Communicate
Connect
Collaborate

STEPHANIE J. COOPMAN

San José State University

JOY L. HART

University of Louisville

cognella®

SAN DIEGO

Bassim Hamadeh, CEO and Publisher
Todd R. Armstrong, Publisher
Michelle Piehl, Senior Project Editor
Sara Watkins, Developmental Editor
Alia Bales, Production Editor
Stephanie Adams, Senior Marketing Program Manager
Natalie Piccotti, Director of Marketing
Kassie Graves, Senior Vice President, Editorial
Jamie Giganti, Director of Academic Publishing

Printed in the United States of America.

3970 Sorrento Valley Blvd., Ste. 500, San Diego, CA 92121

BRIEF CONTENTS

DETAILED CONTENTS

PART II   Interpersonal and Small Group Communication

CHAPTER 7   Interpersonal Communication Processes   142

PART III   Public Speaking

CHAPTER 11   Developing Your Speech Purpose and Topic   240

CHAPTER 15 Speaking to Inform 357

TABLE OF BOXES

LIVE IT LOCAL

PREFACE

We live in a communication environment unlike any other in human history. The internet—the vast global network of interconnected computers—serves as the foundation for the new and powerful ways in which people communicate, connect, and collaborate. More than a communication channel, the internet has changed how people think about communication, enact relationships, and inform and persuade others. Unlike other communication texts that consider technology an "add on," *Communicate, Connect, Collaborate* embraces the everydayness of this heightened connectivity and its implications for human interaction.

Communicate, Connect, Collaborate addresses the central characteristics of the ubiquitous communication era: communication as convergent, communication as connection, and communication as collaborative.

With the convergence of communication tools into the smartphone, frequently transitioning among multiple communication channels has become integrated into the flow of everyday life. The distinction between online and offline communication has moved from static to fluid, with people connecting via text, video chat, phone, email, and in-person across and within relationships. *Communicate, Connect, Collaborate* recognizes that communicators don't have "virtual" versus "real-life" friendships. They just have friendships.

Communicators achieve their goals—in families, with friends, at work, in public presentations—by connecting with others. And today, those connections can occur from nearly anywhere at anytime. The COVID-19 pandemic brought the human imperative to connect into sharp relief. In past crises such as 9/11 and the Great Recession, people could come together in their homes, at work, at school, in their favorite restaurants and coffee shops, in parks—anywhere that people naturally congregate. But with the pandemic and physical distancing, there was no coming together in the ordinary sense. Instead, Zoom, Houseparty, and FaceTime became the go-to apps for students, coworkers, friends, and family to connect. Work, school, weddings, birthdays, and happy hours moved online, as people adjusted to this strange new world of being alone together.

Online tools such as social networking sites, video chat, and Google Docs reflect the collaborative nature of human communication today. College students are hardwired for collaboration, having grown up with TikTok, Instagram, and YouTube, along with learning pods, soccer games, and study teams. Students expect to learn experientially, actively participating in their learning environment through meaningful applications, hands-on examples, and creating novel content. *Communicate, Connect, Collaborate* fundamentally identifies communication as a collaborative endeavor.

In short, *Communicate, Connect, Collaborate* centers on communication's ubiquity, with convergence, connection, and collaboration as the guiding principles animating the fundamentals of communication.

Organization

As the book's title conveys, interwoven throughout the book are the concepts of communicating, connecting, and collaborating. The chapters illustrate these interconnections and encourage student reflection on the underlying principles and overarching commonalities, as well as the specific skills needed to excel across contexts. Understanding communication fundamentals, developing and managing relationships, and preparing and delivering speeches are central components in students' personal and professional lives.

- Part I, "Foundations of Communication," introduces students to the elements of communication, such as the pervasive communication environment model, ethical principles, culture and inclusiveness, identity, self-disclosure, listening, language, and nonverbal communication. These fundamentals provide a springboard for students to develop skills and apply materials in the parts that follow. Through understanding these fundamentals, students are well positioned to achieve communication goals.

- Part II, "Interpersonal and Group Communication," covers interpersonal and small group communication processes, such as how relationships evolve and change, interpersonal and group dialectics, conflict and power, and strategies for productive interpersonal and small group relationships. Understanding the complexities of interpersonal and small group relationships positions students for success in connecting and collaborating with others.

- Part III, "Public Speaking," guides students through the speech-making process. Effectively developing and delivering speeches is essential for professional, personal, social, and civic growth. Whether advocating for community change or presenting a project team's report, students will be called upon to give presentations in a variety of communication contexts. Abilities in both informative and persuasive speechmaking will serve students well beyond their college years.

Features

Each chapter begins with learning outcomes that correspond with the major sections in the chapter, setting the stage for the material to come and ensuring that the chapter's key ideas are foregrounded for students. Key terms and concepts are highlighted, defined, and clearly explained throughout the book and in the glossary, making important ideas clear and facilitating the review process when studying. Each chapter concludes with discussion questions to stimulate additional collaboration and application of essential ideas covered. For example, five of the six public speaking chapters include sample student speech outlines or complete speeches for analysis and discussion.

In addition to these elements, *Communicate, Connect, Collaborate* features ETHICAL QUESTIONS, CONNECT & REFLECT, and LIVE IT LOCAL boxes to further engage students and embed interactivity in the textbook.

Ethical Questions

Underscoring the importance of usability or application, ETHICAL QUESTIONS boxes provide brief scenarios to give students opportunities for reflection on and discussion of ethical issues specific to the various areas of the field. For example, Chapter 3, "Perception and Identity," addresses the ethical challenges of combating stereotypes by illustrating how a White pop singer is called a "bad boy" when he is accused of driving while under the influence and vandalism, but a Black professional football player and Stanford University graduate is called a "thug" when he shouts during a TV interview (page 55). Asking students how they develop a more complex picture of someone or seek out information about a group of people with which they are unfamiliar can help them combat stereotyping.

Reflecting on ethical issues alongside chapter concepts increases relevance and connections. Rather than relegating ethics to a separate chapter, this approach encourages students to consider ethical questions in tandem with chapter information and examples.

Connect & Reflect

CONNECT & REFLECT boxes build on participation and collaboration in communication, presenting in-class activities that can be completed online or in-person that apply the text's content. For example, in Chapter 4, "Listening," one CONNECT & REFLECT box (page 81) invites students to reflect on listening as or to a person with a hearing impairment. In Chapter 5, "Verbal Communication," students consider ways to make language more powerful by rephrasing statements (page 113).

Across the chapters, the CONNECT & REFLECT opportunities fuel consideration of a range of topics. They encourage deeper thinking, reinforce key ideas, and provide flexibility for students and instructors alike.

Live It Local

LIVE IT LOCAL boxes suggest ways students can apply concepts in the various communities in which they participate. Highlighting service learning, the student-scholar, and community engagement, LIVE IT LOCAL boxes connect each chapter's content with strategies for applying course concepts in community-related activities. For instance, in Chapter 6, "Nonverbal Communication," a LIVE IT LOCAL box (page 135) invites students to analyze nonverbal rituals, physical settings, and nonverbal ways to make groups more inviting and welcoming to others. In a LIVE IT LOCAL box (page 185) in Chapter 9, "Small Group Formation," students identify volunteer groups in their community, meet with a few volunteers, and explore why people join such groups.

When students live it local, they collaborate with their local communities, expanding their connections as well as enhancing their knowledge. This feature encourages students to find their voice and visualize the impact they can have on their various communities.

Communicate, Connect, Collaborate's engaging and conversational tone, practical and integrative approach, and reflective exercises resonate with communication majors and non-majors alike, facilitating their understanding and application of communication concepts and theories in their everyday lives. Whether students are taking this course due to interest in the topic, to fulfill a general education requirement, or as the first step toward a communication degree, they will gain a sound introduction to the discipline, preparation to engage with others, and essential communication skills, from evaluating messages to speaking publicly. Effective communication is foundational in addressing problems near and far, small and large. In short, to attempt to solve society's wicked problems—such as hunger, climate change, and income inequity—today's students, our future leaders, must be adept communicators, skilled in connecting and collaborating with others.

Instructor Resources

Instructor resources are available to adopting instructors. For more information on these materials and about adopting this textbook for your classroom, please contact adopt@cognella.com.

Acknowledgments

Like the best communicative endeavors, writing this book was a team effort. We benefitted from many collaborative relationships. Todd Armstrong, our publisher at Cognella, offered practical guidance and key resources over the course of many email messages and Zoom meetings and could always be counted upon for excellent insights and great humor. Michelle Piehl, senior project editor, kept us on track and on time—always in helpful and kind ways. Sara Watkins, developmental editor, provided the line-by-line detailed attention essential to any textbook. Alia Bales, production editor, lent her expertise to creating the book's engaging and inviting layout. Sabrina Baeta, editorial assistant, brought together all the pieces of the instructors' resources. Beyond support from the Cognella team, the book's development was improved by thoughtful and insightful comments from instructors who reviewed early drafts: Jenna Abetz (College of Charleston), Leonard Assante (Volunteer State Community College), Abby M. Brooks (Georgia Southern University), Gina Firenzi (Foothill College), Stefan Hall (High Point University), Andrew F. Herrmann (East Tennessee State University), George Lawson (The University of Nebraska at Kearney), Archie Wortham (Northeast Lakeview College), and Fletcher O. Ziwoya (The University of Nebraska at Kearney).

Foundations of Communication

CHAPTER 1

Communicating, Connecting, and Collaborating

The beginning of the 21st century brought with it an explosion of communication tools for people to interact with each other. People around the globe use social media to stay in touch with family and friends, organize events, find jobs, and join groups. Companies identify employees with the needed skills and knowledge to complete a project—no matter where they live—and link them together via mobile communication technologies. Physicians use video chat to interview and treat patients who are hundreds or thousands of miles away. What do these activities have in common? They are all digital media. But in addition to that, they are forms of human communication—people connecting and collaborating with each other.

You can think of how humans communicate as spanning a historical continuum with four epochs or eras: oral, written, print, and electronic. Storytelling and passing along information through the spoken word characterized the oral tradition. With the invention of the alphabet, people recorded their thoughts and ideas, although only an elite few could write and had access to books. The printing press brought the written word to the public along with the development of common language uses and spelling. Television led to the rise of the electronic age and what McLuhan called the "global village," or the free flow of information across traditional geographical and cultural boundaries.[1] Although earlier epochs cannot include later ones, later epochs include earlier ones. For example, oral communication continues its strong role today even with the addition of written, print, and electronic communication.

An Era of Convergence

Now human communication has moved into an era called **convergence**, where previously separate activities, industries, and mediums come together, collapse into each other, or become effectively indistinguishable. Convergence involves three main components. The first is the convergence of activities that used to take several devices and now take just one. The most obvious example is the smartphone. Originally, cell or mobile phones functioned like landline telephones—you used the phone to talk to people. Now smartphones are minicomputers that allow you to take and send pictures, text, play games, map out directions, watch and record video, read a book, and use a myriad of apps that take the place of other devices (Figure 1.1).

The second is the convergence of the companies that produce and distribute media. Apple is a good example. Originally, Apple just made computers and the software that ran them. Now the company produces the hardware and software for one of the top-selling smartphones in the world—the iPhone. iTunes, another Apple product, is the largest music distributor on the planet. Apple's entrance into the music and mobile phone businesses greatly changed those industries, making it easier to communicate with others and access information. In addition, the iPad and Apple News+ are changing the newspaper and magazine business. Moreover, Apple recently entered the video production and streaming market with Apple TV+.

Third, convergence is about the collapse of media into a single medium. Traditionally, the medium defined media. Radio, television, film, and print (books, newspapers, magazines) were all distinct mediums. The restrictions of each channel defined the content you would get from each one. Almost as important, how companies made money from their content was directly tied to what a consumer could or could not do with it. You know what happened to the music business when you no longer needed to buy physical CDs but could just go online and download the music you wanted—for free or at a low cost. Another example is the TV program. If you want to watch an episode of a network TV show, you can wait for it to come on TV each week, you can record and watch it later, you can download the app and view it on demand on your mobile device, or you can buy

convergence The intersection of two different entities that previously were separate

FIGURE 1.1 With their extensive computing capabilities, smartphones represent the convergence of activities into a single device.

episodes. In addition, some platforms and media outlets make a show's entire season available at once. You also can check out reviews, fan sites, the show's social media, and other online sources. If you can view and interact with a TV program in so many ways, is it still just a TV show? Reflect on how this level of access impacts your life when you no longer are tied to a particular device, time, or place to consume media. That is what happens with convergence.

digital divide The gap between those who have access to new communication technologies and those who do not due to social, economic, educational, and geographic inequities

Not everyone enjoys the affordances of convergence. The **digital divide** refers to the gap between those who have access to new communication technologies and those who do not due to social, economic, educational, and geographic inequities. A study by Microsoft estimated that half of the people in the United States—163 million Americans—do not have reliable high-speed internet access, either because it is not available where they live or they cannot afford it.[2] The Pew Research Center reports that for those Americans who earn less than $30,000 per year, nearly one third do not own a smartphone and nearly half do not have broadband access or a computer at home. In addition, more than one third of low-income children in K–12 do not have home computers with internet access.[3] In addition, rural homes are less likely to have broadband access than those in urban areas and rural residents are less likely to own a smartphone or tablet. Moreover, people in rural areas are less likely to go online than those in urban areas, with 15% of rural residents not using the internet at all.[4] Finally, people of color continue to experience the digital divide in spite of efforts in the United States to increase internet accessibility. About one quarter of Latinx and Black adults rely solely on their smartphones to go online, making tasks that require larger screens inaccessible. In addition, more than 40% of Black library patrons use the library to go online compared with 25% of White library patrons.[5]

The COVID-19 pandemic brought the digital divide in the United States into sharp relief. For example, when K–12 and college classes moved online after governors issued stay-at-home orders, not all students had equal access to class materials. Without such access, and a tablet or computer, students could not watch their instructors' lectures, participate in online discussions, or turn in assignments, exacerbating what education experts call a homework gap.[6] A Pew Research Center study found that nearly 20% of low-income teenagers could not do their schoolwork at home, because they did not have a computer or internet access.[7] Moreover, 60% of low-income parents reported that their school-aged children would have to complete their homework on a smartphone, use public Wi-Fi to do their assignments, or not turn in their homework at all due to lack of internet access.[8] Some school districts, such as Berkeley, California, gave out laptops and created wireless hotspots so students could continue with their studies during the pandemic (Figure 1.2). But not all locales could afford such technology. In Oregon, teachers mailed printed

FIGURE 1.2 To help bridge the digital divide during the COVID-19 pandemic, some K–12 school districts handed out laptops to students so they could continue their studies online while staying at home.

materials to students without internet access. Parents used their phones to take photographs of students' completed assignments for teachers to print out, grade, and mail back to the students.[9] Explore your community's level of digital connection in Box 1.1, Mapping the Digital Divide.

BOX 1.1: LIVE IT LOCAL
Mapping the Digital Divide

How digitally connected is your community? Go to the Federal Communications Commission Broadband Map website at broadbandmap.fcc.gov. Type in your address. How many residential broadband providers are available on your block? Using the zoom-out feature on the map, explore the availability of residential broadband providers in your town, county, and state. Where are there gaps or few choices in coverage? What are the implications of those gaps and the lack of choice for the people who live there? Use the Area Comparison tab to compare high-speed internet availability by state. Are there places with no providers or only one or two? What might be some issues associated with locales that have few providers? For instance, the Alliance for Affordable Internet found that worldwide, lack of competition among providers increases prices and decreases choices for consumers, preventing nearly 50% of the world's population from accessing the internet.[10] Consider your own neighbors, friends, and family members. Do all of them have internet access? What does it mean to function in today's society for people who do not have reliable and affordable internet access? What can be done to bridge the digital divide?

Converging technologies extend the realm of human communication, providing you with more options to tap into the global networked society. Still, humans have been communicating since they first walked the Earth. Now they share their experiences by texting, chatting, blogging, and tweeting. Although the 21st century has brought with it all sorts of new ways to communicate, the basic human desire to connect with other people has not changed.

The Communication Imperative

Consider these responses to advances in various new communication technologies:

- No one will need to remember anything.
- People will be overwhelmed with information.
- It focuses on the superficial and trivial.
- It's an invasion of privacy.
- Children will stop reading and their performance in school will suffer.
- Conversation as we know it will end.[11]

These statements might sound like they are about the internet, social media, and smartphones. Yet these claims were made about older technologies—some of them much older. Here are the statements again, this time with the technology they were referencing:

- No one will need to remember anything. [written language]

- People will be overwhelmed with information. [printing press]

- It focuses on the superficial and trivial. [telegraph]

- It's an invasion of privacy. [telephone]

- Children will stop reading and their performance in school will suffer. [radio]

- Conversation as we know it will end. [television]

FIGURE 1.3 Each new communication technology brings fears that it will alter human communication for the worse.

Each new communication technology comes with dire warnings about the negative changes to humans and humanity that will occur (Figure 1.3). In the ancient text *Phaedrus*, Plato, Socrates's student, set up a dialogue between Socrates and Phaedrus. In part of that conversation, Socrates tells the story of the god Theuth, who invented writing. Theuth extolled the virtues of writing, but another god, Thamus, was skeptical, arguing:

> For this discovery of yours will create forgetfulness in the learners' souls, because they will not use their memories; they will trust to the external written characters and not remember of themselves. The specific which you have discovered is an aid not to memory, but to reminiscence, and you give your disciples not truth, but only the semblance of truth; they will be hearers of many things and will have learned nothing; they will appear to be omniscient and will generally know nothing; they will be tiresome company, having the show of wisdom without the reality.[12]

Here, Socrates argued that writing allowed individuals to avoid committing information to memory and as a result failed to learn. In countering Socrates's claim, you could argue that writing frees your memory to retain essential information and reading what others have written greatly expands what you can learn. More recently, researchers have suggested that the ability to search for information online has negatively impacted human memory processes. As individuals become more reliant on their smartphones and computers to store information, they rely less on their own memory.[13] Called **cognitive offloading**, individuals reduce their mental effort by using an external device to extend their own mental capabilities. From one vantage point, you might conclude that soon people will not be able to

cognitive offloading The use of external devices such as smartphones to reduce an individual's mental efforts and extend mental capabilities

remember anything at all—they will look up anything they need to know on their phones. Yet you will need to recall how to search for and identify information. In that respect, you will work your memory in different ways, with devices such as smartphones extending the information available to you (Figure 1.4).

The **technological imperative** suggests that technology controls human behavior and there is little you can do about it. Taken to its extreme, the technological imperative results in a world where humans are at the mercy of uncontrollable and unstoppable technological advances. This was the world Socrates imagined—one in which writing would lead to the end of true knowledge and intelligence. But his prediction was completely wrong. Writing opened up a world of information and knowledge for thousands and then millions of people.

In contrast to the technological imperative, the **communication imperative** suggests that you are born with a desire to maximize your communication satisfaction and interaction with others. That is, as a human being, you are driven to communicate, both to form social bonds and to express yourself. Your brain is hardwired to communicate. At the most fundamental level, your brain wants to connect with other people. For instance, neuroscientists have found that truly understanding how the brain learns and uses language requires observing people as they interact with others. Because language is inherently a social function, your brain becomes more active when you communicate with those around you versus just watching them. As shown in Figure 1.5, your emotional engagement with others as you interact with them triggers a desire to use and learn language.[14]

You communicate because you are a social being. You strive to reach your communicative goals, such as explaining a point, and you help others reach theirs, such as working together to solve a problem. And you use technology in ways that help you achieve what you have set out to do. Rather than technology determining how and what you communicate, your motivations to communicate drive how you use technology.[15] For example, when mobile phones first included cameras, they were set up for users to take photos of what they could view. But people wanted to take photos of themselves or selfies—with their friends, at a scenic spot, wearing a new outfit. They tried all sorts of ways to hold the phone so they could get that picture of themselves. Rather than letting technology dictate

FIGURE 1.4 External devices, such as smartphones, computers, and books, allow people to extend their information storage capacity and mental abilities.

technological imperative The assumption that technology controls human behavior

communication imperative The inherent social nature of human beings that drives them to communicate with others using the tools available

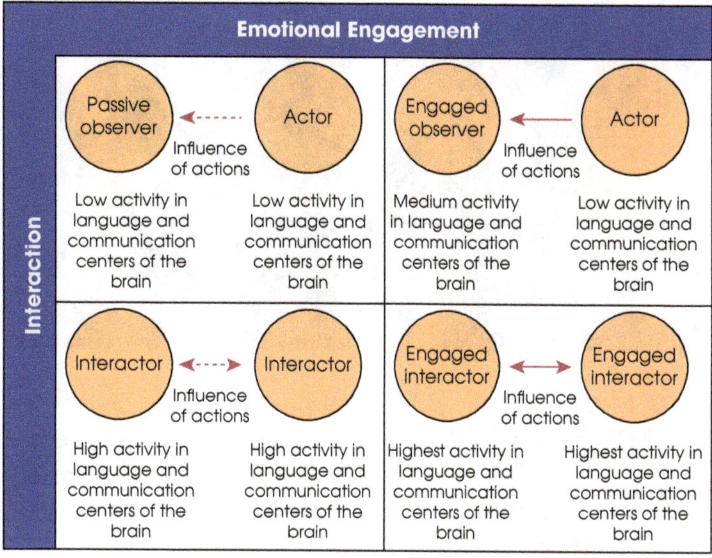

FIGURE 1.5 Your brain on communication.

FIGURE 1.6 Even with social distancing during the COVID-19 pandemic, people found ways to be together.

how they used their phones, people figured out ways to take selfies, even before phones were equipped with the reverse-view camera.

At no time in recent history has the communication imperative been more apparent than during the COVID-19 pandemic. When other traumatic events have happened, such as 9/11 and the Great Recession, people could reach out to each other in-person for support and comfort. Especially before vaccines were available, many countries, states, cities, and towns put in place stay-at-home and social-distancing requirements. While these measures contained the transmission of the virus, they ran completely counter to what humans do in a crisis—come together. Yet people did connect, often in innovative and creative ways (Figure 1.6). Couples had their weddings on FaceTime, seniors from a university in Japan attended their graduation ceremony using robot stand-ins, and churchgoers went to services on Zoom. In addition, people changed how they used the internet, with traffic on Facebook's, YouTube's, and Netflix's websites up and use of the corresponding apps down.[16] As people stayed home, they switched from connecting via their phones to connecting via their computers because that worked best for them.

The resurgence of the Black Lives Matter movement and the global protests against racial injustice that occurred during the pandemic provide additional support for the communication imperative. The widespread protests reflect both the human need to connect in times of social turmoil and to collaborate to address social problems. Using social media platforms such as Instagram and Facebook, protest organizers quickly planned outdoor rallies and marches in large cities and small towns (Figure 1.7). Wearing face masks to protect against the spread of the COVID-

FIGURE 1.7 Social media played a key role in organizing and documenting protests against racial injustice, offering an example of how individuals use technology to achieve their communicative goals.

19 virus, protesters gathered to condemn police violence against Black people, indigenous people, and people of color. Participants recorded the protests using their smartphones and posted videos online, extending the reach and impact of their message. Try out the strength of the communication imperative for yourself in Box 1.2.

The communication imperative does not guarantee you will achieve your goals in using communication technologies. Just as with in-person communication,

BOX 1.2: CONNECT & REFLECT

The Communication Imperative in Action

The communication imperative posits that humans are driven to communicate with each other—that individuals will use all available technology, or create new technologies, to form a bridge between themselves and others. Consider these scenarios, brainstorming multiple ways to solve each communication problem:

1. You are supposed to meet your study group at the campus library in 15 minutes. Unfortunately, you find your bicycle has a flat tire and you are going to be late. You have your phone, but none of the group members' numbers. How do you contact them?

2. You are staying with friends for the weekend and have a paper you must turn in online by midnight. It is one hour before the deadline and your friends' internet service is not working. How do you turn in your paper on time?

3. You are in charge of publicity for your student group's fundraiser to benefit communication programs in local middle schools. How do you get out the word to possible contributors?

Now that you have generated your ideas, reflect on how you developed them. What role did the communication imperative play in coming up with new ways to use communication technologies?

you may use technology both in conventional and innovative ways that you think will lead to the outcomes you expect, but that does not always happen. For example, during the COVID-19 pandemic when K–12 schools moved their classes online, lack of access and unfamiliarity with the technology resulted in frustration for instructors, students, and parents. Even with the best of intentions, plans may go awry and objectives unmet.

Models of Communication

How do you think communication works? Reflect on a recent in-person or video chat conversation you have had. You talked and the other person talked. You gestured and the other person gestured. You thought about what the other person said and did. That person thought about what you said and did. Sketch it out—how did the exchange unfold? How would you explain what happens when humans communicate?

Communication scholars have developed a wide range of models that attempt to depict the communication process. The mathematical, interactive, and transactional model are the most well known and provide a historical context for current understandings of communication. These models and a more recent one, the pervasive communication environment model, which offers a new view of communication in an era of communication mobility, are summarized in Table 1.1 on the next page and discussed in detail in this section.

TABLE 1.1 Comparison of Models of Communication

MODEL	COMMUNICATION AS …	METAPHOR FOR COMMUNICATION	CONTRIBUTION TO THE DISCIPLINE
Mathematical	Action	Conduit	Identified noise
Interactive	Interaction	Lens	Added feedback element
S-M-C-R	Transaction	Bridge	Construed communication as relational
Pervasive communication environment	Convergence	Air	Illustrated communication's ubiquity

Mathematical Model: Communication as Action

In the late 1940s, two mathematicians, Claude Shannon and Warren Weaver, developed a model of communication that attempted to account for differences between the message someone sent and the message the other person received. Their interest in the communication process stemmed from their work in radio and telephone. Also called a transmission model, Shannon and Weaver drew attention to communication as action that involved an information source, a message, a way to transmit the message, and a receiver, as shown in Figure 1.8. Most important, they identified noise as interference between the message sent and the message received.[17]

Transmission models rely on a conduit metaphor, depicting words as containers for information that travels from one person to another.[18] These models make intuitive sense, especially from the sender's perspective. For instance, you text your study group with detailed instructions on when and where to meet you at the library, but somehow they mix up the location. How could that happen? You were so clear. The Shannon and Weaver model would identify noise as the problem—something that interfered with the study group members' reception of your message, such as a problem with your phone.

Although they were groundbreaking at the time, transmission models suffer from several limitations. First, communication is depicted as linear and one-way, from the sender to the receiver. One of the study group members might have texted the group, "Let's meet at the Communication Resource Center instead so we can talk with a tutor," but you had not checked for messages. Second, in the transmission model, only the sender plays an active role in communication; the receiver is a passive recipient of the message. However, you know that people

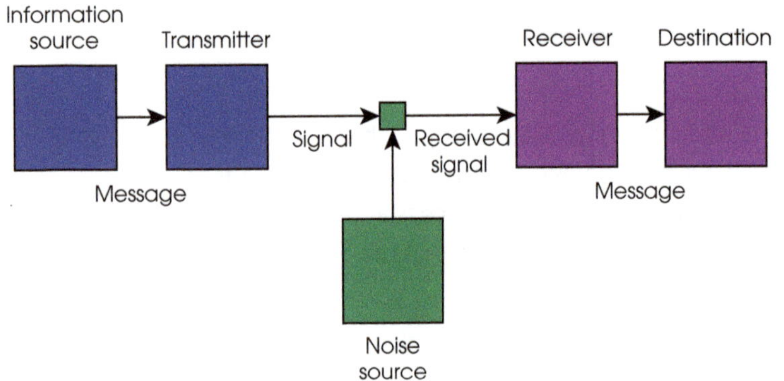

FIGURE 1.8 Mathematical model of communication.

actively interpret messages. The other study group members may have interpreted the message as, "If you're on campus, meet at the library, but if you're not, we'll meet in class tomorrow." Third, the context for communication is ignored. Is the study group meeting during a time the library typically is crowded? Have the meetings usually been at the library or someplace else? Are most group members dedicated to the group? Fourth, communication is characterized as mechanistic, much like transmitting sound via a telephone.[19] But think about some of the communication problems you have had and you will agree that interacting with others is more like the creative exuberance of art than the precise gears of a machine.

Interactive Model: Communication as Interaction

Not long after Shannon and Weaver published their model of communication, Wilbur Schramm offered an alternative model based on his research in propaganda for the U.S. Information Agency. Shown in Figure 1.9, his interactive model included an additional key component in the communication process—feedback. Also, rather than senders and receivers, the model identified interactants as encoders-interpreters-decoders. Communication moved from action to interaction in which individuals exchanged messages in a two-way process. Moreover, Schramm introduced the notion of field of experience, or background, as influencing how communicators formulate and interpret messages.[20]

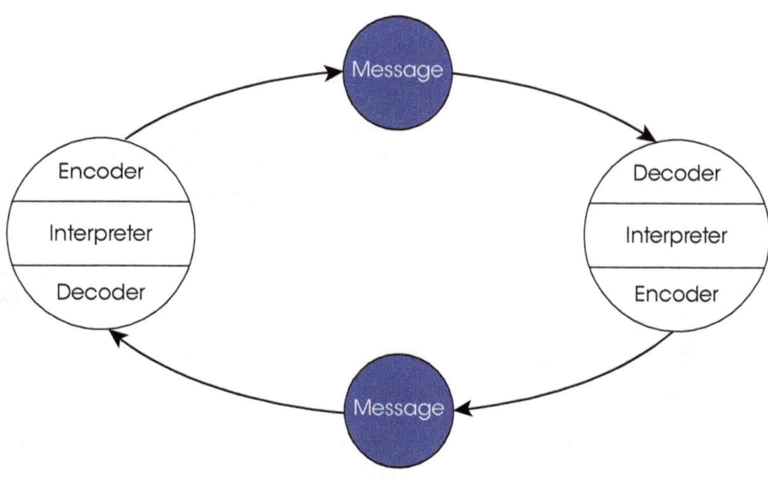

FIGURE 1.9 Interactive model of communication.

With the interactive model, effective communication involves careful message design to produce the results communicators seek. Gathering information about the other person's background and attending to feedback can help communicators develop successful messages. The metaphor for communication is a lens, as communicators search for and screen information.[21] In contrast to the passive receiver in the mathematical model, the interactive model assumes an active, participative audience.[22] Say, for instance, you are trying to convince a friend to watch a movie with you rather than study for an exam. In trying to formulate a convincing argument, you consider what you know about your friend, such as likes, dislikes, and interests. You say something such as, "I know you were planning to study for your chemistry exam, but it isn't until Friday and it's only Tuesday. You'll be in a much better frame of mind to study tomorrow morning if you have a relaxing evening tonight. Then when you wake up you'll feel fresh, rested, and ready to review your notes." You also watch and listen for feedback. If your friend starts to smile, you might go on and say, "I know you think I'm right. You're a morning person—that's the best time to study." In contrast, if your friend starts to frown and turns away, you might say, "How about you study for a couple hours, then take a break and watch the movie? Then you'll get a head start on your schoolwork, but still get in some relaxation time." Feedback and each person's field of experience provide a more complete picture of the communication process.

Still, the interactive model has several shortcomings. First, communication remains a linear process, with messages going back and forth from one person to the other. Second, the model does not take into account the context for communication, such as the relationship between the participants. Third, the model emphasizes the attainment of individual goals rather than mutual goals that might unfold in the interaction.

S-M-C-R Model: Communication as Transaction

The social changes of the 1960s brought with them a new view of communication as something that connected people together. Rather than a conduit for ideas—the mathematical model—or filtering mechanism for information—the interactive model—communication became central to developing human relationships. In 1960, David Berlo, professor and chair of communication at Michigan State University, published his seminal book *The Process of Communication* in which he outlined his **S**ource-**M**essage-**C**hannel-**R**eceiver (S-M-C-R) model of communication, shown in Figure 1.10.

With the S-M-C-R model, Berlo attempted to identify all the elements or ingredients involved in human interaction.[23] The metaphor for communication is a bridge, in that communication links together people. Interactants are interdependent, with each relying on the other to achieve their goals. In this model, empathy—understanding another's perspective—provides the basis for effective communication. More than gathering information to design a message for an audience, the S-M-C-R model stresses experiencing the world as another person would to develop a true understanding of the message that person is trying to communicate. In a nursing class, for instance, students may engage in a role-playing activity to better

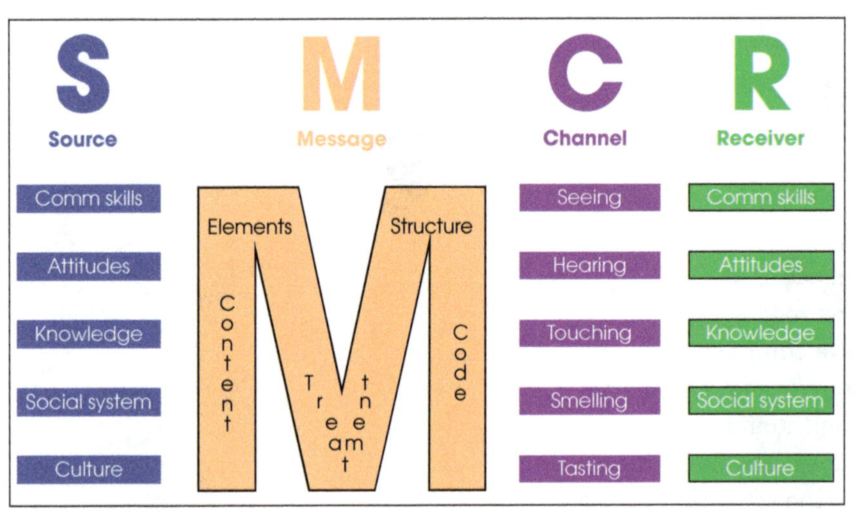

FIGURE 1.10 S-M-C-R model of communication.

understand how patients might experience their first hospital visit. Research on Second Life and other virtual worlds has found that interacting as an avatar increased individuals' sensitivity toward differing cultural perspectives.[24] For example, constructing an avatar with purple skin and green hair can give you insight into how others respond to differences in physical appearance.

Although transactional models provide a more complex view of communication than the transmission and interactive models, the focus remains on a source sending a message via a channel to a receiver. Factors such as cultural and social backgrounds, knowledge, attitudes, and skills influence communicators' effectiveness, but larger institutional and organizational structures are absent from the model. For example, when you are talking with your supervisor, that person's power in the organization's hierarchy impacts the conversation.

Pervasive Communication Environment Model: Communication as Convergence

The mathematical, interactive, and S-M-C-R communication models trace their origins to well before the development of the internet and mobile communication technologies. Why is that important? Because like the printing press, telegraph, telephone, television, and other new technologies that came before, the internet has revolutionized how people connect with each other. But this time, the impact may be even more profound.

The printing press allowed the mass distribution of information from one person or party, such as a newspaper publisher, to many people. With the telegraph, people for the first time could send brief messages almost instantaneously across long distances. The telephone provided voice transmission and a person-to-person connection from people's homes and businesses. The internet gave communicators a wide variety of ways to connect and share information, especially when linked to a mobile phone. In less than a generation, cell or mobile phones have gone from a high-tech toy for the wealthy to the most pervasive mobile computing platform on the planet. Nearly all adults in the United States—97%—own a mobile phone, and 85% own a smartphone. Including tweens, teens, and adults, there are 294 million smartphone users in the country. Worldwide, 5.28 billion people, or 67% of the Earth's population, own a mobile phone, and 3.8 billion of those people own a smartphone.[25] The combination of mobility and more computer power in a cell phone than in the Apollo spacecraft that first visited the moon has caused an upheaval in the communication terrain (Figure 1.11). Now communicators can send text, images, audio, and videos to one person or millions while relaxing at home, attending a sports event, or taking a break between classes.

FIGURE 1.11 Your smartphone has more computing power than the Apollo 11 spacecraft that landed on the moon in 1969.

Today you can communicate nearly anytime, from anywhere, with anyone. This **pervasive communication environment** provides you with multiple access points to an integrated communication structure with text, audio, video, and voice capabilities.[26] So what does this mean for you? The short answer is the ability to access, create, and share information in multimedia anytime, from almost anywhere, with anyone, for any reason. The social impacts are staggering and go far beyond online shopping, texting friends, and turning in your homework online. Mobile devices connected to the internet have played a key role in recent political movements. For example, in the Hong Kong protests against the extradition of criminal suspects to mainland China, activists used mobile phones to communicate and take video of demonstrations they then shared on social media, successfully bypassing the traditional and state-controlled media organizations.[27] As another example, a Pew Research Center study found that just a few months into the COVID-19 pandemic, 87% of Americans reported the internet was important or essential in coping with the adjustments to their daily lives.[28] Because new communication technologies so permeate everyday life, they have formed an increasingly invisible network connecting individuals and groups.

As you probably have experienced, the pervasive communication environment has fundamentally changed how you communicate. Sure, you still interact in-person with friends, family members, coworkers, neighbors, and the like. But you also use email, chat, text messaging, instant messaging, and other mobile communication technologies to exchange ideas, solve problems, make decisions—to connect with others from wherever you are at any particular moment. With the internet linking technologies and people, time and place no longer constrain communicators' activities.

Your ability to access information and communicate with others 24/7 stems directly from convergence. Communication as convergence recognizes the unique contributions of digital technology to individuals' abilities to construct their everyday lives in new and different ways.[29] Because you can connect anytime with anyone from anyplace, the traditional boundaries of home/work and private/public are increasingly blurred. Rather than meeting friends after work at a local coffee house, you can video chat with them from work. Rather than sitting in a classroom on campus, you can listen to a lecture at home in a real-time webinar or later in a video capture of the event. Consider the ways in which smartphones have transformed how you communicate with others and access information. In the past, telephones served one function—transmitting audio from one point to another. Now your phone can do all sorts of things, such as take a photo, record video, help you find a nearby restaurant, pay for groceries, and remind you of an upcoming meeting. In short, camera, video recorder, map, wallet, and calendar—along with many other communication and organizing tools—have converged into a single device, the smartphone. Most Americans have a smartphone, and for about 17% of Americans, their phone is their only access to the internet.[30] Consider your ethical responsibilities in using your smartphone in Box 1.3.

BOX 1.3: ETHICAL QUESTIONS

The Ethics of Smartphone Use

Smartphones greatly extend your ability to communicate with others, but with that ability comes a greater ethical responsibility. Medical personnel, for instance, have found smartphones allow them to more easily document patient care and look up relevant information. However, they must carefully secure patient records to protect privacy and adhere to legal requirements.[31] Similarly, lawyers regularly use their smartphones to record notes, keep records, and file court documents while maintaining client confidentiality and following the law.[32] In your everyday personal use of your smartphone, the ethical standards are not as clear cut. Is it okay to record your instructor's lecture without asking? Should you ask permission before taking a photo of your friend and posting it on a social media site? Is it all right to text while you are having dinner with your family? How do you decide on the ethical and unethical uses of a smartphone?

The pervasive communication environment model identifies and embraces the integration of digital media in human communication. From this perspective, the metaphor for communication is air—ubiquitous, essential for life, made up of multiple components, and varies in quality. Humans are driven to communicate with each other, just like their bodies force them to breathe. Communication forms the essential foundation for human development. The pervasive communication environment model views human interaction as conversations, often fragmented and disjointed, involving multiple goals and possible outcomes.

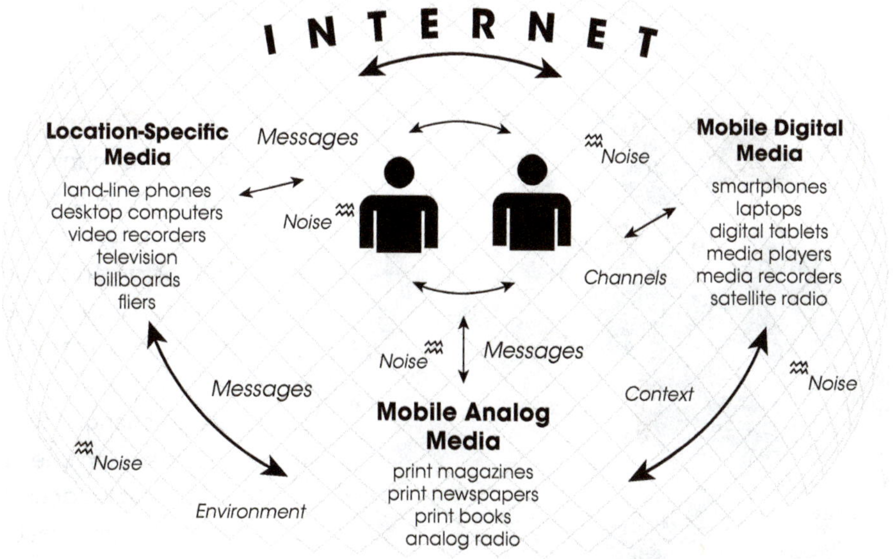

Figure 1.12 represents human communication in a pervasive communication environment. The model includes three major components: the outer ring, the inner ring, and floating elements. The internet provides the infrastructure that links together the various mediated components. Human communication forms the model's center—the conversations that connect people.

FIGURE 1.12 The pervasive communication environment model.

The Outer Communication Ring

The model's outer communication ring includes three broad areas of mediated communication: location-specific, mobile digital, and mobile analog.

Location-specific media remain anchored to a particular place, such as landline telephones, desktop computers, billboards, and posters. Mobile digital media are portable and include cell and smartphones, laptops, tablets, and satellite radio. Mobile analog media are also portable and include print magazines, newspapers, books, and analog radio. However, as the pervasive communication environment evolves, these categories begin to blur. Television, for instance, is a location-specific media—unless you are streaming a Netflix show on your smartphone or other mobile communication device. The connectivity among the outer ring media reveals a complexity that has become possible with the internet and digital communication.

The Inner Communication Ring

The model's inner ring represents communicators engaging in conversations. These conversations may be in-person or mediated, depending on individuals' needs, motivations, interests, objectives, and the like. The communication imperative suggests that people will use communication tools that work best for them. As the center of the model, communicators are both communication producers and consumers, or **prosumers**. That is, you produce messages—written reports, Twitter tweets, YouTube videos, Snapchat posts, in-person comments—and consume messages—reading reports and tweets, watching YouTube videos, viewing Snapchat posts, and listening to in-person comments (Figure 1.13).

prosumer An individual who is both a communication producer and consumer

Floating Communication Elements

Floating communication elements in the model include the environment, context, messages, channel, and noise. Larger external surroundings, such as societal and cultural norms or organizational practices, make up the environment. For example, legal rules and regulations may influence how members of a city government task force communicate. In turn, task force members may have the power to change those rules and regulations.

Context includes both the physical space in which communication occurs and the communication parameters individuals create as they interact with each other. Students may create a "study group" context in which group members follow certain rules and interact in a particular location. Messages may be verbal and nonverbal, mediated and nonmediated. Messages serve multiple purposes, such as developing relationships among interactants, creating the context for communication, exchanging information, and forming identities. Channels are the modes or mediums of communication individuals use, such as in-person, video chat, email, and text. Communicators may leverage multiple communication channels

FIGURE 1.13 As a prosumer, you both produce and consume messages.

simultaneously, as with texting during an in-person conversation. Finally, noise includes anything that interferes with communication at any point in the model, as with communicators not listening to each other or a poor internet connection during a video call.

The pervasive communication environment model offers a fundamental shift in thinking about how people connect and interact with each other. The model's major strength is moving human communication out of the traditional physically co-located or "face-to-face" context and placing interaction within the larger stream of digital media. The pervasive communication environment model avoids privileging one form of communication over another, recognizing the communication imperative that drives individuals to make communication technology work for them. Additionally, this perspective highlights the ubiquity of communication, the ability to connect and collaborate with others, and the unprecedented access to information anytime from anywhere. Finally, the pervasive communication environment model defines communicators as prosumers— simultaneously information producers and consumers.

Several weaknesses are associated with the pervasive communication environment model. First, because the perspective is so new, it lacks the more extensive research associated with the other three models. Second, the pervasive communication environment model attempts to integrate a wide range of elements, which presents a challenge when representing the complexities of human communication. Third, the model relies on some of the terminology from older models, such as noise and channels, which may not accurately represent the human communication process as it unfolds.

Defining Communication

Definitions serve to identify what is included in something and what is not. Just like the boundaries between cities, counties, states, and countries, definitions offer a way to mark the parameters of a concept. For the purposes of our discussion, **communication** involves the exchange of socially shared symbols that represent interactants' thoughts, ideas, and feelings. **Verbal symbols** refer to oral and written language. **Nonverbal symbols** are those that do not involve language, such as gestures, eye contact, body movement, and touch. Although scholars differentiate between verbal and nonverbal communication, people construct and interpret messages as a total package, assigning meaning based on both verbal and nonverbal symbols. For example, as you read this book, you focus on the words, but you also attend to the photographs and other images included in the chapter. In addition, you notice different font types and styles, headings, and boxes. The verbal and nonverbal messages provide context for each other.

communication The exchange of socially shared symbols that represent interactants' thoughts, ideas, and feelings

verbal symbols Representations of thoughts, ideas, and feelings with oral and written language

nonverbal symbols Representations of thoughts, ideas, and feelings without the use of language, as with gestures, eye contact, body movement, and touch

FIGURE 1.14 Is sleeping in class communication or behavior? Is it a symbol or a symptom?

symptoms Behaviors that convey information about a physical state

Some scholars argue that people cannot not communicate, which means that individuals communicate messages all the time, whether they are awake or asleep and whether they are with others or alone.[33] From this perspective, sleeping could be interpreted as communicating the person is tired and being alone could be interpreted as communicating the person is reluctant to interact with others. But other scholars draw a distinction between behavior and communication.[34] Although both convey information, *communication* uses symbols, whereas *behavior* uses symptoms (Figure 1.14). **Symptoms** indicate an individual's physical state, such as yawning when tired or dropping a hot plate when it burns the skin.[35] The actions convey information, but without intentionality.

For communication to occur the sender must intentionally send a message, the receiver must perceive that the sender intended to communicate something, or both, as shown in Table 1.2. For example, suppose you are outside on campus and wave your hand. Is your action communication or a behavior? It depends. If you notice some friends and you are waving to get their attention, then you are trying to send a message—it is communication. Your friends notice you and wave in return. In this case, both parties perceive intentionality, so it is communication. What if you wave at your friends and they do not notice? It is still communication because you intended to send a message. However, if it is a hot summer day, you are waving your hand just to cool off, and no one perceives you as attempting to communicate, then it is not communication. But if someone observes you waving and interprets the behavior as intentional—such as conveying a greeting—then it is communication, even though you did not intend it that way.

TABLE 1.2 **Intentionality and Communication**

	SENDER HAS INTENTION	**SENDER HAS NO INTENTION**
RECEIVER INFERS INTENTION	Communication	Communication
RECEIVER DOES NOT INFER INTENTION	Communication	Behavior

Throughout the day, people engage in many routinized exchanges that do not appear to involve much in the way of intention. But communication always involves intentionality, however implicit.[36] You may start a conversation with comments such as "Hope everyone's having a good week" and "How are things going with you?" You might dismiss this kind of interaction as small talk and trivial. Yet these simple social scripts acknowledge the others' presence, demonstrate a concern for their well-being, and help maintain positive relationships among interactants.

Ethics and Communication

Ethics is about morality—how you tell the difference between right and wrong. Making moral judgments stems in large part from your cultural experiences. Consider how you learned about what you should and should not do. From an early age, adults likely rewarded you when you demonstrated good behavior, such as sharing your toys with other kids. You probably were punished for bad behavior, such as taking another child's toy without asking. But as you have found out over the years, most situations are ambiguous and do not lend themselves to simple determinations of good and bad, right and wrong.

How can you communicate ethically? The National Communication Association (NCA) provides guidance in its *Credo for Ethical Communication*, offering this definition: "Ethical communication enhances human worth and dignity by fostering truthfulness, fairness, responsibility, personal integrity, and respect for self and other."[37] Ethical communication centers on respect: respect for others regardless of gender, age, race, status, or the relationship to the speaker; respect for others' ideas, feelings, and integrity; and respect for evidence and the rational weighing of alternatives.

You can think of ethics in two ways: an ethic of care and an ethic of responsibility.[38] An **ethic of care** is concerned with providing support and not harming others. An **ethic of responsibility** centers on justice and equity. Ethical communication often arises from the process of producing a **public good**, or something beneficial to the larger society or group, like José Andrés's World Central Kitchen that provides free meals in the wake of natural disasters to people around the world as well as programs that promote food security. Communicating both an ethic of care and an ethic of responsibility is important for achieving a common public good. For example, perceived gender discrimination negatively impacts productivity in the workplace, violating an ethic of responsibility. Women who perceive gender inequity are less committed to completing tasks and blame their work group or organization for the discrimination. Interestingly enough, perceived inequity toward *anyone* leads to higher dissatisfaction among men than women, a violation of an ethic of care. In contrast, perceived gender equity improves the well-being of both women and men in the workplace (Figure 1.15).[39] Thus, when organizations adopt an ethic of responsibility, as with gender equity, an ethic of care often accompanies it, as with improved organization member well-being.

ethics Making judgments based on moral principles to differentiate between right and wrong

ethic of care An approach to ethical behavior that involves providing support to others and not causing harm

ethic of responsibility An approach to ethical behavior based on justice and equity

public good Outcomes that benefit everyone

FIGURE 1.15 When people perceive there is gender equity in a workplace, all organization members have greater feelings of well-being.

Both an ethic of care and an ethic of responsibility are included in the nine ethical principles NCA established to promote ethical communication. For example, "truthfulness, accuracy, honesty, and reason as essential to the integrity of communication" and "freedom of expression, diversity of perspective, and tolerance of dissent to achieve the informed and responsible decision making fundamental to a civil society" emphasize the importance of fairness and equity in interactions with others—an ethic of responsibility. Other principles, such as promoting "communication climates of caring and mutual understanding that respect the unique needs and characteristics of individual communicators" and condemning "communication that degrades individuals and humanity through distortion, intimidation, coercion, and violence, and through the expression of intolerance and hatred" suggest an ethic of care. As you read through the National Communication Association Ethical Principles in the box below, consider some of the ways you have applied them in your own communication with others or might apply them in the future.

National Communication Association Ethical Principles[40]

We advocate truthfulness, accuracy, honesty, and reason as essential to the integrity of communication.

We endorse freedom of expression, diversity of perspective, and tolerance of dissent to achieve the informed and responsible decision making fundamental to a civil society.

We strive to understand and respect other communicators before evaluating and responding to their messages.

We promote access to communication resources and opportunities as necessary to fulfill human potential and contribute to the well-being of families, communities, and society.

We promote communication climates of caring and mutual understanding that respect the unique needs and characteristics of individual communicators.

We condemn communication that degrades individuals and humanity through distortion, intimidation, coercion, and violence, and through the expression of intolerance and hatred.

We are committed to the courageous expression of personal convictions in pursuit of fairness and justice.

We advocate sharing information, opinions, and feelings when facing significant choices while also respecting privacy and confidentiality.

We accept responsibility for the short- and long-term consequences for our own communication and expect the same of others.

Still, a code of ethics or list of principles can encourage ethical behavior, but even explicit policies do not guarantee that individuals will follow the rules.[41] On college campuses, hazing is an ongoing problem in sports programs, fraternities, and other student groups.[42] Although nearly all higher education institutions in the United States have antihazing policies, a study of undergraduates found hazing prevalent, with more than half of students who belonged to a team, club, or other campus organization reporting they had experienced hazing. Those doing the hazing often posted photos of the incidents online. Only a tiny minority of students who were hazed—5%—reported what happened to campus officials. Sadly, students tended to view hazing positively rather than negatively.[43] Explore what your campus has done to address hazing in Box 1.4.

BOX 1.4: CONNECT & REFLECT

An End to Hazing

Go to your campus's website and search for your institution's policy on hazing. How is hazing described? What activities constitute hazing? What are the penalties for hazing? How are students encouraged to report incidents of hazing? Also search for training to prevent and avoid hazing. What workshops or programs are available for students and faculty advisors? What topics do they cover? Why do you think hazing continues on college campuses? How might you join in the fight to end hazing?

Communicating With Integrity

You act with **integrity** when you implement ethical practices, communicating honestly, openly, and respectfully. You contribute to an ethical world by facilitating dialogue, demonstrating respect, engaging in honest communication, and welcoming conflict.

integrity The implementation of ethical practices

Facilitate Dialogue

One of the NCA's principles of ethical communication is to "strive to understand and respect other communicators before evaluating and responding to their messages." Facilitating dialogue offers a clear path to achieve that principle. **Dialogue** involves inviting all participants to express their viewpoints while at the same time retaining an openness to other perspectives.[44] When you engage in dialogue with others, you share a genuine commitment to understand what others think *and* feel. You listen to logical and emotional arguments. You value each person's experience as a unique human being. Dialogue requires collaboration—participants must work together to create meaningful communication (Figure 1.16).[45] Communicators depend on each other to clearly articulate their thoughts and actively listen to what everyone has to say. In addition, dialogue means that individuals approach ethical dilemmas together, encouraging a cooperative approach to finding solutions.[46] With

dialogue Communication that invites all participants to express their viewpoints while at the same time retaining an openness to other perspectives

FIGURE 1.16 When people engage in dialogue, they collaborate to create meaningful interactions.

dialogue, you seek a balance between expressing yourself and understanding the perspectives that others bring to an issue. But what happens when some people do not have access to a conversation? Consider the issue of access in Box 1.5.

BOX 1.5: ETHICAL QUESTIONS

Dialogue and the Digital Divide

Dialogue welcomes all participants to the conversation. But what happens to those who cannot participate because they lack access to or an understanding of the necessary technological tools to join in? Recall one NCA ethical principle: "We promote access to communication resources and opportunities as necessary to fulfill human potential and contribute to the well-being of families, communities, and society."[47] Yet in the United States many households still do not have reliable internet access. In addition, about 7% of U.S. adults never go online.[48] Although that number has dwindled substantially in the past 20 years—from nearly 50% in 2000—that 7% still translates to about 18 million people. Most are over 65 and low-income earners living in rural areas. In a recent survey of 34 countries, younger people (ages 18–29) were much more likely to go online than older people (50 and over), especially in countries such as Indonesia, Tunisia, Greece, Philippines, and Ukraine.[49] Moreover, technology and communication experts predict a future with a "tele-everything" world for work, social life, and health care.[50] With the increasing use of social media and other online platforms for dialogue, the digital divide means some people are left out. What are the ethical implications of these missing people, especially older individuals, from online discussions? Some people choose not to go online. Is that okay? What happens when their voices are missing from online conversations? How can people get better prepared for dialogue in a world of tele-everything?

Demonstrate Respect

When you demonstrate respect, you show others that you are interested in their ideas and opinions. You enthusiastically engage with others, exhibit an awareness of them and their thoughts and feelings, actively listen to others, and display a sensitivity toward their experiences.[51] For instance, if you are discussing topics for a group project, you invite all participants to voice their ideas. You wait for them to finish speaking before responding to what they had to say. You ask questions that encourage others to elaborate and clarify their points. You suspend judgment and withhold evaluation. You encourage others to speak freely, ask questions, and share information. You truly want to know and understand others' viewpoints.

Engage in Honest Communication

Communicating honestly with others and encouraging their honest communication with you puts into practice the ethical principle to "advocate truthfulness, accuracy, honesty, and reason as essential to the integrity of communication." This transparency builds trust and positive interpersonal relationships. In addition, honest communication promotes information sharing, an essential component of effective decision making and problem solving. Still, you must balance openness with a sensitivity to personal and private information. Ethical communication involves "sharing information, opinions, and feelings when facing significant choices while also respecting privacy and confidentiality," another NCA ethical principle.

Welcome Conflict

Conflict is an unavoidable part of human existence and a part of daily life. Knowing that conflict will happen prepares you to face it ethically. In addition, conflicts can surface moral dilemmas that might otherwise be ignored. Working through disagreements can promote understanding and bring about positive change. Approach contentious issues with an open mind and indicate you are receptive to new ideas and opinions.[52] Recognize and assess your own biases, assumptions, and predispositions. Once you are more aware of your own thought processes, you are more open to considering alternative ways of approaching and solving a problem (Figure 1.17).

FIGURE 1.17 Communicating with integrity involves facilitating dialogue, demonstrating respect, engaging in honest communication, and welcoming conflict.

How a Communication Class Benefits You

Why take a class in communication? You may be asking yourself that question. After all, you have been communicating all your life. But even the best communicators know there is room for improvement. **Communication competence** requires both effectiveness in achieving your goals and appropriateness in adapting to norms and contexts. As you successfully work your way through this class, you will become a more competent communicator by deepening your knowledge about communication, increasing your motivation to communicate well, and putting what you know into practice. Improving your communication with others benefits your self-concept, interpersonal relationships, academic success, career, and community.

communication competence Occurs when an individual demonstrates effectiveness in achieving goals and appropriateness in adapting to norms and contexts

Communication Competence and the Self

How do you rate your own communication competence? Do you regularly achieve your goals when you communicate with others? Or do interactions often not go as you have planned? Communication competence and how you view yourself are intertwined. For example, if you view yourself as a shy person, you also will view yourself as a less competent communicator. In contrast, if you view yourself as outgoing, you are likely to think of yourself as a more competent communicator.[53] How you communicate with others and how they respond to you influence how you view yourself. When you have more successful interactions that demonstrate your communication competence, your self-confidence increases and you are more likely to feel more positive about communicating with others. Your communication class is designed to help you develop your knowledge base of communication concepts, enhance your positive attitudes toward communication, and sharpen your communication skills. Improving your communication competence will strengthen your confidence as an effective communicator and reduce your anxiety about communication.[54]

Communication Competence and Your Interpersonal Relationships

Developing effective communication skills improves your relationships with others. Research shows that romantic partners with higher levels of communication competence also report higher levels of relationship satisfaction and commitment.[55] Although communication is not a panacea for all relational problems, developing effective strategies for talking with family, friends, and romantic partners provides a basis for addressing difficult issues. For example, when parents divorce, children often feel caught in the middle and pressured to side with one parent or the other. A recent study found that parents with greater communication competence were able to reduce children's feelings of ambiguity about relational closeness and avoid choosing sides in the conflict.[56] These examples demonstrate that communication competence has a direct influence on all your interpersonal relationships (Figure 1.18).

FIGURE 1.18 Improving your communication skills improves your interpersonal relationships.

Communication Competence and Your Academic Success

Communication competence translates directly into academic success. College instructors evaluate students with better classroom communication more positively than students who are poor communicators.[57] Effective oral and written communication skills are essential in any college class. You write papers, take essay exams, give oral presentations, and answer questions in class. You also interact with your classmates and instructors. Yet you may evaluate your classroom communication as better than how others would rate you. For example, in a study of college students using nagging to persuade their instructors, researchers found that while students thought various nagging strategies were effective, instructors disagreed. When students barraged an instructor with requests ("Can we please get out of class early? Can we please?") or tried to strike a deal with the instructor ("If we finish our work quickly, can we leave class early?"), thinking that such a strategy was persuasive, the instructor was not convinced at all.[58]

Communication Competence and Your Career

Employers expect employees to demonstrate effective oral and written communication skills. A recent study by an international recruiting firm identified interpersonal and communication skills, team management and leadership skills, and organizational skills in the top 10 global skills that are in short supply.[59] That is, such skills are in high demand, but companies are having problems finding employees who can effectively build relationships with others, cooperate and collaborate with team members, and efficiently organize their time. Another study that examined the skills needed for entry-level positions—those filled by new college graduates—found interpersonal and oral communication skills the ones managers identified as most important.[60]

Even if you are majoring in a discipline that does not seem directly related to communication, you will need strong communication skills to do your work and advance in your career. For example, you might think scientists just have to worry about communicating with each other. However, scientists' effective communication with the public can help people understand key scientific information needed to make informed policy decisions, such as regulations that keep our food and water safe. The Alan Alda Center for Communicating Science at Stony Brook University in New York recognized the need for scientists to explain their research in ways nonscientists can comprehend. Recently, the organization hosted a conference on applied improvisation where attendees learned how improv can improve their ability to explain complex scientific and health care problems.

Communication competence has far-reaching effects in the workplace. For instance, supervisors can make their subordinates' work lives happy—or not. You probably have experienced both situations and know that how your supervisor communicates influences how you view your job. Research supports the link between supervisor communication competence and subordinates' feelings about the organization. Subordinates who evaluated their supervisors as sharing and responding quickly to new information, actively listening to others, interacting effectively across organizational levels, and using multiple communication channels reported higher levels of job satisfaction, communication satisfaction, motivation, and commitment to the organization.[61]

FIGURE 1.19 Collective communication competence makes for more productive and happier teams.

collective communication competence Occurs when individuals share an ability to interact effectively and appropriately, adapting to each other and various contexts

Many workplace tasks require teamwork, and collective communication competence provides the foundation for teams to achieve their goals (Figure 1.19). With **collective communication competence** group members share an ability to interact effectively and appropriately, adapting to each other and various contexts. Shared laughter, actively listening, engaging in collaborative learning experiences, and reflecting on group processes help build collective communication competence. In contrast, making sarcastic jokes, vying for power, and expressing boredom and disinterest create barriers to developing collective communication competence.[62] Workers who facilitate collective communication competence are more likely to feel happy, do well, and get ahead than those who engage in nonproductive behaviors that hinder teamwork.

Communication Competence and Your Community

Communication and community share the same etymology or origin, developing from the Latin word *communis*, meaning public or shared by all.[63] Communicating provides the foundation for forming and developing community, and effective communication is essential for you to participate in your community. Whether it is joining a student club, serving as a representative in your campus's student government, writing a letter to the editor of your local newspaper, blogging about issues in your neighborhood, or documenting an art project at an elementary school, the contributions you make to your community require you to articulate your ideas coherently and cogently. Box 1.6 asks you to identify ways you might apply your communication skills in contributing to your community.

Successfully completing this class will improve your communication competence, benefit you in a wide variety of contexts, and help you achieve your personal, relational, academic, professional, and community goals.

BOX 1.6: LIVE IT LOCAL

Contributing to Your Community

Communicating with others in your local community provides you with opportunities to create connections with others and contribute to their well-being, along with your own. Make a list of the communication skills you have that would benefit others. Do you have excellent writing skills? Are you a good listener? Do you find it easy to talk one-on-one with others or small groups? Do you enjoy expressing yourself creatively, such as drawing or performing? Now consider how you could apply those skills in your local community, such as working with middle school children to create skits about their experiences, helping English language learners feel more comfortable in everyday conversations, and interviewing long-time residents about the area's history. These are just a few suggestions of how you might apply your communication skills in your community—you likely will imagine many more.

CHAPTER REVIEW

- **Explain the components of the era of convergence in human communication.**

  First, communication activities that once required several devices now need only one, as with the smartphone. Second, companies now both produce communication hardware and distribute media, as with Apple manufacturing the iPhone and distributing music via iTunes. Third, media have collapsed into a single medium, as with the online availability of TV shows and print news.

- **Discuss the implications of the communication imperative.**

  As social beings, humans communicate as a natural part of everyday life. The communication imperative drives individuals to figure out the best ways to achieve their communication goals. From this perspective, you determine how you want to use technology in your communication with others, rather than technology making that decision for you.

- **Identify and describe historical models of human communication.**

  Three traditional models of communication informed early understandings of the process. The mathematical or transmission model depicted communication as action—something people do. The interactive model added a critical dimension to the communication process—feedback. The Source-Message-Channel-Receiver (S-M-C-R) model provides a more detailed view of human communication and attempts to identify all the elements needed for communication to occur.

- **Discuss the key components of the pervasive communication environment model and its implications for human communication.**

  The pervasive communication environment model depicts communication as convergence—the bringing together of elements that previously were separate. The smartphone provides the best example, with a camera, web browser, alarm clock, calendar, telephone, and many more communication tools, all in one device. The pervasive communication environment model foregrounds communicators' ability to access information and interact with others anytime from almost anywhere.

■ **Explain the definition of communication.**
Communication involves exchanging socially shared symbols that represent interactants' thoughts, ideas, and feelings. Oral and written language are verbal symbols. Nonverbal symbols are all those that are not linguistic.

■ **Recognize the elements of ethical communication.**
Ethical communication involves demonstrating an ethic of care and an ethic of responsibility. Ethical communicators display integrity by facilitating dialogue, demonstrating respect, engaging in honest communication, and welcoming conflict.

■ **Identify the benefits of studying communication.**
Successfully completing a class in communication increases your communication competence, which improves your self-confidence, academic success, interpersonal relationships, career options, and participation in your community.

DISCUSS AND APPLY

1. How have you experienced the era of convergence? For example, what are some ways you use your smartphone in addition to making phone calls?

2. The communication imperative suggests that individuals will make technology work for them. How do you use technology to communicate with each of the following?

   a. friends
   b. family
   c. instructors
   d. businesses

3. How does what you want to communicate influence your choice of communication channels?

4. How does your relationship with the person or group influence your choice of communication channels?

5. What is your response to the pervasive communication environment model? How does the model help you better understand communication? What do you think would improve the model?

6. How would you define communication? In what ways does your definition differ from that offered in this chapter?

7. Some communication scholars argue that communication does not have to be intentional—that one cannot *not* communicate. What do you think? Is everything you do communication?

8. Go to the TED Talk website (ted.com) and watch José Andrés's TED Talk: "How a Team of Chefs Fed Puerto Rico After Hurricane Maria." How did his team apply an ethic of care? How did they apply an ethic of responsibility? Go to his organization's website (wck.org), and review the World Central Kitchen's mission and programs. How is food an ethical issue?

9. Why are you taking this class? What do you hope to get out of it? What are your goals for this class?

CREDITS

IMG 1.0: Copyright © 2019 iStockphoto LP/gremlin.

Fig. 1.0: Copyright © 2020 Pexels/Andrea Piacquadio.

Fig. 1.1: Copyright © 2020 Pexels/TUBARONES PHOTOGRAPHY.

Fig. 1.2: Copyright © 2020 Unsplash/Ahmed Hindawi.

Fig. 1.3: Copyright © 2016 Pexels/Pixabay.

Fig. 1.4: Copyright © 2020 Pexels/Vlada Karpovich.

Fig. 1.5: Adapted from Leonhard Schilbach, et al., Your Brain on Communication, from "Toward a Second-Person Neuroscience," Behavioral and Brain Sciences, vol. 36, no. 4, pp. 396. Copyright © 2013 by Cambridge University Press.

Fig. 1.6: Copyright © 2020 Depositphotos/Kzenon.

Fig. 1.7: Copyright © 2020 Unsplash/Julian Wan.

Fig. 1.8: Wanderingstan, "Shannon Communication System," https://commons.wikimedia.org/wiki/File:Shannon_communication_system.svg, 2007.

Fig. 1.9: Wilbur Lang Schramm, "Interactive-model," The Process and Effects of Mass Communication, pp. 8. Copyright © 1954 by University of Illinois Press.

Fig. 1.10: David Kenneth Berlo, "SMCR-model," The Process of Communication. Copyright © 1960 by Houghton Mifflin Harcourt.

Fig. 1.11: NASA, "Apollo 11 Liftoff from Launch Tower Camera," https://commons.wikimedia.org/wiki/File:Apollo_11_liftoff_from_launch_tower_camera.jpg, Cambridge University Press, 1969.

Fig. 1.13: Copyright © 2020 Pexels/Ono Kosuki.

Fig. 1.14: Copyright © 2013 Depositphotos/Wavebreakmedia.

Fig. 1.15: Copyright © 2020 Pexels/Andrea Piacquadio.

Fig. 1.16: Copyright © 2020 Pexels/Cliff Booth.

Fig. 1.17: Copyright © 2020 Pexels/William Fortunato.

Fig. 1.18: Copyright © 2018 Pexels/mentatdgt.

Fig. 1.19: Copyright © 2019 Pexels/Jopwell.

CHAPTER 2

Culture and Communication

When Disney sought to trademark "Día de los Muertos" in conjunction with an animated movie the company produced, Facebook pages, Twitter feeds, and other social media became the tools used to protest the request. *Día de los Muertos*, or Day of the Dead, is a traditional Latinx holiday in which participants honor loved ones who have died. Disney was accused of cultural insensitivity, questionable ethics, and trying to trademark a holiday like Christmas or Easter.[1] Disney also faced criticism and an online petition for trademarking the phrase "hakuna matata," words from Swahili meaning "no worries" used in *The Lion King*.[2] The petition urged Disney to halt attempting to trademark what it did not invent, including words, phrases, and holidays.

These are not Disney's only cultural missteps. Arab Americans and others criticized song lyrics in the Disney movie *Aladdin* for stereotyping Arab people. Additionally, Disney was questioned about the underlying voodoo theme in *The Princess and the Frog*, which featured a Black lead character.[3] Disney drew criticism for cultural appropriation and misrepresentation in its film *Moana*.[4] And although Euro Disney managed to improve its image and popularity since first opening in 1992, initial criticisms cited the company's lack of attention to local culture. For example, the theme park was alcohol-free and served primarily fast food in contrast to predominant cultural norms that include beer or wine with meals and a more leisurely approach to eating.[5]

Disney learned from its mistakes and did its homework before opening theme parks in Hong Kong and Shanghai. In Hong Kong Disneyland, for instance, the company consulted a feng shui specialist in designing the park's layout and hired employees who speak the predominant Chinese languages, Cantonese and Mandarin.[6] Still, even with greater attention to cultural differences, Disney has encountered problems when it tries to export its approach to entertainment to locations outside the United States. An ethnographic study of Hong Kong Disneyland found that the company did not take into account residents' love of photographing their experiences, even though those actions may disrupt rides and interfere with employees' work.[7]

Because of globalization, you live and communicate in a world that extends far beyond your physical space and culture.[8] The pervasive communication environment makes it much more likely you will interact and build relationships with people from a number of other cultures.[9] Like Disney, you may encounter surprises and problems as you interact with others whose backgrounds differ from yours. So, as Disney did in launching its more recent theme parks, you will want to develop ways to bridge and embrace cultural differences.

Defining Culture in a Pervasive Communication Environment

Every day, or at least nearly every day, you interact with others across a variety of contexts. For example, you text or FaceTime your friends, meet classmates for coffee, and email your professors. How do you know what you are supposed to do in these interactions? How do you know what is expected? How should you interpret what others say? Culture provides the basis for your communication, and you learn about culture through your communication with others. In other words, communication both creates and reinforces culture. Although **culture** is defined in many ways, Manual Castells offers a brief and useful definition: "a set of values or beliefs informing behavior [and a] collective construction that transcends individual preferences."[10] Let's examine that definition more closely.

culture A set of values or beliefs informing behavior [and a] collective construction that transcends individual preferences

"A set of values or beliefs informing behavior" refers to the similar interpretations or worldviews members of a culture share and how those interpretations influence actions. For example, historical analysis of U.S. culture suggests several primary beliefs associated with work: Everyone should work, workers must get some benefit from their efforts, and physical labor is good.[11] These beliefs shape how we talk about work, how we do our work, and the expectations we hold for how others should work. Do you think of your college classes as work? Is volunteering in your community work? Do you consider parenting a job? Answering these questions depends on your beliefs about work.

FIGURE 2.1 Men in traditional wear.

FIGURE 2.2 What cultural beliefs and values does this photo illustrate?

norms Rules that are socially enforced

"Collective construction that transcends individual preferences" means that you produce culture in your interactions with others and that cultural views differ from individual perceptions. Developing, maintaining, and changing culture depends on individuals communicating with each other, so culture goes beyond what just one person may think or believe. You construct culture every day as you go about your life. You learn about culture from obvious sources, such as your family, friends, instructors, and institutions. You also learn about culture through less transparent sources, such as proverbs, folk tales, legends, myths, art, and the media (Figure 2.1).

You create culture through communication, such as telling a story to someone at work or in class or by posting a photo on Facebook or Snapchat. Culture also influences how and what you communicate, such as the particular story you decide to tell and the specific photo you choose to share. That is, culture provides a framework both for the messages that you produce and how you interpret messages others produce (Figure 2.2). For example, suppose you observe people crying at a wedding. You likely would interpret those tears as tears of joy. Yet if you observe the same people crying at a funeral, your interpretation would likely be very different—you probably would view those tears as tears of sorrow. Finally, if you viewed the individuals crying after getting fired from their jobs, you probably would think they felt rejected, sad, or angry. The behavior is the same, but culture influences how you interpret the behavior in these three situations.

Often, people do not choose to share just one story or one photo. How is it that people choose to represent themselves through communication, and what guides those choices? For example, what types of pictures do your friends usually share via social media? What common themes do you see across these posts? Are there pictures that you and your friends would not post? Culture surrounds you with recurring patterns of behavior (e.g., posting photos of one's food in restaurants or one's activities while on a beach vacation), and these recurring patterns serve to create norms or ways of behaving. **Norms** are socially enforced rules or guidelines, often implicit and unstated. For example, in a recent class discussion on culture and norms, one of our students observed that when people in the United States board a train, they generally seek out seats in empty rows, spacing themselves away from others. A student from Japan remarked that this seemed odd. She said that in Japan, even if there is only one person in a train car, it is considered rude not to go and sit next to that person.

Because you are born into and deeply enmeshed in your own culture, it is often difficult for you to identify the cultural values and beliefs that guide your own behaviors. You simply accept ways of doing and saying things as natural and normal—just how it should be. For example, the saying that "fish can't see the water in which they swim" conveys the idea that when you are enmeshed in a culture, it is difficult to see how it shapes your behavior and to recognize the taken-for-granted assumptions. When you attend in-person class meetings, what cultural norms guide your behavior? Where and how do you sit? What types of clothing might you wear? Conversely, what types of clothing would you never wear? What are example of differences in online class meetings? How do you know when to talk, when not to talk, and how long to talk? Explore these ideas further by considering the examples and questions in Box 2.1.

Creating and Detecting Culture

Five **cultural indicators** help define key aspects of culture: vocabulary, practices, rituals, artifacts, and stories. **Vocabulary**, or language, is one way cultural members define themselves, build identity, and identify who is in the group and who is not. Using specialized language or speech codes can build group identity.[12] Consider the language you need to know as a student at your college. At San José State University, *MySJSU* is the website students log in to for their class schedules and fee payment notices, a class greensheet is a syllabus, and BSAC is the advising center for business students. What are examples of similar vocabulary used on your campus? When joining a new group or organization, part of the socialization process is learning the language system. For example, if you joined the military or went to medical school, you would learn new a new vocabulary, and despite some commonalities, that vocabulary would differ somewhat based on where you were stationed or the hospital where you practiced. When entering a group or organization, newcomers might feel excluded when they do not know the terms being used.

Practices, or how tasks are accomplished, also reveal important cultural information. Practices may or may not coincide with the rules or guidelines for

cultural indicators Help define key aspects of culture; several important indicators are vocabulary, practices, rituals, artifacts, and stories

vocabulary One way cultural members define themselves, build identity, and distinguish who is and is not in the group

practices Ways tasks are accomplished; may or may not coincide with the rules or guidelines for getting things done or how things work

rituals Cultural events that happen regularly; some are grand occasions that mark major transitions, such as a college graduation ceremony, and others are daily activities, such as how you greet people when you get to class and how the instructor ends the class each day

getting things done or how things work. A campus workout facility, for instance, may have rules about how long members may use a piece of equipment, such as a treadmill. But those rules may be enforced only when the gym is busy. You have probably observed how practices develop in your various college classes. In one class, students may talk quite a bit and frequently ask the instructor questions. In another, the instructor might lecture, with students remaining quiet. Some classrooms have signs that read, "No eating, drinking, or smoking." Despite these signs, are there differences in cultural practices? For example, smoking is likely viewed differently culturally than it was when those signs were originally posted. But has the practice of carrying water bottles changed drinking in class at your school? Are there differences in practices based on the type of classroom (e.g., standard classroom, computer lab)?

Rituals refer to cultural events that happen regularly (Figure 2.3). Some rituals involve grand occasions that mark major transitions for cultural members, such as first-year orientation, honors convocation, and graduation at a college or university. Others are daily activities, such as how you greet people when you get to class and how the instructor ends the class each day. You get accustomed to these rituals and notice when they are changed. For example, if your instructor always reserves 5 minutes at the end of class for students to ask questions and skips this ritual one day, you might wonder why.

Artifacts might make you think of items found during an archeological dig, such as parts of a building, pottery, buttons, or tools. But artifacts can also be found all around you (Figure 2.4). They are simply objects associated with a culture, such as buildings, artwork, and clothing. For example, uniforms and dress codes provide an easy way to identify who is part of some cultural groups and who is not. When you go to a hospital, you can tell who works there by their clothing and other artifacts, such as scrubs, lab coats, stethoscopes, and ID badges. Many restaurants require employees to follow a narrowly defined dress code, such as white shirts, black pants, and dark shoes. Wearing a uniform can give cultural members a greater feeling of unity. Elementary and secondary schools often have a dress code to increase identity with the school and reduce individual identity. When you visit professors' offices, what types of artifacts do you commonly see on display? What are examples of artifacts from your school's sports teams?

FIGURE 2.3 In this greeting ritual, what is communicated about culture?

artifacts Objects associated with a culture, such as build-ings, artwork, and clothing

FIGURE 2.4 If you went on an "archeological dig" in this classroom, what artifacts would you find particularly meaningful? What do these artifacts tell you about the culture?

Stories and myths transmit cultural values and beliefs in implicit ways (Figure 2.5).[13] You learn about a group's culture when members recount how the group was started, reveal anecdotes about group members, or relate tales of overcoming adversity. Sometimes group members tell stories to reinforce their connectedness, such as when your high school friends retell the story of the big game or people you work with retell the story about the crazy customer, even though everyone already knows it. The stories group members tell, however, do not always transmit positive values, as in the case of hate groups. Hate groups promote hostility and often violence against others based on ethnic background, religion, sexual orientation, and similar social categories. The Southern Poverty Law Center documents more than 830 hate groups in the United States and increases in actions by such groups, such as extremist flyers placed across many college and university campuses.[14]

Hate groups increasingly use the internet to communicate with each other, disseminate their ideas, and persuade people to join the group.[15] Research has found that adolescents find stories told on White supremacist websites more persuasive than when the site states the group's position explicitly. When values are embedded in a story, you are less likely to critically examine them. Another study of four hate group websites identified three fantasy themes the groups shared: "a separate state provided as fair compensation," "resurrection of the righteous," and "the original people of the Earth are called to give new meaning to race." Although information included on these sites is factually wrong and clearly one-sided, the use of narrative and myth to convey their message softens the tone and makes the hate groups seem less threatening. Yet there is no doubt these groups have developed a culture of hate toward people who are not like them.[16] Additionally, the COVID-19 pandemic and new online platforms have made hate and extremism more challenging to track, making assessments of cultures of hate and their associated cultural indicators (e.g., vocabulary, practices, stories) more difficult.[17]

FIGURE 2.5 How have stories influenced your views on your school, your workplace, or your friends?

stories and myths Transmit cultural values and beliefs in implicit ways; may recount how the group was started, reveal anecdotes about group members, or relate tales of overcoming adversity that let others learn about the culture; also used to reinforce connectedness

Culture and Cocultures

Individuals are rarely part of a single culture. Rather, they are part of **cocultures**, or groups with behaviors or beliefs that differentiate them from the broader culture in which they are embedded.[18] Within the larger U.S. culture, you probably belong to several cocultures based on your ethnic background, religion, gender, education, sexual orientation, and hobbies and other interests. For example, if you are a Black woman raised in the Baptist church majoring in computer science, you may consider yourself part of cocultures associated with each of those groups. You rarely identify with only one coculture, but likely find yourself involved in several cocultures, often overlapping, but sometimes quite dissimilar.[19]

coculture A set of people with a distinct set of behavior and beliefs that differentiate them from a larger culture of which they are a part

speech code Socially constructed system of rules, terms, and meanings that guide communication within a particular group

Often, different cocultures have distinctive speech codes. A **speech code** is a socially constructed system of rules, terms, and meanings that guide communication within a particular group.[20] Speech codes let you know who can talk when, what they can talk about, and how others should respond. For example, a study of a Lebanese community in the United States found that participants developed communication practices centered on traditional ways of preparing and serving food. Deference was paid to elders—especially women—in cooking and baking the food. Men generally were not found in the kitchen, except for physically demanding tasks, such as lifting large, heavy pots. When working together to prepare food for the annual food festival, participants told stories that reinforced the community's traditions.[21] Some coculture members develop considerable fluency in **code switching**, which means that they can go back and forth between ways of speaking and behaving to adapt to different cocultures.

code switching Moving back and forth between ways of speaking and behaving to adapt to different cocultures

Professions develop speech codes as well. A study of nanotechnology engineers and scientists identified implicit rules associated with how they described their work. Talk about their research centered on the benefits to society, the desire to pursue knowledge, and the altruistic nature of their work. In contrast, they referred to corporate work in nanotechnology as motivated by profit and consumerism. They also differentiated scientific research in nanotechnology from science-fiction plots portrayed in the popular media. For the engineers and scientists, talk about nanotechnology should focus on how this aspect of science will help people lead better lives.[22] Another example is the ways that employees of a police department and the American Civil Liberties Union might use different speech codes when discussing the racial justice movement. Box 2.2 provides a real-life example of what happens when speech codes are violated.

BOX 2.2: ETHICAL QUESTIONS

When Cocultures Clash

News coverage of Hurricane Katrina demonstrates what happens when speech codes are violated.[23] This 2005 natural disaster, which killed nearly 2,000 people in the southeastern United States, caused extensive damage in South Florida, Louisiana, and Mississippi. Tens of thousands of people had to abandon their homes. As the news media reported on the events, journalists started using the term "refugees" to refer to those displaced by the storm. Yet those fleeing their flooded neighborhoods were not crossing national borders. Using the term suggested that the storm's victims were not United States citizens and were separate from the rest of the country. Many of those most affected by the storm were Black, raising questions of racism in the use of term "refugees."

1. How would you define a refugee? What do you think of when you see or hear the term "refugee"?

2. As a coculture, journalists often stress the need for objectivity and an unbiased view in reporting events. How does the use of refugees in identifying victims of Hurricane Katrina fit with those values?

3. Why do you think Black people were particularly outspoken about this issue?

4. What have you learned about cocultures and language based on this example?

Cocultures form in a variety of places. For example, four hierarchical layers, or cocultures, contribute to internet culture.[24] The first layer is the techno-meritocratic coculture with roots in academia and science. This coculture values peer review and sharing knowledge, with members believing in the inherent good of science and technology as key components in the development of humankind.

Next is the hacker coculture, which builds on techno-meritocratic culture. Members of the hacker coculture believe in open source knowledge and merit earned through quality work. However, hacker culture distains the institutional base of the techno-meritocratic culture and instead values technological excellence, creativity, and freedom.

The techno-meritocratic and hacker cocultures have given rise to the communitarian coculture, which values open communication and the ability to create online groups and communities. All the various ways that people organize themselves online, from promoting causes to planning parties, stem from the communitarian commitment to providing new ways for individuals to connect with each other. Such connections were especially pivotal during the recent pandemic.

Finally, the entrepreneurial culture developed from the work of the techno-meritocratic, hacker, and communitarian cultures. The entrepreneur uses the structure and ideas of the other cultures and turns them into profit and products that support the larger internet culture. Social networking sites such as Twitter, Facebook, and Instagram provide examples of companies that took the communitarian coculture to the entrepreneurial level.

This analysis of internet culture illustrates how different cocultural forces interact and influence each other, as well as how cocultures collaborate to make up the larger society. These internet cocultures are depicted in Figure 2.6.

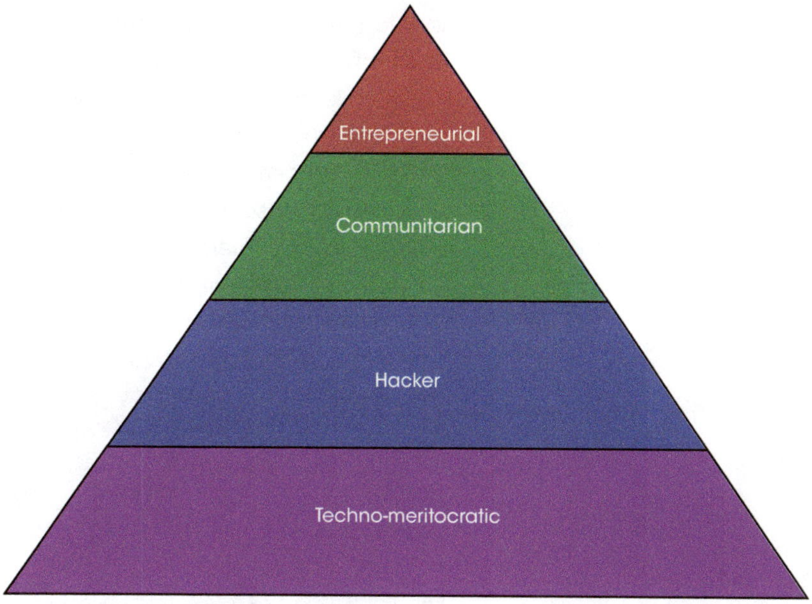

FIGURE 2.6 Internet cocultures.

Collectivist and Individualist Cultures

Research suggests that cultures range on a continuum from individualist to collectivist. Broadly speaking, Western cultures like the United States, Australia, and Europe lean toward individualism, while Eastern and Southern cultures such as China and Guatemala focus more on collectivism. The labels for each reflect the

individualist culture Prizes self-attainment, competition, and personal growth

collectivist culture Prizes loyalty to the group and cooperation

emphases. Self-attainment, independence, competition, and personal growth are highly prized in **individualist cultures**, whereas loyalty to the group, compliance, and cooperation are valued in **collectivist cultures**.[25] Of course, these descriptors are broad generalizations that oversimplify how people from different regions behave as well as some individual personality differences; however, these categories do provide information on generally held beliefs and ways of behaving. Most cultures include a mix of attributes.

From a collectivist perspective, members of a culture or group define their own interests and values in reference to the collective good. A collectivist perspective assumes what is good for the group is good for the individual. Thus, behavior that prioritizes group needs is valued. Examples include other-orientedness, self-sacrifice, and conforming. In contrast, an individualist perspective views success as achieved through the pursuit of individual objectives that fit with the culture's or group's goal. From the individualist perspective, what is good for the individual is good for the group. Using this lens, behavior that maximizes individual goals is prioritized. Examples of such behaviors include a focus on individual excellence, independence, and uniqueness. In your opinion, how might educational or health care systems differ in collectivist and individualist cultures?

A collectivist orientation typically results in a higher degree of cooperation and concern with a group's goals. In contrast, an individualist orientation leads to more competition between people. However, individualist cultures are more accepting of dissent and disagreement, which can fuel innovation. Collectivist cultures are less tolerant of dissent, which may stifle creativity.[26]

Table 2.1 summarizes the basic attributes of individualist and collectivist cultures.

TABLE 2.1 Individualist and Collectivist Cultural Attributes

INDIVIDUALIST CULTURES	COLLECTIVIST CULTURES
Emphasis on "I"	Emphasis on "we"
Identity grounded in the individual	Identity grounded in the group
Promote competition	Promote cooperation
Reward self-attainment	Reward group loyalty
Foster dissent and disagreement	Foster social harmony
Encourage independence	Encourage interdependence

Your culture also influences how you view yourself. Researchers have found that if you are a member of an individualistic culture, you likely see yourself as independent from others and your self-descriptions emphasize that separateness. In contrast, if you are a member of a collectivist culture, you are likely to hold an interdependent view of yourself, viewing yourself through relational connections.[27]

For example, if you are from an individualistic culture, when asked you describe yourself, you might focus on your individual traits: "I'm an independent thinker who is resilient, straight-forward, and creative." Alternatively, if you are from a collectivist culture, you might focus on your connection with your social group: "I'm from a large, close-knit family, and I enjoy family gatherings and making meals together."

Collectivist and individualist orientations frame how you interpret behaviors—what you think ways of behaving mean. In many collectivist cultures, for instance, silence and listening are signs of leadership and wisdom. Most individualistic cultures, however, associate verbal dominance with leadership and silence with weakness or indecisiveness. Therefore, those from more individualist cultures might be suspicious of someone who talks little, and those from more collectivist cultures might be suspicious of those who talk a lot (Box 2.3).[28]

BOX 2.3: CONNECT & REFLECT

The Sounds of Silence

Consider the role of silence in your everyday interactions. For example, how do you evaluate your classmates who do not say much in class or who are silent? What about those who talk a lot? Reflect on a recent interaction with friends. How would you describe those who did not talk much? What about those who talked a lot? How do you think culture influences your evaluations? How might culture influence their conversational participation in these two settings?

Research for several decades in a number of nations identified key differences in individualist and collectivist cultures that impact communication across a wide range of contexts.[29] At school, for example, students in individualist cultures are expected to express themselves in class, asking questions, offering ideas, and seeking others' opinions. In contrast, students in collectivist cultures are expected to remain quiet and speak only at certain times. At work, employees in individualist cultures focus on getting the job done, whereas in collectivist cultures, relationships typically are given a higher priority than tasks. Table 2.2 summarizes some key differences between individualist and collectivist cultures in education and the workplace.

Although research that examines individualist and collectivist cultures continues to find distinct differences between the two, social media may be blurring those lines. For example, a study on organizations' Facebook pages found that they often varied from cultural expectations. That is, some organizations from more collectivist cultures promoted their competitiveness, and some from more individualist cultures highlighted their collaborative approach to problems.[30] Further, the digital and media networks are largely owned by an

TABLE 2.2 Individualist and Collectivist Cultural Differences at School and Work

INDIVIDUALIST	COLLECTIVIST
Individual students are expected to speak up in class	Students speak up in class only when sanctioned by the group
The purpose of education is learning how to learn	The purpose of education is learning how to do
A college degree increases economic worth and/or self-respect	A college degree provides entry to higher status groups
Occupational mobility is higher	Occupational mobility is lower
Employees are "economic persons" who will pursue the employer's interest if it coincides with their self-interest	Employees are members of in-groups who will pursue the in-group's interest
Hiring and promotion decisions are supposed to be based on skills and rules only	Hiring and promotion decisions are based on being part of the in-group
The employer–employee relationship is a legal contract between parties	The employer–employee relationship is basically moral, like a family link
Management is managing individuals	Management is managing groups
Management training teaches the honest sharing of feelings	Management training teaches that direct appraisal of subordinates spoils harmony
Every customer should get the same treatment (*universalism*)	Customers in your group get better treatment (*particularism*)
Tasks prevail over relationships	Relationships prevail over tasks

Adapted from: Hofstede et al. (2010, p. 124)

increasingly small set of multinational conglomerates, contributing to increased blurring of cultural distinctions.[31]

Several points above give reasons to appreciate and learn from cultures that differ from your own. But there is a great difference between cultural appreciation and cultural appropriation. **Cultural appreciation** is demonstrated by wanting to learn about another culture, build intercultural relationships, and increase understanding. Conversely, **cultural appropriation** refers to taking something from within a culture and using it for your benefit, often with little understanding of its cultural significance. In other words, cultural appropriation typically takes a narrow view of the culture (e.g., focuses on an article of clothing to mass produce, a photo that may get lots of "likes" from friends), whereas cultural appreciation takes a broad view of the culture (e.g., seeks to understand the origins of the culture's practices, the meaning of norms, and the relations between people).[32] In Box 2.4, the issue of cultural appropriation in regard to Halloween costumes is explored.

cultural appreciation
Demonstrated by wanting to learn about another culture, build intercultural relationships, and increase understanding; broad view focused on understanding

cultural appropriation
Taking something from within a culture and using it for your benefit, often with little understanding of its cultural significance; narrow view (e.g., interest in one item or aspect, rather than understanding culture more fully)

BOX 2.4: CONNECT & REFLECT

"We're a Culture, Not a Costume"

In response to racist Halloween costumes,[33] Ohio University students created a communication campaign, including posters, to educate their peers about problems with some costume choices and encourage thoughtfulness. Their campaign, "We're a culture, not a costume," went viral.[34] Later, two University of Utah students, Amerique Phillips and Alexis Baker, addressed cultural appropriation, cultural appreciation, and key questions regarding Halloween costume choices.[35] A question to consider is whether the costume represents a culture that is not your own, perpetuates stereotypes, includes race-related or ethnicity-related accessories, sexualizes the sacred, or includes packaging with words such as "authentic" or "tribal." If you answer yes or are not sure about an answer to one or more of these questions, you should reconsider.[36]

Are there other questions that you would add to the list? What can colleges and universities do to encourage costume choices that are not offensive? How would you respond to Halloween or other costumes that appropriate culture? How should colleges and universities respond?

Diversity and Communication

Diversity is most often associated with differences such as race, sexual orientation, and gender (Figure 2.7). But thinking about the concept more broadly helps you understand **diversity** as individuals with different attributes contributing their unique experiences to a communication encounter. With this definition, you can consider diversity in a wide range of areas, such as socioeconomic status, religion, dis/ability, political affiliation, age, education level, and personal experiences. For example, in the United States your chances of earning a bachelor's degree are more than five times higher if you are from a rich family than if you are from a poor one. Further, fewer students from low-income families go to college than students from middle- and upper-income families.[37] So in many colleges and universities, the student population based on socioeconomic background is fairly homogeneous, even when there is diversity in other areas, such as ethnicity and religious affiliation.

diversity Individuals with different attributes contributing their unique experiences to a group

Why be concerned about diversity and communication? First, diverse groups of people usually are more successful than homogeneous ones. Racially and ethnically diverse groups report a stronger belief in their combined abilities to get tasks done at the end of their projects than groups lacking diversity. A study of a Fortune 500 information-processing company found that when teams included members with a wide range of personal and professional experiences, they worked together better than teams made up of people with

FIGURE 2.7 What are your observations about this photo?

similar backgrounds.[38] In several other studies, diverse groups have been found to perform better for several reasons, including decisions rooted in facts, better overall information processing, broad range of views, and increased innovation.[39]

Second, humans have a natural tendency to identify similarities in others. You try to link others' experiences with your own and often assume you have more in common than might be the case. This assumed similarity leads to assuming others think like you as well. When you interact with others, you typically experience the **common knowledge effect**, where you overemphasize common knowledge everyone has and underemphasize knowledge only certain individuals have.[40] So even when you are interacting with a diverse group of people, you might miss out on the unique information each person has to offer because you assume everyone is drawing from the same knowledge base. Overcoming the common knowledge effect requires actively seeking everyone's input to leverage all possible knowledge, skills, or contacts.

Third, you also tend to be attracted to and want to interact with people you perceive as similar to you. It is natural to feel hesitant when interacting with others you think are not very much like you.[41] Yet having a diverse relational network can help you learn how to better handle conflict and disagreements.[42] Moreover, a study of African Americans and European Americans working together found that when faced with racial humor, the presence of an African American resulted in a higher degree of cultural sensitivity among the European Americans.[43] Twitter, Facebook, Instagram, YouTube, and other social media provide opportunities to expand your interaction circle and get to know how a wide variety of people experience their worlds. Stepping outside of your usual comfort zone and following someone on Twitter you ordinarily would not encounter in your everyday life or searching for a YouTube video about a culture with which you are not all that familiar are just two examples of ways to develop a more diverse network. Interacting with a wide range of people with varying backgrounds and interests increases awareness and acceptance of differences. Actions like the above examples can help reduce cultural myopia—nearsightedness that makes one's own culture seem best, appropriate, or comfortable across situations—and encourage a broader view of cultural perspectives. Box 2.5 invites you to explore experiencing diversity firsthand.

common knowledge effect Occurs when you overemphasize common knowledge held by a group of people and underemphasize knowledge held by certain individuals

---

BOX 2.5: LIVE IT LOCAL
Experiencing Diversity

Attend a local religious or cultural event sponsored by people outside the usual group of individuals with whom you interact. Observe the indicators of culture, including vocabulary, practices, rituals, artifacts, and stories, you can detect. What do these indicators tell you about the culture or coculture in which the event is embedded? Is the culture or coculture more collectivist or individualist? What led you to that conclusion? What have you learned about your own cultural background based on this experience? What have you learned about your local community and diversity?

Gender and Diversity

Some years ago, John Gray wrote the book *Men Are From Mars, Women Are From Venus*.[44] Gray's argument that women and men innately communicate in distinctly different ways persists even today. His view is appealing because it removes any responsibility you have for less-than-ideal communication with the opposite sex/gender. But Gray's work is not based on empirical research, and it actually contradicts what gender researchers have found. These researchers identify differences between how men and women communicate based on socialization and social structures. Because these differences are not innate, you can adjust, adapt, and modify your communication to better interact with others.[45]

In the dominant U.S. culture, men tend to use talk as a way to negotiate and maintain status and independence, whereas women talk to negotiate relationships and build connections. For example, research on social support has found that women are better able and more motivated to attend to supportive messages and situations.[46] Although the many reasons for these differences are too complex to address here, you can focus on how you can practically negotiate women's and men's communication styles in your everyday inter-actions. For example, research suggests that women embrace digital media in more and different ways than men. Research on college students has found that women prefer social networking sites, video chat, and text messaging more than men and use these forms of mediated communication more frequently than men.[47]

Still, traditional interaction styles can persist. A study of how teens present themselves with words and images in chat rooms found that self-presentations generally fit gendered stereotypes.[48] However, differences in gendered communication styles reinforce the importance of embracing diversity across all communication situations (Figure 2.8). Ethical communicators recognize the positive role diversity plays in human communication. Consider also how moving beyond a gender binary might influence communication (Box 2.6).

FIGURE 2.8 Although the people featured in this photo are not from Mars or Venus, are any gender differences evident?

BOX 2.6: CONNECT & REFLECT

Beyond the Binary

For many years, most research examined gender as binary. Only relatively recently have researchers begun to incorporate a fuller array of gender identities into their work; thus, there are few scientific findings regarding different gender identities and culture. Consider gender identities, including man, woman, both, and neither. As researchers move beyond the binary, what new findings do you anticipate? How might diversity in gender identities contribute to group and workplace performance?

Promoting Inclusiveness

When you promote inclusiveness, you communicate in ways that indicate you accept others' perspectives and ideas as important and valuable, even if you disagree with them. Inclusiveness makes others feel welcomed and part of activities or events; exclusiveness makes others feel unwelcomed, distrusted, and not accepted for who they are.

Promoting inclusion involves three parts: awareness, knowledge, and skills (Figure 2.9).[49] First, you must be aware of your own cultural frame of reference, biases, thought patterns, norms, and ways of doing things. Second, you need to learn about cultures and cocultures different from your own so you have a better understanding of how other people experience the world. Third, you must develop skills for interacting with a broad range of people in a variety of contexts. The actions listed below are ones you can practice to promote inclusiveness across your interactions with others:[50]

FIGURE 2.9 Beyond signs, what are other ways to promote inclusion?

- *Build an awareness of your own personal biases, prejudices, and stereotypes*: When you find yourself evaluating someone, consider the source of your evaluation. Are you basing it on a general assumption or your personal knowledge of the individual? How might your preconceived ideas affect your interpretations?

- *Suspend your first impressions*: You make judgments about people within the first few seconds of interacting with them. Based on limited and many times incorrect interpretations, these first impressions can shape your subsequent observations. Recognize that first impressions often are wrong and give yourself time to get to know others before drawing conclusions about them.

- *Take on the role of observer*: Consider your own and the other person's interaction from the perspective of a third-party observer—someone outside the conversation. This distance will help you gain another view of the situation, especially your own actions and behaviors. That is, you want to think about how another person observing the interaction would view what you say and do.

- *Demonstrate empathy*: Empathy involves trying to understand the world from the other person's point of view, setting aside your own emotions and perspective. This idea is demonstrated by the saying that "you should walk a mile in the other person's shoes," indicating that you need to try to see things from the other's perspective and experience the world from that vantage point. When you demonstrate empathy, you listen to what the other person has to say rather than focusing on getting your own ideas across. You let the other person know you are interested in their perspective and ideas and that you truly want to understand.

- *Engage in active listening*: Active listening goes hand-in-hand with empathy. When you actively listen to others, you give them your full attention, mentally work to comprehend what they are saying, and give them relevant feedback. Active listening is communication work invested to understand another.

- *Show respect for perspectives that are different from your own*: More than simply tolerating others' viewpoints, you want to show that you respect those differences. Demonstrating an interest in attitudes, values, and beliefs that are different from you own will indicate your respect for those differences.

Engaging in these practices will not guarantee a more inclusive communication climate, but such practices will go a long way toward building more effective relationships across cultures. Consider the information in Box 2.7 and Box 2.8 in terms of fostering inclusion, especially as it relates to values and ethics.

BOX 2.7: ETHICAL QUESTIONS

Create Classroom Inclusion

Have you ever taken a class that had a code of conduct? Some instructors include one in the syllabi for their classes. Have you ever thought about what you would include in a classroom code of conduct that would promote effective and inclusive communication across cultures? Even if your instructor has provided a code of conduct for the class or something similar, creating a class code for inclusive communication will give you an opportunity to consider what aspects of inclusive communication you and your classmates think are most important in the classroom. Follow these steps in creating a code of inclusive communication for your class:

1. Develop a list of the five to 10 most important values associated with effective communication across cultures in the classroom, such as listening to others' ideas, giving everyone the opportunity to contribute to discussions, and appreciating differences. For each value, give at least one example you have observed that supports that value.

2. Develop a statement for each of the values you have identified. Review and refine your statements so they are clear, concise, and practical. That is, you want classmates to understand each part of the code and be able to implement it.

3. Working with a group of other students in the class, compare the codes of inclusive communication you and your classmates have developed. Where are their commonalities? Where are the differences? Why do you think there are differences? As a group, come to an agreement on the five to 10 statements that will form the basis of your class code of inclusive communication, again revising for clarity, conciseness, and pragmatics.

4. Finally, as a class, compare the codes of inclusive communication each group developed, noting both convergence and divergence. Work together to reach a consensus on the five to 10 most important ethical statements, revising and rewriting as needed.

5. Should such codes be included as part of course syllabi? Why, or why not? Would they be useful in achieving more inclusive classroom cultures?

BOX 2.8: ETHICAL QUESTIONS

Cancel Culture

A relatively new form of ostracizing people is to "cancel" them for their ideas or behaviors. Through the "cancel culture" or "call-out culture," the person is blocked from social media or social settings. Some suggest that calling in, rather than calling out, can lead to greater understanding and acceptance.[51] For example, Loretta Ross, in her work on calling people in instead of out, addresses the challenging questions of how to confront unfairness and injustice in ways that encourage self-analysis and growth in others.

What are your views on the cancel or call-out culture? What are its advantages and disadvantages? Is this approach an ethical and acceptable one? What do you see as the potential benefits and pitfalls of "calling in," rather than "calling out," people after their offensive behavior?

CHAPTER REVIEW

■ **Explain the relationship between culture and a pervasive communication environment.**

Culture provides the basis for your communication, and you learn about culture through your communication with others. A pervasive communication environment provides many more opportunities to learn about different cultures and interact across cultures. These intercultural interactions present opportunities both for increased understanding and increased misunderstanding.

■ **Discuss the relationship between culture and cocultures.**

Culture involves values and beliefs that inform behavior, collectively constructed by the group, transcending the individual's specific perspective. Individuals rarely interact within a single culture; most people are part of multiple cocultures within a larger culture. Cocultures are groups with behaviors or beliefs that differentiate them from the broader culture of which they are a part. Cocultures typically are formed based on ethnic background, religion, gender identity, education, sexual orientation, and specific interests.

■ **Define and describe differences between collectivist and individualist cultures.**

Self-attainment, competition, and personal growth characterize individualist cultures. Loyalty to the group and cooperation are valued in collectivist cultures. Most cultures are a mix of individualism and collectivism.

■ **Discuss the role of communication in understanding and promoting diversity.**

Diversity includes socioeconomic, experiential, religious, dis/ability, race, and gender identity differences. Embracing diversity improves communication for all involved. Strategies that promote inclusiveness include building an awareness of your own personal biases, prejudices, and stereotypes; suspending first impressions; taking on the observer role;

demonstrating empathy; engaging in active listening; and showing respect for individuals whose views are different from your own.

DISCUSS AND APPLY

1. Identify two to three norms that students follow in this class. How do you know what the norms are? What happens when students do not follow a particular norm?

2. Identify and give examples of the elements of culture on your campus:

    a. vocabulary
    b. practices
    c. rituals
    d. artifacts
    e. stories and myths

3. Is the culture with which you primarily identify more collectivist or more individualist? How do you know? What examples support your conclusion?

4. What are some of the cocultures to which you belong? In what ways do they overlap? In what ways are they different? How do they influence your communication?

5. What are your experiences with diversity in your everyday life? What have you learned about communicating with people from backgrounds that are different from your own?

6. What role do social media play in the cancel or calling out culture? What historical forerunners of cancel culture can you identify?

7. How have social media platforms shaped practices of inclusion?

CREDITS

CHAPTER 3

Perception and Identity

Three White students called their Black roommate "three-fifths," in reference to the original clause in the United States Constitution that defined an enslaved person as three-fifths of a free person. When the student objected, they called him "fraction." The White students hung a Confederate flag and Nazi symbols in their dorm suite. They wrote the n-word on a dry-erase board in the suite. They put a bicycle lock around his neck and locked him in his room. The harassment went on for months, until the student's parents visited the campus and reported the incidents to campus housing. The White students admitted their activities, but claimed it was all in fun.[1] How could this happen at San José State University, one of the most diverse college campuses in California?

Many students on campus asked themselves that question, including communication majors, who launched a campaign to fight racism, sexism, ableism, and other isms that engender intolerance and hate against groups of people. The "No Kidding/It's Never Funny" campaign provided a platform for the campus community to talk openly about stereotyping, discrimination, prejudice, hate speech, and hate crime (Figure 3.1). The students developed a website with diversity-related resources, created a postcard to promote inclusiveness, and organized a contest to design an "honor diversity" badge. They encouraged faculty to spend time in their classes creating safe spaces for students to discuss diversity issues. The students' goals were to change perceptions of what constituted joking around and encourage people to speak out against disparaging comments about others, even when those comments are framed as "all in fun."

Protests for racial justice and the resurgence of the Black Lives Matter movement underscore the widespread institutional and systemic racial inequities people of color face today. Why do these inequities exist? Why do people engage in disparaging talk about and take negative and even violent actions against those whom they perceive as different from themselves? The tendency to stereotype is rooted deeply in how the human brain perceives and processes information. Those perceptions, in turn, provide the basis for how you communicate and act with others and how you view who you are. In this chapter, you will learn about the relationships among the perception process, identity, and a pervasive communication environment.

FIGURE 3.1 Students at San José State University launched a campaign to increase awareness of hate speech and encourage the campus community to speak out against racism.

Perception in a Pervasive Communication Environment

Although the mechanics of perception have not changed in a pervasive communication environment, the availability, number, and range of stimuli within the human perceptual field have changed tremendously. The internet did not exist before 1973, the World Wide Web went public in 1991, and the first web browser was launched in 1993. Facebook is not quite 20 years old, with Twitter and YouTube a few years younger.[2] Previously, people engaged in most of their social interactions in-person or over the phone; now so many more options exist. In addition, information from far-off locales, once the province of large news media organizations, now is available readily on social media sites. Before the advent of the internet, the stream of messages individuals encountered in their daily lives seems more like a small creek compared with the river of information through which people now wade. Because you are almost always connected via your phone, tablet, laptop, or smart TV, your brain must process so much more than in the past, much of it demanding your attention as you try to make sense of it all.

Sorting Out the Genuine From the Fake

Arguably, the greatest challenge to communicators' perceptual skills is determining what information is genuine and what is fake. Trickery and shams are as old as humans, from forged paintings to counterfeit money. Stories with fanciful embellishments and imaginative plots have entertained people throughout the ages. Still, you expect authenticity when listening to the news or watching a documentary. That expectation is what primes your susceptibility to pseudo news and deepfakes.[3] **Pseudo news**, also called fake news, refers to false information that purposefully mimics the form of news stories, but the content lacks accuracy and credibility.[4] Recent examples of the top pseudo news posted on Facebook include that Fred Trump, President Trump's father, was a Ku Klux Klan member; Congresswoman Alexandria Ocasio-Cortez proposed a ban on motorcycles; and Catholic altar boys put marijuana in the church's incense burner.[5] Disseminating disinformation is not new, as propaganda long has been a tool of politicians and governments. Spreading false narratives for political gain dates back at least to Aristotle's time. In the 13th century, the first Western legal code concerned the prohibition of fraudulent stories.[6] However, the ability for anyone with an internet-connected communication device to create and widely distribute completely fabricated news arose with the pervasive communication environment and social media.[7] Pseudo news became an especially serious problem during the COVID-19 pandemic, with false stories claiming 5G cell towers caused the virus and other stories reporting so-called cures, such as drinking bleach, that were in fact dangerous.[8]

Research on pseudo news demonstrates how the mechanics of the human perceptual process predisposes you to believe this form of false information. Because the human brain organizes information through recognizing repetitive patterns, seeing or hearing the same fake news headline over and over will reduce your ability to identify it as false.[9] That is, if the same headline keeps showing up in your Twitter feed, that repetition will increase your perception of the information's accuracy. However, the content must seem at least somewhat plausible. No matter how many times you read "The Earth is flat" on Facebook, you will not believe it, because you know the statement is not true (Figure 3.2). In addition, social media offer pseudo news a degree of credibility, as you read a story someone in your network has posted. The human brain likes what is familiar because what is known is easier to interpret than what is unknown. Therefore, your brain takes a shortcut, assuming the story must be accurate because you know the person who shared it.[10] For example, a study of 357 Facebook users across 44 countries found that trust in a story's source—the friend who had posted it—was the main reason for believing the message's accuracy and then sharing it.[11]

pseudo news Also called fake news, these messages are disinformation that purposefully mimics the form of news stories, but the content lacks accuracy and credibility

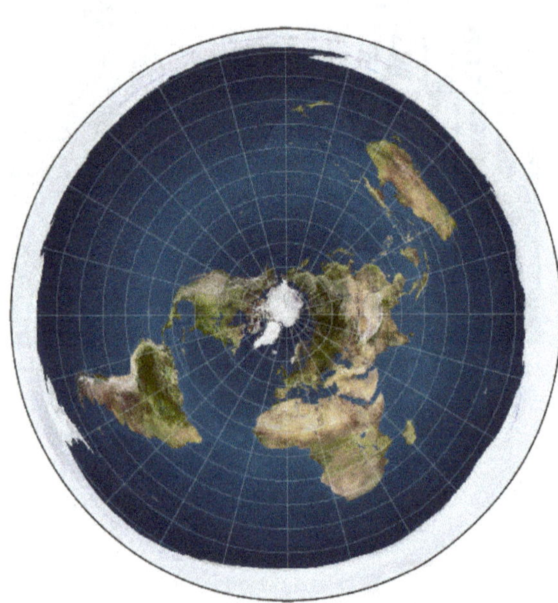

FIGURE 3.2 If a statement seems plausible, viewing it repeatedly will increase the likelihood you will believe it. However, if a statement is implausible, as with "The Earth is flat," you will judge it as inaccurate no matter how many times you view it.

Deepfakes are videos and other digital content that have been manipulated using artificial intelligence (AI) or similar technology to create false images that appear authentic.[12] Deepfakes often feature celebrities or politicians. For example, in a deepfake of Mark Zuckerberg posted on Instagram, the Facebook founder appeared to claim that the company owned its users. A deepfake of President Obama had him warning about the technological capabilities to create deepfakes.[13] This form of disinformation presents an even greater challenge to human perceptual processes than fake headlines and stories, as the image quality often makes it difficult to detect that the images are fabricated. A recent study in the United States involving 1,404 participants found pseudo news stories that relied on images and text were perceived as more credible than those with just text.[14] The tendency of the human brain to work as efficiently as possible means that current perceptions rely on previous perceptions. Because deepfakes appear to be so like images you have seen before, your brain concludes they must be real. In fact, deepfakes can seem so authentic that viewing them can create false memories.[15]

While a pervasive communication environment facilitates the proliferation of pseudo news and deepfakes, your ability to access information anytime from anywhere means that you can check on the authenticity of the information you encounter, especially stories on social media. In addition, a recent study found that training participants on the characteristics of pseudo news and deepfakes, such as the use of emotional language and discrediting opposing views, increased the ability to identify fictious information posing as genuine information.[16] You will learn more about evaluating information sources and spotting fake news in Chapter 12.

deepfakes Videos and other digital content that have been manipulated using artificial intelligence (AI) or similar technology to create false images that appear authentic

The Perception Process

perception Sensing, selecting, organizing, interpeting, and recalling verbal and nonverbal messages

Perception involves sensing, selecting, organizing, interpreting, and recalling verbal and nonverbal messages. Summarized in Figure 3.3, the process happens quickly, and you probably do not even notice all the steps that go into perception.

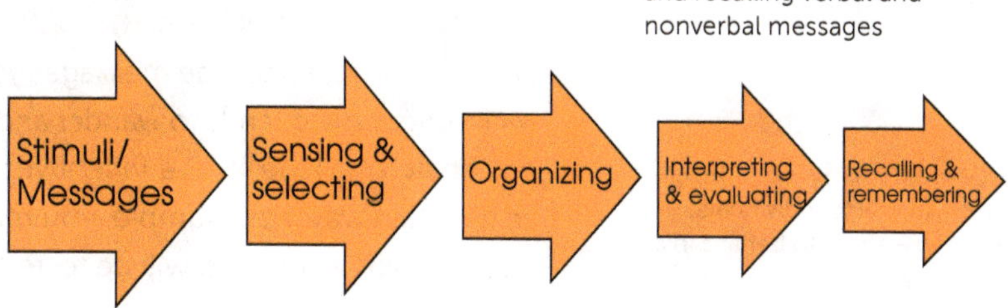

FIGURE 3.3 Information is lost at each stage in the perception process.

Sensing and Selecting

You are bombarded with messages daily. The average American spends about 10 hours each day engaged in communication activities, such as interacting in-person, listening to the radio, talking on the phone, reading magazines and books, emailing, texting, and viewing websites.[17] Worldwide, individuals send and receive nearly 300 billion email messages every day, with that number expected to grow to almost 350 billion by 2023.[18] In the United States, individuals send about 26 billion text messages every day, with each person on average receiving just under 100 messages.[19] And researchers estimate that on average Americans encounter 5,000 advertisements each day, 10 times the number from 50 years ago.[20]

You seek out messages as well. A recent survey found that 71% of mobile phone users keep their phones close by when they sleep so they do not miss a call or text message. On average, smartphone users check their phones more than 60 times each day. In addition, the vast majority of people—86%—will check their phones when talking in-person with family and friends.[21] The majority of Instagram users check the app once a day, with more than 40% checking multiple times a day.[22] Similarly, 60% of Snapchat users open the app every day, with regular users averaging 20 times a day.[23] The COVID-19 pandemic prompted people to engage in increased message-seeking behaviors, as they searched for the latest information on the virus. For example, visits to local and well-known news sites, such as the *San Francisco Chronicle*, *Seattle Times*, and *Boston Globe*, skyrocketed as the virus spread across the United States.[24] And of course you regularly seek out friends, family, classmates, coworkers, and others to talk with daily.

Sensing and selecting stimuli is the first stage in perception and involves all your senses—touch, taste, smell, sound, and sight. For example, you might touch the arm of friend while offering verbal reassurances. When meeting classmates at a local café, you taste your mocha chai and smell the coffee's aroma. In chatting with your parents on FaceTime, you note their gestures, facial expressions, and the like. As you sense stimuli, you must decide whether or not to continue processing them. This selection process often occurs unconsciously. Based on past experience, your brain determines almost automatically whether or not the message requires further attention, blocking out irrelevant noise and other distractions.

Motivation and Message Reception

Motivation influences the messages you sense and receive.[25] When you lack motivation, your mind tends to wander and you pay less attention to your surroundings. **Extrinsic motivators** are incentives external to you that cause you to behave in a certain way. For example, you might focus closely on an instructor's lecture because you know you will be tested on the content later. **Intrinsic motivators** drive you to take action based on internal forces, such as being attentive in a presentation because you are interested in the topic. Identifying what motivates you can help you avoid missing important messages. If you know external rewards bring out your best behavior, then remind yourself of those rewards when you are communicating with others. A 3-hour work meeting, for instance, can become a positive experience when you consider the praise you will receive for providing a concise summary of everyone's contributions. If internal motivators provide

extrinsic motivators Incentives external to individuals that cause them to behave in a certain way

intrinsic motivators Incentives that encourage people to take action based on forces internal to the individual

the incentive you need to focus on message reception, figure out how to best activate them. Listening to a relative's lengthy story can become more enjoyable when you think about how much you like the person. If you are motivated to attend to messages, you are more likely to physically receive them.

The Myth of Multitasking

The human brain has the capacity to attend to several things at the same time. To keep track of incoming stimuli, humans have executive control processes that act as managers or supervisors, prioritizing information and allocating the brain's resources. When you encounter relatively routine situations, the executive control processes easily sort out what needs to be done and how to do it. However, resources quickly become limited with more complex tasks, and the executive control processes cannot keep up.

When you switch back and forth between tasks, no matter how simple or complex, you become less efficient, as the executive control processes must keep revising mental priorities and resources. In addition, the executive control processes often default to emotional needs over cognitive ones. For example, if you are watching a video or listening to a podcast while studying, the media content you are consuming will have a higher priority because it is more entertaining and fun. An example of noise in the pervasive communication environment model introduced in Chapter 1, multitasking is a distraction that interferes with the mental processes needed to produce and interpret messages.[26] You lose time when you multitask, rather than gain time (Figure 3.4). Multitasking also leads to mental fatigue, slower reflexes, and memory problems.[27] One study found that college students who let their minds wander during a class lecture remained engaged and learned the material. In contrast, those who tried to multitask—such as texting with a friend during the lecture—disengaged, missed key points, and performed poorly on the class assessment. And if your multitasking involves a conversation, you are losing time and messages as you shift your attention back and forth.[28]

FIGURE 3.4 Multitasking is less efficient than focusing on one task at a time.

Organizing

Think of your brain as a huge filing cabinet. After you have received and selected a message, you need to determine where it belongs. One way you organize information is through **schemata**, cognitive structures representing basic categories of information grouped together by differences and similarities.[29] Schemata provide

schemata Cognitive structures representing basic categories of information grouped together by differences and similarities

TABLE 3.1 *Types of Schemata Used in Organizing Messages*

SCHEMATA TYPE	BRIEF DEFINITION	EXAMPLES
Self	View of own beliefs, attitudes, goals, emotions	I am shy. I am outgoing. I am practical. I am carefree. I am motivated.
Role	Social categories	Instructor–student; supervisor–subordinate; patient–health care provider
Person	People we know and types of people	My friend, Erika; my dad, Bob; unfriendly people; happy people
Relational	Types of relationships	Parent–child relationship; acquaintance relationship; sibling relationship; friendship
Event	Common occurrences	Movie night with friends; lunch with coworkers; classroom speech

a way for you to organize information about yourself and others. Table 3.1 summarizes the five types of schemata directly related to human communication.

Self Schemata

self schemata Cognitive structures associated with your view of self, including your beliefs, attitudes, goals, and emotions

Self schemata refer to those structures associated with your view of self, including your beliefs, attitudes, goals, and emotions.[30] Although you might think of the self as a single unit, the self is multidimensional. For instance, you likely project a somewhat different image with your family than you might at school or work. That is, how you represent yourself alters depending on the context and the relationship you have with other communicators. Generally, you do not undergo complete personality changes when you interact with different people and in different situations. Core aspects of the self will remain important to you in any context.[31] For some people, shyness may be a central category associated with the self. For others, maintaining a positive outlook may be a major feature of their identity. You organize the messages you receive and select in part based on how you think of yourself. If demonstrating good leadership skills is an important part of your identity, then you select and include messages in your self schemata file that support that view and may dismiss messages that appear to counter that view.

Role Schemata

role schemata Cognitive structures that provide information about how you and others should behave based on social categories

stereotyping Categorizing people based on the social group to which an individual thinks people belong

Role schemata tell you how you and others should behave based on social categories, such as ethnic background, class, gender, age, dis/ability, and occupation. For example, you likely have a different schema for "physician" than you have for "car salesperson" and interact with people who fit those categories in different ways. That is, you expect your physician to ask you about your health and the salesperson to ask about your driving habits. However, although schemata provide predictability, they also can make your brain lazy. Schemata provide a mental shortcut for organizing information. You develop expectations of others based on role schemata and can be surprised when others do not fit categories. **Stereotyping** involves categorizing individuals solely based on the social group

BOX 3.1: ETHICAL QUESTIONS

Fighting the Brain's Default to Stereotype

In arguing that "thug" has become a racial slur, *Miami Herald* columnist Leonard Pitts Jr. observes that a White pop singer is called a "bad boy" when he is accused of driving while under the influence and vandalism. But a Black pro-football player who graduated from Stanford University is called a "thug" when he shouts during a TV interview.[32] There is no doubt that stereotyping others is unethical. Yet combating stereotyping can prove challenging because your brain is hardwired to categorize incoming information. Still, you can choose how you file that information and how you label it. Perceiving ethically means attending to information that challenges your existing schemata and revising schemata based on that information. The more complex your ways of categorizing people, the less likely you are to stereotype. Think of someone you know well. What are all the words you would use to describe that person? Now think of someone you have interacted with on only a superficial basis. How could you get to know that person better? How could you develop a more complex picture of that person? How can seeking out information about a group of people with which you are unfamiliar help you combat stereotyping?

to which you think they belong. Although a natural product of your cognitive structures, stereotypes become a problem when you do not revise your role schemata based on new information. For example, part of your role schemata for people in their 70s (categorizing by age) may be that they do not know much about social media. Then, when you meet or learn of people in that age group posting on TikTok, Twitter, and YouTube, you must revise your role schemata. If you do not, then you are stereotyping. Box 3.1 highlights the role of language and role schemata in stereotyping.

Stereotyping directly affects your behavior and even can prevent you from engaging in healthy activities. For example, getting people to wear face masks in Western countries during the COVID-19 pandemic was challenging at first due to cultural resistance and negative stereotypes.[33] However, celebrities and others soon were showing off their face masks on social media. Images of a mask-wearing Statue of Liberty, planet Earth, Mona Lisa, and other popular culture icons appeared on Facebook, Instagram, and Twitter (Figure 3.5). Wearing a face mask became fashionable, commonplace, and a symbol of solidarity in the fight against the virus.[34]

FIGURE 3.5 Images of popular culture icons wearing face masks offered one strategy for overcoming negative stereotypes during the COVID-19 pandemic.

Person Schemata

Person schemata are concerned with specific people you know and particular types of people. For example, you might have different person schemata for each of your siblings, such as "Andre is hard working, conservative, quiet, easy going, and an excellent listener," and "Samarah is ambitious, outspoken, smart, funny, and athletic." You also have schemata for person types, such as "creative people," "friendly people," and "industrious people." The "friendly person" schema might

person schemata Cognitive structures concerned with specific people you know and particular types of people

include "smiles often," "always says hello to others," "laughs easily," "helps when needed," "thinks of others' feelings," and "speaks positively."

Relational Schemata

relational schemata
Cognitive structures that provide a frame for romantic, kinship, friendship, work, and other relationships

Relational schemata are associated with romantic, kinship, friendship, work, and other relationships.[35] You likely have a relational schema for "friend" that suggests your expectations of friendships and frames your interpretations of others' actions. Your friend schema likely addresses what you expect a friend to say and do and what you think you should say and do as a friend. You probably expect friends to help you when needed, engage in fun activities with you, and respond quickly when you text or call. The COVID-19 pandemic affected individuals' relational schemata, as they found themselves redefining their expectations of friends, coworkers, neighbors, partners, and others within and outside their everyday communication network.[36] With quarantines, social distancing, and mask requirements, common rituals and routines were upended. For example, your "friend" schema might have shifted to someone you expected to wear a mask and not hug you when meeting in-person; fun activities may have moved online, as with streaming movies at the same time but in different physical locations; and help might have been a FaceTime study session.

Event Schemata

event schemata Cognitive structures that provide scripts for common occasions and situations

Event schemata help you anticipate common occurrences. For example, you have scripts for "dinner party" with a group of friends and how it should unfold and "project meeting" with your classmates and how it should be conducted. These event schemata assist you in predicting and comprehending what will happen and in what order. At a project meeting, whether in-person or online, you might expect everyone in the group to greet each other and engage in small talk at the start of the meeting. Then each group member might report on their part of the project, with the other group members making suggestions. Finally, you would expect some sort of adjournment pronouncement at the meeting's end. Box 3.2 encourages you to consider how naming an event influences your ethical judgment of it.

Interpreting and Evaluating

After you have sensed, selected, and organized a message, you interpret and evaluate it. Interpretation and evaluation occur quickly—in nanoseconds. Yet this part of the perception process is quite complex and multidimensional, as you are both assigning meaning to and appraising the message. Although the words the speaker chooses are a key factor in this step, you take many other variables into account. For example, you consider nonverbal cues, such as tone of voice, posture, gestures, and body movement. You also frame

BOX 3.2: ETHICAL QUESTIONS

Protest, Rebellion, or Riot?

On December 16, 1773, more than 100 American colonists snuck onto privately owned ships and tossed 45 tons of English tea into the Boston Harbor—cargo worth about $1 million today. While the colonists considered this action a protest, the governor of Massachusetts called it a riot.[37] The British interpreted the act as a rebellion and quickly enacted sanctions and punishments, such as rescinding the Massachusetts constitution and ending local judicial authority.[38] The two different groups applied different labels—and interpretations—to the same event. Fast-forward to the 2020 demonstrations across the United States and worldwide against racial injustice. Although for the most part the gatherings remained peaceful, some were marred by violence, flipping media descriptions of the demonstrations from protest to riot. Moreover, once government officials declared a demonstration a riot, police were allowed to use nonlethal weapons, such as pepper spray and stun grenades, to disperse participants (Figure 3.6). Conduct your own research on the racial justice marches that occurred during the summer of 2020. Were they protests, rebellions, or riots? How does taking the different perspectives—participant, police officer, government official, community member—influence how you label the events? Review the National Communication Association Ethical Principles in Chapter 1. Which principles are especially relevant to the racial justice demonstrations? How can those principles help guide all stakeholders in interpreting these events?

FIGURE 3.6 How you name an event—protest, rebellion, or riot—influences how you define it, act in it, and talk about it.

the message within your own past experiences and your relationship to the other person. You interpret and evaluate differently what friends and what acquaintances say to you.

Empathy in Message Interpretation and Evaluation

Your ability to interpret and evaluate another's message depends largely on **empathy**, in which you try to understand others' messages from their perspective. Three types of skills make up empathy: perceptive, cognitive, and behavioral.[39] Perceptive skills involve using all your senses to understand the other person's message. You interpret not only what the person says but also how the person says it. Your friend might say, "I guess my idea wasn't all that great. Let's just forget it." However, you note the person's downcast eyes and dejected tone of voice. She may *say* "forget it," but your assessment of her nonverbal cues suggests that she wants to take more time to discuss her idea.

Cognitive skills refer to your ability to think about something from someone else's viewpoint. When you share similar backgrounds and experiences with others, you probably find it easier to put yourself in their place. Empathy

empathy Understanding a message from the sender's perspective, or when you try to understand the world from another person's point of view proves much more challenging when you are less familiar with others, yet still is possible if you are willing to get to know them. Developing cognitive skills associated with empathy requires you to suspend judgment when communicating others. When a coworker says, "I didn't get my section of the report done, because my daughter was sick yesterday," you try to consider the message from the parent's perspective. You might think, "My coworker didn't finish their part of the report, but from the viewpoint of a parent, an ill child is a serious circumstance."

When you show others verbally and nonverbally that you empathize with them, you exhibit behavioral skills. That is, you perceive the person's message, you think about it, then you respond. For example, in the case of the coworker with the daughter who was ill, you might respond, "I'm sorry your daughter was ill yesterday. I'm sure that was stressful for you. I hope she's doing better and that you're okay. How can we best handle the work that needs to be done now?" Empathy requires trying to understand others' thoughts and feelings. When you interpret and evaluate a message, you consider both the obvious content of the message and the emotions that might underlie it.

Emotional Intelligence in Message Interpretation and Evaluation

emotional intelligence
The ability to perceive emotions, link emotion to thought, recognize the complexities of emotion, and manage your own and others' emotions

Emotional intelligence helps you interpret the emotional components of a message. Emotional intelligence involves four interrelated abilities: perceiving emotions, linking emotion to thought, recognizing the complexities of emotion, and managing emotions, as shown in Figure 3.7.[40] In perceiving emotions, you attend to others' tone of voice, facial expressions, posture, and other nonverbal cues as well as the words they use. When perceiving emotions in mediated text communication, such as email and texting, you rely more on linguistic cues, such as specific word choice. But you also consider nonverbal cues, such as emojis, the length of the message, and how long it takes the other person to respond.

In linking emotion to thought, you are concerned with how emotion might contribute to cognitive processes, such as problem solving and decision making. For example, your passion for an idea might spur you on to develop a creative, innovative plan that addresses key issues in a project you have been developing. Rather than ignore emotions, you identify ways those emotions productively advance your and other communicators' goals.

Although you may use simple terms to describe your emotions, such as "happy" or "sad," emotional intelligence involves recognizing the complexities of emotion. You seldom experience emotions in a one-dimensional way. When a class ends, for instance, you may feel relief, joy, and sorrow all at once—relief that you completed the class, joy that you did well, and sorrow that you may not see your classmates and instructor again.

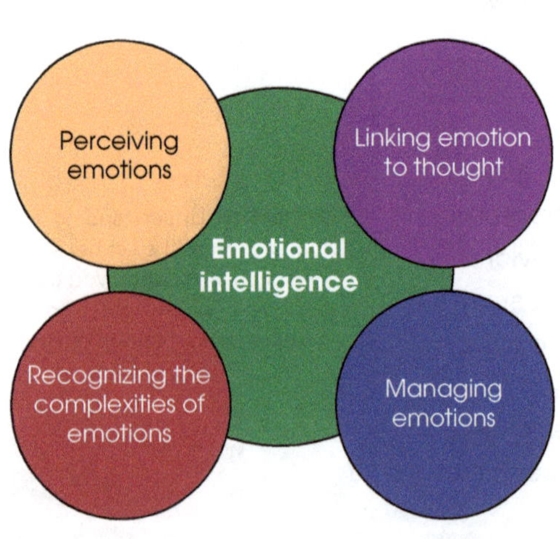

FIGURE 3.7 The interrelated abilities of emotional intelligence.

Finally, emotional intelligence involves managing your own and others' emotions. In some settings, such as at school and work, emotions are often hidden away or denied. Managing emotions means using feelings to achieve productive outcomes for yourself and others. You may become angry when a classmate disagrees with you, for instance, and interpret the message as a personal attack. Managing that anger requires reframing the feeling so that you and your classmates can have a productive discussion. In contrast, if a classmate expresses anger at something you have said, then you would want to consider ways to address those feelings in a positive manner. Box 3.3 invites you to assess your own emotional intelligence using an online quiz.

BOX 3.3: CONNECT & REFLECT

Testing Your Emotional Intelligence

How emotionally intelligent are you? The Greater Good Science Center at the University of California, Berkeley (greatergood.berkeley.edu), offers a quiz on emotional intelligence. Go to the website or type the keywords "test your emotional intelligence" in your favorite search engine to take the quiz for free. What did you think of the quiz? How useful was the feedback on each question? How did your score match with your own view of your emotional intelligence? What are some of the drawbacks of the quiz based on what you have learned about emotional intelligence in this chapter? In reflecting on the quiz, what have you learned about yourself that you could apply in your interactions with others?

In emotionally intense conversations, you probably have had the experience of regretting something you said right after you said it. One advantage to asynchronous communication, such as email or texts, is the time lag between responding to messages. Although you might feel you need to respond right away, waiting to answer a contentious message gives you time to think over—and likely temper—what you want to say. Often, writing the message and not sending it allows you to express your feelings and then move on. So before firing off that sarcastic email, save it in the draft folder and then do something else. Once you have gained some perspective by waiting, then formulate a more measured response.

Recalling and Remembering

The messages you recall, and those you do not, depend on what has happened in the previous stages of the perception process—receiving, selecting, organizing, and interpreting. At each stage you lose or modify the message, affecting how you recall it.

Recalling messages involves three memory systems: immediate memory, working memory, and long-term memory.[41] **Immediate memory** gives you just enough time to determine if you should continue processing the information,

immediate memory
Memory you retain for just enough time to determine if you should continue processing the information

use it immediately, or discard it. For example, if someone tells you the address for a website, it stays in your immediate memory until you have typed it into your web browser. Then the address is gone, as no additional processing was required. Similarly, if you hear someone call out a name and you realize it is not yours, the information goes no further than your immediate memory.

working memory Also known as short-term memory; involves the temporary storage and processing of information

Working memory, also known as short-term memory, is the next step in remembering. Working memory involves the temporary storage and processing of information. Without conscious rehearsal, the time span for information to stay in the working member is quite brief—only about 30 seconds. However, if you pay close attention to your perception habits, you can lengthen the time messages remain in the working memory, increasing the likelihood they will get stored in your long-term memory.

Repetition, chunking, and identifying logical patterns are the primary ways to retain information in your working memory. You have used repetition before—it is how you learned basic things like the alphabet, numbers, and colors. For instance, when listening to others, you repeat what they have said to yourself. This strategy is especially effective when learning people's names. Chunking involves grouping together pieces of information, as with remembering phone numbers and email addresses. For example, when someone says their email address, you split the address into two groups: before the @ and after the @. Finally, identifying logical patterns helps you store messages for later recall. Patterns are elements that repeat in predictable ways, as with the spirals found in nature or the grid formation of a city's streets. The groundbreaking children's television show *Sesame Street* used the idea of identifying logical patterns to teach young children all sorts of lessons through rhyme and song. The patterns of lyrics and melodies make it easier for children to recall the information, such as recognizing the object that is different from the others. You listen for patterns as an adult, which is why advertisement jingles stick with you. Patterns also are how you identify people. For instance, you often can recognize a friend by tone of voice because you know their pattern of speaking. Identifying patterns is essential to engaging in play with others. Observe children at a playground. If they are 5 or older, they have learned how to take turns and engage in reciprocal play because they have figured out the pattern (Figure 3.8).[42]

FIGURE 3.8 Children learn the rules of play by identifying patterns of interaction, such as turn-taking.

long-term memory The main storage area for incoming messages

Long-term memory provides the main storage area for incoming messages. Once messages have passed through the immediate and working memory filters, you store that information for later retrieval. In comparison to your working memory, your long-term memory has storage capacity but functions more slowly and inefficiently. With working memory, your concern is with hanging on to the information so you can move it to your long-term memory. With long-term memory, retrieval is the main challenge.

Five strategies increase the effectiveness of your long-term memory. Association, where you link a message to something familiar, provides an easier path for your brain to follow when you try to retrieve a bit of information. When listening to a neighbor give directions to a new restaurant, you might connect the instructions with landmarks you already know to better recall the route you should take. Categorization, or consciously placing a message in a class of things that go together, aids recall as well. For example, as your instructor describes a writing assignment, you might put it in the "writing college papers" category. Visualization involves using your imagination to create mental images of what you are listening to. If a friend is telling you about a trip to Indonesia, you try to picture the scenes the person describes. Mnemonics combine association and visualization by linking a message to something familiar and visually vivid. Say a classmate has asked you to email a news article to her after class. You associate "article" with an article of clothing—a hat, for instance. Then you imagine the most outlandish hat you can think of—bright pink and green with a yellow brim and a fake purple ostrich feather. Last, memory movement links messages and motion. You are better able to recall information that involves some sort of action. When someone in your neighborhood describes a proposal for a new park, you imagine yourself walking through the park, climbing steps, and sitting on a bench.[43] Actively developing your memory abilities enhances your skills in recalling others' messages, as summarized in Table 3.2.

TABLE 3.2 Improving Your Working and Long-Term Memory

	STRATEGY	WHAT YOU DO	USEFUL FOR RECALLING …
WORKING MEMORY	**Repetition**	Repeat something over many times	Snippets of information, such as people's names
	Chunking	Group together bits of information	Large amounts of information, as with a long conversation
	Identifying logical patterns	Recognize recurring elements	How to do something
LONG-TERM MEMORY	**Association**	Connect new information to something familiar	Novel or unfamiliar information
	Categorization	Assign something to a class or category	How something is related to a larger class of things
	Visualization	Create mental images	Complex ideas
	Mnemonics	Combine association with visualization	Out-of-the-ordinary processes and procedures that are different from your usual routine
	Memory movement	Connect messages with motion	Spatial information or things that involve activities

Identity in a Pervasive Communication Environment

identity An individual's view of the self and the image of the self presented to others

FIGURE 3.9 Your identity is multidimensional and complex, not just a single self.

Identity, or the self, refers to how you perceive yourself and the image of yourself that you present to others. You might think of your identity as something that is just about you, apart from other people. But your identity is social in that you create who you are in your relationships with others as you communicate with them. How you present yourself and how you perceive yourself depend on the people you interact with every day, as well as those from your past and those you will interact with in the future. Thus, your identity is complex and multidimensional, going far beyond a single self (Figure 3.9). For example, you might perceive and present yourself as a classmate, a sibling, a coworker, a friend—all various identities that make up who you are.[44] In addition, your identity is connected to how others present themselves. For example, a study of Facebook users found that viewing the messages posted by family and friends in the users' news feed had a negative impact on the participants' self-esteem.[45] That is, Facebook users unfavorably compared their own identities to the identities their Facebook friends presented. A study of adult women who used Facebook found that those who more often compared their own appearance to photographs of their Facebook friends also were more concerned about their body image.[46] Your identity and your perceptions of the world are closely intertwined.

A pervasive communication environment has changed the ways in which individuals create and present their identities. New communication technologies provide venues and platforms for distributed identities that cross time and space as you present images of yourself in social media, video conferences, and text messages, for instance, as well as in-person.[47] Even the choice of ringtone on your phone reveals an aspect of your identity. Is the ringtone bright and cheery? Did you download a song snippet for your phone's ringtone? Or did you leave the ringtone on the default setting? And if you assign different ringtones to different callers, those ringtones reflect your perceptions of their identities.[48]

Communicating Identity Online

Online environments allow you to try out different identities, as with choosing an avatar for a game or selecting specific photos on Instagram. While you make choices about how you want to present yourself in-person, such as your clothes or hairstyle, you have many more options for creating identities online. The hyperpersonal model of online communication, depicted in Figure 3.10, suggests that communicators have greater control over how they present themselves online than they do in-person. The heightened ability to manage your identity online means you can highlight the more positive aspects of your identity and downplay the more negative ones. This leads to the notion of hyperpersonal relationships in which communicators feel more affection for and view each other as more attractive than similar communicators report for in-person situations.[49]

The hyperpersonal model of communicating identity online includes four elements: selective self-presentation, idealization, channel management, and feedback. Selective self-presentation refers to your ability to control what information you present about yourself. Individuals strategically use social media and other forms of mediated communication to enhance more positive aspects of their identities and downplay less appealing ones.[50] With social media such as Instagram and Snapchat, you decide which photos to post and which hashtags to include in your message. When using Zoom, Skype, FaceTime, or other video chat platforms, you likely make choices about how close you are to the camera and objects visible in the background, especially with people you know less well or are meeting for the first time.[51] Similarly, users of online dating apps curate what information they include and exclude. For example, people particularly sensitive to rejection prefer to use online dating sites because they can determine carefully what and how much information to reveal about themselves.[52] Teens consider their audience when deciding what content to share on Facebook and Instagram, sometimes consulting with their friends before posting.[53]

Cultural norms influence what you decide to post on social media and how you frame the information. For instance, researchers compared status updates posted by Facebook users in the United States and Turkey. Those in the United States reflected an individualistic culture, expressing positive attitudes and describing their accomplishments as a sole effort. In contrast, those in Turkey reflected a collectivistic culture, expressing more negative attitudes, as with complaints, and attributing their achievements to a group effort, as with help from family and friends.[54] These choices also affect if and how others interact with you. Research on Facebook found that college students were more likely to initiate friendships with people who included attractive photos in their profiles than people who either included unattractive photos or no photo at all.[55]

Idealization involves applying generalizations related to what you know about groups of people to complete the impression you have of an individual. That is, with online interactions, you often have limited information about a person, such as text from an email and possibly a few photos or videos posted to a social media site. You fill in the gaps of your knowledge with what you know about people who might be similar to the person with whom you are communicating.

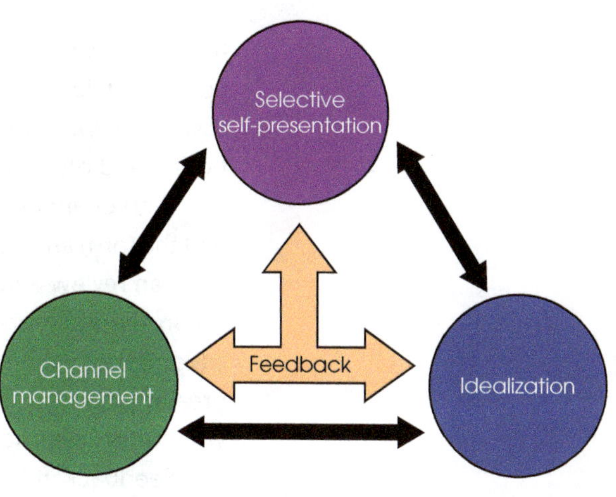

FIGURE 3.10 The hyperpersonal model of communicating identity suggests that you have more control over your presentation of self online than in-person.

If you know, for instance, that the person grew up in a small midwestern town, you would develop an idealized image of that person based on your experiences associated with people from that region of the country as well as related films, books, and other mediated images.

With channel management, you determine when you interact with others and the form and content of the message. For example, when you send an email, you can review and revise it before sending the message. College students, for instance, used more complex language when emailing a professor than emailing a high school student about the same topic.[56] In addition, college students feel greater affinity for instructors who quickly responded to email than instructors who rarely respond.[57]

Feedback influences the other three components of the model. When you post a message on social media, others respond in ways that affirm or disconfirm your presentation of self. For instance, romantic couples viewed their partners as communicating their satisfaction with and commitment to the relationship when they changed their Facebook profile photo from an individual one to one showing the couple together.[58] Changing the profile photo affirmed both the individual's presentation of self and the couple's presentation as a dyad. Similarly, if you change your profile picture on LinkedIn, thinking that the new photo is more flattering, your view is confirmed when those in your network compliment you on it. In contrast, if the responses are neutral or negative, you likely will revise your assessment of the photo or even of yourself. **Identity shift** occurs when feedback from your online self-presentation leads to changes in your identity. Say you post a photo on Instagram that you think shows you as friendly and fun. You only get a few comments on the photo, and they are not all that positive. You reflect on these responses and wonder if you really are a friendly and fun person. Your online experience altered your view of yourself, causing a shift in your identity. In a similar way, feedback can influence idealization, leading you to update your view of others based on new information. Finally, feedback can alter channel management by reinforcing or disconfirming your choices.

As you know from living in a pervasive communication environment, your connections with others are not so neatly divided into online or in-person interactions. You text a friend to meet for coffee, you phone your partner that you will be late for dinner, and your email your boss the report you will present at an in-person meeting later in the week. In fact, research on college students in dating relationships found that those who interacted both in-person and online experienced higher levels of self-disclosure and intimacy.[59] In another study, undergraduates who communicated with on-campus friends in-person and also via Instagram reported feeling more socially adjusted to college life than other students.[60] As you go about your day, you present and negotiate your identity in multiple communication contexts with both the same and different people.

Whenever you engage in self-presentation, you learn from that experience and reflect on who you are.

A Communication Perspective of Identity

From a communication perspective, identity involves four intertwined and interrelated layers: personal identity, enacted identity, relational identity, and communal identity.[61] Figure 3.11 shows the relationships among the four layers of identity.

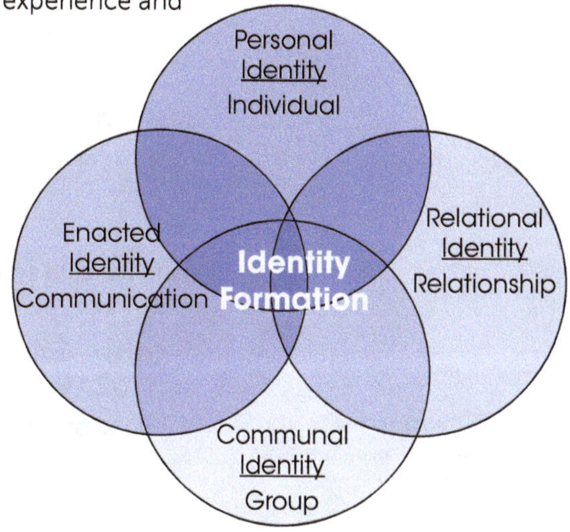

FIGURE 3.11 The four layers of identity are intertwined and interrelated. (*Adapted from* Weaver et al., 2021, p. 307.).

personal identity An individual's self-concept

Personal Identity

The first layer, **personal identity** is your self-concept. The individual forms the central focus of this layer. When you think about yourself and who you are, you are reflecting on your personal identity. Gender, ethnicity, and health are key aspects of this layer. For instance, you may think of yourself as a well and healthy person. But if you experience an injury or illness, such as breaking your leg or contracting the flu, you may revise the health aspect of your self-concept. As your health improves, your self-concept will change, but the injury or illness will remain a part of how you define your level of health. In Box 3.4, consider the intersection of language in how people think about gender identity.

BOX 3.4: CONNECT & REFLECT

The Language of Gender Identity

Up until recently, nearly all questionnaires and surveys included only two categories for gender: female and male. But for some—if not many—people, those categories do not accurately reflect their personal gender identity. Social media sites have increasingly recognized the problems with a binary view of gender, providing additional options, such as agender, pangender, transgender, and cisgender, as well as an option to customize the user's gender. Why is it important to have a wider range of choices in representing your gender identity online? What are the advantages of designating your gender in your social media profiles? What are the disadvantages of doing so? How does the idea of moving from two gender categories to a spectrum of choices influence how you think about gender in your everyday interactions with others?

Your personal identity influences many of the choices you make in life. Individuals with stronger ethnic identities—feeling a sense of belonging to an ethnic group or groups—tend to engage more in civic and political activities, such as joining community groups, serving in volunteer organizations, and participating in protests.[62] Ethnic identity also influences college students' career choices, in part because of the cultural activities in which they engage that give meaning to their lives.[63] In addition, ethnic identity can help students pinpoint and overcome career barriers, as a study on Latinx college students revealed. The researchers found that a strong ethnic identity provided a foundation for students to cope with and develop strategies for obstacles they encountered in making career choices.[64]

Enacted Identity

enacted identity The aspects of the self an individual expresses to others when interacting with them

The second layer, **enacted identity**, is what you express to others when you interact with them. Identity as communication is the center of this layer. Here, you perform or present yourself to friends, family, acquaintances, and the other people with whom you come in contact every day. You might tell others about working with a team in a recent Global Game Jam to demonstrate your creativity and knowledge of game development. Or you could relate a story from your semester studying abroad in Chile to present your commitment to intercultural experiences. Breast cancer survivors, for instance, may participate in events to raise funds for breast cancer research (Figure 3.12).[65]

FIGURE 3.12 Breast cancer survivors may enact their identities by participating in events related to breast cancer awareness, research, and funding.

Your voice—your vocal intonation, tempo, pitch, accent, and dialect—is unique to you and provides one way you enact your identity. Yet in the United States alone, 2.5 million people do not have the physical ability to use their voices at all or in ways others can understand. For many, the solution is speaking with a generic computerized voice. Professor of computer science at Northeastern University Pupal Patel's groundbreaking research provides those without the ability to speak a voice that is uniquely theirs. Employing new technology in voice analysis, her team has developed synthesized voices that reflect the people using them. The technology merges the donor voice of a person whose characteristics closely match those of someone who is unable to speak with the vocal qualities of whatever sounds the target person can generate. Professor Patel and her team create a personalized vocal identity that is then integrated with an assistive communication device. For the first time, people who have had to use one of a few synthesized voices can have a voice of their own.[66]

Enacted identity varies with the context in which you are interacting. Research on online avatars has found that individuals will enact the gender of their avatar, regardless of their own gender. For example, men who were represented online by female avatars used gender-typical language consistent with their avatar, such as references to emotions, making apologies, and expressions of uncertainty. Similarly, women with male avatars made fewer references to emotions, apologies, and uncertainty. In this way, the study participants performed their gendered identities on the screen, rather than their offline gendered identities.[67] As another example, although people tend to create avatars that reflect their identities, those who create avatars more attractive than they perceive themselves are more extroverted in online interactions.[68]

Recent recognition of gender as fluid and nonbinary has led to a closer examination of the pronouns used to refer to yourself and others. You might have noticed—or use yourself—email signature lines that include the self-referential pronouns senders prefer, as with they/them, ze/hir, ze/zir, she/her, or he/him. In meetings, group members may introduce themselves by both their names and personal pronouns. Indicating your personal pronouns is one way of enacting your gender identity, and that enactment can have important consequences. Research on transgender and diverse gender people found that when health care providers affirmed their patients' personal pronouns, patients felt respected, had greater trust in providers, and experienced more positive health outcomes.[69] Similarly, asking about LGBTQIA+ patients' personal pronouns for the intake form used in a youth health clinic offered a key opportunity for providers to support patients' enacted gender identity.[70]

You communicate your enacted identity with others, who may or may not support the self you present. In her TED Talk "The Danger of a Single Story," writer Chimamanda Ngozi Adichie tells the story of meeting her roommate her first semester of college in the United States. Because her roommate knew Adichie was from Nigeria, she assumed Adichie was poor, listened to "tribal" music, and did not speak English. Yet Adichie is from a middle-class Nigerian family, listens to American music, and of course speaks English, Nigeria's official language. Adichie's roommate created an impression of her based on a one-dimensional view of Nigerians—a single story. Adichie found she had to negotiate her identity with her roommate, presenting an alternative story of herself and Nigeria that more closely aligned with her personal identity. In her talk, Adichie argues that individuals' identities are comprised of multiple intersecting stories.[71] What are your intersecting stories? Explore them in Box 3.5.

Enacted identity underscores the ever-developing, -negotiating, and -changing nature of identity. Although basic aspects of your identity remain constant over time and across contexts, other aspects of your identity are more mutable and change in your interactions with others. Research on Facebook status updates provides a useful example.[72] When you update your status on Facebook, others may post a comment on it, like it, or link to it. These actions all become part of the narrative associated with your status update and your enacted identity. You might post, "Just registered for classes in my study abroad program!!" and include a link to the school. Friends post their congratulations and similar comments, others click on "like," and still others share the link on their own Facebook pages. Your enacted identity becomes a larger networked narrative, a story of you tied together with those who have in some way participated with your status update. In addition, status updates often spill over into interactions outside of Facebook, as you talk with others about their recent Facebook posts and they comment on yours. Examples such as this underscore the blurring of boundaries between social media and other venues for connecting with others.

BOX 3.5: CONNECT & REFLECT

The Power to Tell Your Own Stories

Chimamanda Ngozi Adichie observes that you gain power when you tell the stories of you and give others power when you allow them to tell your stories (Figure 3.13). That is, if others define who you are, you lose control over the narratives of you. While identity always involves a negotiation process with others, you have the power to correct flawed narratives, as Adichie did with her college roommate. However, to exert this power, you first want to trace your own multiple intersecting stories. Consider the narratives of your life that make up how you view yourself today. You might think about stories from the various familial roles you fulfill, as with sibling, child, parent, cousin, and spouse. Reflect on your other social roles, such as student, coworker, colleague, and friend, too. Also consider stories associated with turning points or major events that influenced your life. What are some of those stories? How have they intersected to create the you of today? How are some of those stories still unfinished and developing? How might you tell others about the stories of you?

FIGURE 3.13 Writer and speaker Chimamanda Ngozi Adichie argues it is up to you to tell your stories.

Relational Identity

relational identity The aspect of identity that evolves in the relationships individuals form through communication with others

The third layer, **relational identity**, develops in the relationships you form with others through your communication with them. The center for identity in this layer is relationships. Relational identity has three dimensions. First, your interpretations of how others view you become part of your relational identity. If others tell you that you are smart and well-organized, you will come to think of yourself as smart and well-organized. Second, the relationships you have with other people, such as sibling, student, and coworker, contribute to your relational identity and influence each other. That is, you experience many identities simultaneously as you go about your day. While you are in class as a student, you remain a friend, sibling, child, coworker, and all the other identities in which you may engage. Third, a relationship itself can form an identity, as with a group of friends, a work unit, or a family. For example, research on refugees who resettled in the United States found that family served as an important relational identity, offering both a cultural anchor and a mechanism for coping with a new environment.[73]

Communal Identity

In **communal identity**, the fourth layer in a communication approach to identity, identity is focused in the group. This layer refers to the identity groups or networks people form over time, as with the members of a church, a sorority, a social club, or other organization. Such groups develop a history and common experiences that are reflected in shared characteristics. For example, Google employees value innovative and creative thinking; they refer to themselves as "Googlers." When people feel strong identification with their work group, they experience low levels of emotional exhaustion and depersonalization and a high level of personal accomplishment.[74] Communal identity helps you feel connected with something larger than yourself and provides you with a context for understanding your contributions to the larger environment (Figure 3.14). Box 3.6 encourages you to explore the groups to which you feel especially connected.

communal identity The identity groups or networks people form over time

FIGURE 3.14 Your communal identity gives you the sense that you belong to something greater than yourself.

BOX 3.6: LIVE IT LOCAL

Uncovering Your Communal Identity

Consider the social networks in which you are embedded. To what groups do you feel the most connected? Reflect on your interactions for the past week. Which people did you interact with the most? What groups do they represent? Develop an in-depth analysis of one of the groups, documenting its history, values, and activities. How is this group part of who you are? What have you learned about your identity in researching this group?

Ideally, these four layers work together harmoniously to present a unified and coherent set of identities. For instance, in the personal layer, you view yourself as friendly and well-liked. In the enacted layer, you are outgoing and personable. In the relational layer, you have many friends who enjoy spending time with you. In the communal layer, you belong to several groups and organizations that value friendships and social connections. However, tensions can exist between and among the four identity layers. When such a discrepancy happens, an **identity gap** occurs. That is, there is disjuncture between or among the layers, as with viewing yourself as outgoing (personal identity) yet avoiding interactions with others (enacted identity). For example, refugees in the Midwest found that while their personal identity often was grounded in their interpretations of U.S. cultural norms, their relational identity—how their families viewed them—was framed within the cultural norms of their country of origin.[75]

identity gap A disjuncture between or among the layers of identity

When individuals experience identity gaps, especially in personal-enacted identities and personal-relational identities, they are less satisfied with their communication with others, feel more misunderstood, have a more difficult time engaging in conversations, and report symptoms associated with depression. In addition, people who are more assertive have greater alignment between their personal and enacted identities than people who are less assertive, likely because they feel more comfortable expressing their self-image. In contrast, people with higher levels of communication apprehension tend to have larger gaps between their personal and enacted identities, probably stemming from discomfort in revealing their self-concept to others.[76] For example, a study of college students found that those with larger identity gaps were more apprehensive talking with their instructors and less satisfied with their communication experiences with their instructors than students who had more integrated identities.[77] Still, not all identity gaps produce negative outcomes. Research on people actively involved in social protests found that identity gaps motivated some study participants to facilitate dialogue between opposing groups. That is, their more complicated identities provided a foundation for them to view the issues from multiple perspectives.[78] Similarly, young farmers in Indonesia who encountered gaps in their personal and relational identities took steps to connect with other farmers, become involved with farming-related groups, and develop cooperative relationships to improve farming techniques.[79]

Self-Disclosure: Revealing Your Identity

self-disclosure The information individuals reveal about themselves to others

A communication perspective of identity underscores the ways in which who you are arise from your connections and interactions with other people. In those interactions, you present, defend, negotiate, and manage your and others' identities. **Self-disclosure** refers to the information individuals reveal about themselves to others. People self-disclose for many reasons, such as deepening a relationship, seeking help, developing trust, and demonstrating similarity with the other person (Figure 3.15). In the health care context, for instance, patients view providers who self-disclose experiences similar to theirs as more trustworthy and likeable. In addition, the patients are more satisfied with the health care provider and more willing to continue the relationship when the health care provider self-discloses relevant information.[80]

FIGURE 3.15 Self-disclosure can deepen a relationship.

There also are reasons not to self-disclose, such as concern with hurting the other person or the relationship, wanting to maintain privacy, feelings of shame, possibility of stigma, and lack of appropriateness to the relationship.[81] For example, social media users balance privacy concerns with **social awareness**—the recognition of connections with others in shared online experiences—in decisions to self-disclose.[82] That is, self-disclosure serves to increase social awareness or online personal connectedness yet at the same time can impinge on individuals' privacy boundaries. For instance, members of online support groups for people trying to quit smoking who chose to reveal demographic information—such as age, gender, and marital status—felt a stronger connection to the group and were better able to avoid smoking than those who remained completely anonymous.[83] However, the researchers noted that disclosing demographic information is not without risks, especially for stigmatized groups.

Your mood also influences what and if you self-disclose. If you are feeling happy, you usually disclose more information and it is usually positive. If you are feeling sad, you usually disclose less information and it is more negative.[84] However, if you are feeling highly stressed, you likely will disclose more information and that information will be more personal.[85] You also may choose not to self-disclose based on how you think the other person might respond. In a study of young adults who were cancer survivors, their decision to talk with friends about the various aspects of their experience with cancer—diagnosis, treatment, effects, emotions—was grounded in how the friend had responded in the past as well as how they thought the person might react. For instance, study participants reported not self-disclosing because they did not want to upset others or out of concern for being treated differently after the disclosure.[86]

How do you decide what to tell people about yourself? When you disclose personal information to someone else, you take into consideration four factors: the relationship, risks and benefits, relevancy, and the degree to which the other person will reciprocate.[87]

Self-Disclosure and Relationships

First, the relationship you have with the other person influences what and how much information you will disclose. Generally, people disclose more superficial information to strangers and more personal information to friends and family.[88] Even more specifically, you likely disclose different information based on the type of relationship, such as parent, sibling, coworker, close friend, or spouse/partner. Research on college students has found that they disclose personal information more often to same-sex friends and dating partners and least often to parents.[89] Not surprisingly, parents and adolescents who reported a warm relationship also reported higher levels of adolescent self-disclosure and more accurate parental knowledge of their teens' daily activities.[90] Self-disclosure also varies based on gender. For example, one study found that women were more likely than men to self-disclose to friends and offer supportive listening responses to the other person's self-disclosures, especially with other women.[91]

FIGURE 3.16 Self-disclosing online may serve different functions than self-disclosing in-person.

online disinhibition effect The tendency to self-disclose at higher levels when interacting online than when interacting in-person

Self-Disclosure Risks and Benefits

Second, you consider how much risk is involved in the disclosure and how much it will benefit you and the people with whom you are interacting. The most apparent benefit of self-disclosure is that it provides a critical foundation for developing and maintaining friendships and other interpersonal relationships. Research on bloggers, for example, revealed that higher levels of self-disclosure were associated with making more new friends and increasing the overall quality of friendships.[92] A study of illness-related support groups on Facebook found that self-disclosing within the groups helped participants develop their individual illness narratives and better interact with the health care system.[93] Interestingly, online and offline self-disclosure may serve different functions for different people (Figure 3.16). For instance, a recent study of adolescents in France found that those who cope with loneliness in a more passive way, as with distracting themselves, self-disclose on Facebook to manage their loneliness. In contrast, adolescents who use more active coping strategies, as with strengthening social bonds, rely on in-person self-disclosure to lessen feelings of loneliness.[94]

Online self-disclosure can pose different risks and benefits than in-person disclosures. Research on online communication has found that anonymity leads to higher levels of self-disclosure, a phenomenon called the **online disinhibition effect**. When an individual's self-disclosures are anonymous, the cost of disclosing is low because there is little fear of personal consequences.[95] The benefits are high because people can express themselves freely and even try out different identities. For example, individuals who blog about health issues that are embarrassing or stigmatized, such as HIV/AIDS and cancers associated with lifestyle choices, disclose more personal information when their posts are anonymous.[96] In addition, anonymous blogging provides people with stigmatized illnesses an opportunity to enact their identities in ways they might not feel comfortable with when interacting in-person. Anonymous posting also allows people to self-disclose actions or thoughts about which they might feel guilty. For example, a study of posts to a question-asking site found that the primary motivation for choosing answers was guilt-relief.[97] Analysis of online dating profiles also has found that people disclose highly personal information when their screen names are not their real names, lowering the risk factor.[98] The benefit in revealing this information is the possibility of identifying a match. Interestingly, people who post photos of themselves but not their real names tend to disclose more than those who do not include any photos.[99] In this way, the level of self-disclosure via visual imagery parallels the level of written self-disclosure.

Self-Disclosure and Relevancy

A third consideration when deciding to disclose information is how relevant it is to the topic and context. Public speakers often disclose personal information relevant to their topic to demonstrate their experience with it. For example, college instructors disclose relevant personal information to better relate to their students and explain important concepts. College students express greater student–instructor relationship satisfaction when instructors disclose personal information relevant to the class, which in turn leads to improved comprehension of course material and greater satisfaction with the class. In addition, students more actively engage in a class, such as voluntarily interacting with the teacher or asking a question, when an instructor discloses relevant personal information.[100] In online classes, instructors who appropriately self-disclose give students a greater sense of the instructor's presence, enhancing the student–instructor relationship and increasing students' knowledge of the topic.[101]

Self-Disclosure and Reciprocity

Fourth, you consider the likelihood of reciprocity when you self-disclose. That is, when you reveal personal information to others, you assess the degree to which they will reveal similar sorts of information. Police officers strategically disclose personal information during an interrogation to demonstrate that they understand a suspect's situation and encourage the person to self-disclose—and admit to the crime.[102] A study of bloggers in 32 countries found that their levels of self-disclosure were strongly tied to their intended audience. Bloggers engaged in reciprocal self-disclosure with other bloggers they considered part of their audience.[103] Members of online breast cancer support groups demonstrated reciprocal self-disclosure both in what they discussed and the writing styles they used in their posts.[104]

If you self-disclose and the other person does not respond in kind, you likely will evaluate the response negatively and you are unlikely to continue disclosing. The norm of reciprocity in self-disclosure is why you might feel uncomfortable when people disclose excessive or deeply personal information for which you are unprepared. However, when individuals are engaged in a conversation, they work to respond in positive ways in the face of another's excessive self-disclosure (Figure 3.17). Even if someone unexpectedly discloses something especially personal, you will do your best to put them at ease. In contrast, those observing the conversation will judge a violation of the reciprocity norm more harshly.[105]

FIGURE 3.17 When one person in a conversation unexpectedly discloses deeply personal information, the other person will attempt to respond in positive ways even if they feel uncomfortable.

74   PART I: Foundations of Communication

Self-disclosure always involves balancing expression with privacy.[106] As a social being, you are motivated to express yourself and share who you are with other people. Yet defining the self as separate from others requires a degree of privacy and keeping some aspects of the self to yourself. In addition, individuals may keep information private out of concern others will judge them in a negative light. Women who have been abused by their partners, for example, often feel more comfortable disclosing the abuse with counselors and in support groups rather than with friends and family members, due to the stigma associated with the abuse.[107] Job interviews are another context in which individuals may be hesitant to self-disclose. In a study of recruitment interviews, interviewees were less willing to self-disclose if they were concerned about the privacy of their personal information.[108]

CHAPTER REVIEW

■ **Explain how a pervasive communication environment influences perception.**
A pervasive communication environment has greatly increased the availability, number, and range of stimuli within the human perceptual field. Social media has contributed to the rise of pseudo news and deepfakes, both forms of disinformation.

■ **Identify and describe the stages of the perception process.**
The process of perception includes four stages: sensing and selecting stimuli, organizing information, interpreting and evaluating, and recalling and remembering. At each successive stage, some information is lost.

■ **Explain the relationship between identity and a pervasive communication environment.**
The hyperpersonal model provides insight into how a pervasive communication environment has changed the ways in which individuals present the self and view their identities. The hyperpersonal model posits that people have more control over what and how they present themselves online as well as the channels used for self-presentation and interpreting how others present themselves. Feedback allows people to revise and adapt their self-presentation strategies.

■ **Discuss a communication approach to identity and its elements.**
A communication approach to identity includes personal identity, enacted identity, and communal identity. These layers of identity are all part of an individual's identity. Disjunctures between and among the identity layers result in identity gaps.

■ **Describe the relationship between identity and self-disclosure.**
You reveal your identity through self-disclosing information to others. Four considerations influence your decision to self-disclose: the relationship, risk-benefit analysis, relevance, and reciprocity.

DISCUSS AND APPLY

1. What are some examples of pseudo news you have encountered on social media? What examples of deepfakes have you viewed? How do you sort out genuine stories from fake ones?

2. What motivates you to pay attention to a message? Do you find intrinsic or extrinsic rewards more motivating? Give some examples.

3. How would you describe your relational schema for "friend"? What words would you use to describe what you expect from a friend? How does your relational schema for friend influence how you interact with your friends?

4. Review Table 3.2, which suggests strategies for improving your working and long-term memory. Try out two or three of the strategies, such as chunking (working memory) and association (long-term memory). How effective were these strategies in improving your memory? How might you use these strategies in your interactions with others?

5. Go to the VocaliD (vocalid.ai) website to find out more about the importance of voice in an individual's identity. How would you feel about donating your voice or receiving a VocaliD?

6. View Chimamanda Ngozi Adichie's TED Talk, "The Danger of a Single Story" on the TED website. Give an example of a time when someone stereotyped you based on a single story. How did you feel when it happened? How did you respond? What stories would you tell others about yourself to provide a more complex picture of who you are? Have you ever categorized someone based on a single story? How can you avoid the danger of a single story in the future?

7. The hyperpersonal model of online communication suggests that individuals engage in selective self-presentation. How do you control the image you present online? For example, how do you decide which photos to post and which not to post?

8. Analyze your own identity. How would you describe your personal identity? How do you enact your identity with others? What relational identities can you identify? What communities or groups do you consider part of your identity? How do the four layers of identity fit together for you?

9. How do you decide what information to self-disclose and what information to keep private when you interact with other students in your classes? How do those decisions change—or not change—over time?

CREDITS

Fig. 3.0: Copyright © 2018 Pexels/Brett Sayles.

Fig. 3.2: Trekky0623, "Flat Earth," https://commons.wikimedia.org/wiki/File:Flat_earth.png, 2008.

Fig. 3.4: Copyright © 2019 Pexels/Jopwell.

Fig. 3.5: Copyright © 2020 Pexels/cottonbro.

Fig. 3.6: Copyright © 2020 Pexels/Life Matters.

Fig. 3.8: Copyright © 2015 Pexels/Lukas.

Fig. 3.9: Copyright © 2019 Pexels/Bran Sodre.

Fig. 3.12: Copyright © 2014 Depositphotos/Wavebreakmedia.

Fig. 3.13: Copyright © by Carlos Figueroa (CC by-SA 4.0) at https://commons.wikimedia.org/wiki/File:Congreso_Futuro_2020_-_Chimamanda_Ngozi_Adichie_02.jpg.

Fig. 3.14: Copyright © 2019 Pexels/fauxels.

Fig. 3.15: Copyright © 2018 Pexels/nappy.

Fig. 3.16: Copyright © 2020 Pexels/Zen Chung.

Fig. 3.17: Copyright © 2018 Pexels/mentatdgt.

CHAPTER 4

Listening

What are the key issues students attending historically Black colleges and universities (HBCUs) face? To address that question, the U.S. Congressional Black Caucus (CBC) launched a listening tour and set out to visit more than 100 HBCUs. The CBC members asked students about their perspectives on a range of topics, such as student loan debt and voter suppression. In addition to finding out about students' concerns, the congressional representatives designed the campus visits to spur students' interest in civic engagement and motivate them to take action on issues especially relevant to them.[1] Similarly, colleges, universities, corporations, government agencies, and organizations in the United States and worldwide sponsored listening sessions, often in conjunction with Black Lives Matter, to promote open and thoughtful discussions of racial injustice and actions to promote racial equity.[2]

When you think of yourself communicating in nearly any context, you probably focus on speaking and what you want to say. Listening tours and sessions offer a reminder of the other side of communication—the listening side. In initial interactions, good listeners are rated as friendlier and more respon-sive than poor listeners. College professors who demonstrate excellent listening skills provide a more satisfying classroom environment for students. In the workplace, listening requires your attention throughout the day, from meetings with coworkers to a project update from your boss. The people you work with expect you to listen to them and evaluate your performance in part on how well you seem to listen.[3] Effective listening skills increase employee productivity and well-being, improve teamwork, and raise workers' morale.[4] The more positively employees evaluate their supervisors' listening skills, the less likely employees will leave an organization or suffer from emotional exhaustion and the more likely they will experience less stress, feel greater happiness, express satisfaction with their job, and exhibit higher levels of creativity.[5] Employees with better lis-tening abilities generally receive stronger performance evaluations. Customers and clients are more likely to remain loyal to an organization when they view members' listening skills as effective.[6]

Effective listening skills provide a central founda-tion for effective human communication and success in achieving your academic and professional goals. A study of college students' metaphors for listening found that participants recognize the importance of listening in connecting with others. Students' descriptions of listening centered on understanding, attentiveness, and building and maintaining relationships through listening.[7] Students who listen well tend to do better in school, while those with poor listening habits struggle to succeed (Figure 4.1). Top-rated physicians report that listening is central to understanding and sharing decision making with their patients. In a survey on communication skills essential for work groups, personnel directors rated effective listening as the most important. When choosing a leader, people who listen well top the list. Effective listening also lowers your stress and increases feelings of well-being at work.[8] To a large extent, effective listening forms the basis for success in life and your connections with others.

FIGURE 4.1 Students with good listening habits usually excel in their studies.

In spite of listening's importance in everyday life, only a tiny fraction of people receive any training in listening. Most colleges and universities offer no courses in listening. Listening often earns just brief attention in courses associated with communication skills, such as persuasive presentations and interviewing. Yet extensive research has found that listening is ranked consistently as one of the most important soft skills in the workplace.[9] In contrast, nearly all college courses place a much greater emphasis on speaking than listening.[10]

In this chapter, you will learn about the importance of listening in a pervasive communication environment, misconceptions about listening, and the process of listening.

Listening in a Pervasive Communication Environment

Studies of human communication going back nearly a century have found that people spend most of their communication time listening. The exact amount varies from 15% to more than 50%, but there is no doubt that listening dominates your daily communication activities.[11] When you compare the four primary communication skills—listening, speaking, reading, and writing—listening is the first skill you learn and the one you use most often throughout your life. As an infant, you listened well before you talked. As an adult, you listen to friends, family, coworkers, and others as you go about your daily routine. You tune in to talk shows on the radio and listen to music on your phone.[12]

FIGURE 4.2 With the ability to communicate nearly anytime with anyone from anywhere, effective listening skills have never been more important.

Effective listening skills never have been more important in your interactions with others (Figure 4.2).[13] Listening plays a key role in sorting through all the information you encounter in a pervasive communication environment—an environment in which communication opportunities have greatly expanded. You may listen in real time or synchronously to others in-person, via phone, or in video chat, such as FaceTime, Houseparty, Skype, or Zoom. You can talk with others in real time who may be in different time zones and different countries. You must adapt to diverse cultures and diverse listening styles. Asynchronous listening may occur when you receive a voicemail or download a podcast. In addition, while you can access information anytime from anyplace, the communication tools you use—phone, tablet, laptop, and the like—can distract you from listening to others.

The 21st century has brought with it an expanded notion of listening, too. In most cases, texting, email, and social media share more characteristics with oral communication than written communication.[14] For example, when you text a friend, you write as you would speak—informal, short sentences, personal. Interpreting those messages involves listening more than reading, as you consider the sender's "voice" and how you will respond. Who or what is listening has changed as well. For instance, voice-based assistants (VAs), such as Siri, Alexa, and Bixby, have become commonplace on digital devices. Research suggests that users often view VAs in human-like terms and engage with them following conversational norms. In other words, individuals expect VAs to listen and respond to them as a person would.[15] VAs can become part of a family and even play a role in facilitating positive relationships among family members, as one study found when parents and children shared an Alexa. The common experience of interacting with the VA provided a familial bond.[16] However, VAs are not a perfect substitute for human listeners. When job applicants were told their responses would be evaluated by an artificial intelligence (AI) program, interviewees were less likely to use deceptive communication strategies and gave shorter answers than those who were told a person would assess their performance. Without a human to listen to their responses, applicants were more honest but missed the human connection that would motivate them to elaborate on their answers.[17] In a study of adults in their mid-60s to mid-80s, how Alexa responded—or did not respond—was central to participants' descriptions of the VA. When Alexa gave relevant and supportive responses, users viewed Alexa as more like a friend and someone to keep them company. When there was no response or the response seemed irrelevant or dismissive, then Alexa became an object. Still, while the researchers found that interacting with Alexa alleviated times of momentary loneliness for these older adults, connecting with people was essential to lessening social isolation.[18]

Misconceptions About Listening

Because you engage in listening as an everyday practice, you might get lulled into thinking you know everything about it. Five misconceptions about listening interfere with developing an accurate understanding of the listening process and improving listening skills.

People Are Born Good (or Bad) Listeners

You might assume people are born good (or bad) listeners. To the contrary, you learn listening habits from your family, peers, and others you interact with regularly. How well you listen, or do not listen, depends on your response to a particular situation, rather than general personality traits. For example, more social, likeable, and outgoing people are not necessarily better listeners.[19] Reflect on your own listening habits. When do you listen well? When are you not as attentive as you might be? In what ways might you improve your listening skills?

Intelligent People Are Better Listeners

You might think that intelligent people are better listeners. But having a high IQ does not automatically make someone a good listener.[20] Rather, better listeners usually have a better vocabulary, are better notetakers, are more empathic, and are more adept at asking useful questions than those whose listening skills are lacking. Increasing your vocabulary, improving your notetaking, demonstrating greater empathy, and asking good questions are all skills you can develop—they are not necessarily the product of a higher IQ.

Listening Is Easy

listening fatigue Occurs when communicators become physically and mentally weary due to a listening situation

You might assume listening is easy. But if you ever have taken detailed notes in a class lecture or listened to a friend's problems, you know listening well requires energy and hard work. **Listening fatigue** occurs when you feel physically and mentally weary due to a listening situation.[21] Listening to the information itself might drain your energy, as with lengthy instructions for a new task at work. Distractions, such as outside noise, can make listening tiring. For instance, listening to someone in a crowded, noisy restaurant with poor acoustics takes much more energy than the same conversation in a quiet café. The topic and setting both contribute to listening fatigue. In the workplace, supervisors may experience listening fatigue and emotional exhaustion when they listen to employees' concerns, even when the comments are presented in constructive ways.[22]

FIGURE 4.3 Communicators are especially likely to experience listening fatigue when video conferencing.

Video conferencing with Zoom, FaceTime, Skype, or similar software presents its own problems that lead to listening fatigue (Figure 4.3). Because you are not in the same physical location with other communicators, you must focus more diligently on the screen and ignore distractions in your surroundings. In addition, with the camera always on, you are under more pressure to perform active listening behaviors—making eye contact, saying "uh huh," and nodding your head—at all times. During in-person conversations, you can take a break, look away from the person, let your mind rest a bit. But with the camera unnaturally close to participants' faces—much closer than when interacting in-person—you feel obligated to appear completely engaged in the conversation.[23] Finally, sound often is not as clear in such mediated contexts, forcing communicators to work all that much harder to understand what a speaker is saying.

listening anxiety Occurs when communicators feel inadequate in receiving and processing messages

Another indication that listening is not easy is when you experience **listening anxiety**, which occurs when you become anxious due to feeling inadequate in receiving and processing messages.[24] High levels of listening anxiety interfere with your ability to process information and result in lower levels of **listening fidelity**, the degree of similarity between the sender's intended message and the receiver's interpretation of that message.[25] Those who experience listening anxiety find it difficult to concentrate on what others are saying, feel self-conscious about their listening skills, become confused when receiving important information, become tense when learning about new ideas, and feel less motivated to listen.[26]

listening fidelity The degree of similarity between the sender's intended message and the receiver's interpretation of that message

listening The process of receiving, interpreting, understanding, evaluating, and responding to messages

Listening and Hearing Are the Same

hearing The physiological reception of aural stimuli or sounds

Listening occurs when you receive, interpret, attempt to understand, evaluate, and respond to messages.[27] **Hearing** is the physiological reception of aural stimuli

BOX 4.1: CONNECT & REFLECT

Listening and People With Hearing Loss

Although hearing typically is considered the foundation of listening, Deaf people or others with hearing loss engage in listening, as with listening to sign language. For a Deaf person, listening to hearing people can be especially stressful, even with an interpreter present, as the person speaking often talks to the interpreter rather than the Deaf person.[29] When the hearing person attends only or primarily to the interpreter, the identity of the Deaf person is lost or diminished in the interaction. For people with some hearing loss, listening to hearing people can result in listening fatigue due to the additional focused attention needed to understand what a speaker is saying.[30] Schools, companies, and other organizations that embrace accessibility provide qualified interpreters and text-based communication options for Deaf or hard-of-hearing people.[31] If you are a hearing person, what are your experiences communicating with Deaf people or those with hearing loss? How did you listen to them? How did you speak with them? If you are a Deaf person or have a hearing loss, what are your experiences communicating with hearing people? How did you listen to them? How did you speak with them?

or sounds, whereas listening involves applying mental processes to figure out the meaning associated with the sounds you have detected (Figure 4.4). When you listen, you use all your senses to interpret, understand, and respond to a message. Sometimes listening does not involve the physical act of hearing at all. Research suggests, for instance, that asynchronous discussions in online classes have more in common with oral than written communication.[28] That is, when you reply to a friend's text message, you are listening to that person—interpreting, attempting to understand, and responding to the message. In addition, Deaf people or those who have hearing loss listen to messages in a variety of ways, as you will explore in Box 4.1.

Listening Just Happens

A third misconception about listening is that it just happens. But as with speaking, you must prepare for and actively participate to listen effectively. The **thought-speech differential** can help you listen well or lead to poor listening habits. The average person speaks between 120 and 180 words per minute, yet people process speech at about 400–500 words per minute.[32] During the extra processing time caused by the thought-speech differential, you might think about what you want to say when it is your turn to talk. But the thinking you do in preparation for expressing yourself is different than how you think in preparation for listening to messages.[33] As you consider what you are going to say, you focus on the main points and the general idea of the message you want to convey. For example, you associate many different thoughts with the first time you met the people in this class. To put those thoughts in a manageable form, you condense them to a few key ideas and leave out the tangential information. In telling the story of

FIGURE 4.4 "Hearing" is more than just sounds reaching your ears.

thought-speech differential The difference between the rate senders can produce messages and receivers can process those messages

your first day in class, you focus on the basics, such as your initial impressions of your classmates. In contrast, when you are listening to others, you add your own interpretations, thoughts, feelings, and related information. In **thinking to speak**, you reduce the topic; in **thinking to listen**, you expand it (Figure 4.5). Listening does not just happen—you must think to listen.

FIGURE 4.5 When you think to speak, you narrow your focus on the ideas you want to convey in your message. When you think to listen, you expand your thoughts, adding your interpretations and meanings for the topic.

thinking to speak When communicators think to speak, they reduce the topic, identifying the key points they want to convey in their message

thinking to listen When communicators think to listen, they expand the topic, adding in their own interpretations and thoughts on what the other person is saying

Types of Listening

Different situations call for different types of listening. You adjust your listening based on the context and your goals.[34] For example, if a friend is telling a story about a recent cycling trip, you likely want to simply enjoy the narrative. In contrast, if a salesperson is trying to convince you to purchase a gym membership, you likely want to carefully evaluate the reliability of the information. Becoming a more effective listener requires you to match your listening approach to the requirements of the situation. Table 4.1 summarizes the five types of listening researchers have identified: appreciative, content, empathic, critical, and dialogic.[35]

TABLE 4.1 Types of Listening

TYPE	BRIEF DEFINITION	WHEN TO USE
Appreciative	Listening for enjoyment	Enjoy what others are saying
Content	Listening for the essence of the message	Understand the sender's main idea
Empathic	Listening for emotion	Feel as the other person feels
Critical	Listening to evaluate	Assess or critique a message
Dialogic	Listening to logic and passion, and for culture, power inequities, dialectics	Facilitate dialogic communication

Appreciative Listening

appreciative listening
A type of listening that involves listening for enjoyment

When you listen for enjoyment, you practice **appreciative listening**. This type of listening often relaxes you and puts you at ease with others. Appreciative listening is especially important in initial interactions when you are just getting to know someone. In the classroom, for instance, informal conversations after the class has ended provide important venues for appreciative listening. In online classes, introductory discussions in which students engage in light-hearted small talk can bridge the physical distance between classmates. Listening with appreciation to a joke, humorous anecdote, or funny story in-person or online can help build positive and productive interpersonal relationships, as students feel more comfortable with each other. Appreciative listening can bring people together in other ways, too. A recent study found that when people listen appreciatively to music, they feel a sense of shared experience, identification with others, emotional closeness, and belonging to a social group.[36]

Content Listening

In **content listening**, you are interested in the sender's main ideas. Also called comprehensive or informational listening, content listening involves understanding the essence of the person's message. Listening to instructions on how to do something, a lecture on a topic, or a docent giving a museum tour are examples of content listening. Much of your listening at school and work involves content listening as others share their ideas, report on assignments they have completed, or explain a problem that needs attention.

content listening A type of listening in which you want to understand the essence of the person's message

Empathic Listening

With **empathic listening** you engage in role taking and attempt to feel as the other person does by putting yourself in that person's place. You imagine how that other person might respond to something you or others say.[37] Sometimes called therapeutic listening, engaging in empathic listening means withholding judgment, advice, suggestions, and evaluation. When others express happiness, sadness, joy, and frustration, empathic listening helps you better understand their feelings. Empathic listening signals to the others that you care about and feel affection for them.[38] This type of listening often requires that you attend to more than just the other person's words. You try to discover what the person might be saying nonverbally, as with a loud voice, animated gestures, slow response to an email, or an abrupt text message. You express sensitivity to how the other person feels and focus on understanding their emotions. This may involve guesswork as you try to determine the other's feelings. If appropriate, you check your interpretation by asking the other person. Your goal is to understand how the other person feels and avoid misunderstanding the person's emotions. Box 4.2 suggests one approach to achieving empathic listening.

empathic listening A type of listening that involves withholding judgment, advice, suggestions, and evaluation

---

BOX 4.2 ETHICAL QUESTIONS

Listening Being

Central to responding empathically and ethically is suspending your judgment of others and considering the world from their perspective. That means setting aside your own way of thinking and how you evaluate your environment and experiences. *Listening being* offers a way to work toward this goal of understanding the other person's message from their perspective.[39] With listening being, you define yourself as listening, rather than as speaking. You quiet your own thoughts so you are open to what others have to say. Listening being encourages you to avoid jumping to conclusions and rushing to judge others. Instead, you exist—at least for that moment—as someone whose reason for being is to listen. Say this aloud: "I am listening." What does that mean to you? How might you apply this way of thinking about listening when you need to listen with empathy?

Empathic listening in the workplace can prove especially challenging. For example, research on hairdressers has found that part of their work involves listening with empathy to clients, requiring a balancing act between demonstrating their skills as a stylist and providing emotional support to patrons. The hairdressers engage in **emotional labor** in which they must manage and suppress their own feelings while on the job.[40] That is, while the hairdressers are expected to listen empathically to their clients, their clients are not expected to reciprocate. Still, empathic listening at work can have positive outcomes. Supervisors who listen empathically increase subordinates' engagement with their work and dedication to their jobs.[41] In addition, employees who received training in empathic listening reported feeling less anxiety in social situations, engaging in greater self-reflection, and viewing topics in more complex ways.[42] Thus, empathic listening can produce benefits for both listeners and speakers.

emotional labor Occurs when workers must monitor and control their own emotions as part of their job

Critical Listening

When you evaluate what the other person says, you use **critical listening**. Effective critical listening depends on effective appreciative, content, and empathic listening. You must first appreciate and understand the other person's ideas and feelings before you can offer any critique. Without clearly understanding the other person's message, you are unable to offer an informed evaluation. In critical listening, you provide an assessment based on a reasoned examination of the information presented, rather than a judgment. Central to critical listening is avoiding personal attacks, name calling, derogatory comments, and other inappropriate listening responses.

critical listening A type of listening in which the goal is to evaluate or assess the other person's message

dialogic listening A type of listening that involves four components: listening with heart and mind, listening for and to culture, listening for power inequities, and listening for dialectics

Dialogic Listening

Dialogic listening integrates and extends on appreciative, content, empathic, and critical listening. As shown in Figure 4.6, four elements make up **dialogic listening**: listening with heart and mind, listening for and to culture, listening for power inequities, and listening for dialectics.[43] First, dialogic listening involves attending to logic and passion in what others say—the heart as well as the mind. In that respect, dialogic listening requires that you listen for the content and emotion in a message.

Second, dialogic listening means listening for and to culture. You listen to detect cultural differences and similarities between yourself and others—listening *for* culture. And you listen for how culture influences your communication with others—listening *to* culture. So first you have to discover cultural differences and similarities you have with others, then you identify the role various aspects of culture play in the listening

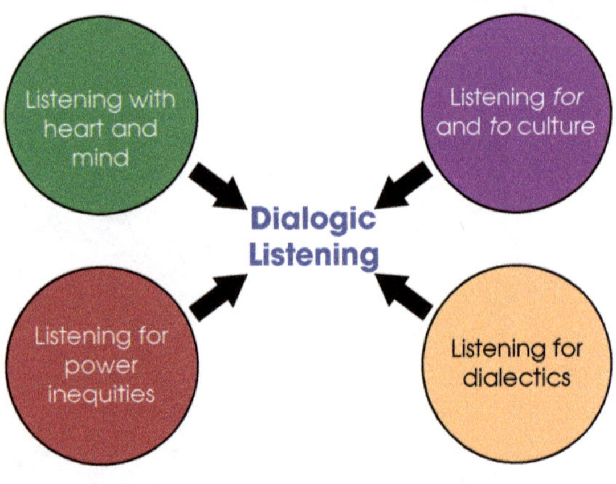

FIGURE 4.6 The four elements of dialogic listening.

process. Listening for culture requires a commitment to interrogate your own cultural assumptions and remain open to listening to culture—the differences and similarities other communicators express.[44]

As you learned in Chapter 2, how communicators use silence differs across cultures. Americans tend to be uncomfortable with silence. When you encounter a pause in a conversation, you probably want to fill it with talk. In many situations, such as classroom discussions, you may view silence in a negative way. Silent classmates may be viewed as nonassertive, overly compliant, unintelligent, and even lazy.[45] But many cultures prize silence and consider silence an important way to demonstrate respect for others. In addition, self-imposed silence can improve your listening skills and make you more aware of your own and others' listening behaviors.[46] Silence allows you to reflect on what speakers say. Scientists on interdisciplinary teams, for instance, are encouraged to use silence to consider the perspectives their colleagues bring to the group. That is, rather than immediately jumping into a conversation and offering a critique, the scientists contemplate what was said before deciding how to respond.[47]

Culture also affects how you listen to messages. For example, many Eastern cultures view listeners as the key to successful communication, whereas many Western cultures place the burden of success on the speaker. Research has found that while listeners across cultures generally agree on the functions of listening, those from Eastern cultures place more of an emphasis on building relationships through listening than those from Western cultures.[48] Listeners from Eastern cultures tend to listen more attentively than those from Western cultures. On the other hand, speakers from Eastern cultures tend to provide less detail and repetition, assuming listeners will fill in the missing information. Speakers from Western cultures recount information in a more linear fashion and with greater specificity.[49] Being sensitive to culture in dialogic listening means recognizing differences in how others listen and speak.

Although you might think of emotions as universally shared, individuals' expressions and interpretations of emotions are grounded in cultural experiences (Figure 4.7).[50] Some cultures, particularly in the West, view emotions as arising from an individual's physiological state. Other cultures think of emotions as created and shared between people or among a group. As you learned in Chapter 2, culture influences the norms or rules associated with when to express certain emotions and how they should be expressed. For instance, families develop their own rules for expressing and interpreting emotions. When you listen with your heart in dialogic listening, you recognize your own cultural filters as well as others'.

Third, dialogic listening requires detecting inequities in power among communicators.[51] Differences in power often are difficult to identify. Equality is a core value for Americans and their identity. Yet power differences based on ethnicity,

FIGURE 4.7 What emotion do you think the person in this photo is expressing? Your interpretation depends on your own cultural background and experiences.

economic status, sex, sexual orientation, dis/ability, and similar factors exist in U.S. society. The increasing activism of social justice organizations, such as Black Lives Matter, National LGBTQ Task Force, National Disability Rights Network, National Congress of American Indians, The Elder Justice Coalition, and Code Pink, highlights systemic inequities in power across U.S. demographic groups. Those differences in social power influence your interactions with others, particularly when communicators have diverse backgrounds. Explore issues of power in the classroom in Box 4.3.

BOX 4.3: CONNECT & REFLECT

Recognizing Power in Classroom Discussions

Carefully observing how students interact in a class can reveal power inequities. Closely examine how students interact in one of your classes, either in-person or online. For example, during a class discussion, keep track of who talks the most, whose ideas are listened to, who talks the least, and who is ignored. Then consider why some students exercise more power in controlling the direction the discussion takes and the topics discussed. Is everyone really getting an equal opportunity to contribute? Do you listen with equal diligence to what each person has to say? If the instructor also participates, what effect does that have on the discussion? When you carefully scrutinize the flow of conversation in the class, you get a better idea of how power is distributed.

Finally, dialogic listening recognizes the tensions or dialectics of human experience and rejects either/or thinking. As you will learn in Chapter 7, some common dialectics are autonomy-connection, openness-closedness, and predictability-novelty. Individuals often feel pulled in two opposite directions at the same time, such as wanting things to stay the same while also wanting things to change. In dialogic listening, you seek to avoid thinking in ways that promote one end of a dialectic over another, such as viewing autonomy as better or more important than connection. Instead, you realize that people experience both sides of the dialectic simultaneously.

Identifying dialectics in dialogic listening also means understanding that you can listen for appreciation and content, for empathy and to critique, and for content and empathy. Dialogic listening may seem overwhelming as it involves complex skills in hearing, understanding, remembering, interpreting, evaluating, and responding to others' messages. You will not apply dialogic listening all the time. But when you have reached the point where you are as eager to listen to others' ideas as you are to share your own, then you have achieved **dialogic virtuosity**.[52]

dialogic virtuosity Occurs when communicators are passionate about listening to and understanding the other person's perspective as well as clearly articulating their own

The Process of Listening

Listening is a continuous process that involves four stages: anticipating a listening situation, applying the basic components of listening, assessing listening effectiveness, and aligning your listening goals and processes.[53] These stages, summarized in Figure 4.8, are explained in detail in the following sections.

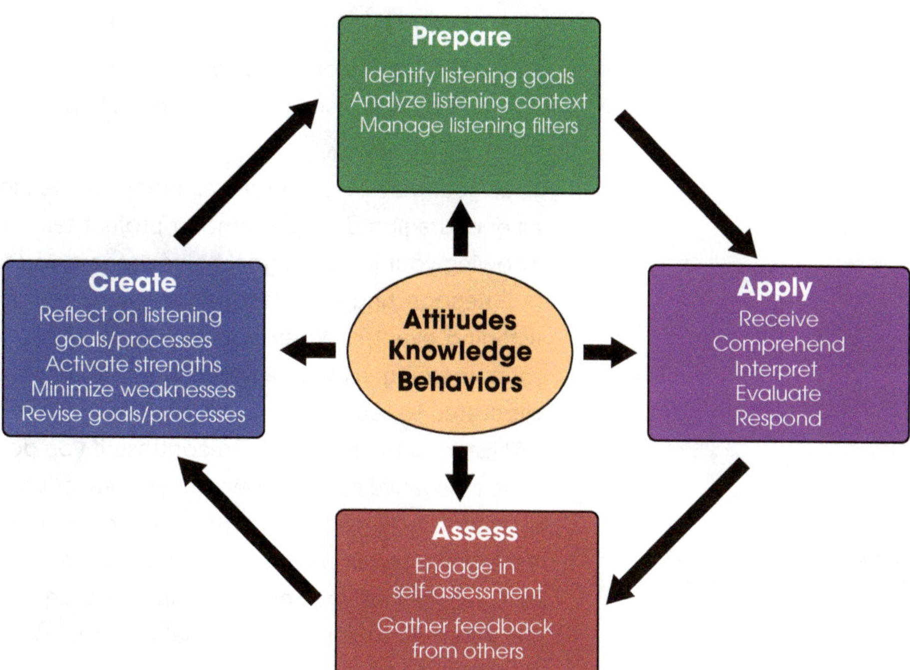

FIGURE 4.8 A process approach to listening.[54]

Anticipate Listening

Just like you anticipate and prepare for important events in your life—a job interview, an exam in a class, a team presentation—you get ready to listen. To begin, identify the participants, the occasion, and the purpose to analyze the listening context. Determine who will be presenting information and who will be listening. Consider the relationships among communicators. Find out about the setting for the listening event and reflect on how it will influence your ability to listen. Finally, identify the purpose for the interaction. Are communicators trying to solve a problem or decide on a solution? Is this just a casual conversation among friends or a problem-solving discussion among neighbors? What outcomes do participants hope to achieve? Having a sense of the listening context in advance will help you identify your listening goals and better prepare you to respond appropriately.[55]

Second, identify your listening goals. Review the types of listening and determine if you are listening for enjoyment (appreciative listening), the essence of the message (content listening), emotion (empathic listening), evaluation (critical listening), logic, passion, culture, power inequities, dialectics (dialogic listening), or some combination. Some example listening goals are:

- Enjoy meeting new people at school [listening to other students at a college social function]

- Understand what I need to do [listening to the instructor at a training session]

- Recognize how my friend is feeling [listening to a friend whose family member is ill]

- Evaluate the utility of a project [listening to a project proposal at work]
- Facilitate open and engaged problem solving [listening to a diverse group of people discuss a contentious issue]

Your goal may be as simple as enjoying the story a friend tells or as complex as encouraging dialogue among project team members with opposing views. Knowing your listening goal helps prepare you to listen.

Recognizing motivations for listening is another factor in anticipating a listening event.[56] When you lack motivation, you fail to prepare adequately for something, as with a meeting or videoconference. For example, if you are motivated to listen and have a positive attitude toward the speaker's topic, you will listen more effectively. In contrast, if you do not feel motivated to listen and have a negative attitude toward the speaker's topic, you likely will not pay much attention. **Extrinsic motivators** are incentives external to you that cause you to behave in a certain way. For instance, you might be predisposed to listen carefully to a member of your project team because completing your task depends on the information you get from that person. **Intrinsic motivators** drive people to take action based on internal forces, such as listening attentively to a presentation because you are interested in the topic. Identifying what motivates you can help you get ready to listen and avoid missing important messages. If you know external rewards bring out your best behavior, then remind yourself of those rewards as you anticipate listening to others. A 3-hour student club meeting can become a productive listening experience when you consider the praise you will receive for providing a concise summary of everyone's contributions. Similarly, if internal motivators provide the incentive you need to listen well, figure out how to best activate them. Imagining how you will listen during that 3-hour meeting can become more enjoyable when you think about how much you like working with the other club members. If you are motivated to listen, you are more likely to anticipate the occasion in a positive way.

Apply Listening Components

The basic act of listening involves receiving, comprehending, interpreting, evaluating, and responding to messages, shown in Figure 4.9. When you receive a message, you respond to stimuli, which typically are aural but often include all the senses—hearing, taste, touch, sight, and smell. In a pervasive communication environment, you can receive messages from nearly anyone, at anytime, anywhere. But communication technologies do fail or work intermittently. Mobile phone connections drop, video streams freeze, emails bounce back, and voice mail messages get erased accidentally. Technological innovations can interfere with hearing messages as well. Ringing phones, animated advertisements, text alerts, and other digital media features can cause distractions that interrupt your focus and result in not hearing messages.

---

extrinsic motivators Incentives external to individuals that cause them to behave in a certain way

intrinsic motivators Incentives that encourage people to take action based on forces internal to the individual

Distractions or extraneous **noise** interfere with message reception. With distracted listening, you focus on the noise rather than the speaker. Some distractions are within the listener.[57] **Internal distractions** include physical sensations, emotions, and thoughts. If you did not get enough sleep the night before an important class session and you are tired, you might have a difficult time focusing on the discussion. Having a great weekend with friends may leave you feeling so happy on Monday that concentrating on what a coworker is talking about at the moment is the last thing you want to do.

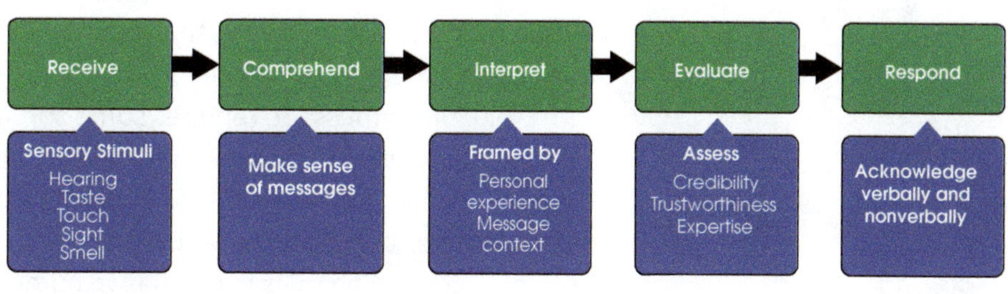

FIGURE 4.9 The components of listening.

noise Distractions that interfere with the listening process

internal distractions Physical sensations, emotions, and thoughts that interfere with listening

external distractions Anything in the environment that interferes with listening

External distractions refer to anything in the environment that interferes with listening, such as your phone ringing, traffic outside a classroom window, or an overly warm room. These distractions serve as mental as well as physical barriers to listening. Even when you can hear a message, external noise can distract your brain, leaving you with less cognitive capacity to process the message.[58] Video conferences and chats are especially susceptible to external distractions, as communicators are in different locations with their own distractions, multiplying the possible stimuli to which the brain might attend. The practice of using background images in Zoom meetings and other video conferencing platforms presents an additional visual distraction, as participants' heads seem to bob, float, and weave over an often-cluttered image.

Comprehending means that you understand or make sense of a message. Listening comprehension occurs internally as you think about the message you hear and try to figure out what it means. The degree to which you understand a message depends on your vocabulary and familiarity with the topic. For example, if your advisor suggested that you attend a guest lecture on your campus about big data and business informatics, you would want to do some research on those subjects in advance to help you better comprehend the speaker's message.

In interpreting a message, you assign meaning to it based on your personal experiences and the context of the message.[59] Although the words the speaker chooses are a key factor in interpreting the message, you take many other variables into account. For example, you consider nonverbal cues, such as tone of voice, posture, gestures, and body movement. Even when you only can hear the person's voice, as with a phone call, you attend to vocal qualities such as pauses, speech rate, and tone. You also interpret the message based on your relationship with the other person. You interpret what your instructor says differently from what your classmates say to you, for instance. Your past experiences play an important role in interpreting a speaker's message.

How much you know about the topic will influence your interpretation of the message.

Evaluating a message requires some sort of judgment about it. Evaluation occurs regardless of the listening context—appreciative, content, empathic, critical, or dialogic. As you listen to others, you evaluate their credibility, trustworthiness, and expertise. Evaluation is most important when you apply your critical listening skills in persuasive communication situations.[60] Research on listening has found that closely listening to a speaker's verbal messages rather than focusing on nonverbal cues increases your likelihood of detecting when someone is lying.[61] Reflect on your critical listening skills and ethical practices in Box 4.4.

BOX 4.4: ETHICAL QUESTIONS

Developing Ethical Critical Listening Skills

Not all interactions require careful evaluation and critical listening, but some situations call for critical listening as an ethical responsibility, such as when listening to a persuasive speaker. The National Communication Association's *Credo for Ethical Communication*, introduced in Chapter 1, asserts that ethical communicators "strive to understand and respect other communicators before evaluating and responding to their messages."[62] You might equate critical listening with a negative response. But as the NCA credo suggests, understanding and respect provide the basis for ethical critical listening. In fact, ethical listeners evaluate others' messages, rather than simply accepting what others say without question. Reflect on a time in which you were an ethical critical listener. Why was critical listening important in that situation? How did you respond to the speaker? In what situations are ethical critical listening skills especially important?

Responding is the behavioral component of listening that lets others know you are paying attention to them. You may respond nonverbally, such as smiling and nodding. You may respond verbally, asking questions and commenting on what the person has said. Responding is how you tell others, "I'm listening to you." Appropriately responding to what others say also affirms them as individuals. When you respond by paraphrasing what the speaker has said or asking for additional information, the person feels more understood than if you simply nod your head in agreement. In addition, such active responses reflect positively on you, as the speaker is more satisfied with the conversation and finds the listener more socially attractive, supportive, accepting, and emotionally aware.[63] Active responses also help you avoid **pseudolistening**, where you appear to be listening but are not fully paying attention to the message. Your nonverbal and verbal messages, such as nodding your head and saying "Uh huh," suggest to the speaker that you are listening. However, because you are

pseudolistening Occurs when communicators appear to be listening but are not fully processing the message

not fully attending to the speaker, you miss the complete message (Figure 4.10).

Assess Your Listening Skills

When you assess your listening skills, you learn from your experience and determine what went well and what you want to improve on in the future. For this stage, reflect on your own actions and behaviors and seek feedback from others on their views of your listening abilities. Focus closely on how you responded to others as you listened to them.

FIGURE 4.10 Responding appropriately with verbal and nonverbal messages lets others know you are engaged in the conversation.

One way to evaluate your effectiveness as a listener is to assess the degree to which you listened defensively or supportively. When you listen defensively, you focus on yourself rather than the person to whom you are listening. **Defensive communication** discourages free expression and interferes with listening by personally attacking or threatening others.[64] Name calling, derogatory comments, and purposefully offensive language all serve to dampen others' willingness to speak up and close off opportunities to listen. Defensive responses when listening to others reflect judgment, control, strategizing, indifference, certainty, and superiority. Judgment messages involve "you" statements that give an evaluation of the other person. Although evaluation is part of the listening process, you want to avoid personal attacks. Consider the difference between "You are an idiot" and "I'm not sure how that would work. Would you tell me more about it?" The first response presents an immediate negative evaluation of the speaker as a person. In the second, the listener expresses some skepticism in the plan but requests more information before making a final assessment.

Controlling messages are challenging in nature, suggesting that the interaction is a competition over who dominates the conversation. Abrupt topic shifts, tangential comments, facial expressions that convey exasperation—all indicate the listener's attempt to control the discussion.

Strategizing messages stem from deception and concealed motivations. When listeners strategize, they react in ways that serve their own interests, rather than what is best for the other person. These **hidden agendas** promote the individual's plans or objectives, often at the expense of others' goals. For example, Jamie might respond positively to Pat's ideas only as a way to gain Pat's support for Jamie's own proposal. Feigning interest to make others think you are listening, manipulating others to achieve your own goals, and appearing to be spontaneous are all strategizing responses.

Indifferent messages indicate a lack of concern for the other. These responses ignore the humanness of communication and have a detached, generic quality.

defensive communication
The use of personal attacks and similar unproductive communication strategies that discourages individuals from freely expressing themselves

hidden agenda An individual's concealed motivations or plans that underlie the person's actions

FIGURE 4.11 Indifferent responses when listening to others often are expressed nonverbally, as with purposefully avoiding eye contact with the other person.

supportive communication
Messages that encourage others to freely express their ideas

It is like getting a form letter or junk email sent to thousands of addresses when you were expecting a personal message. Often, indifference is conveyed nonverbally, as with a lack of vocal enthusiasm or not making eye contact with others (Figure 4.11).

Communicating with certainty means you portray your view as the best or correct way to think and act. You suggest you are right and that those who do not agree with you are wrong. Certainty messages typically involve taking a position and expressing an unwillingness to explore alternative ideas or solution, as with "We need to choose my plan; no other plan will work."

Messages suggesting superiority leave others feeling inadequate. When you communicate in a superior way, you are telling others you think you are better, more skilled, or more important than they are. Statements such as "Be quiet when you don't know what you're talking about," "I'll explain it since you obviously don't have a clue," and "If you'd prepared for the meeting, your presentation wouldn't have been so disorganized" display superiority and lead to defensiveness in others.

In contrast, **supportive communication** encourages others to freely express their ideas and creates spaces for effective listening. Supportive responses acknowledge the other person as a unique individual and indicate a respect for that person's thoughts, perspectives, and experiences. Research shows that supportive communication has long-lasting positive effects.[65] For first-generation college students, supportive communication both in-person and on social media helps them adjust to college life. In addition, as these students become more comfortable on campus, they in turn serve as sources of supportive communication for newer first-generation college students.[66] Supportive communication is a key component of active listening.[67]

Supportive messages are characterized by description, problem orientation, spontaneity, empathy, provisionalism, and equality. Descriptive messages offer sincere comments on what the other person has said. In description, you make "I" statements, as with "I disagree with your position and here's why." You describe how you think, feel, or believe. You focus on the topic or issue and avoid judging the other as a person.

In a problem orientation the listener responds in ways that encourage a collaborative approach to communication. Appearing genuinely interested, smiling, brief agreement phrases (e.g., "I see," "Good point," "Yes, exactly"), and other affirmative behaviors let others know you are listening and cooperating in the conversation with them. For instance, research on racial microaggressions found that Black women express both individual support for their friends who are the targets of these commonplace hostilities as well as collective support for the group's experience with racial discrimination. The women listened to their friends' stories, agreeing both verbally and nonverbally with the personal experience, then shifted their discussion to addressing racial microaggressions as a larger problem for all Black women.[68]

FIGURE 4.12 Spontaneous messages reflect authenticity and a commitment to understanding others' perspectives.

Spontaneous messages arise directly from the situation and are free of hidden motives and deception. They originate from a genuine commitment to understanding others' thoughts and feelings (Figure 4.12). Responding spontaneously takes into account all communicators in the conversation. When you respond spontaneously, your responses are appropriate to the situation and the other participants.

Empathic messages recognize and legitimate others' emotions. You acknowledge others' perspectives and demonstrate a respect for who they are as individuals. For example, research on cyberbullying has found that the most effective responses are ones that offer emotional support to those who are the targets of online intimidation. In contrast, messages that suggest the cyberbully might change and improve their behavior were not viewed as effective, because the focus shifted away from the victim's concerns and lived experience.[69] Similarly, when listeners responded to friends describing a difficult life event by affirming the speakers' identity, indicating agreement, and displaying kindness, speakers viewed listeners as more competent. Listeners who appeared uncaring or expressed doubts about their friends' stories, especially if the events were particularly upsetting, were viewed as less competent.[70]

Provisionalism means you qualify what you say with words and phrases such as "in my experience," "it seems like the best course of action would be," and "based on our discussion so far." When you respond in provisional terms, you acknowledge that things are not worked out completely or certain and you are interested in others' ideas and viewpoints. You also indicate that your perspective is just one of many.

The last type of supportive communication, equality messages, fosters respect and trust. Statements like "Let's work together on this," "I like your idea. If you and I divide up the tasks, neither one will be overly burdened," and "You've said what I bet all of us are thinking" let others know you are listening and express respect for what they said. Table 4.2 summarizes the two types of communication listeners may use in response to others.

TABLE 4.2 Defensive and Supportive Listening Responses

DEFENSIVE MESSAGES	BRIEF DEFINITION	EXAMPLE LISTENING RESPONSE	SUPPORTIVE MESSAGES	BRIEF DEFINITION	EXAMPLE LISTENING RESPONSE
Judgment	"You" statements	"You're not explaining yourself clearly at all."	**Description**	"I" statements	"I'm not quite clear on what you mean."
Control	Attempts to dominate the conversation	"Enough of that. Let's move on to more important things."	**Problem orientation**	Collaborative approach to interaction	"Your idea seems really interesting. Please tell us more about it."
Strategizing	Deception to achieve own goals	"Sure, I'll help you out with your project" while thinking, "Now you owe me a favor."	**Spontaneity**	Free of hidden motives	"I'd be happy to help you out with your project" while thinking, "My friend just needs a little assistance."
Indifference	Apathy, lack of enthusiasm	"Who cares about your excuses?"	**Empathy**	Recognize others' emotions	"This sounds like a difficult time for you."
Certainty	Unwilling to consider alternatives	"Clearly, this is the only way to proceed."	**Provisionalism**	Consider new information	"From what we know so far, we have three options."
Superiority	Indicate being better than others	"Here's a much better way to assign household chores."	**Equality**	Foster respect and trust	"Let's figure out a way to divide up the work fairly."

Align Your Listening Goals and Skills

Once you have completed your evaluation, you can identify the ways in which you will improve on your listening skills. For this stage, reflect on the outcomes from the listening event, determine your listening strengths, acknowledge your listening weaknesses, and develop strategies for better aligning your listening skills with your goals. Develop a chart such as the one in Table 4.3 to map out a plan for updating your listening goals and skills. The next section of the chapter discusses specific strategies to improve your listening skills.

TABLE 4.3 Documenting Your Listening Goals and Skills

ORIGINAL LISTENING GOAL	OUTCOME	LISTENING STRENGTHS	LISTENING WEAKNESSES	STRATEGIES FOR IMPROVEMENT
Evaluate my group's idea for the class project, and provide appropriate feedback	Provided feedback, although some group members seem unhappy with my evaluation	Listened to the entire presentation before offering feedback; made eye contact with all group members; provided verbal responses to indicate I was listening	Let my mind wander a few times; forgot to silence my phone; engaged in some defensive communication when responding	Manage internal distractions; silence phone or leave in my office; commit to creating a supportive communication climate

Improving Your Listening Skills

Although there are many strategies for improving your listening skills, some common ones, summarized in Table 4.4, are listen mindfully, commit to listening reciprocity, facilitate a supportive communication climate, engage in discovery listening, be present, and put technology to work. Use the ones that will best help you align your listening goals and outcomes.

TABLE 4.4 Strategies to Improve Listening

STRATEGY	USEFUL FOR …	IMPLEMENT
Listen mindfully	Overcoming habitual listening habits	Take an active interest in the speaker's topic
Commit to listening reciprocity	Listening to others when you do not agree with them	Listen for differences as well as common ground
Facilitate a supportive communication climate	Reducing stress and promoting trust	Use "I" statements
Engage in discovery listening	Taking a learning approach to listening	Develop an improvement plan based on your own listening strengths and weaknesses
Be present	Focusing the context for the interaction	Actively manage distractions
Use technology strategically	"Relistening" to a conversation	Record interactions with participants' permission

Listen Mindfully

Making a conscious decision to listen is the most important step you can take to improve your listening skills. **Mindful communication** occurs when you actively process information as you interact with others.[71] Once you commit yourself to listen to others, you will put your energy toward understanding what they have to say. Listeners who practice mindful communication are more empathic, promote a supportive communication climate, and listen more actively.[72] Listening

mindful communication
Involves the active and conscious processing of current information during interactions with others

mindfully applies in mediated contexts as well. For instance, individuals who practice mindful communication when texting report more positive emotions, less stress, and greater well-being when messaging with their friends than those who are less mindful.[73] Thus, when you respond to others in a more deliberate and thoughtful way, you enjoy benefits from your mindful listening. In contrast, when your listening habits become overlearned, habitual, and routine, you experience **mindless communication** and respond to others automatically with minimal attention to the situation.[74] You hear a family member start telling a story you have heard many times before and without fully thinking about it, you tune out the person. The response is automatic; you immediately select out what that person has to say.

There are several ways to become more mindful of your listening. First, closely examine your general attitudes toward listening. Do you view some people or communication situations as boring and let your mind wander? Simply reframing your mindset to have a more positive attitude toward listening will help you be more mindful. Second, taking an active interest in the topic will increase your curiosity and make you more aware of your listening habits. Even if the subject is one you usually tune out, working to identify how you might apply information to your own experiences will increase your conscious mental processes. Third, use mental imagery to develop a clear picture in your mind of what others are saying.[75] For example, if a friend recounts a recent vacation, use your imagination to reconstruct the events. Engaging in the mental work of listening will increase your mindfulness and help you become a better listener.

mindless communication
The outcome of habitual and routinized interaction involving minimal processing of new information

listening reciprocity
Occurs when individuals listen to each other openly, completely, and thoughtfully, withholding judgment until the speaker is finished

FIGURE 4.13 Listening reciprocity occurs when communicators listen thoughtfully to each other even when they have different viewpoints.

Commit to Listening Reciprocity

Most people expect reciprocity in expressing themselves, assuming that everyone will get a chance to contribute their ideas. Through listening, you invite others to articulate their viewpoints. **Listening reciprocity** occurs when communicators withhold judgment and listen to each other openly, completely, and thoughtfully (Figure 4.13). You engage in listening reciprocity when you listen to others even when you do not agree with them or find what they say repetitious and boring. Moreover, listening reciprocity requires listening for differences as well as common ground and welcoming perspectives that you might not share with others.[76] Try out listening reciprocity in Box 4.5.

BOX 4.5: LIVE IT LOCAL

Listening to Your Community

To find out the health concerns of Native Hawaiian elders and their family caregivers, public health researchers at the University of Hawai'i organized a series of listening meetings.[77] The researchers asked only eight questions during each 90-minute listening meeting, allowing elders and caregivers to voice their concerns at their own pace and in their own words. Committing to listening reciprocity allowed the researchers to withhold judgment and listen openly, completely, and thoughtfully. Go to a local senior center or other place where older people in your community gather. Ask them a few questions about themselves or the history of the community. Invite them to talk about their viewpoints and ideas. What did you learn about your community in these listening meetings? What did you learn about yourself? What did you learn about listening reciprocity?

Facilitate a Supportive Communication Climate

Creating a supportive communication climate helps all participants listen more productively. Supportive communication involves using "I" statements when responding to others, viewing conversation as collaboration, remaining free of hidden motives, acknowledging others' emotions, taking new information into consideration, and promoting respect and trust among communicators. Supportive messages reduce participants' stress levels and allow them to actively engage in listening.[78] For example, in the workplace, a supportive communication climate makes employees more likely to listen and respond positively to constructive criticism of their performance.[79]

Letting others know you are listening to them is especially important in online interactions. When you respond appropriately and in a timely fashion, you achieve co-presence, or the sense that communicators are interacting together, regardless of differences in time and space.[80] Whether the interactions are synchronous, asynchronous, text, audio only, or video, let others know you have received their messages. In video interactions, responding involves many of the same behaviors as when conversing face to face—head nods, smiles, encouragements (such as "Uh huh" and "I understand"), as well as lengthier comments. In audio-only interactions, such as phone conferences, clearly vocalize your responses, as with "Okay," "Go on," and "Good point," in addition to responding in greater detail. For text messages, acknowledge what the other person said before giving your own views.

Engage in Discovery Listening

Discovery listening helps you take a learning approach to listening. With **discovery listening**, you identify your own listening problems, determine their causes, and develop specific ways to address them.[81] If you have trouble focusing on a speaker

discovery listening
A process in which communicators identify specific listening problems, determine what causes those problems, and develop ways to address those problems

in a noisy environment, try to find a quieter place or work to block out extraneous sounds. If you miss a speaker's main points because you are daydreaming, figure out what you can do to concentrate on the present. In discovery listening, you develop a plan tailored to your listening strengths and weaknesses. Not everyone has the same listening habits, so individual plans will differ. Reflecting on and honestly assessing your listening abilities provide the essential basis for discovery listening.

Be Present

You think faster than others speak or type, so it is easy for your mind to wander when you are listening. You make plans for the future, you reminisce about the past. Often you do everything except stay in the present. But research shows how well you listen depends on how attuned you are to the context of the interaction.[82] When you are not present, you miss out on how others may be responding, important nonverbal cues, and the conversation's general flow.

Being present and actively participating in a communication event will help you manage internal and external distractions that can take your focus off the present interaction (Figure 4.14). Being present means recognizing internal distractions and setting them aside while you interact with others. Sometimes these distractions are under your control, while other distractions you must just try to ignore. Taking notes during a classmate's presentation, for instance, will keep your attention on the speaker's main points. Contributing to a group discussion not only helps you avoid distractions but also increases the likelihood you will recall what others said.[83]

FIGURE 4.14 When you are present, you manage internal and external distractions, keeping your attention on the unfolding interaction.

Use Technology Strategically

While video conferencing greatly extends your ability to connect with others, it also becomes wearing. For example, during the COVID-19 pandemic, people reported experiencing Zoom fatigue, a cognitive overload from participating in often daily online video conferencing with coworkers, supervisors, family, and friends.[84] Researchers found communicators were more worn out from a Zoom meeting than they ordinarily would be from an in-person conversation due to the increased mental work listening required. When you plan to schedule a video conference with someone, consider if it might be best to have a phone conversation. Not every meeting requires video and screen sharing. Sometimes a simple phone call is enough. Alternately, turning off the video during a video conference and just focusing on the audio can reduce listening fatigue and allow you to concentrate on the speaker's voice.[85]

Whether you are interacting with someone in-person or online, minimize technology distractions. If you are talking online, keep your own space free from distractions. Close unnecessary programs and browser windows, go to a place where you can avoid interruptions, and turn off other communication devices, such as smartphones. The same rules hold true for in-person interactions. Identify a space where you will be undisturbed, silence your phone, and focus your efforts on being with the other person. Keeping your phone out of reach is especially important. When you engage in **phubbing**, or gazing at your phone rather than the people you are with, not only do you miss out on what others are saying, they also feel less connected to you (Figure 4.15). Moreover, research on phubbing suggests that speakers evaluate listeners more negatively when they look at their phones than when they look at printed material. Speakers view phubbing as devaluing the relationship between communicators.[86]

In addition to phubbing, a pervasive communication environment can interfere with effective listening, as communicator rely on digital recording. Sound specialist Julian Treasure observes that the ability to record, especially audio and video recording, has reduced our ability to listen.[87] Rather than staying in the moment and attending to what others are saying, you depend on the recording to listen for you. While there may be times to record a listening event with participants' permission, as with an instructor's lecture, setting aside your device often is the best practice for achieving your listening goals.

FIGURE 4.15 Phubbing disconnects you from other communicators.

phubbing Gazing at your phone rather than attending to the people you are with

CHAPTER REVIEW

- **Discuss the importance of listening in a pervasive communication environment.**
  A pervasive communication environment allows you to listen to others synchronously, in real time, as well as asynchronously, nearly anytime and anyplace. New communication devices, especially mobile phones, can cause listening distractions. Texting, email, and social media posts often share more in common with spoken language than written language, broadening the notion of what counts as listening.

- **Identify misconceptions about listening.**
  Dispelling misconceptions about listening provides a start on improving your listening skills. First, listening and hearing are not the same. Second, people are not born good or bad listeners. You learn your listening habits—so you can change them. Third, you might assume listening is easy. To the contrary, good listening depends on hard work, energy, and a commitment to listen

well. Fourth, you might think listening just happens, yet you must prepare for listening, just as you prepare for speaking. The fifth misconception is that intelligence is related to listening. Research shows just the opposite: IQ and listening are independent of each other.

■ **Recognize and explain the types of listening.**
There are five types of listening: appreciative, content, empathic, critical, and dialogic. Appreciative listening involves listening for enjoyment. You focus on others' main ideas when you listen for content. In empathic listening, you are primarily concerned with detecting others' emotions. Critical listening requires evaluating what others have to say. Dialogic listening includes four elements: listening with heart and mind, listening for and to culture, listening for power inequities, and listening for dialectics.

■ **Explain the four stages in the process of listening.**
The process of listening includes four stages: anticipation, application, assessment, and alignment. In the first stage, anticipation, you analyze the listening context, identify your listening goals, and determine your own motivations for listening. In the second stage, application, you apply the listening components—receiving, comprehending, interpreting, evaluating, and responding. In the third stage, assessment, you engage in self-assessment and gather feedback from other participants. In the fourth stage, alignment, you reflect on the outcomes of the interaction, use your strengths to work on your weaknesses, and revise your goals and listening processes in anticipation of another listening opportunity.

■ **Identify strategies to improve your listening skills.**
Key strategies to improve your listening skills include listen mindfully, commit to listening reciprocity, facilitate a supportive communication climate, engage in discovery listening, be present, and use technology strategically.

DISCUSS AND APPLY

1. How does a pervasive communication environment influence your own listening habits? Think back to a recent in-person conversation you had. What role did communication technologies, such as your phone, play in your listening experience?

2. Give an example of each type of listening you have experienced: appreciative, empathic, content, critical, and dialogic. How do you know when each is appropriate? In what situations have you found yourself switching from one type of listening to another?

3. Is there ever a time you can listen too mindfully? Are there times when mindless listening is appropriate? Provide examples to support your answers.

4. How do you know when others are actively listening to you? How do you feel?

5. What do you do when you perceive that someone is not listening to you?

6. What are some strategies you plan to try out to improve your listening skills?

CREDITS

CHAPTER 5

Verbal Communication

As you think about verbal communication, consider calls to change the names of some sports teams. For example, relatively recently, Washington's NFL team decided to change its name. Until a new name is decided upon, the team will be called the "Washington Football Team." Despite the team using its earlier name for decades (1933–1936 in Boston and since 1937 in Washington),[1] many individuals and groups demanded change. Given that the team's name had not changed in years, why the calls for change?

Advocates for change asserted that the team name was derogatory and racist. They emphasized that such language reinforces stereotypes of Native Americans, perpetuating misunderstanding and discrimination, and is linked to years of exploitation. Calling the "R-word" a racial slur, advocates for change invoked issues of civil and human rights.[2]

The team's owner, Dan Synder, however, maintained that the name recognizes and honors Native Americans (e.g., celebrating positive attributes, like strength, bravery, and determination). Other supporters of retaining the name suggested that, historically, Native Americans used the term to signify group membership, and White people used it to recognize Native Americans' achievements.[3]

Across several years, considerable public attention focused on this issue and debate grew. In the fall of 2013, the Oneida Nation launched a campaign for a name and mascot change, featuring pickets and advertising. By spring of 2014, momentum expanded, with support from Indigenous groups, civil rights organizations, sports figures, journalists, politicians, and religious leaders (Figure 5.1).[4] After years of resistance, Synder agreed to change the name in mid-2020. By the end of 2020, Cleveland's baseball team announced an upcoming name change, and despite no formal announcements, other teams, such as Atlanta's baseball team, Kansas City's football team, and Chicago's hockey team, were believed to be considering name changes.[5] What would you do if you were in the decision-making position for one of these or a similar team?

In an earlier chapter, you examined the relationship between stereotypes and perception. Language, or verbal communication, influences our interpretations and ultimately the judgments we make about people and issues. The language that you choose affects other people, and their language affects you. Certainly, such issues come into play when you consider names, such as those of the teams above.

FIGURE 5.1 Protesters with signs labeling the team name and mascot offensive demand change.

Verbal Communication in a Pervasive Communication Environment

Verbal communication and nonverbal communication are interrelated. For example, you often reinforce what you say with nonverbal gestures, such as smiling when you give a compliment in a face-to-face interaction or adding a smiley face emoji when you text a compliment to a friend. In the next chapter, you will explore nonverbal communication in more detail. In this chapter, the focus is on what you say—or your verbal communication. Essentially, **verbal communication** is defined as the use of language. **Language** is the system we use to convey our meaning through symbols, and language relies on a key type of symbol—**words**.

verbal communication Using language to communicate

language The system we use to convey our meaning through symbols (words)

words Key type of symbol on which language relies

Language Foundations

Four foundational principles underlie understandings of language. The four foundations of language are that language is arbitrary, abstract, ambiguous, and evolving.

FIGURE 5.2 The above dog, Raven, lives with one of your authors, Stephanie Coopman. But is this photo really of a dog? River could be a *kalb* (Arabic), *cicing* (Balinese), *jakay* (Cambodian), *allcu* (Ecuadorian Quechua), *chien* (French), *hundur* (Icelandic), *sag* (Farsi), *ogday* (Pig Latin), *perro* (Spanish), *mbwa* (Swahili), *áso* (Tagalog), *ntsa* (Tswana), or *inja* (Zulu), depending on where she and Stephanie lived.

FIGURE 5.3 Meanings do not reside in words. Meanings reside within you.

message production
Choosing the best words to get your ideas across to others

message interpretation
Processing what your conversational partner communicated

Language Is Arbitrary

Think about the language you use: "Let's get together and study," "Cold pizza is my favorite snack," "I play ultimate frisbee." How and where did you learn these words? And what meanings do you associate with them?

A common communication myth is that meanings are in words. Certainly, across your life, you have come to associate particular meanings with specific words. If you ask your study partner to borrow their pencil for a moment, you do not expect them to hand you a wrench. But the words "pencil," "wrench," and "pizza" are simply agreed-upon conventions used to try to convey our meanings to others. These agreements are recorded in dictionaries where you can look up meanings of words you are not sure about or do not know. But keep in mind that these recorded definitions are simply social agreements on how we use an abstract symbol, in this case a set of letters, in a particular way.

Consider the word "dog." What does this word mean? Your definition for this common word may be quite similar to the definitions of your classmates (Figure 5.2). But your "dog" could be someone else's *perro* (Spanish for "dog").

The combination of the letters "d," "o," and "g" to make the word "dog" is an arbitrary arrangement. The same animal can be referred to with the letters "d," "o," and "g" (English) or the letters "m," "b," "w," and "a" (Swahili), and each arrangement can be used to refer to one's pet canine. Alternately, rather than "dog" in English, we could have used the letters "t," "r," "e," and "e" (tree) to refer to our favorite canines. If so, you would be introducing your new pet tree, Rover, to others at the beach or along the hiking trail.

In other words, language is arbitrary because there is generally not a natural relationship between our words and the concepts that they come to represent. Any number of combinations of letters could have been selected to represent the word dog, but the language you are reading in now just happened to select the letters "d," "o," and "g." In short, words are symbols that we use to represent ideas when we communicate.

Another facet of the arbitrariness of language is that rules vary across languages regarding how to convey thoughts. For example, different languages use different rules for where adjectives are placed in relation to nouns. In English, you would say, "I have a brown dog." But in Spanish, you would say, "Tengo un perro marrón," with the adjective "brown" following the noun.

So, if meanings are not in words, where do they reside? Within YOU (Figure 5.3). You bring your understandings of words to bear in interactions when you try to choose the best words to get your ideas across to others (**message production**) or when you try to process what your conversational partner said (**message interpretation**).

Such message processing is also influenced by the abstract nature of symbols.

Language Is Abstract

Symbols, such as words, are abstract and help you make connections between objects and ideas and your understandings of them. The relationship between objects or ideas, words, and understandings is explained by the **meaning triangle**.[6] The meaning triangle has three components (Figure 5.4).

Ogden and Richards' meaning triangle represents the relationship between (a) a concept or entity (called a **referent**), such as an iPad; (b) the **symbol** you use to communicate about that referent, such as the word "iPad"; and (c) your thoughts about the referent (called **reference**), such as your views and opinions of an iPad. The referent is linked to **denotative meaning** (the formal, dictionary definition of a word), whereas the reference is linked to **connotative meaning** (our personal or informal meanings associated with words). Thus, the relationship between a symbol, reference, and referent is different for everyone. Although the word "iPad" (symbol) and the actual item iPad (referent) may be the same, your understanding of (connotative meaning) an iPad may differ from another person's. You may view your iPad as an aid for relaxing because playing music helps you unwind and tune out the hectic pace of the day. Your romantic partner, however, may view an iPad as a distancing device, one that you use when you do not want to interact. Thus, a simple question like "Where's my iPad?" may trigger different responses. As we saw in the chapter's opening, the names of some sports teams have different connotative meanings, ones that have resulted in change and others that are fueling pressure for change.

Of course, if even concrete, routine words like "iPad" can result in differing interpretations, imagine what happens as symbols grow more abstract. Your reference, or connotative meaning for words, is more likely to differ from views held by others as words increase in abstraction. Concrete words are associated with specific and actual referents, such as particular objects (e.g., your iPad or your skateboard) or people (e.g., your Aunt Jill or your math professor). In contrast, abstract words refer to broader ideas or categories. Hayakawa's ladder of abstraction depicts movement from concrete to abstract concepts (Figure 5.5).[11]

At the lowest rung of Hayakawa's ladder are concrete words or symbols, such as "Professor Allen," your math teacher. As you climb each rung of the ladder, the level of abstraction increases. So you might go from "Professor Allen" on the first rung, to "college professors" on the second rung, and then to "teachers" on the third rung. Alternately, you might go from "SARS-CoV-2" on the first rung, to "coronavirus" on the second, to "virus" on the third, and to "microscopic parasites" on the fourth. With each step up the ladder, your thinking focuses on broader conceptualizations. In many respects, it is easier to be clear when you

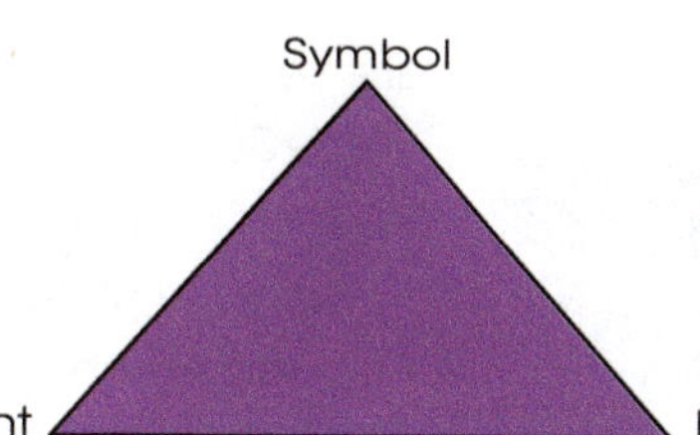

FIGURE 5.4 Ogden and Richards's Meaning Triangle has three interrelated parts—a referent, a symbol, and a reference.

meaning triangle Explains the relationship between objects or ideas (referent), words (symbols), and understandings (reference/thought)

referent A concept or entity, such as an iPad; one of three parts of the meaning triangle

symbol Representations, such as words and signs, used to convey meanings; symbols are arbitrary; one of three parts of the meaning triangle

reference Your thoughts about a referent, such as how your regard your iPad; one of three parts of the meaning triangle

denotative meaning The explicit, formal, or "dictionary" definition of a word or other symbol; related to the referent in the meaning triangle

connotative meaning Personal or informal meanings that you associate with a word or other symbol; related to the reference in the meaning triangle

FIGURE 5.5 Hayakawa's ladder of abstraction: You can think of the lowest rung of the ladder as the most concrete words. As you climb higher with each rung, you move to increasing levels of abstraction.

are communicating about concrete objects or people, like Professor Allen or SARS-CoV-2, than when dealing with broader categories of information, like all teachers or all illnesses.

Beyond representing specific entities or ideas, like iPads or Professor Allen, words also symbolize broad, abstract concepts, like freedom, justice, and democracy. Although you and your romantic partner or friends might have different views of iPads that influence your communication, imagine how your viewpoints on freedom might contribute to understandings or misunderstandings. As you move up Hayakawa's ladder, increasing abstraction, be sure to strive for clarity in language.

To be a successful communicator, you need to carefully determine when more concrete words are needed to help others understand your message. You need to choose appropriate words based on your meaning triangle and your understanding of the other person's meaning triangle to get your message across. This word choice should vary in level of abstraction based on how much you think your conversational partner knows about the topic. For example, there are various levels of understanding of photosynthesis, Higgs boson, and string theory, and of course, there are different connotative meanings for even more common words. Also, consider the information in Box 5.1 in terms of word use and understandings.

BOX 5.1: ETHICAL QUESTIONS

If You Were a Sports Journalist or Commentator …

If you were a sports journalist or commentator, would you use team names in your blogs, in newspaper or magazine articles, in tweets, or on the air that some groups find derogatory and racist? As you recall from the chapter's opening, the team's owner asserts that the team name is not meant in disrespect but rather is intended to honor Native Americans. However, many individuals and groups are offended by the term, and even many who are not suggest that what matters is that others interpret it as racist and derogatory. Clearly, different views and connotative meanings are associated with the team name.

Prior to the Washington Football Team's name change, some sports journalists, like Peter King at *Sports Illustrated*/TheMMQB and John Smallwood at the *Philadelphia Daily News*, stopped using Washington's team name.[7] King commented:

> I have no idea if this is the right thing to do for the public, or the politically correct thing to do … I can just tell you how I feel: I've been increasingly

bothered by using the word, and I don't want to be a part of using a name that a cross-section of our society feels is insulting. …

> Here's what it came down to for me: Did I want to be part of a culture that uses a term that many in society view as a racial epithet? The answer kept coming back no—and now that I have been charged to run a website, I thought I would finally do what felt right to me.[8]

Some publications, like the *San Francisco Chronicle* and *Slate*, also quit using the team's name.[9]

What do you think of King's and Smallwood's decision? What are your views on entire publications banning the use of the names of other teams? What would you do if you worked in this context? What about in talking with your friends while watching the team play? Calls also have been made for Jeep to rename some of its SUVs.[10] If you made advertising decisions, would you run ads using the SUVs' names? What ethical considerations guide your decision?

Language Is Ambiguous

Related to our discussion of language being abstract is another foundational principle: Language is ambiguous. Words, phrases, or sentences can have more than one meaning or be unclear (Figure 5.6). For example, when Groucho Marx said, "One morning, I shot an elephant in my pajamas," the message seems relatively clear. However, when he adds, "How he got in my pajamas I don't know," your interpretation likely changes somewhat. Imagine that someone at work asks you to "get the thingamajig that we use for this stuff, stat." Is it clear what the person wants and when?

FIGURE 5.6 Expressions like "so close but so far" are ambiguous, leaving room open for message interpretation.

In other instances, you can interpret what is said in different ways. Consider, for example, the interaction between Milo and Officer Short Shrift in the following passage from *The Phantom Tollbooth*, a popular book that has much to say about language:[12]

OFFICER SHORT SHRIFT: Now then, would you like a short sentence or a long sentence?

MILO: Well, I suppose a short one, if I have a choice.

OFFICER SHORT SHRIFT: How about "I am"? It's the shortest sentence I know. [Writes "I AM" on his pad and hands it to Milo]

MILO: It's very kind of you to give me … such a short sentence.

OFFICER SHORT SHRIFT: And when do you think you can go to prison and start serving it?

Clearly, Officer Short Shrift is having fun with the meaning of "a short sentence."

Ambiguity occurs simply because words can have multiple meanings as well as be combined in multiple ways, which can result in differing (and sometimes problematic) interpretations. Despite some critiques of language ambiguity,[13] this capacity of language likely serves useful purposes.[14] For example, it allows us to reuse words across different contexts, thus making our language system simpler and more efficient. But these multiple meanings also sometimes result in confusion or misunderstanding. If you invite Shelly to meet you at the corner restaurant at seven o'clock, you walk there from your house, and she does not show, you will likely wonder why—just as Shelly may be wondering why you have not shown up at The Corner Restaurant. Clearly, you meant the restaurant on the corner near you, and Shelly interpreted the phrase to mean a restaurant named The Corner Restaurant. Words (e.g., *mean, run, draft,* and *pitch*) can have multiple meanings. Some words have hundreds of meanings. These multiple meanings as well as level of abstractness can lead to miscommunication. Box 5.2 offers you an opportunity to create your own list of words that have multiple meanings and to consider the implications of this feature of language.

BOX 5.2: CONNECT & REFLECT

Words and Meaning

How many definitions can you think of for the following words: *mask, extra, set, solution, reservation, date, fan,* and *plane*? What are other common words that have multiple definitions? Make a list and write as many definitions for each word as you can. Then ask a classmate to do the same. Where are the areas of overlap in your lists? What distinct meanings appear on one person's list but not the other's list? Given these multiple definitions, how do you try to get your meaning across in ways that do not confuse others? What features of language help us with processing such words? How can language ambiguity lessen our ability to get our message across to others? How and when can ambiguity be useful?

Whether a word has one definition, two definitions, or more in the dictionary, it can always be used in new ways and acquire new definitions. One example is "tight," which added a new meaning a few years ago reflecting its use as a slang term that means "looks good." More recently, "slay" has taken on additional meaning, such as in the statement "Marcos slayed the exam." These examples also refer to the remaining foundation: Language is evolving.

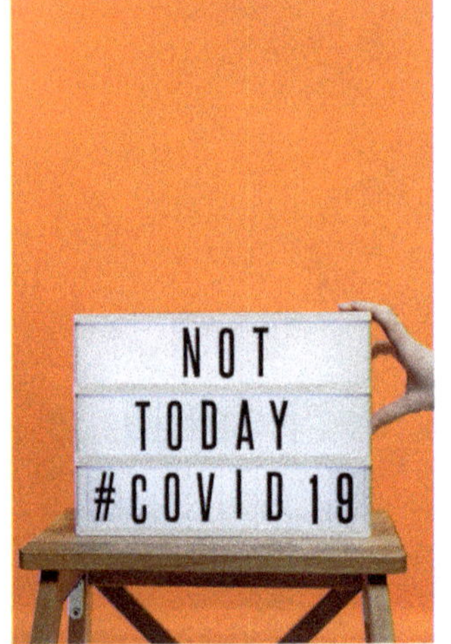

FIGURE 5.7 A few years ago saying, "Not today #COVID-19" would have had no meaning. But using hashtags to promote phrases via social media, such as Twitter, where hashtags gained fame, and COVID-19, which became a household word for coronavirus disease 2019, are examples of the evolution of language.

Language Is Evolving

Language is constantly undergoing change.[15] Consider our iPad example above. When your authors were in college, the word "iPad" did not exist. Each year new words are added to dictionaries. This fact is more evidence that meanings are not in words and that words are simply combinations of letters that we have agreed to use to reflect particular ideas. What words can you think of that did not exist just a few years ago? "Twitter," "Google it," "Frankenfood," "vape," "Uber," "cringey," and "COVID-19" are just a few examples (Figure 5.7).

What about acronyms, such as FOMO and JOMO? Chances are that you understand these phrases. Do you know people who likely would not? As languages change, some people learn the new words quickly, while others are slower to learn these expressions, and some may not incorporate them into their usage at all or even be aware of them.

Some language changes take place slowly, and others move quickly.[16] A friend's parents moved to the United States from Lithuania. He grew up speaking Lithuanian at home with them and English at school. Years later, as an adult, he visited Lithuania to attend a professional conference held there. He discovered people loved to hear him talk. Language in Lithuania had slowly evolved, but since he and his parents were living abroad, they retained the original use and expressions. Thus, his word choices and speech reminded people of an earlier version of the language. Language changes for several different reasons, including exposure to new people, such as by travel, immigration, or colonization, and the needs of speakers, such as to describe new experiences and use new technology.[17]

Recently, technologies such as Twitter, IM, and texting have influenced language. Considerable debate has addressed the influence of such technologies on language.[18] For example, although some people suggest that texting is harming individuals' abilities to produce language (e.g., corroding spelling and grammar), others suggest that text is simply a new form of language, one designed to convey points just as well as other forms and that may have several benefits.[19] Thus, if you understand "CU 2moro," then is that expression any different than "See you tomorrow," even if it does not use standard English? John McWhorter's TED Talk, discussed in Box 5.3, presents an interesting view—that texting is not a decline in language but rather a new form of language: "fingered speech." What are your views? How have tweeting, IM, and texting changed your language? How have texting conventions, like LOL and the slash symbol, changed over time? What other new technologies have influenced your language or that of your friends?

BOX 5.3: CONNECT & REFLECT

McWhorter: "Txting Is Killing Language. JK!!!"

In a TED Talk, titled "Txting Is Killing Language. JK!!!," John McWhorter[20] addresses the relationship between texting and language. Despite many criticisms on how texting is damaging language, McWhorter argues the opposite. In fact, he asserts that texting is "miraculous." Rather than categorizing texting as a type of writing, McWhorter suggests that text is more aligned with speech. Consequently, he believes that we should think of it as "fingered speech," an important, emerging type of communication. Instead of texting resulting in a language decline, McWhorter emphasizes the growing complexity of this relatively new language form and likens it to being bilingual. Watch his TED Talk (http://www.ted.com/talks/john_mcwhorter_txtng_is_killing_language_jk) to find out more about his position. What do you think of McWhorter's views on texting? Can you think of ways that text has evolved over time? How does the title of McWhorter's talk convey his overarching point? Can you think of any downsides of texting in terms of language use? What positives do you associate with texting language?

Now that you have explored the four foundations of language, let's examine meaning in more detail. Several processes influence meaning-making in verbal communication.

Semiotics:
Meaning and Verbal Messages

Semiotics is the study of how you make meaning from, or interpret, communication.[21] It focuses on the signs and symbols from which we construct meanings.

semiotics The study of how you make meaning from, or interpret, communication

In this section of the chapter, you will examine three areas of semiotics associated with language: semantics, syntactics, and pragmatics.

Semantics

semantics The relationship between words and their meanings

Semantics focuses on the relationship between words and their meanings. The interpretations that you make are linked to your understandings of words, both their denotative and connotative meanings, as well as how the words are used together in sentences (or syntax, which we will discuss next). Thus, semantics deals with the relationship between the signs and symbols used when we communicate.[22] If your friend says, "My boss is a cheesecake," you know not to employ your typical definition of cheesecake in interpreting this message. In short, semantics examines how we string words together in our utterances and how the relationship between these words contributes to interpretation.

Syntactics

syntactics The rules within a language that govern word order

Syntactics refers to the rules within a language that govern word order.[23] The order of words in phrases, clauses, and sentences influences meaning. Of particular interest is how sentences are properly constructed in a language as well as the variation in sentence structures across languages.

Where words are placed influences ease of interpretation as well as the interpretations that are made. For example, the following two sentences use exactly the same words:

Chris loves Pat.

Pat loves Chris.

But because of word placement, they do not have the same meaning. Also, consider how shifting one phrase changes the meaning of the following utterances:

The dog with the broken leg curled up in the chair.

The dog curled up in the chair with the broken leg.

If you are a Star Wars fan, you and your friends might have devoted some time to trying to talk like Yoda. "Yoda speak" involves different syntactic rules than regular English. For example, you would typically say, "I'm going to school." Yoda, however, would say, "Going to school, I am." You would say, "I can't fly." Yoda would say, "Fly, I cannot." Notice the difference between typical word placement in English and how Yoda places words within a sentence.

When you are learning a new language, understanding syntax is critical to being easily understood by native speakers. For example, if someone said, "Let's sing Grandmother's house, through woods, over river, go we," it might take you a few seconds to realize that the lyrics the speaker intends are, "Over the river, and through the woods, to Grandmother's house we go."

Although it is important to know how to structure words in a sentence according to semantic and syntactic rules, it is also important to use the language system appropriately.

Pragmatics

Pragmatics is concerned with using signs and symbols correctly, which enhances the likelihood that your messages will be received in the way that you intend them. The chief concern of pragmatics is constructing sentences that are socially and culturally appropriate, so the focus is on how we use language.[24] For example, imagine your new boss stops you in the hallway and asks, "How has your first day here been?" Responding with, "Horrible. My coworkers aren't nearly as fun as the ones at my last job, I don't like the cafeteria food, and the employee manual is ridiculous" might not be considered culturally appropriate and may result in some perceptions or consequences that you did not desire. Several types of conversational appropriateness fall under pragmatics. These types include taking turns talking, using language that fits your speech purpose (e.g., requesting an employee do something versus directing an employee to do something), and adapting to situations and people (e.g., speaking differently to kindergarteners than to your peers or speaking differently at a football game than at a funeral).

When trying to learn a new language, you are often confronted with semantic, syntactic, and pragmatic concerns. In his bestselling book *Me Talk Pretty One Day*, David Sedaris[25] describes his attempts to learn French:

> On my fifth trip to France I limited myself to the words and phrases that people actually use. From the dog owners I learned "Lie down," "Shut up," and "Who shit on this carpet?" The couple across the road taught me to ask questions correctly, and the grocer taught me to count. Things began to come together, and I went from speaking like an evil baby to speaking like a hillbilly. "Is thems the thoughts of cows?" I'd ask the butcher, pointing to the calves' brains displayed in the front window. "I want me some lamb chop with handles on 'em."

As he interacted with others in France, Sedaris learned more about the meanings of words (semantics), the order in which he should put them together (syntactics), and how he should use them in different contexts (pragmatics). Learning a new language, such as David Sedaris did, can also teach you more about your native

pragmatics How to construct messages that are socially and culturally appropriate

language as well as about cultural appropriateness. Our culture shapes what is pragmatically appropriate. Next, we examine language and culture in more detail.

Language and Culture

From the discussion of culture in Chapter 2, you learned that culture influences communication—both messages you produce and how you interpret the communication of others. Culture is highly important in shaping your language choices, as well as the meaning you attach to the language of others. For example, in Chapter 2 you learned that people from individualist cultures are more likely to use certain verbal patterns, such as "I" language and promoting competition and independence. In contrast, people from collectivist cultures are more likely to use "we" messages and to stress cooperation and interdependence.[26] Consider the following language choices:

> "I will take the lead on the group project. I am the best student and the best writer."

> "By working as a team, we can gain a lot from everyone's strengths and produce a better project, which will help all of our grades."

These messages differ in the dimensions discussed above. But how are people with different cultural backgrounds likely to interpret each other's messages? To someone with a collectivist orientation, the first message may be interpreted negatively, as it does not work to build or maintain positive relationships within the group. To someone with an individualist background, the second message may be problematic because it does not allow any one person to stand out from the overall group.

As you learned above, meanings are in people, not in words. And the meanings that you hold are largely shaped by your culture or cocultures (Figure 5.8). For example, a friend of one of your authors recently began dating a man from Nigeria. She is African American, and he is African. She was surprised one evening when his friends began teasing him about having a White girlfriend. This expression seemed odd to her because in U.S. culture she would not be seen as White. However, he explained later that in his culture the use of the term "White" meant that she was foreign.

FIGURE 5.8 Through collaborative work, group members take on different roles.

Also, think about situations in which one stands out as exceling. In the United States, talking in class to answer the teacher's question is regarded as good, a way to show your intelligence. Students who stand out in this way often receive several types of rewards, such as teachers and peers perceiving them as intelligent and socially competent. In New Zealand, standing out, however, might result in you being labeled a "tall poppy," with the resulting response being to cut you off so that you will fit in with the group and not stand out above others. Such cultural understandings influence the types and frequency of your verbal communication.

In addition, because language and culture have a reciprocal relationship, language also influences culture over time. What examples can you describe of how language has changed culture? Consider the prompts in Box 5.4 and then expand the examples.

BOX 5.4: CONNECT & REFLECT

Language Changing Culture

How has language changed culture? For example, adopting communication technologies, especially when adoption becomes widespread, influences language; in turn, this changing language can influence culture. What are our current cultural expectations about how quickly news should be available and updated? What language do we use that has resulted in these cultural expectations? What about language choices related to communicating over distances? How has language surrounding the ease of staying in touch changed cultural attitudes on geographical separation? What other examples can you think of regarding how language has changed culture?

Because we are members of multiple cocultures, we often move seamlessly between them, not even realizing that we are changing our language. In much the same way that someone fluent in English, Spanish, and Quechua can transition back and forth between these languages without even realizing it, you transition back and forth between using language appropriate to your various cultures. For example, think about the language that you use at work, especially when you are trying for a promotion or are hoping that your boss will give you a raise. How does this language differ from the way you talk with your friends at a party or when debating politics?

Direct and Indirect Speech

Your ability to switch back and forth between language systems in these contexts shows that you understand what is considered culturally polite or acceptable. For example, cultures vary in the degree to which they prefer direct or indirect speech. **Direct speech** is specific, detailed, and forthright. **Indirect speech** relies on subtlety and the listener understanding social conventions to infer meaning.[27] Members of cultures like the United States, which tend to prefer direct speech, pride themselves on being open, upfront, and candid. They tend to regard indirect

direct speech Language that is specific, detailed, and forthright

indirect speech Speech that relies on subtlety and the listener understanding social conventions to infer meaning

---

BOX 5.5: LIVE IT LOCAL

The Activist in You

Think about your community. What binds people together as a group? What blocks group cohesion or understanding? How does language work to facilitate understanding or to fuel separation? For examples, examine how local news issues are framed, how groups in opposition talk about their positions, and how activists describe situations. What steps can you take to increase understanding and acceptance? What new language choices might help clarify local issues and build positive solutions?

---

speech as vague and puzzling. In contrast, members of cultures that tend to prefer indirectness, like Japan, tend to see themselves as empathetic and tactful. They tend to regard direct speech as blunt and impolite.

As the interpretations above illustrate, culture also influences language that is considered impolite or hurtful. Although your parents may have taught you expressions like "Sticks and stones may break my bones, but words will never hurt me" to help you through the teasing often associated with youth, the last part of this expression is far from the truth. You can, no doubt, think of times that someone's words have hurt you as well as times that your words hurt another. Commenting on the *San Francisco Chronicle*'s decision not to use the nickname of Washington's NFL team, Audrey Cooper, the managing editor, said, "Words are powerful, and so is how we choose to use them. ... Our long-standing policy is to not use racial slurs—and make no mistake, 'redskin' is a slur."[28] Although there is debate regarding this label's origins and cultural use over time, many Native Americans today find it an insulting, hurtful term. As discussed earlier in the chapter, increasing support suggests that the team name is culturally inappropriate and should be changed. Consider the questions in Box 5.5 in terms of deepening understanding and acceptance.

Hate Speech

Have you ever considered how language is used to promote hate and separate people into cultural groups? Some work in communication has analyzed hate speech and the means by which people are recruited into communities of hate. **Hate speech** is language that attacks people based on perceptions that they belong to a specific group, such as a particular race, sexual orientation, or religion. For example, work by Michael Waltman and John Haas has examined how hate is communicated and, in particular, how hate "comes alive" through language and other symbols.[29] They assert that beyond hurting others, hate speech may have several intended functions. These functions include language designed to reach goals such as intimidation, perpetuating cultural in- and out-groups, and encouraging ethnoviolence. By analyzing racist discourse and hateful messages, Waltman and Haas hope to promote antihate discourse and embracing difference.

hate speech Language that attacks people based on perceptions that they belong to a specific group, such as a particular race, sexual orientation, or religion

In short, language can be used to bring people together, to increase understanding of cultural differences, and to embrace and celebrate our different heritages or it can be used to reinforce negative attitudes (Figure 5.9). As you read in Chapter 3, Chimamanda Adichie focuses on the power of narrative—the stories that we weave with our words. Near the end of her TED Talk, she emphasizes that "stories have been used to dispossess and to malign. But stories can also be used to empower, and to humanize. Stories can break the dignity of a people. But stories can also repair that broken dignity." Think about your language, and make language choices that move toward furthering understanding and acceptance.

FIGURE 5.9 A poster with "Hate has no home here" written in multiple languages is designed to be welcoming and convey acceptance.

Language and Gender

Similar to work in language and culture, researchers have long studied how one's gender influences language choices. Attention has been devoted to understanding why language varies across genders as well as particular language differences. A number of gender differences in language have been found. For example, research has examined difference in turn-taking, topic changes, interruptions, tag questions, intensifiers, giving directives, politeness, self-disclosure, hesitation markers, and verbal aggression. Let's examine a few examples below.

Power and Tentativeness

Robin Lakoff's 1975 book *Language and Woman's Place* played a key role in laying the foundation for subsequent work in the area of language and gender.[30] Lakoff observed that women used a variety of tentative language forms more often than men. Tentative language forms include:

- **Tag questions** (e.g., "It's a beautiful day out, isn't it?" and "Chad is such a nice guy, isn't he?") function to encourage conversation and agreement.

- **Intensifiers**, such as "truly pleased" and "so crazy," function to amplify the conversational thrust of a statement.

- **Hesitation markers**, such as disclaimers (e.g., "I'm probably not right, but I seem to recall …" or "It's probably not important now, but the last …") and hedges (e.g., "like," "kind of"), function to lessen the impact of language.

Lakoff's work focused on cultural or societal power, associated with traditional social roles, and how power influenced women's and men's speech. Her conclusions suggested that men engaged in more powerful speech because they

tag questions Function to encourage conversation and agreement

intensifiers Function to amplify the conversational thrust of a statement

hesitation markers Function to lessen the impact of language

held dominant positions in U.S. culture and society and that women's speech reflected their subordinate status. She argued that the cycle of power is reinforced by women's continued use of less powerful language patterns and men's continued use of powerful ones. Research over the years has continued to examine the use of powerful and powerless speech in a variety of contexts, such as workplaces and relationships.[31] **Powerful language** conveys social power, such as clearly stating a position on a topic or taking credit for an accomplishment, as with "I support the proposal because it is the most comprehensive" and "I instituted the first staff appreciation day on this campus." **Powerless language** conveys a subordinate social position and typically includes qualifiers, such as "That's the best proposal, don't you think?" and "I guess you could say I started that program." During your interactions, examine differences in powerful and powerless language and see if you notice themes in who tends to use each type. In Box 5.6, examine ways to make statements more powerful and consider the advantages and disadvantages of such rephrasing.

powerful speech Language that conveys social power; often attributed to men

powerless speech Language associated with a subordinate social position; often attributed to women

---

BOX 5.6: CONNECT & REFLECT

Pour on the Power!

Consider the following statements that use powerless language and rewrite them so the language is more powerful:

It was probably one of the first instances.

In my opinion, we should set Friday as the deadline.

Although I may not recall every detail, I think that *The Matrix* was a better movie.

I am so very truly excited to be here.

I have been thinking of this event for so, so long.

I sort of think we should.

It would be good to finish before the first of the month, don't you agree?

I know that not everyone agrees, but we should prioritize finishing the report early.

If it's ok with the group, let's set deadlines for next week.

I'm wondering if we should divide topic areas by person.

As you rewrote the statements to use language more powerfully, what themes did you see in the changes? What are the likely advantages and disadvantages of using more powerful language? What factors influence these advantages and disadvantages?

---

dual-culture approach Suggests that gender differences in language occur because women and men are members of different cocultures and thus have different goals in interaction

Rapport and Report Talk

Other work on language and gender examines life experience and cocultural membership.[32] More specifically, this approach to examining gender differences in language use, often called the **dual-culture approach**, suggests that women and men intend to achieve different objectives with their verbal messages—in other words, to *do* different things. From this perspective, women are seen as working to build closeness and solidarity, whereas men are interested in independence. These differing communication goals result in differing language choices. For example, Deborah Tannen suggests that women often engage in "rapport talk" while men tend to engage in "report talk." **Rapport talk** seeks connection and agreement with others; **report talk** emphasizes individuality and talents.[33]

Furthermore, some scholars have questioned the interpretations associated with particular language devices, such as tag questions. For example, some tag questions could be interpreted as less powerful or as a mechanism to build consensus. Conversely, consider the following tag questions that seem to serve neither of these purposes: "You aren't going out with the guys again this weekend, are you?" and "You are out of money in your bank account, aren't you?" Researchers have labeled these types of tag questions as confrontational and softening.[34]

Whether you believe that gender differences in language are shaped more by social roles or cocultural memberships and communication goals, it is important to recognize that many factors shape such language differences. For example, General Motors CEO Mary Barra's language may be influenced more by her organizational role, rather than her gender. In summary, language differences are influenced by many factors other than gender, such as social class and age, and even when they do exist, identifying their function is not clear cut.[35]

rapport talk Seeks connection and agreement with others; often associated with women

report talk Emphasizes individuality and talents; often associated with men

---

BOX 5.7: CONNECT & REFLECT

Pronouns and Inclusion

Much of the research on language and gender has categorized gender as binary, generally classifying gender as female or male. Thus, much work remains to be done to better understand the influence of other genders on communication. One example of ways that binary views of gender have been expressed in the past is through the predominant use of female and male pronouns (she/her and he/him). In the past few years, use of other pronouns (e.g., they/them used singularly; ze/hir) to be more inclusive has become more frequent. Also, stating one's pronouns, such as when introducing oneself or including pronouns in email signature lines, has become more common. What are additional ways that verbal communication can be used to be inclusive? What other language use reflects binary views of gender? How might that language be changed?

CHAPTER REVIEW

■ **Identify the key foundations of language in a pervasive communication environment.**
The four foundational principles are that language is arbitrary, abstract, ambiguous, and evolving.

■ **Discuss semiotics, especially language and meaning-making.**
Semiotics is the study of how you make meaning from, or interpret, communication. It focuses on the signs and symbols you use to construct meanings. Three areas of semiotics—semantics, syntactics, and pragmatics—are especially important.

■ **Describe the relationship between language and culture.**
Meanings are in people, not in words, and the meanings that you hold are largely shaped by your culture or cocultures. Culture shapes language, such as preferences for inclusion or individualization and direct or indirect speech, and is also influenced by language over time.

■ **Explain the relationship between language and gender.**
Research has identified a number of gender differences in language use, such as use of intensifiers and interruptions. One perspective suggests that such differences occur due to social roles and power differences. Another viewpoint suggests that, based on cocultural memberships, women and men often have differing communication goals that result in language differences. Whichever perspective you find most persuasive, it is important to recognize language differences are influenced by many factors other than gender, such as social class and age.

DISCUSS AND APPLY

1. How does a pervasive communication environment influence your language choices? What ways can you think of that texting or tweeting has influenced your conversations with friends? What role have other new communication technologies played in your interactions?

2. What frequently used new words can you think of that are related to communication technology use? When and with whom do you tend to use these expressions?

3. Given that language is arbitrary, abstract, and ambiguous, how are you able to understand your conversational partners?

4. Consider this quote by Nathaniel Hawthorne: "Words—so innocent and powerless as they are, as standing in a dictionary, how potent for good and evil they become in the hands of one who knows how to combine them." How can you relate this quote to our discussion of where meaning resides? How does it fit with the work of Waltman and Haas on hate speech?

5. Bestselling author Jodi Picoult wrote: "Words are like eggs dropped from great heights; you can no more call them back than ignore the mess they leave when they fall." How does this quote fit our discussion on pragmatics and/or culture?

6.   Explain what is meant by the reciprocal relationship between language and culture.

7.   What types of gender differences in language use have you experienced in your friendship groups? At work? At home? With family members?

8.   What factors might explain gender difference in language use? Which perspective on examining gender difference in language do you find most helpful? Explain your perspective.

9.   During and after the COVID-19 pandemic, what language changes did you experience? To what extent do the new words surrounding the virus in Spring 2020 remain in common use today? What factors have shaped this use?

CREDITS

Learning Outcomes

- Identify the key functions of nonverbal communication in a pervasive communication environment.

- Examine the types of nonverbal communication.

- Describe the relationship between nonverbal communication and culture.

- Explain the relationship between nonverbal communication and gender.

CHAPTER 6

Nonverbal Communication

The weather is lovely, and lots of people are out as you walk across campus. If you did not have to be in class shortly, it would be a good day for people watching and connecting with friends. But the walk provides you with some insights, even though you are not talking with anyone.

In the distance, you notice your friend Kavonte walking toward you. Before he turns on a different sidewalk, he makes eye contact, smiles, and waves his hand at you. Looking in the direction of that sidewalk, you see Carolina and Eduardo walking very close together and laughing, and you think that perhaps Tracy was correct when she speculated that they were interested in each other.

Two people exit a building and slap hands in what appears to be a high five. "Did they just finish a test?" you wonder. As you get closer, you hear one ask the other, "Who you do think will win the game?" The response is simply a shrug of the shoulders. The question asker then replies, "Agreed. Toss up." On a nearby bench, two couples sit, one holding hands and one with their arms around each other.

After class, you walk to the library for your study group. Some people are participating online, so you log in. As you greet people in the session, Lilly cups her hand to her ear, making you realize that again you have forgotten to unmute your microphone. Once you unmute and begin to talk, Lilly gives you a thumbs up. When group members have their videos on, you can observe their surroundings. Siobhan must be a serious basketball fan given all of the team items on her wall. Millie appears to be into Broadway musicals, as she has used playbills to decorate her room. From the images in Alejandro's tattoos, you get the impression that he is religious. When Elisa joins the session and flashes the peace sign, you know that it is time for the meeting to start.

During your campus walk and your study group session, you made note of several examples of nonverbal communication, and your impressions were shaped by how you interpreted these messages. In earlier chapters, you learned about how perceptions are formed, including how these are influenced by beliefs that you hold, stereotypes, and language that communicators use. Just as our previously held views and understandings of verbal communication influence our interpretations, so too do our understandings of nonverbal communication (Figure 6.1). In the examples above, you will notice several inferences that were made based on nonverbal communication.

FIGURE. 6.1 If you walked by Adrian on the way to your library study group, would you stop to say hello? Why or why not?

Nonverbal Communication in a Pervasive Communication Environment

body language Nonverbal communication

nonverbal communication Ways we communicate beyond words; sometimes called body language

As discussed in the previous chapter, nonverbal communication and verbal communication are interrelated. With verbal communication, you use symbols—words—to try to convey your ideas to others. Through nonverbal communication, you may enhance what you say, detract from it, or not need to use words at all. In this chapter, you will examine nonverbal communication. Sometimes called **body language**, **nonverbal communication** includes all of the ways that we communicate beyond words. Nonverbal communication can occur alongside verbal communication or separately from it and involves gestures, mannerisms, and other physical behaviors (Figure 6.2). In this chapter, you will first explore how nonverbal communication functions—the array of relationships that nonverbal communication has with verbal communication. Then you will consider the types of nonverbal

FIGURE 6.2 What does this expression convey? Are words needed?

communication, which range from eye contact and facial expressions to posture and use of physical space. Finally, you will look at the influences of culture and gender on nonverbal communication.

Functions of Nonverbal Communication

Nonverbal communication functions in several ways. As discussed earlier, nonverbal and verbal communication are interrelated;[1] thus, the functions of nonverbal communication often are examined in terms of their relationship to verbal communication. Nonverbal communication functions include repeating, substituting, complementing, accenting, contradicting, and regulating our verbal messages.[2]

Nonverbal Communication as Repeating

repeating Using nonverbal communication to repeat your verbal statements; to reinforce

One function of nonverbal communication is to repeat your verbal message. Through **repeating**, your nonverbal communication reinforces or underscores what you say. For example, when saying "yes" to answer a question, you might nod your head up and down and smile, which adds reinforcement to your positive verbal answer. If you are describing your sore elbow from playing tennis to your friend, you might rub your elbow while talking, which underscores the idea that it hurts.

substituting Using nonverbal communication instead of verbal communication; to replace

Nonverbal Communication as Substituting

At times, nonverbal communication can take the place of—**substitute** for—verbal communication (Figure 6.3). Following the example above, if your friend later asks whether your elbow feels better and you hold up your elbow and grimace, these responses likely will substitute for a verbal answer. If a new friend asks about whether you would like to go to the ballet and you shake your head side to side while frowning, many people would consider your answer "obvious," although you did not say a word.

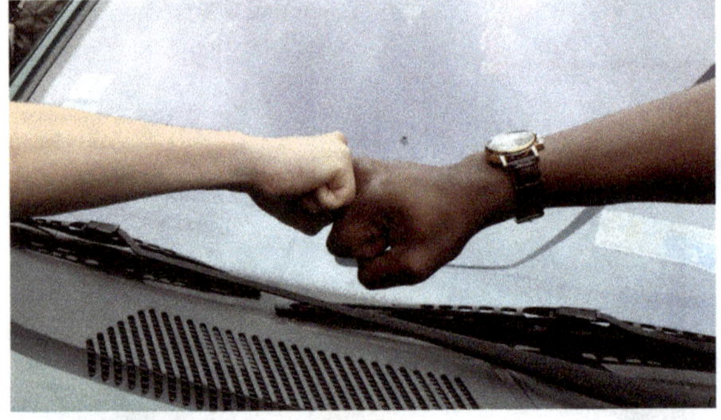

FIGURE 6.3 What is communicated by touching fists? For what words might this gesture substitute?

Nonverbal Communication as Complementing

complementing Aligning nonverbal and verbal communication; to support

Another function of nonverbal communication is to **complement** your verbal message. When a nonverbal message complements a verbal message, the messages are aligned and easier to interpret. For example, if your friend is feeling a bit down because she is not sure how well she did on her anthropology test, you might touch her arm and pat it a few times while you reassure her that she is an excellent student who studied hard and is likely to get a good grade. Sometimes when you tell people that you love them ("I love you so much!") or when you are saying how happy you are that they got an A on the test ("Brilliant work! You are a superstar!"), you simultaneously smile and hug them. In each of these cases, this type of nonverbal communication complements the verbal statements.

Unlike nonverbal messages that substitute for verbal ones, nonverbal messages that complement verbal responses do not make sense or are not clear without the accompanying verbal response. In other words, these behaviors need the accompanying words. For example, if you pat your friend's arm a few times or smile at your friend without a verbal message, it is not exactly clear what you mean.

When your nonverbal communication complements your verbal communication, messages are easier for others to understand, to believe, and to remember. Because the messaging is aligned, others can more easily process the agreement between the two types of communication.

Nonverbal Communication as Accenting

Related to complementing a verbal message is **accenting** a verbal message. When nonverbal communication accents, it serves to stress parts of the verbal communication, placing emphasis in specific areas. For example, if you say, "I am just so very, very upset with you right now" and you place your face in your hands when saying "upset" or you raise your voice when saying "very, very upset," then you accent those facets of your verbal message.

accenting Using nonverbal communication to stress or emphasize parts of verbal communication; to accentuate

Nonverbal Communication as Contradicting

In contrast to the above examples, not all nonverbal messages complement verbal ones. In fact, the opposite also occurs—nonverbal messages may **contradict** verbal messages. In these instances, people might be accused of "saying one thing but meaning another." Let's revisit two examples above. When telling your friend that you are happy about the A that they made on the test, what if your tone of voice did not sound happy, you did not smile, and when you hugged the person, you did not really embrace them? Have you ever asked someone whether they were mad at you and gotten a "No" response that was loud and delivered without a smile or eye contact? If so, how did you interpret this message?

contradicting Misaligning nonverbal and verbal communication; to deviate

When someone's nonverbal message contradicts their verbal statements, such as when your friend says she is not mad but does not smile and seems to raise her voice, how do you make sense of this discrepancy? In other words, do you believe the nonverbal message or the verbal one? Research suggests that people tend to believe the nonverbal messages in such situations. In many cases, you are processing several nonverbal messages (e.g., the tone and volume of her voice, her facial expression, her posture, etc.) alongside her words, so that may influence believability.[3] Additionally, research indicates that nonverbal communication is perceived as harder to control than verbal communication.

Of course, sometimes the contradiction between verbal and nonverbal messages is intentionally used as sarcasm or humor. If you miss an easy layup on the basketball court and then proclaim, "I am the GOAT!," your friends will likely know that you are joking.

Nonverbal Communication as Regulating

regulating Using nonverbal communication to start, indicate speaking order and length, and end verbal conversations; to synchronize

Nonverbal communication also serves to **regulate** verbal communication. How do you know when it is time to begin or end a conversation? How do you know when it is your turn to talk, and how you do indicate to others that you would like them to talk? Conversational beginnings, turn-taking, and endings are all typically regulated by nonverbal communication.

Until relatively recently, looking at your watch, especially if more than once, was interpreted as signaling that you would like a conversation to end or had another pressing commitment. Now that fewer people wear wristwatches and instead rely on smartphones for keeping track of time, what cues that another needs or wants to end a conversation have you observed?

When working in group settings, how do we negotiate who talks when? Several aspects of nonverbal communication typically play a role in turn-taking as well as in indicating when we would like to speak.[4] For example, eye contact, body position, and vocal pitch can be used to attempt to regulate conversational flow. When taking online courses, either planned or in adapting to the COVID-19 pandemic, or participating in online meetings, what similarities and differences have

---

BOX 6.1: CONNECT & REFLECT

Nonverbal Communication and the Pandemic

Consider the coronavirus image (Figure 6.4). What other images or symbols can you think of that represent the COVID-19 pandemic? How have your views of seeing people wearing masks changed? What examples can you give of ways that the pandemic has changed nonverbal communication? Consider both small adjustments and larger, longer-term changes. How might these changes vary by group or place (e.g., college students, health care workers, people living in rural or urban areas)?

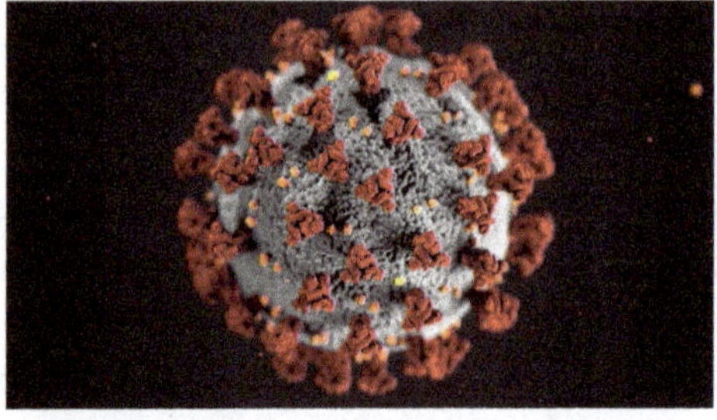

FIGURE 6.4 When do you recall first seeing this or a similar image? What words do you associate with the image?

TABLE 6.1 Overview: Nonverbal Communication Functions

FUNCTION	DEFINITION	EXAMPLE
Repeating	Using nonverbal communication to repeat your verbal statements; to reinforce	Shaking your head side to side while saying "no"
Substituting	Using nonverbal communication instead of verbal communication; to replace	Shaking your head side to side or making a thumbs-down gesture while saying nothing
Complementing	Aligning nonverbal and verbal communication; to support	Smiling and hugging someone while saying "I love you"
Accenting	Using nonverbal communication to stress or emphasize parts of verbal communication; to accentuate	Talking faster and louder when saying "VERY angry" as part of a sentence
Contradicting	Misaligning nonverbal and verbal communication; to deviate	Saying that you are not mad at someone with a loud voice, crossed arms, and a frown
Regulating	Using nonverbal communication to start, indicate speaking order and length, and end verbal conversations; to synchronize	Checking the time on your watch or phone to cue the end of a meeting

you observed in how nonverbal communication regulates verbal communication? Both modalities offer options like raising one's hand to ask a question, but have you observed any differences with using eye contact to regulate conversations?

Now that you have considered the ways that nonverbal communication can function (Table 6.1), let's consider how it might change (Box 6.1) and the types of nonverbal communication. As you read the next section, think about how the types of nonverbal communication may work together (or fail to do so) in accomplishing the functions discussed above.

Types of Nonverbal Communication

Nonverbal communication takes place in a variety of ways. The types of nonverbal communication include the face and eyes, kinesics, proxemics, haptics, paralanguage, chronemics, and artifacts. Rather than conveying information in an isolated fashion, the types of nonverbal communication often function together to shape our interpretations, which can range from understanding (when messages reinforce each other) to confusion (when messages appear to be mixed). In other words, we consider different types of information (e.g., facial expressions, words, posture) when we form impressions of others and their communication.

Face and Eyes

It is impossible to talk about nonverbal communication without discussing the **face and eyes**, which are areas that people most often think of when nonverbal communication is mentioned. The eyes, frequently said to be the "window to

face and eyes Types of nonverbal communication that include eye contact and facial expressions; frequently used to communicate interest and emotion

our souls," both provide information to others that they use in forming impressions of us as well as take in information for shaping the impressions that we form of others. The ways that we look at others may be interpreted as conveying interest in what they are saying, liking or attraction to them, dislike, anger, or unfriendliness. Further, as discussed in the section above, the ways that we look at others may signal that we want to talk or to leave a conversation, serving to regulate interactions.

People pay close attention to eyes and eye contact and use this information in attempting to understand others.[5] For example, whether assessing blinking, pupil dilation, or direct eye contact, others use our eyes to shape their understandings. More specifically, rapid blinking may be interpreted as signaling discomfort, and infrequent blinking might be regarded as attempting to hide information.[6] Additionally, although pupil dilation is affected by indoor and outdoor lighting, it also can be affected by our emotions (e.g., interest in another person may be indicated by increases in pupil dilation).[7] In particular, the amount of direct eye contact made is an important feature of interactions. Eye contact may be interpreted as indicating that someone is interested or paying attention or, especially if prolonged, as threatening. Little eye contact or failure to make eye contact may be interpreted as indicating that someone is uncomfortable, disinterested, or dishonest.[8]

Of course, the eyes, including the eyebrows, are only one part of our overall facial expressions. When interacting with others, you likely pay considerable attention to their faces because facial expressions are believed to convey emotions. Some facial expressions are described as universal, meaning that these are the same across cultures, such as expressions of happiness, surprise, sadness, fear, anger, contempt, and disgust.[9] However, the willingness to display these

microexpressions Facial expressions that occur rapidly, may go unnoticed, and unconsciously reveal information

---

BOX 6.2: ETHICAL QUESTIONS

Microexpressions

Beyond our normal facial expressions, some researchers, such as Paul Ekman and David Matsumoto, examine "microexpressions."[13] These **microexpressions** are faster than our usual macroexpressions and, according to these researchers, unconsciously reveal information. Because they occur so rapidly, many people miss or cannot interpret these expressions. However, understanding these subtle cues can be helpful, especially when you want to accurately understand how someone feels. Although such understanding may be helpful to anyone, these researchers suggest that the ability to interpret such communication is especially important for people in particular areas of work, such as antiterrorism, security, and criminal justice.

Can you think of ways that correctly identifying microexpressions would be useful in other fields? For example, how might such information be useful in teaching young children? For employers? For health care providers? For attorneys? In what other areas of work would skills in this area be helpful? How might an ability to interpret microexpressions be useful in daily life beyond work?

Is it ethical to train others to "read" such expressions? How should we decide who would be trained first? Are there groups of people that you would not recommend training (if so, please describe these groups)? Why, or why not?

expressions may vary across culture, which will be examined later in this chapter. In addition, there are a variety of other facial expressions, such as excitement and confusion, that you may try to discern from others' faces.

Whether or not you believe or trust another may be linked to your interpretation of their facial expressions, including microexpressions (Box 6.2). For example, displaying a slight smile and slightly raised eyebrows is likely to be interpreted as self-assured, forthcoming, open, and trustworthy, thus leading to perceptions of believability.[10] As discussed earlier, our nonverbal communication, such as facial expressions, may reinforce what we say or contradict it. For example, you might say you are having a great day and smile while sitting with relaxed posture, underscoring what you said, or you might make that same statement but frown and have downcast eyes, thus suggesting otherwise. How might examining eye contact and facial expressions contribute to experiences such as Zoom fatigue (Box 6.3)?

Other facets of facial expression include the mouth and nose. Smiling is often discussed as a means of conveying happiness. Although that may be the case, a smile can also conceal or divert as well as reveal scorn or criticism.[11] Other examples include covering your mouth, which might be a polite gesture (e.g., to hide a yawn or a cough) or an attempt to conceal a negative expression (e.g., a smirk), and biting your lips, which might indicate concern or stress. The nose is less frequently discussed in examinations of nonverbal communication; however, holding one's nose or crinkling it are relatively common expressions. Also, beyond our interpretations based on sight, it may be that smell influences us in many ways. Some research examines chemosignals, information processed via smell, and how these shape interpretations.[12] For example, some findings suggest that we may smell fear based on chemicals others give off and become fearful ourselves.

Kinesics

In the section above, we focused on nonverbal communication using the eyes and facial expressions. In each of these areas, some movement is involved, and this section extends the examination of movement, or kinesics, to gestures and posture. People often use hand and arm gestures and body positioning to send messages to others.

Previous research has classified gestures into one of five categories: emblems, illustrators, regulators, affect displays, and adaptors. As we examine each one, consider how they relate to the functions of nonverbal communication discussed above.

Emblems have culturally agreed-upon meanings that represent words or phrases.[15] For example, raising your middle finger to "flip someone the bird" or placing your index finger vertically across your lips has a specific meaning. However, if you raise your index and middle fingers into a V shape and bring your other fingers and thumb together in front of your palm, you have a gesture that differs from each of the above. Whether you are "flipping someone off," shushing

kinesics Use of movement to communicate

emblems Gestures with culturally agreed-upon meanings that represent words or phrases

BOX 6.3: CONNECT & REFLECT

Zoom Fatigue: Causes and Solutions

During the COVID-19 pandemic, much of school and work life transitioned to online spaces. You likely had courses and meetings online via Zoom or similar platforms, and you might have experienced Zoom fatigue. Researchers at Stanford University explored some of the differences with in-person communication, especially related to nonverbal communication, that may result in this exhaustion.[14] These video meeting differences include more extensive, close eye contact, continual exposure to one's own face (and the potential for evaluating one's appearance), decreases in types of movement possible in face-to-face meetings (e.g., restricted space to stay centered with the computer's camera), and extra cognitive processing (e.g., participants engage in more work to both send and interpret nonverbal messages).

How often do you participate in online class sessions or meetings now? How common is Zoom fatigue for you and others that you know? What ideas might be used to lessen Zoom fatigue? Consider both changes to the technological platform as well as changes that you could implement in your meetings.

illustrators Gestures that illustrate what is said; typically do not have clear meanings distinct from the verbal messages that they accompany

regulators Gestures that help to regulate conversation

someone, or making the peace sign, you are using an emblem to communicate—a gesture that can substitute for its verbal equivalent.

Illustrators serve to illustrate what you are saying. These frequently used nonverbals do not have clear meanings distinct from the verbal messages that they accompany. Perhaps one of the most common examples of an illustrator is the "big fish" story—saying that you caught a huge fish and spreading your arms to illustrate its length. Of course, you could make the gesture without the fish tale but the meaning of the gesture would not be clear on its own. Illustrators help maintain or increase attention, add clarity, and assist in remembering.

Regulators are nonverbal gestures that assist in the function of regulating interaction, as discussed in an earlier section of this chapter. As discussed in the chapter on listening, you do not just "sit back and listen." Instead, listening is an active process where you convey to the speaker, through behaviors such as eye contact, head nods, posture (e.g., leaning forward or slumping), and verbalizations (e.g., "uh-huh"), your level of interest and attention, interest in turn-taking, and willingness to continue engaging in conversation. For example, how do you nonverbally convey to a speaker that you do not understand or want the person to slow down? Your textbook authors have noted a common regulator when they lose track of time and keep lecturing or holding a discussion at the end of the scheduled class period. When this happens, you and others in the class likely engage in nonverbal leave-taking behaviors (e.g., fidgeting in your seat, starting to close notebooks, picking up backpacks, pointing at or looking at the clock or your watch or phone, etc.) to cue us to the time and to encourage us to conclude.

Affect displays are behaviors that share emotions. Whether you are feeling fearful, happy, or surprised, you convey your emotions to others via both facial and body gestures or movements. You might share your surprise or excitement by jumping up and down, your anger by putting your hands on your hips, or your sadness by slouching and hanging your head. Sometimes we are aware that we are sharing these emotions, such as when we hug someone who made us happy with a thoughtful gesture. At other times, people may notice our nonverbal gestures and ask us about feelings that we did not know we were conveying or wanted to hide (e.g., when someone asks whether you are bored although you had hoped to feign interest).

affect displays Gestures that share emotions

Adaptors are more related to your own comfort and less centered on communication with others. However, the impressions that others form about us can be influenced by these behaviors. Adaptors are used to help manage stress, discomfort, nervousness, lack of control, excess energy, or similar feelings. Examples of adaptors include playing with items (e.g., twirling your pencil, twisting the strings on your hoodie), twisting your hair, and organizing and reorganizing the items on your desk.

adaptors Gestures related to comfort and less centered on communication

Beyond specific gestures, we also communicate through our posture. For example, when observing someone who is seated, our impressions might vary based on whether they are sitting in an erect manner and keeping their arms and legs close to their body or sitting at an angle and sprawled out across their area with their legs stretched as far out as possible. In some instances, people interpret leaning forward in one's chair as expressing interest and leaning backward as expressing disinterest. What examples can you think of that illustrate how posture might be an indication of friendliness or unfriendliness, interest or disinterest, or formality or informality?

Proxemics

Although it might not be talked about as often as some forms of nonverbal communication, such as eye contact and facial expressions, proxemics—how we use space to communicate—shapes our interactions and impressions. Throughout your day, you use space to send many messages to others. For example, which people do you sit close to and which ones further from? How do you position your belongings relative to your roommate's? Maybe you recall sibling squabbles from childhood about who gets to ride in the front seat of the car or to claim the larger bedroom in a new apartment.

proxemics Use of space or distance to communicate

These examples show how space is used to send messages and how our views of space shape our impressions. Factors such as the distance between people, who is afforded the most space, and use of objects to influence space (e.g., being seated at the head of the table versus at a roundtable) can be revealing.[16] Through our use of space, we communicate closeness, liking, dominance, and aggression. For example, during an argument if someone

public interaction zone Used in situations such as public speeches or presentations; 12 or more feet between communicators

social interaction zone Used in contexts such as workplaces; usually 4–11 feet

personal interaction zone Used for interacting with family and friends; usually 1.5–3 feet

intimate interaction zone Used for interactions with individuals we are very close to (romantic partners); typically under 1.5 feet and may involve touching

"gets in your face," they are asserting their dominance in an aggressive fashion. Space can also be an indicator of social position, with powerful people taking up more space. The next time that you are in a large meeting, pay attention to how people in the room use space—who sits in the most expansive way, how people spread their possessions out on the table, how close the chairs are to each other, and so forth.

Although we may take space for granted, we are reminded of its importance when violations occur. If you are the only person in an elevator and someone else enters, where is each of you "supposed" to stand? How would you feel if the person entered the elevator and stood right beside you? The next time that you are at dinner with a friend seated at a small table for two, try the following space experiment. Each time that you take a drink from your water glass, put the glass down further and further across the table's center line. Use the same behavior to place other objects (e.g., a book you brought along, your dirty dishes, etc.) on the friend's "side" of the table. What reaction are you likely to get?

Researchers have examined four interaction zones, each defined by space expectations based on our relationships with others. The four zones are: **public**, **social**, **personal**, and **intimate**.[17] The public zone involves the greatest distance, likely 12 or more feet, between communicators and would be used in situations such as a classroom speech or public presentation. The social zone, usually estimated at 4–11 feet, is used in general interactions in workplaces or across campus. The personal zone, from approximately 1.5–3 feet, is comfortable for interactions with family members and close friends. The intimate zone, typically viewed at under 1.5 feet and may involve touching, is limited to very close relationships, such as those we have with romantic partners. When our space expectations are violated, either by someone invading what we perceive as our space or distancing too much, discomfort is common.

Of course, these interaction zones emerged from research prior to the COVID-19 pandemic. The physical distancing recommendations during the pandemic are in contrast with these expected interaction zones. For example, you would usually interact with close friends in the personal zone; however, following the pandemic, physical distancing requirements would involve distances of at least 6 feet, moving into the social or public zone. How might these differences affect your communication or your relationships? In what ways might the pandemic experience alter these interaction zones in the future (Figure 6.5)?

FIGURE 6.5 This graphic reinforced the physical distance needed during the COVID-19 pandemic in a fun way. How do you think that the pandemic will influence our physical interactions in the future?

Haptics

Another type of nonverbal communication is touch, also known as **haptics**. Touching can range from hugs and kisses to punches and kicks and from high fives to handshakes (or footshakes during the pandemic). We use touch frequently to convey our feelings, such as liking, encouragement, sympathy, and anger (Figure 6.6).

Researchers describe seven categories of touch: playfulness, liking/positive emotion, control, ritual, hybrid, task-related, and accidental.[18] Playful touching acts to make interactions less serious and more enjoyable. Touching that conveys liking/positive emotion may show support or appreciation or convey romantic interest. Touch also is employed to try to control others or show dominance. Some touch is a ritual practice, such as shaking hands before starting and upon ending a meeting. Hybrid touch has multiple purposes; for example, it can both show liking and serve as a ritual (e.g., hugging someone before departing). Task-related touch involves performing a task, such as helping someone put on a lifejacket or off a boat. Accidental touch refers to inadvertent brushes that do not serve as intentional communication.

Several factors about a touch affect how it may be interpreted.[19] For example, the location of a touch is especially important; touching someone on the arm is likely to be perceived quite differently than touching their buttocks. Additionally, the length of the touch (e.g., was it a lingering caress or a fleeting pat?) and the regularity (e.g., did it occur once or multiple times?) will influence perceptions.

Paralanguage

Have you heard the expression that "it's not what you say but how you say it"? If so, you are already familiar with paralanguage. "Para" refers to alongside or with, so **paralanguage** means nonverbal communication that is associated with your voice or how you say what you say.[20] Also called vocalics, paralanguage includes verbalizations such as "uh," "um," "uh-huh," and clearing your throat as well as other factors associated with the voice, including how pauses and silence are used.

Your use of paralanguage can function in several ways. It can help increase the accuracy of your verbal statements by reinforcing or emphasizing key parts. For example, if you were trying to convince your partner that you preferred a small, rather than a large, birthday party, how might your paralanguage play a role in your success? Paralanguage also can reveal your attitudes and emotions.[21] When someone says, "I'm so surprised" or "I enjoy presenting in class," what types of paralanguage might reveal their true attitudes or emotions? Laughing or crying while talking are examples of the ways in which paralanguage can reveal emotions. In addition, paralanguage can reinforce or undercut your words.[22] If you respond "Fine" when someone asks how are you, what types of paralanguage might reinforce this answer? What might call it into question? What might contradict it? Paralanguage can even reveal information that you did not intend to share.

haptics Use of touch to communicate; includes playfulness, liking/positive emotion, control, ritual, hybrid, task-related, and accidental

FIGURE 6.6 How would you describe communication via touching in this photo? What categories of touch are illustrated?

paralanguage Use of features associated with the voice, but not words, to communicate; also called vocalics

Paralanguage can be divided into two types: **vocal characteristics** and **vocal inferences**. Vocal characteristics include your speech volume (soft or loud), rate (slow or fast), pitch (low or high), quality (easy to listen to/appealing or hard to listen to/unappealing), articulation (clear or unclear), resonance (vocal depth and strength), and accent (way of pronouncing). These vocal qualities intertwine and shape how you perceive others and their messages. For example, if you encounter a person who is speaking rapidly and loudly, what might be your first impression?

Vocal inferences are utterances that we use when searching for what we want to say or attempting to delay communicating. These utterances are also called verbal fillers, as they serve to fill conversational space. Examples include "uh" and "um." Utterances, such as "um," if used frequently in a conversation, might lead others to think that one is insecure, uninformed, uncertain, or attempting to deceive. In fact, we often pay attention to paralanguage when trying to decide whether someone is telling the truth. In attempting to assess truthfulness, you might consider use of verbal fillers, length of pauses, and higher vocal pitch, among other nonverbal cues.

Chronemics

Our understanding of time shapes our communication and influences our relationships, and this area of study is referred to as **chronemics**. It is through communication that we develop an understanding of time, and then that understanding of time shapes our subsequent communication.[23] Chronemics explores how we believe that time functions, how we use it to communicate, and its meaning across our interactions (Figure 6.7).

FIGURE 6.7 Why do we place clocks on walls? On buildings? Beside our beds? Why do we wear watches on our wrists? Why do our computers and phones display time?

Whether you were raised in the United States or came here more recently, you have likely frequently heard a number of expressions that emphasize the value and importance placed on time.[24] In phrases such as "big waste of time," "spend your time wisely," "running on borrowed time," "lost time," "no time to spare," and "time is money," what is communicated about time? What do these phrases suggest about how we should view time and use our time?

The ways that we spend our time communicate information, such as our interest in others and how we see ourselves in relation to them. For example, we often let others know their importance to us and our wish to develop or continue close relationships by spending time together.[25] When you engage in frequent conversation with someone, make time to get together, and show that you are willing to work on the relationship, you indicate your dedication and commitment to that person and relationship. If someone tells you how important you are to them but rarely has time to talk with or see you, how would you interpret their level of interest?

Social status and power are also often conveyed via use of time. For example, if you have a restaurant reservation and an appointment with your physician at noon, would you be equally comfortable waiting an hour beyond the scheduled time for each? If you were running 15 minutes late to meet your best friend for lunch or for a job interview, how might your stress level differ, and why? Individuals in certain positions (e.g., organizational leaders, celebrities, etc.) are often allowed to talk more, interrupt others, keep others waiting, and set the agenda for meetings or conversations (i.e., determine what and how long topics will be discussed). What differences have you observed in who interrupts whom, who needs to be on time, and who needs to schedule an appointment, and what do these differences tell you about social relationships?

In addition, the ways that each of us views time shape facets of our interactions, such as how long we might wait for others, be willing to listen, and allot to activities as we plan our daily schedules. It might also influence factors such as our rate of speech. Electronic calendaring systems send reminders so that we do not forget scheduled activities, emphasizing the importance of adhering to a planned schedule, and arrive on time, underscoring the importance of punctuality.

Time can be divided into four categories—biological time, personal time, physical time, and cultural time.[26] Each category of time has important implications for communication. Biological time refers to the circadian rhythms that shape our life patterns, such as when we sleep. Considering biological time can influence your approach to communicating with others; for example, different strategies might be needed to secure and maintain interest from your classmates for a presentation in an evening course versus an early morning course. Personal time is centered in how you experience time. For example, when does time seem to "fly by" for you, and when does it "go on and on"? These perceptions are influenced by your interests and emotions. You also may have a greater orientation to the past or to the future, which influences how you enjoy spending time. If you tend to be past-oriented, you may enjoy keeping up with long-term friends and sharing stories about earlier adventures. If your focus is more future-oriented, you might like to discuss your anticipated career or plans with others or strategize about accomplishing these goals. Physical time involves the calendar and seasons. Because our emotions can be affected, such as through less access to light during winter, our communication may also be influenced by physical time. Cultural time refers to the way that our culture shapes our perceptions and use of time. For example, some cultures are characterized as monochronic and others as polychronic. People in the United States tend to have a monochronic approach to time. "Mono" emphasizes the focus on one—the view that tasks should be accomplished one by one. From this vantage point, you might plan your schedule in small blocks of time and be disappointed if the plan is altered. In contrast, polychronic time embraces engaging in multiple actions simultaneously, which is conveyed by "poly," meaning "many." For example, someone operating from a polychronic perspective might participate in an online meeting, answer emails, and brush their dog.[27] People employing a polychronic view of time tend to see time as expansive, prefer unscheduled time, and be less bound by planned activities. In a later section of this chapter, you will learn more about the ways that culture influences nonverbal communication, including our perceptions and use of time.

Artifacts

The physical objects that we wear or surround ourselves with are called **artifacts**. Often, these are items that we choose to represent ourselves (e.g., our attitudes, social status, personality) to create an impression that we desire (Figure 6.8). Look at the buttons on your backpack or the stickers on your laptop or water bottle or how your friends have adorned their possessions. What do these artifacts communicate?

There are many different types of artifacts. Some are found on the body. For example, you may choose to wear your hair a certain way or to wear makeup or nail polish. You also may choose to get tattoos or body piercings, and if so, you may choose the types, locations, and colors. In addition, you add other artifacts, such as clothing, jewelry, watches, glasses, and sunglasses, to your body. In some situations, you carry additional artifacts with you. These artifacts include items such as a backpack, purse, briefcase, pen, organizer, book, tennis racket, food, and phone and its case.

You also may surround yourself with other objects to create an impression. For example, artwork, furniture, and photographs may be carefully selected and displayed. When you visit your professor's office, what do the shelves lined with books and academic journals convey? What does the sports memorabilia covering the walls of your friend's apartment suggest? Why do many managers display pictures of their family members in their offices?

In addition, your automobile or home can lead others to form particular impressions. If automobiles simply provided transit from one place to another, they might all be the same. Instead, we have an array of different types and colors. How might purchasing a particular auto shape how you are viewed by others? What types of impressions might you make about another person based on their home (e.g., type, location, size)? In part because impressions are important to people and accessories may shape impressions, professionals may be hired to "stage" a home before potential purchasers view it. Using artifacts, the stager creates an environment that will make it easier for a house hunter to envision the identity conveyed by living in the space.

The list of artifacts is long, and new possibilities emerge. For example, virtual reality headsets are moving from science-fiction movies closer to our daily lives. Your dog, cat, hedgehog, or pig (and its collar or bandana—and whether you wear a matching one) also may influence how people perceive you. From designer clothing brands to designer pets and maybe, ultimately, to designer babies, many people devote considerable attention to cultivating a desired image. In some

FIGURE 6.8 Notice the artifacts in this picture. How do they shape your impressions of the people?

social media circles, strategizing about one's "brand," the image one wishes to cultivate, is a hot topic, and artifacts are carefully selected to enhance the desired image.

Beyond communicating one's desired image, artifacts are also important in providing information that may be useful as we try to adapt our messages to others. You may discover common interests with a classmate or your professor based on a water bottle sticker or book. Noticing that someone is interested in climate change from the buttons on their backpack may give you information to draw upon in getting them to sign your petition for more vegetarian food on campus or be useful in planning your speech. Artifacts can signal group membership (e.g., favorite sports team, sorority affiliation, etc.), relational status (e.g., wedding ring), socioeconomic status, and personality, and we pay close attention to such information, as it may be useful in our communication.

Certainly, although people may attempt to create particular impressions through their artifacts, there is no guarantee that they will get the desired result. Your friend, for example, may wish to convey her good taste and sense of style by decorating her apartment walls with shopping bags from expensive, specialty stores. Though some people may hold this interpretation, it is possible that others will see her as privileged and showy. Just like beauty is in the eye of the beholder, impressions are left up to individual perceivers.

In the sections above, you have considered several types of nonverbal communication and the ways that each might shape impressions and thus communication. Although each has been discussed separately, it is important to note that all or many take place together. Thus, our overall impressions are informed by the full set of nonverbal and verbal messages. Accuracy in interpreting others' messages is enhanced when we pay careful attention to many types of nonverbal messages, especially noticing whether these align with or contain discrepancies from each other and verbal messages. How can nonverbal communication be used to create more welcoming and inclusive spaces (Box 6.4)?

BOX 6.4: LIVE IT LOCAL

Your Community, Your Communication

Consider your membership in community groups and organizations. Select one of these groups or organizations for assessment. With the challenges of the times, such as racial injustice, political division, and economic disparities, how can we communicate in more welcoming and inviting ways to foster civil dialogue and seek solutions? In particular, consider the forms of nonverbal communication discussed above. Often, when we consider ways to promote conversation, we think of verbal communication.

However, nonverbal communication is powerful, setting the stage for comfort, receptivity, and sharing. In the group or organization that you are analyzing, what nonverbal rituals currently take place? What is the physical setting like (e.g., how is it arranged)? What nonverbal communication changes could you or others make to help people feel more welcome and comfortable? How successful would you anticipate such changes to be?

Nonverbal Communication and Culture

In Chapter 2, you learned about the many ways that culture influences communication. It affects both the messages that we generate for others as well as our interpretations of what others say and do. Your culture is foundational in shaping your nonverbal behaviors and the lens through which you interpret the nonverbal communication of others.

All the types of nonverbal communication discussed earlier in this chapter—face and eyes, kinesics, proxemics, haptics, paralanguage, chronemics, and artifacts—are influenced by one's culture. Our understandings of how each type of nonverbal communication works and what specific behaviors can achieve is rooted in our culture. The meanings that we glean from the nonverbal behaviors of others are informed by our culture (Figure 6.9). Assessments regarding what is an appropriate greeting, what counts as standing too close, what types of touch are acceptable, how much eye contact should be made, whether a home is small, what being on time means, and how one pays attention are all based on cultural interpretations. This fact, of course, means that your understanding of what is normal, appropriate, or expected may differ from the understandings of others.

Earlier you learned about the four interaction zones—public, social, personal, and intimate—and the distance associated with each zone. However, those distances are based on U.S. cultural expectations. In other locations, space expectations for a zone may differ. For example, if you have spent time in Latin America, you may have experienced people standing closer to you when talking. These differences often contribute to misunderstanding and, at times, unexpected or undesired outcomes. If you perceive that someone is standing too close to you while talking, you might take a step away. However, by stepping away to increase the distance so that you are comfortable, you may create discomfort for your conversational partner because they have different space expectations. To alleviate their discomfort, they seek to decrease the space and take a step closer to you. Now, you once again feel that your conversational partner is encroaching upon your space. In this situation, several impressions are possible. Your conversational partner may start to perceive you as unfriendly or distant, as you keep stepping away. You may perceive them as aggressive, as they keep invading what you perceive as your space. Both of you may perceive the interaction as odd and uncomfortable.

As discussed earlier, you know that U.S. culture places value on time, often equating it with money.[28] Other cultures, however, regard time differently. These differences may be evidenced in arrival times. For example, if you are invited to dinner at 7:00 p.m. in the United States, what time should you arrive? How might your answer differ if the invitation were issued in France, Japan, Germany, Ghana,

FIGURE 6.9 What is your attention drawn to in this photo? What are your impressions?

Spain, or Ecuador? Of course, if you are invited for dinner in some places, like the southern United States or Ireland, that might mean an early afternoon meal.

Another type of culture is workplace or organizational culture, and such cultures also influence communication. Considering chronemics again, let's examine the military. A popular phrase from military life that you may have heard is "If you're on time, you're late." This phrase stresses the importance placed on time, and because being late is simply not acceptable, one must always arrive early to be considered on time.

Although gesturing occurs across cultures, there are key differences.[29] For example, you learned about emblems and illustrators earlier in this chapter. Across different cultures, using illustrators is common; however, that use differs in frequency and form. Emblems, which replace words or phrases, are typically culture-specific (i.e., defined within a culture). Thus, these differ across cultures. For example, you may use a thumbs-up gesture to signal approval or agreement in the United States; however, in Iraq, using this gesture is the equivalent of raising your middle finger in the United States.[30] Despite these differences, increasing numbers of emblems are becoming widely recognized due to globalization.[31]

The old adage "When in Rome, do as the Romans do" is important because it recognizes the significance of culture and differences in communication (Box 6.5). What may be polite in one culture may be rude in another. As you interact with people from other cultures and travel to new places, be sure to consider how culture shapes your nonverbal messages as well as the interpretations that you make.

BOX 6.5: CONNECT & REFLECT

No Dictionary for Nonverbal Communication

Although it is sometimes called "body language," nonverbal communication cannot be characterized into a dictionary of definitions. With words, you can consult a dictionary for agreed-upon meanings. Look up "cardinal," "wombat," "glove," or "sequoia," and you will find definitions. However, despite attempts to catalogue nonverbal behaviors and their associated meanings, specific behaviors in this system of communicating cannot be classified with individual definitions. Why? As discussed above, there are several functions and types of nonverbal behavior, and they act in tandem with each other, and in some cases verbal communication, to produce meaning.

People sometimes suggest that crossing one's arms in front of one's body is a "closed" posture that communicates a lack of interest or receptivity to others—that crossed arms indicate that a person is trying to block or protect themselves from interaction with others. However, is this meaning the only possible one? What if the person is smiling and making eye contact with you? Could crossing one's arms in front of one's body occur for other reasons? Perhaps the person is cold or finds this posture comfortable.

What are other examples of nonverbal behaviors that might be misunderstood or characterized too quickly?

Nonverbal Communication and Gender

Often alongside work examining cultural influences, researchers have explored the influence of gender on nonverbal communication (Figure 6.10). In general, this work has identified gender differences in specific areas of nonverbal behavior, sometimes also examining differences in one country or area compared to another.

Across the years, several findings have generated interest and have been discussed in terms of their importance in specific contexts, such as the classroom, workplace, families, friendships, and romantic relationships, as well as what these differences potentially reveal about gender overall. For example, women have been found to engage in more frequent smiling and head nodding as well as to use more expressive gestures.[32] They also are better at recalling nonverbal information and correctly interpreting nonverbal behavior. In addition, women use less space in positioning themselves in interpersonal conversations.

Conversely, men more often interrupt conversation, talk loudly, fidget, and spread their bodies out physically. They use fewer hand gestures and indicators of listening, and men cry in public less often. Differences such as these examples have been reported in numerous studies. What do these findings mean? Further, what accounts for these differences?

On their face, these findings may appear to illustrate differences based on sex or gender. However, some researchers have suggested instead that these behaviors may be shaped or driven by factors such as psychological state (e.g., mood, comfort, etc.), group expectations, and cultural norms.[33] For example, some suggest that differences may be evidenced because people understand how to enact gender in interaction with others (i.e., how to "do gender").[34] Some nonverbal behaviors are generally viewed as associated with femaleness or maleness; thus, communicators may behave in ways that they believe will help to accomplish their interactional goals (including as one dimension of the presentation of self, their presentation of gender or display of culturally relevant or accepted gender behaviors). For example, crying has long been associated with femaleness; thus, if men cry, the risk of negative impressions is greater due to violating expectations.[35] Although shifts are occurring in how crying is viewed, as well as more generally in how gender is understood, the risk of negative impressions may explain why men have traditionally cried less often in public. In short, because it has been more accepted for women to cry, women may have felt more comfortable and anticipated fewer negative social sanctions in crying than men have perceived.

According to a recent review, the findings on nonverbal gender differences are better contextualized by considering a host of influencing factors, such as age, desire to fit in, and situation.[36] Rather than being the singular or main cause of behavior, gender and one's understanding of gender roles and associated behavior may be only a portion of what shapes our actions. Further, if one's identity extends beyond the binary, then nonverbal behavior may be less likely to adhere to particular sets of expectations. Thus, as we seek to interpret nonverbal behavior, considering context and situation as well as individual differences and preferences is likely to result in a better informed understanding of nonverbal communication and gender.

CHAPTER REVIEW

- **Identify the key functions of nonverbal communication in a pervasive communication environment.**
  The six key functions of nonverbal communication are repeating, substituting, complementing, accenting, contradicting, and regulating.

- **Examine the types of nonverbal communication.**
  The seven types of nonverbal communication include the face and eyes, kinesics, proxemics, haptics, paralanguage, chronemics, and artifacts.

- **Describe the relationship between nonverbal communication and culture.**
  Culture shapes our nonverbal behavior as well as the lens through which we interpret the nonverbal communication of others. Culture influences each type of nonverbal communication.

- **Explain the relationship between nonverbal communication and gender.**
  Studies have identified nonverbal behaviors that are more commonly exhibited by women as well as ones more commonly exhibited by men. Recent work, however, suggests that several situational and individual factors may have greater influence than gender identify. Future work also needs to incorporate gender identities beyond the binary.

DISCUSS AND APPLY

1. How does a pervasive communication environment influence nonverbal behavior? How have social media shaped nonverbal communication? What influence have other relatively recent communication methods and technologies had on nonverbal behavior?

2. When nonverbal communication and verbal communication are in conflict, how do you decide which to believe? Why? What are examples of key behaviors that would shape your conclusions? Can you think of situations where your nonverbal behavior led another to think that you agreed with them when you had not said so verbally? If so, what are the advantages and disadvantages of this assumed agreement?

3. Determining truthfulness and detecting deception are concerns in many interactions. What nonverbal behaviors might be especially helpful in these areas? To what extent is it easy or challenging to lie nonverbally?

4. Consider your membership in cocultures. What differences in nonverbal communication do you notice in these cultures?

5. Which types of nonverbal behavior seem most important in trying to understand others? Explain why these types seem most revealing or informative.

6. What limitations do you see in the literature examining gender as binary? How will a broader conceptualization of gender identity increase understanding?

7. How easy is it to control your nonverbal behavior overall? What areas are easiest to control? What areas are hardest?

8. Have you noticed any new forms of nonverbal behavior or past nonverbal behavior that have new meanings recently? If so, what factors contributed to these changes or developments?

9. Consider the following saying: "Actions speak louder than words." Do you agree or disagree? Why?

10. Before interviewing for employment or delivering a speech in class, you likely spend considerable time planning what you will say. How much time should you devote to planning your nonverbal communication?

11. According to Robin Marantz Henig, "Lies can be verbal or nonverbal, kindhearted or self-serving, devious or bald-faced; they can be lies of omission or lies of commission; they can be lies that undermine national security or lies that make a child feel better." Across these types of lies, how might nonverbal communication differ?

CREDITS

PART II

Interpersonal and Small Group Communication

CHAPTER 7

Interpersonal Communication Processes

It is your typical Thursday. You get up and join one of your roommates for breakfast in the kitchen. She shares information about her schedule for the day, discloses issues that she is having with a team at her part-time job, wants your advice on her mom's birthday gift, and asks whether you are interested in meeting for happy hour with her and a couple of other people. While talking with her, you respond to a text and a ping from GroupMe.

You head to your first class. On the sidewalk near the building, you see Jax. It has been awhile since the two of you talked. There have been a few texts and a short phone call, though that is far less communication than usual, and you still feel awkward about that last interaction. Still, Jax is waving and smiling as you all approach each other, and you are pleased to run into her. After a quick hug, you promise to get together soon.

You grab your usual seat in the classroom, and you and Adele start talking as usual. When she gets a text and picks up her phone to respond, that reminds her to show you something on Pinterest. While the two of you are catching up on weekend plans, she is checking social media feeds. Jules pulls up a chair and tries to persuade you both to join the evening study group. You recall how much you like gaming with Jules but that you did not find the study group productive.

Once the professor arrives, class begins. You are learning about child development, and the professor shares a few stories about his children's relationships with their grandparents. The stories are entertaining and also meaningful in making his points as well as helping you recall information. His stories also make you think about your relationship with your grandparents, and you vow to reach out to each of them before the weekend is over. You really want to tell your grandmother again how much you appreciated the "talking" birthday card that she sent and her lovely message.

Next, you are off to your part-time job at the law firm. It has been a good place to learn a bit about the legal system, but even after two years, you are not really close with anyone there. Why? The answer is not exactly clear. Sometimes, you think about the types of conflict that you see across the cases and wonder whether emotion blocked people's skills in resolving problems. So many divorce cases cite irreconcilable differences. How do partners with enough love to marry move to a space where they cannot see paths to improve or solve conflicts? You are reminded though that you have also seen neighbors file cases over where debris is placed or barking dogs. Sometimes these disputes could be addressed through out-of-court mediation, and you have admired the mediators' communication skills.

On your break, you scan website headlines and skim articles of interest. The constant barrage of which celebrities might be dating, which are dating, which might be breaking up, and which might be getting back together as well as who might be mad at whom is both frustrating and interesting. Why does our culture seem to care about the relationships of people we do not really know? At a minimum, it conveys interest in interpersonal interaction! You wonder, though, whether the time would be better spent trying to develop one's own relationships rather than gaining insights into celebrity ones. You also wonder how factors such as fame, paparazzi, and life "under a microscope" would shape relationships.

Ducky texts you a picture, but no other information. So you are left to make sense of it and why he might have sent it to you. You will ask for details at happy hour if you go and he is there, but in the interim, it is interesting to analyze possibilities.

After work, you decide to skip the happy hour gathering and head home to study. Hunter wants to talk with you about a problem, so you spend your time listening and giving advice until the two of you head to dinner. At the cafeteria, you and Hunter join Lezza and Jon; all of you tell stories from the pandemic. Jon's high school had a Minecraft graduation, and you learn more about how he built facets to resemble the usual graduation venue, the school's football field. Lezza talks about her virtual reality course this term and the ways that VR could be used if graduations had to be done remotely again. You describe a news story about a university that used robots with students' faces displayed on their screens for graduation recognition. Beyond graduations, you all give examples of other ways people adapted during physical distancing to try to preserve facets of usual events. For example, there were creative COVID-19 weddings, online visits with relatives in assisted living or the hospital, and birthday car parades with signs and streamers. In each of these cases, it is clear that people improvised to try to preserve togetherness, recognize accomplishments, and maintain relationships.

You think that you should call Mario, but should that feel less like pressure? When you first started dating, you would talk for hours on end. You still want to see him and like being a couple, but sometimes you would also like space. The weekend arrives soon, and you are torn between planning something with him, doing something with friends, or just being on your own for a bit. Is this feeling a bad sign for your relationship or a normal one?

The day comes to a close. You check social media again, and you fall asleep while texting with your mom.

Looking back at this hypothetical day, what observations do you have about interpersonal communication and relationships? How similar are facets of the above to your typical day? What interpersonal examples would you add? In this chapter, you will explore interpersonal interactions in more detail.

Interpersonal Communication in a Pervasive Communication Environment

Interpersonal communication is ubiquitous—as you observed in the opening example, you likely are surrounded by this type of communication and participate in it frequently (Figure 7.1). Whether interacting with a friend or small group of friends, talking with a family member, exchanging Facebook messages with your grandfather, or texting a coworker, you are communicating interpersonally. The pervasive communication environment has increased the number of tools that you have at your disposal to connect interpersonally with others. Of course, sometimes you employ multiple tools simultaneously, nearly simultaneously, or in close sequence. For example, you might be at dinner with your roommate, answer a phone call from someone you are dating, and respond to a text from your mom.

During the COVID-19 pandemic, many people sought even greater interpersonal communication with others, developed new ways of connecting, and creatively used earlier communication methods. When presented with the need for social distancing, the need for social connection was foregrounded—people may be able to maintain a physical distance of 6 feet, but they face challenges with "social" distance. Considerable

FIGURE 7.1 Based on the communication cues in the photo, how would you describe this event and the people's relationships?

concern exists about the ongoing mental health effects of social isolation[1] that some experienced during the pandemic (e.g., anxiety and depression).[2]

In this chapter and the following one, you will examine interpersonal communication and interpersonal relationships. In this chapter, you will become familiar with the ways that such relationships develop and evolve as well as several factors that influence their success and outcomes. For example, how did you and Sally use communication to move from being acquaintances to becoming close friends? When you ended your romantic relationship with Roger, he said that he had knew you were about to take this action. What communication behaviors led him to that knowledge? Also, how have competing wants and conflict shaped your relationships? After you consider these areas in this chapter, the following chapter will examine common contexts for interpersonal relationships (e.g., family, friends, romantic, workplace) and the ways in which these contexts influence your relationships. In other words, you will become familiar with interpersonal processes in this chapter and then examine relational contexts, such as friendships, in the next chapter.

FIGURE 7.2 When you hear the word "relationship," what do you envision?

But before discussing interpersonal communication and relationships in more detail, let's look at what these terms mean (Figure 7.2). **Interpersonal communication** involves message exchanges (message production and interpretation) between two people or a small group of people who are interconnected with each other. In other words, their connections involve some degree of interdependency (e.g., family, coworkers in same department, etc.).[3] For the communication to be considered interpersonal, there must be **mutual awareness**. If someone sets up hidden cameras to monitor your behavior, the video feed is not interpersonal in nature, because you are not aware that you are being monitored and thus are not tailoring your behavior or message for a particular person or audience. Conversely, if your roommate downloads new software on your computer so that the two of you can have online conversations while you are both on winter break, those online feeds are interpersonal communication—you are both aware of the other and can adapt your communication to each other. Interpersonal communication involves both verbal and nonverbal messages, may take place online or in-person, and is shaped by cultural norms. People who have well-developed interpersonal communication skills are likely to experience several benefits, including better mental health, higher relational satisfaction, and greater job and career mobility.[4]

Interpersonal relationships exist across an array of communicative contexts, such as family, friendship, romantic, and education/work. They can vary in duration as well as level of self-disclosure and intimacy (e.g., professor–student compared to parent–child). They are characterized by a common connection (i.e., mutual goal),

interpersonal communication
Message exchanges (message production and interpretation) between two people or a small group of people who are interdependent and mutually aware of each other; can be online or face-to-face and involve verbal and nonverbal communication

mutual awareness All interactants are aware of the other and can adapt their communication accordingly

interpersonal relationships Exist across contexts (e.g., family, friendship, etc.); can vary in duration as well as level of self-disclosure and intimacy; are characterized by a common connection, interpersonal communication, reciprocal benefits, and adherence to or renegotiation of cultural expectations which can be formal or informal (e.g., legally married spouses, friends from a book club, etc.); as ongoing (e.g., continuing indefinitely or perhaps shaped by defined boundaries); by reliance on interpersonal communication; as having reciprocal benefits; and by adherence to or renegotiation of cultural expectations (e.g., what friends can and cannot expect from each other).[5]

Having a better understanding of interpersonal communication and relationships can be useful in your interactions and relationships across the life span from your family to your career. In short, such understanding is foundational to success. To explore these areas, you will first explore how such relationships are developed and shaped across several possible phases.

Developing Relationships

As you think across all of your interpersonal relationships, you may remember how some began, and for others, you may not—they may always have been in place or been in place for so long that you do not recall their beginnings. For example, you may remember meeting your friends in college and how you began spending time together, but not recall exactly when you became closer with a cousin you have known since birth. In this section, we will examine how relationships develop, paying particular attention to how we begin, maintain, change, and end interpersonal relationships and the collaboration involved throughout relationships.

It is important to keep in mind that relationships are dynamic, which means that they continually undergo change. They can grow at different rates, be close and then feel less intimate, and improve or erode. Transitions in relationships are often described as stages, with several models describing these stages. Although some models propose different numbers of stages, Mark Knapp developed an influential one that has been used for many years to describe how we build or escalate relationships and how we de-escalate or terminate them. Table 7.1 presents his 10-stage model, where the first five stages describe escalation and the last five stages describe de-escalation.[6] As you review the stages, keep in mind that progression varies, movement is not linear, and relationships cycle back and forth.

The first stage is initiating. As its name conveys, this stage is characterized by an interest in connecting with one another. In the initiating stage, you get to know another person at the surface level. The second stage, experimenting, is where you begin learning a bit more through small talk. Sometimes the importance of small talk is downplayed, and it is regarded as simply "chitchat." However, although small talk may not appear to accomplish key goals or seem particularly important, it involves communication skill and influences relational development. As many have noted, "Small talk is big talk" because it refines the perceptions that others develop and shapes their decisions about whether they want to stay in this stage or move forward with a deeper relationship. Many relationships stay in the experimenting stage for months or years and never escalate to a deeper level. The third state is intensifying. This stage signifies that an interpersonal relationship has been established. It is characterized by factors such as private jokes and "we" language (i.e., focusing on the relational identity rather than separate, individual identities). In integrating, the fourth stage, the relational partners are clearly defined as a unit, such as best friends or a couple. They are interdependent, have routines and rituals, and have

TABLE 7.1 Knapp's Interpersonal Relationship Stages

	STAGE	FOCUS	EXAMPLE COMMUNICATION
COMING TOGETHER OR "ESCALATING" STAGES	**1. Initiating**	Interest in meeting; connecting	"It's such a pleasure to meet you. Are you in school here?"
	2. Experimenting	Getting to know each other; small talk	"I'm studying communication and am really interested in human resources. What are you studying?"
	3. Intensifying	Interpersonal relationship has begun; use of "we" language	"Why don't we see a movie tonight? We always enjoy comedies. How about seeing the new one?"
	4. Integrating	Are recognized as "a unit" (e.g., a couple, best friends)	"Let's get an apartment together."
	5. Bonding	Public gestures reinforce relationship status	"I'm planning your birthday party. It's going to be the biggest and best ever!"
COMING APART OR "DE-ESCALATING" STAGES	**6. Differentiating**	Seeking space and separateness while maintaining relationship; using "we" language, but reclaiming "I" language	"I need to do something just for myself. I'm going on a vacation with my high school friends."
	7. Circumscribing	Communication wanes; less demonstration of commitment and interest in the other and the relationship	"It's not open for discussion. I've made my decision."
	8. Stagnating	May keep up usual behaviors but with less feeling; relationship feels "hollow"	"We always go to the birthday brunch, so there's no reason not to this time. I don't know that it'll be fun, but we'll be there."
	9. Avoiding	Creating distance; lessening contact	"I've taken a new position with more travel."
	10. Terminating	Declaring or indicating that the relationship has ended	"Here's your key and the tools that you loaned me."

Adapted from Knapp & Vangelisti. (2009). *Interpersonal communication and human relationships.* Pearson.

merged their other networks (i.e., have a shared social circle). Bonding is the fifth stage and the final one of relational escalation. This stage typically confirms the defined relationship in a public fashion. For example, an engagement ring would symbolize this public confirmation (Figure 7.3).

The sixth stage, and the first one that involves de-escalation, is differentiating. In this stage, you work to reclaim some of your independence. While maintaining the relationship, you also want to carve out space for yourself as an individual. Use of "I" language may increase, and more discussion of individual differences may take place. Circumscribing is the seventh stage. This stage typically involves less frequent and less developed communication, and relational

FIGURE 7.3 Weddings signal that relationships are bonded. What are other examples of this relational stage?

FIGURE 7.4 What might the space and posture convey about relational stage?

social exchange theory
Examines relationship rewards and costs to determine likelihood of continuing relationship and satisfaction

FIGURE 7.5 How does one weigh relationship benefits and costs?

partners may convey less interest in each other. In this stage, disagreements may be more challenging to resolve and the lack of resolution may result in partners withdrawing from each other. Unless these behaviors change, they can spur the beginning of the termination process. The eighth stage is stagnating. Just as the name conveys, this stage is characterized by feelings of "autopilot," rather than engagement. Relational partners may keep repeating the same routines but have limited or no excitement or interest in the routines. Once in this stage, the relationship feels "hollow" and without positive momentum. Although a brief amount of time may be spent in this stage in some cases, it is possible that a relationship may remain stagnated for months or years. If the relationship continues to de-escalate, its next stage is avoiding, which is stage nine. In this stage, one or both partners use barriers to create distance. Distance can be physical, emotional, or both. The tenth stage is terminating. This stage lays to rest the current relationship, may embed reasons for the termination and/or possibilities for a future type of relationship (e.g., becoming friends after breaking up), and usually makes clear that the previous relationship is ending. Whether using Knapp's 10-stage model or one of the shorter models, examining usual relational patterns can help you better understand the ways in which we work to build and dismantle interpersonal relationships (Figure 7.4). In part, your decisions to pursue or leave a relationship may be based on whether or not your goals are being met—or your assessment of the rewards and costs associated with the relationship.

Beyond our basic needs for affiliation with others, we seek to achieve goals through our relationships. **Social exchange theory** provides one way to assess our interpretation of goal achievement by looking at relational costs and rewards.[7] Although rewards and costs vary across time, social exchange theory examines our assessments of the benefits we gain compared to the costs of staying in the relationship (Figure 7.5). Costs are considerations such as the time and energy that you put into your relationships. Rewards are benefits that you receive from relationships, such a help with tasks, social support, and companionship. Although a number of factors influence decision making, if you believe that the rewards you receive in a relationship exceed the costs, then you are likely to want to continue that relationship. Alternately, if you judge the costs of the relationship as exceeding its rewards, you may choose to leave or redefine the relationship. Of course, relational type and context influence such choices too. Social exchange theory focuses less in these areas, but certainly if the relationship is with your parent or sibling or in an organization where you want to continue working, then your assessment may include factors beyond the relationships itself. As you read the information below on beginning, maintaining, changing, and ending relationships, consider whether the ideas from social exchange theory might influence your choices and behavior (Box 7.1).

BOX 7.1: ETHICAL QUESTIONS

Weighing Relationship Benefits and Costs

Consider the ideas from social exchange theory on relationship benefits and costs. How does one weigh each and assess balance? From this perspective, when benefits outweigh costs, the scale tips toward staying in the relationship. When costs outweigh benefits, the scale tips toward leaving the relationship.

What other factors influence choices to remain or leave a relationship? For example, extensions of this theory suggest that relational expectations and perceived availabilities also shape decisions. If you are in a relationship where you believe the benefits outweigh the costs and the benefits exceed what you would expect or believe that you would find in another relationship of this type, then you are likely to be satisfied and stay in the relationship. Alternately, if the relationship costs are greater than the benefits and you believe that you can find greater rewards in other available relationships, then you are likely to leave the relationship and be happy to pursue other options. If the relationship costs are greater than the benefits but you do not believe that you have other options for a relationship with greater benefits, then you may stay in the relationship, but you are likely to experience dissatisfaction.

In what ways do you think that social exchange ideas are useful? What are the limitations of these ideas, and what considerations are not covered by this thinking? To what extent is employing an economic model to explain interpersonal relationships a good approach? Is using such a model fair? What ethical considerations guide your thinking?

With this backdrop of relational stages, let's now examine four key periods in the course of relationships. Across your interactions with others, you engage in collaborative communication to begin relationships and to maintain them. In some cases, you wish to change or end relationships. As you read the following sections, think about examples of past or current relationships.

Beginning Relationships

Have you ever thought about how you go from meeting someone to having a relationship with them? What processes shape relationship formation? Considerable research has examined such communication processes, and one way of explaining relational development is filtering. From this perspective, you use a series of filters that people need to pass through to move toward a relationship or greater intimacy. Of course, you will need to pass through others' filters too. Through communication, especially reciprocal self-disclosure and impression formation, you determine your interest in pursuing particular relationships.

According to the **relational filtering model**, developing relationships involves assessments and selections on the part of all participants. In general, there are four levels of filters, with each narrowing the set of potential relational partners.[8] These four levels are appearance, communication, roles, and attitudes/values. You use these filters to progressively assess compatibility and make decisions about relationship status.

relational filtering model Developing relationships involves assessments and selections; four filters through which potential partners must pass

appearance filter Your interpretations of how a person "appears" to you; largely physical appearance

communication filter Using others' communication to try to assess what they think and feel; their internal thinking

roles filter Attention to the behaviors the person uses in performing the formal and informal roles assigned to them or that they choose to fulfill; expand your impressions

attitudes/values filter Concerned with the psychology underlying the person's behavior, their worldview, and shared psychological similarities the two of you have

The first filter, **appearance**, involves your interpretations of how a person "appears" to you. Generally, this filter is concerned with physical appearance. For example, what are characteristics that you might observe when meeting new people? Factors such as their gender presentation, weight, clothing, physical attractiveness, body piercings, tattoos, age, and race might shape your first impressions. Based on these and similar perceptions, you may be interested in talking with the person or less inclined to talk with the person. Certainly, first impressions have limitations, but they also influence people's behavior. What first impressions would make you more likely to talk with someone?

The second filter, **communication**, includes both verbal and nonverbal communication. If a person clears your appearance filter, then you begin employing the communication filter. With this filter, you use their communication to try to assess what they think and feel, their internal thinking. In other words, if someone sends supportive texts to their friends before exams, you may infer that the person is thoughtful and kind. If the person plays games with children at the park, you may believe that they enjoy children and are interested in becoming a parent someday. As you form these impressions, you are likely to assess them in terms of your relational goals and likely outcomes (i.e., the expectations about the rewards of a relationship).

If the relationship continues to escalate, the third filter, **roles**, comes into play. When applying this filter, you pay attention to the behaviors the person uses in performing the roles assigned to them or that they choose to fulfill. These roles can be formal or informal. Examples of formal roles include heading a student organization and managing a retail store. Examples of informal roles include taking initiative to support a group and to mediate conflicts on a team project. As a person engages in formal and informal roles, you develop additional impressions of what the person is like. For example, have you considered how someone you are developing a friendship with treats their other friends? Is the person willing to spend time listening and providing good insights, or are they more focused on their own concerns? Does the person keep their commitments to friends or prioritize their own interests? Your answers to these questions might provide clues to how the person will treat you. Additionally, have you considered how that person behaves as a leader? If the person is assigned to lead an initiative on campus but does not welcome input and feedback from other students, what impressions would you have of their leadership style and confidence level? How might such behaviors play out in friendship situations? In short, as you observe people fulfill formal and informal roles, you gain additional insights and deepen your knowledge base, expanding or revising earlier impressions.

If the person passes through the roles filter, then the fourth filter, **attitudes/values**, becomes the focus. This filter is concerned with the psychology underlying the person's behavior and with shared psychological similarities the two of you have. In applying each filter, the focus has been on trying to understand the other person, assess similarities, and determine compatibility, all of which are helpful in predicting whether you are likely to achieve the relational benefits that you desire. This fourth filter centers in trying to understand the other person's worldview or life philosophies. When another's views are similar to your own, the way that you see the world is reinforced. You may believe that this type of similarity would result in making similar decisions or approaching problems in similar ways—applying the same values to address them.

Forming impressions of others occurs quickly, and typically, one does not focus on the filters used to understand others. However, through such processes, you develop impressions of others, and those impressions guide your communication with them. Such impressions also lead you forward in pursing some relationships and foregoing other opportunities (Figure 7.6).

Maintaining Relationships

Once you have developed a relationship, you need to use communication to maintain it at its current level or it will begin to de-escalate. Maintaining a relationship falls somewhere between escalation and de-escalation and, because relationships are not static, involves times where you may feel closer or less close to the other person. Further, maintaining relationships involves spending time together and availability to the relational partner (Figure 7.7).[9] Several communication strategies can be involved to maintain relationships. One typology suggests that five key behaviors contribute to maintaining relationships.[10] These behaviors are positivity, openness, assurances, social networks, and sharing tasks.

Positivity centers in choosing to be positive and refraining from being negative. In particular, you would behave in upbeat and fun ways. Additionally, you would focus on constructive, rather than critical, messages. Openness includes sharing your views on the relationship's status, health, and future. You can disclose your hopes and information on how you envision the relationship in the future. Related to positivity and openness, assurances provide support for the relationship and the partner. This form of communication emphasizes your commitment. Social networks include engaging in activities with others that reinforce your relationship (e.g., indicate that you are a couple). For example, double dating or attending family gatherings together publicly illustrates and reinforces the status of your relationship. Additionally, continuing to merge your social networks (e.g., having the same friends) also serves a relational maintenance function. Sharing tasks also serves to maintain relationships. For example, if you live with your best friend and neither of you likes

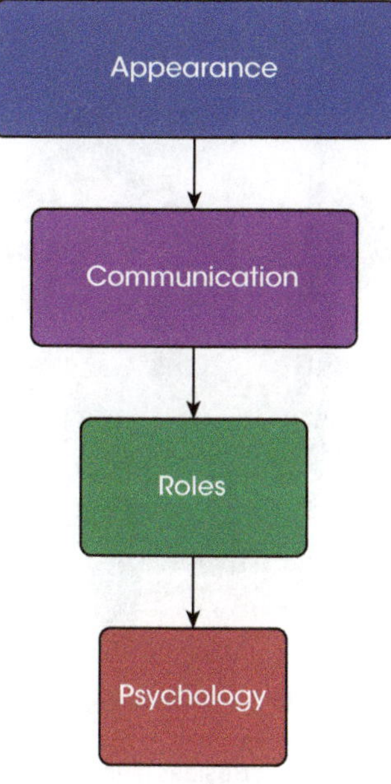

FIGURE 7.6 Using relationship filters, your set of potential partners decreases at each step.

FIGURE 7.7 How would you describe this relationship?

FIGURE 7.8 Besides laundry, what other joint tasks might equalize relationships and promote maintenance?

to clean the kitchen, determining a plan to share the responsibility equitably can act as a relational maintenance strategy (Figure 7.8).

Communication behaviors used to maintain relationships shape factors such as relationship satisfaction, commitment, and stability.[11] Willingness to employ relational maintenance strategies is higher when partners believe that relational rewards are equal; in other words, when you perceive that there is fairness, you are more likely to constructively engage in maintenance strategies.[12]

Of course, it is also possible that rather than maintain your relationship at about the same level, you will wish to change it—to escalate or de-escalate it. Next, we will examine the ways in which communication can accomplish relational change.

Changing Relationships

After maintaining a relationship at a particular level, you may assess its benefits and your feelings and decide to make a change. When changing a relationship, you are either escalating it to a deeper, more personal level, or de-escalating it to a less personal level. For example, have you ever transitioned an existing friendship to a closer friendship or started to date a close friend? In each of these cases, you would have engaged in communication to change the relationship.

Just as you begin and maintain relationships through communication, you also change them through your communication. If you wanted to escalate a relationship, how might you do that? What if you wanted to de-escalate a relationship (Box 7.2)? For example, how might your nonverbal communication change? What topics might you bring up or avoid? Change is negotiated through your verbal and nonverbal messages, which may align with the level where you would like the relationship to move.[13] Given that communication is a cooperative endeavor, your relational partner can embrace or resist these changes, which will influence the ease of the transition or the uncertainty and/or conflict surrounding it.

Ending Relationships

Beyond the earlier discussion of changing relationships, one type of relational change is to end a relationship. In most cases, relationships fade away gradually as communication patterns change. At other times, they may end abruptly with

BOX 7.2: CONNECT & REFLECT

A Change Is Coming

Consider the following scenario. You have been dating Adrian for about 3 months. The relationship has been fun, and you like Adrian, but it is clear to you that this relationship does not have long-term potential. Still, you and Adrian get along well, have a lot in common, and travel in the same social circles. You wonder how to keep Adrian as a friend but end the romantic part of your relationship.

Considering the above information, what communication strategies would you suggest? What approaches might be most helpful in preserving the likelihood of a friendship? What approaches might you want to avoid? What observations from your own or your friends' lives do you have about such relational transitions? Why can some former lovers be friends and others face difficulties with such a transition?

sudden death.[14] In relationships that gradually fade away, relational partners may become less interested in each other and what their relationship has to offer and thus spend less time together. They may also engage in fewer relational maintenance behaviors when they are together. Over time, then, they simply seem to drift farther and farther apart. In sudden death endings, an event typically transpires that violates the boundaries of acceptable behavior in the relationship and destroys established trust. Examples of such behaviors could be partner abuse, romantic involvement with another person, or fraud.

Because ending most relationships takes place over time, researchers have tried to understand the processes involved. One of the best-known explanations for ending relationships is the **relational dissolution model**.[15] This model contains five phases: intrapsychic, dyadic, social, grave-dressing, and resurrection. Triggered by dissatisfaction, the **intrapsychic phase** is an individual one where you consider the costs and benefits of the relationship and try to determine whether you want to remain in the relationship or leave it. During this phase, your partner is not aware that you are contemplating these concerns but may notice that you have pulled back from the relationship. Ultimately, you figure out how to repair the relationship or you move to the next phase of dissolution.

In the **dyadic phase**, you make your partner aware of your relational dissatisfaction and assessment of the relationship. This phase may be characterized by uncertainty, airing of complaints, and honest discussions.

If dissatisfaction is not resolved in the dyadic phase, then the partners may move to the next phase, the **social phase**. During this phase, people outside the relationship learn of the concerns and that the relationship may end—the concerns go from private to public. Because there are shared social networks, this phase can result in surprise and/or conflict as well as provide social support and advice. This sharing outside the relationship can give others time to process the likely breakup, setting the stage for future action.

In the **grave-dressing phase**, the partners develop explanations for the relationship's end (Figure 7.9). These explanations may be developed separately,

relational dissolution model Contains five phases—intrapsychic, dyadic, social, grave-dressing, and resurrection

intrapsychic phase of dissolution Individual phase where you consider the costs and benefits of the relationship and try to determine whether you want to remain in the relationship or leave it

dyadic phase of dissolution Make your partner aware of your relational dissatisfaction and assessment of the relationship

social phase of dissolution People outside the relationship learn of the concerns and that the relationship may end; concerns go from private to public

grave-dressing phase of dissolution Develop explanations for the relationship's end

FIGURE 7.9 What are common stories to explain why relationships ended? How do these stories "dress the grave" of the relationship? What functions do they serve?

resurrection phase of dissolution Takes place once healing has occurred; reflect on what they have learned

but if they are jointly agreed upon, more positive feelings may result. Through crafting particular explanations, partners can deflect blame and illustrate their viability in future relationships. Like at a gravesite, they can pay their respects to a past life and move to a new one. How might approaches to relational dissolution, such as breaking up by text (Box 7.3), influence grave-dressing?

The **resurrection phase** is the final one. This phase takes place once healing has occurred. At this point, past partners can reflect on what they have learned about themselves and relationships as well as figure out how to benefit from this information in future relationships.

Some studies have examined ending relationships in specific contexts. For example, considerable work explores dissatisfaction and potential dissolution of marriages. One approach describes the "four horsemen" that ride in before relationships are ended.[16] The names of the four horsemen are criticism, contempt, defensiveness, and stonewalling. Each infuses negative communication into the relationship, likely resulting in increased relational dissatisfaction and increased likelihood of relational dissolution.

Criticism differs from complaints and thus is likely interpreted differently. Complaints focus concern on a behavior that is perceived as problematic (e.g., staying out too late or not being on time) and open the possibility for behavior change. In contrast, criticism focuses concern on the person—that person's character or personality (e.g., declaring that your partner is unkind, unfair, or selfish). When a partner's essence is criticized, negative responses are likely.

When criticism becomes an intentional action to offend or insult another, the action is labeled "contempt." Behaviors that display contempt range from rolling your eyes to humor or disparaging remarks that are designed to humiliate your relational partner. This type of behavior undermines respect and escalates relational dissatisfaction.

Partners may engage in criticism and contempt toward their partner and simultaneously engage in defensiveness in response to their partner. Behaving in a defensive manner can take many different forms or involve combinations of behavior. For example, you might deny any responsibility, meet any complaint with another complaint, or refuse to acknowledge the other person's viewpoints.

The term "stonewalling" describes the action of blocking communication by turning into a "stone wall." As a stone wall, one would not engage with their partner or would minimize any responses. An example is not listening or paying attention to what your partner says or does.

From these brief descriptions, you can probably see why the four horsemen would likely be associated with relational apocalypse. Although this work examined marriages, such behaviors can be enacted in other relationships as well. Each type of horseman behavior has negative relational outcomes and, when used in progression, greatly increase the costs of staying in a relationship and the likelihood of ending the relationship. In the next sections, you will explore tensions and conflicts in relationships and ways of dealing with these challenges.

Dialectics in Interpersonal Relationships

Have you ever found yourself wanting to talk with a friend about a concern but also not wanted to share that information? Have you ever been in a relationship where you cared deeply about the other person and wanted to spend time together but also felt a bit smothered and wanted to spend time apart? If you answered yes to either or both of these questions or can think of a similar situation, then you have experience with **relational dialectics**. These dialectics are competing wants or needs that exist simultaneously in relationships and pull us in multiple directions (Figure 7.10).[17]

Although we may tend to see relationships as stable, they are constantly evolving and changing. Phrases like "and they lived happily ever after" do not capture the full reality and complexity of even the happiest relationships (and, of course, the level of happiness varies across time). Even in happy, stable relationships, dialectical tensions exist. A dialectical tension is not an indicator that something is wrong or amiss in a relationship but, rather, is an indicator that we have multiple goals and limited time. It is a pull in multiple directions. Dialectical tensions embody the idea that we do not make either/or choices (e.g., you either want to be on your own or in a relationship) and that we are interested in both/and arrangements (e.g., you want to be in a committed relationship but also have time for yourself). Although dialectical tensions occur across relational types, they are especially frequent in close relationships, such as romantic ones.

Some dialectical tensions are external to the relationship, such as wanting to spend holiday time together as a couple rather than with family members and also wanting to meet family expectations and uphold traditions. Others are

relational dialectics
Competing wants or needs that exist simultaneously in relationships, pull us in multiple directions, and play out through discursive struggles

FIGURE 7.10 How does the Chinese yin and yang symbol represent ideas similar to relational dialectics? What hope might it give to people experiencing such relational tensions?

autonomy versus connection dialectic
Competing desires to be both independent and interdependent

openness versus closedness dialectic Tensions surrounding how much information to share in a relationship

predictability versus novelty dialectic Wants for both stability and excitement

internal to the relationship, and we will focus our attention on common examples of internal dialectical tensions—autonomy versus connection, openness versus closedness, and predictability versus novelty.[18]

The **autonomy versus connection dialectic** conveys competing desires to be both independent and interdependent. For example, you may desire to grow as an individual and perceive a need for space to do that growth. And you may simultaneously want to be an integral part of a unit (e.g., family, marriage, etc.) and also grow through that experience. Certainly, your experience with this dialectic may vary across time and be influenced by relational stage. As you are developing a new romantic relationship, you may want to spend considerable time with the other person and not view your independence as in question. But later, once the relationship has been established and is comfortable, you may long for the days when you had greater independence. In short, this dialectic conveys the contradictions embedded in desiring connection with others and carving out space for yourself.

The **openness versus closedness dialectic** expresses tensions surrounding how much information to share in a relationship. You develop closeness and deepen relationships by sharing information about yourself and learning about the other person, but self-disclosure also involves some risks and can make you feel vulnerable. You may also have different privacy needs than others, and such differences can lead to misunderstanding in relationships. For example, your partner may be comfortable sharing details about past relationships, but you may prefer not to divulge much information in this area. When confronted with this dialectical tension, you will need to decide what to share and what to keep to yourself.

The **predictability versus novelty dialectic** indicates wants for both stability and excitement. In relationships, you appreciate consistency. In part, the filtering process allows assessment of whether a partner is consistent in behavior across time, which can increase confidence and trust. As you are escalating relationships and getting to know people, you are experiencing new, novel information. But once the relationship has been established, relational maintenance is underway, and behaviors are more predictable, you may miss the newness and novelty and long for more excitement. These tensions between wanting both consistency (sameness) and excitement (newness) comprise the predictability versus novelty dialectic.

Across relationships, dialectical tensions are normal, and experiencing them is not a sign that you are having relational problems. However, it is important to keep in mind that you and your relational partners may experience these dialectics at different times, may experience different dialectics, and may view the opposing tensions differently. The ways in which you handle such dialectical tensions and negotiate them in relationships can influence relational outcomes. Our points of view, positions on issues, and overarching worldviews as well as culture influence our communication, creating discursive struggles where relational dialectics are in play.[19] Through these discursive struggles, meaning is created, and these

TABLE 7.2 Overview of Relational Dialectics

RELATIONAL DIALECTIC	BRIEF DEFINITION	COMMUNICATION EXAMPLES
Autonomy vs. connection	Desires to be both independent and interdependent	"Sometimes I just want to do my own thing."
Openness vs. closedness	Decisions to share information or keep it private	"You always ask me these questions, and I don't see why you need to know."
Predictability vs. novelty	Interests in both consistency and excitement	"That's our usual place. Why don't we ever do anything new and fun?"

understandings shape subsequent behavior and relational decisions (Table 7.2). Because these are common, expected feelings and frequent discursive exchanges, you should devote attention to better understanding them and managing them throughout the course of your relationships.[20]

Managing Dialectics

The ways in which you manage the dialectical tensions that you experience as well as the ones your relational partners experience will shape your relationships in important ways, including the levels of conflict, closeness, and satisfaction. Beyond the reflection and balance needed to address competing needs and resolve discursive struggles, keep in mind that you and your partner may experience these tensions at different times and that tensions may emerge or be heightened as one partner attempts to wrestle with dialectical tensions. For example, if your partner is experiencing competing wants for autonomy and connection, they may pull back a bit from the relationship or verbalize their interest in spending time with others. As they engage in these behaviors, your interests in and needs for connection may be heightened; thus, you may enact behaviors to encourage greater connection. Without an awareness of these competing demands, you may create additional relational tension or provoke conflicts.

Certainly, one key factor in managing the discursive struggles surrounding relational dialectics is the willingness to discuss them with relational partners. Through such communication, you and your partner may come to better understand each other and develop mutually agreeable ways to handle the tensions. Ignoring or deprioritizing dialectical tensions can lead to misunderstanding and conflict. For example, strategies such as denying the concerns or devoting attention to only one facet do not deal robustly with these tensions and may damage relationships. Consider the information in Box 7.4 in terms of how the pandemic may have affected dialectics and relationships.

Conflict and Deception in Interpersonal Communication

As you manage your interpersonal relationships, you will inevitably experience conflicts and wonder about the potential for deception. Each of these areas has received considerable

BOX 7.4: CONNECT & REFLECT

COVID-19 Hibernation, Pods, and Microgroups

With the physical distancing required to contain the spread of COVID-19, people adapted in different ways. Some essentially hibernated, interacting solely or mainly with only their immediate families or households. Some set up pods that still restricted in-person contact but allowed interaction beyond those living in one's home. Others created microgroups, expanding the boundaries a bit wider but still imposing limits on interaction.

Considering relational dialectics, what are the likely outcomes of these communication practices? What dialectics are most likely to have shifted? What do you expect the effects on relationships to be? How do you imagine these will change after the pandemic or in the face of a similar health challenge?

---

conflict management styles Discussed in terms of how much concern is shown for two types of interests—your own and your relational partner's

research attention, and you may have opportunities to take an upper-level course focusing on one or both or participate in trainings in these areas.

Just as dialectics and discursive struggles are a normal part of interpersonal relationships, so is conflict. What shapes relational satisfaction is not the absence of conflict but how you deal with conflict. Below, different **conflict management styles** are examined to help you learn more about your approach to conflict as well as the approaches that others may use. Understanding differences in these approaches can be useful in helping you learn through conflict, more effectively achieve your goals, and maintain relationships. The five conflict management styles are competing, accommodating, avoiding, compromising, and collaborating.[21] These styles are often discussed in terms of how much concern is shown for two types of interests—your own and your relational partner's.

competing style Defines conflict as a competition and seeks to gain—goal is to win; shows high concern for your interests and low regard for your partner's interests

The **competing style** defines conflict as a competition and seeks to gain—the goal is to win. In other words, you compete with your relational partner to maximize your self-interests, which shows high concern for your interests and low regard for your partner's interests. Suppose that you and you partner are planning a vacation together. You prefer an expensive resort in an exotic location, but your partner prefers backpacking and cooking freeze-dried food over campfires. You have been discussing these two options for days but cannot come to a resolution. Given that your income is higher and you have been working a lot to earn extra money, you believe that you should "win" this battle. Thus, you insist that you two take the vacation to the resort. If you employ this style repeatedly, your partner may grow resentful of the lack of accommodation to their needs.

accommodating style Acquiesces to let the other person win; shows little concern for your own interests but high concern for your partner's interests

Opposite the competing style is the **accommodating style**, which acquiesces to let the other person win. When using the accommodating style, you show little concern for your own interests but high concern for your partner's interests. In the vacation example above, once the different opinions on vacation choices became a conflict, you would give in and let your partner's vacation preference be the destination. Although using this style sometimes may be okay, it may create bitterness in the long-run, as your interests are not fulfilled.

Between the two approaches above is the **compromising style**, which shows some degree of concern for both your interests and your partner's interests. Although not fully achieving either your goal or your partner's goal, compromise results in each of you getting part of what you wanted—the goal is to ensure each of you benefits some. Returning to the vacation example, you could suggest a compromise involving 3 days of wilderness backpacking at a national park followed by 3 days at a nice hotel near the park. This compromise allows you to spend time as a couple and do activities that you each enjoy, though each for less time than originally considered. Reaching a comprise may take some time, but it also may avoid resentments created over time with competing and accommodating styles.

The **avoiding style** seeks to not have the conflict—the goal is no conflict. This style demonstrates little concern for individual interests—your own and the other's—and involves maneuvers to avert conflict situations. If you are employing this style, once there is a disagreement on the vacation, you might refuse to discuss the matter. Although some conflicts may be best avoided, frequent reliance on this approach does not allow you to learn from conflict, develop methods of working though problems, or achieve your goals, and refusing to engage in conflicts can result in dissatisfaction from your partner.

With a **collaborating style,** you work with your partner to try to maximize the outcomes for both of you. When employing this style, you show high concern for both your interests and your partner's interests—you want to find an answer where both of you win. Collaboration takes considerable time, as you work together to invent unique possibilities that are beneficial to all parties. Bringing a collaborative style to the vacation example, you might discuss available vacation time and finances with your partner to try to examine broad solutions. Through these conversations, you might discover that if you traveled when their school was not in session and your workplace was closed, you could be away for 2 weeks. Then, you could choose to combine both your preferences into a 2-week vacation. Approaching conflict collaboratively can result in unique, highly satisfying resolutions, but the process takes considerable time and energy; thus, it may not be the ideal style to employ in all conflict situations.

Ideally, you would become adept at using several conflict styles and employ the most appropriate one based on the situation. In some situations, your preferences may not be strong ones, so it is fine to accommodate to your partner's stronger preferences. Some conflicts will disappear if given time, so it may be preferable to avoid those conflicts. Others involve important issues with significant people in your life; thus, these conflicts may warrant the process of seeking collaborative solutions.

compromising style
Each participant gets part of what is wanted—goal is to ensure each benefits some; shows some degree of concern for both your interests and your partner's interests

avoiding style Seeks to not have the conflict; demonstrates little concern for individual interests (i.e., one's own or the other person's)

collaborating style Tries to maximize outcomes; shows high concern for both your interests and your partner's interests; wants both to win

FIGURE 7.11 Consider the chapter on nonverbal behavior. How would you assess nonverbals in this photo?

FIGURE 7.12 Being truthful in relations can lead to positive outcomes. Is it possible to be too truthful? Is some information best left unshared?

In conflict situations and sometimes outside them, deception is a consideration when assessing communication and relationships. A number of researchers have examined verbal and nonverbal differences that may be associated with telling the truth or deceiving others, and many people are interested in honing their deception detection skills (Figure 7.11). However, people may try to deceive you for an array of different reasons, and you might evaluate these reasons very differently (Figure 7.12).[22] For example, some lies are categorized as prosocial ones, which are designed to be helpful to others (e.g., assuring your friend that others did not overhear her conversation so that she will not worry). Others are protective ones that are designed to shelter you, either by covering up information or attempting to maintain positive evaluations (e.g., denying that you took a negative action). Some lies are antisocial ones, which function in the opposite way as prosocial lies. With antisocial lies, the goal is to hurt another person (e.g., circulating false information about a classmate).

If suspected or discovered, deception can have negative effects on interpersonal relationships. Generally, prosocial lies may not be as harshly evaluated as protective and antisocial ones. With protective ones, the reasons underlying the protective cover may influence how the lies are evaluated. For example, some lies to avoid embarrassment may be easier to understand than lies that hide wrongdoing. Antisocial lies typically result in negative sanctions if discovered. Several negative relational outcomes have been linked to deception. For example, deception is associated with lower satisfaction with and commitment to a relationship.[23] If deception discovery results in considerable negative responses, the chances of relationship dissolution are higher.[24] Conversely, honesty and genuine self-disclosure are associated with a number of positive relationship outcomes.[25] Thus, skillfully managing conflict and cultivating honest and effective communication are important communication skills (Box 7.5).

BOX 7.5: LIVE IT LOCAL

Relationships and Conflict

Based on the information in this chapter, what recommendations would you make to others about creating and sustaining healthy and successful relationships? Identify a conflict in your community, at your workplace, or in your school. The conflict could be one that you are aware of through news or social media or one related to a group in which you are a member. What actions would you suggest for addressing this conflict? Why do you believe that these actions are likely to be helpful ones? What is the likelihood of successful resolution?

CHAPTER REVIEW

■ **Define "interpersonal communication" and "interpersonal relationships."**
Interpersonal communication involves message exchanges between two people or a small group of people who are interconnected and where mutual awareness exists. Interpersonal relationships exist across an array of communicative contexts and are characterized by a common connection, ongoing nature, reliance on interpersonal communication, reciprocal benefits, and adherence to or renegotiation of cultural expectations.

■ **Explain considerations for beginning, maintaining, changing, and ending relationships in a pervasive communication environment.**
Decisions to begin, maintain, change, and end relationships are influenced by communication and impression formation. Research provides insights into the relational stages involved in escalating and de-escalating relationships as well as systems for evaluating relationships (e.g., social exchange's costs and benefits). Filters are used to determine whether to begin and continue escalating relational development. Specific communicative behavior is associated with successfully maintaining relationships (e.g., openness, assurances, etc.).

■ **Describe dialectics affecting interpersonal relationships.**
Relational dialectics are competing wants or needs that exist simultaneously in relationships, pull us in multiple directions, and play out through discursive struggles. Common dialectical tensions are autonomy versus connection, openness versus closedness, and predictability versus novelty.

■ **Discuss conflict and deception in interpersonal communication.**
Conflict and deception can lead to negative relational outcomes. In relationships, conflict is inevitable, and conflict styles can shape how behavior is evaluated. Conflict styles include competing, accommodating, avoiding, compromising, and collaborating. If deception is discovered, relationships can be severely affected, though consideration may be given to the type of lies involved, such as prosocial, protective, and antisocial ones.

DISCUSS AND APPLY

1. How have increasing options to communicate affected relational development? Have newer media facilitated relational development and satisfaction, impeded it, or had no effect?

2. Interpersonal communication does not need to take place in-person. What, then, makes it interpersonal?

3. Why is grave-dressing important in relational dissolution?

4. What are the limitations of the social exchange model? What types of relationships are less influenced by your assessment of costs and benefits?

5. Which conflict style do you tend to employ? What are the strengths and weaknesses of this style?

6. In your opinion, which relational dialectic is likely to influence the largest number of relationships? Why do you believe that this one is so influential?

7. How can conflict be positive and contribute to relational growth?

CREDITS

CHAPTER 8

Interpersonal Relationships in Context

J. R. R. Tolkien's *The Lord of the Rings* has it all—a noble quest, interesting lands, a powerful antagonist, and a spellbinding good-versus-evil plot. It also has a number of interesting relationships (Figure 8.1).

The work features family relationships, including their tensions and complexities. Consider the relationships between Éomer and Éowyn, Bilbo and Frodo, and Elrond and Arwen. How would you describe each of these family relationships, and what can be learned from them?

Next, think about the friendships, such as those between Gandalf and Bilbo, Aragorn and Legolas, and Frodo and Sam. In your assessment, what do these friendships communicate about the nature of being friends or the communication that takes place between friends?

Now, think about the work relationships. Perhaps these relationships are best illustrated by the *The Lord of the Rings: The Fellowship of the Ring*. When

FIGURE 8.1 Consider the interpersonal relationships in *The Lord of the Rings*.

Frodo commits to the task of taking the ring to Mordor, he gains several team members in this work:

> GANDALF: "I will help you bear this burden, Frodo Baggins, as long as it is yours to bear."

> ARAGORN: "If by my life or death I can protect you, I will. You have my sword."

> LEGOLOS: "And you have my bow."

> GIMLI: "And my axe."

> BOROMIR: "You carry the fate of us all, little one. If this is indeed the will of the Council, then Gondor will see it done." ...

> ELROND: "Nine companions. So be it. You shall be the Fellowship of the Ring."

What does the above passage convey about work relationships? Are there other work relationships in Tolkien's work that you can identify? If so, how would you describe those relationships?

Finally, consider the romantic relationships. Romances between Sam and Rosie, Éowyn and Faramir, and Aragorn and Arwen are a few examples featured in Tolkien's works. What observations do you have regarding communication surrounding these romances or how they are depicted? How do these observations inform us about romance?

Interpersonal Relationships in a Pervasive Communication Environment

In the last chapter, you examined interpersonal communication and relationship development, including beginning, maintaining, changing, and ending relationships. Additionally, you considered factors, such as dialectical tensions and conflict, that can influence relational satisfaction and commitment. In this chapter, you will explore several contexts in which relationships take place and the characteristics of these contexts. Because you have many relationships that you manage at once, these relationships overlap and offer a number of opportunities to connect with others. Behavior in relationships is shaped by understanding of overarching and specific cultures, contexts, and type as well as individual differences.

As you move through a typical day, you likely engage in an array of relationships. How would you describe these relationships? In addition, how do you decide which topics to discuss with which people? For example, are there disclosures that you would make to your parent but not to your coworker? Are there topics that you would talk with your best friend about but not bring up with your parent? What guides these choices, the assumptions that you make about what a particular relationship involves (e.g., its possibilities and boundaries), and decisions about escalation and de-escalation? Throughout this chapter, you will become familiar with several influences as you consider relationships in context.

Relational Contexts

Although you may develop relationships in other contexts, four contexts are especially important and will be examined in the sections below. These contexts are family, friendships, work, and romance. In each area, you will become more familiar with key considerations. For example, some relationships are voluntary ones, and others are involuntary. Also, the level of independence and interdependence likely varies by type and across time. Although there are cultural differences, many people in the United States choose their friends and romantic partners; however, you do not choose, but rather are born into or adopted by, your family.

involuntary relationships
Relationships in which you do not select your relational partner or partners; for example, you do not select your family members

Family

For many people, relationships with family members are the first relationships formed. Not surprisingly, then, these relationships are important learning grounds and shape our views of and communication in other relationships that we form. According to most views of what constitutes a family, you are born or adopted into a family. Although, as you grow up, you can attempt to change some dimensions of your family, you cannot change families in the same way that you can change jobs.

In addition to being **involuntary** (i.e., you do not select your family members), family relationships are generally seen as enduring, continuing throughout your lifetime.[1] Due to their enduring nature, communication may be used to negotiate changes across the years (e.g., changes to high school curfews when you visit home while in college, open financial discussions when your parents reach later life, etc.). Thus, while you may not be able to change the members of your family, you can try to change family communication in positive ways.

Considerable research describes the many ways that family relationships influence family members (Figure 8.2). For example, family relationships significantly

FIGURE 8.2 What has your family taught you about relationships?

influence one's opinions and behavior, can facilitate goal achievement and meeting needs (or, if dysfunctional, can fail in these areas), shape one's communication skills, and enhance or erode self-esteem.[2]

Family relationships can be fraught with complications and problems, or they can be safe spaces. They can also move back and forth between these poles across time, or they can be infused with elements of each, simultaneously being complicated and safe, problematic and comfortable (Box 8.1). For example, strained relations during adolescence, as you assert your independence and push back against family rules, may turn into a deeply respectful relationship as you raise your own children. In other words, you may see your family of origin differently once you have your family of procreation, terms used in the academic literature to describe the evolution of families.[3]

One way of conceptualizing family communication uses two dimensions: **orientation to conformity** and **orientation to conversation**.[4] With conversation, the focus is on whether families encourage an open exchange of ideas—that is, the degree to which family members can freely share their views and opinions. In high conversation families, there is considerable sharing, comfort with disagreement, and openness to conflicting thoughts. In low conversation families, less priority is placed on engaging in conversation to maintain family relations; thus, family members may talk less and address fewer subjects. With conformity, interest centers on the degree of overlap in family members' worldviews (e.g., opinions, values, etc.). In high conformity families, priority is given to hierarchy (e.g., following the hierarchy), interdependent functioning, and smooth interactions (e.g., not engaging in conflicts). In low conformity families, emphasis is placed on equal relationships, individuality, and independent thought and behavior. In other words, how families use these orientations shapes their communication. If conversation is emphasized, interest is placed on learning about family members' viewpoints and developing understanding. If conformity is emphasized, agreement and shared opinions will be prioritized.

orientation to conformity A dimension used to conceptualize family communication; conformity focuses on the degree of overlap in family members' worldviews (e.g., opinions, values, etc.)

orientation to conversation A dimension used to conceptualize family communication; conversation focuses on whether families encourage an open exchange of ideas

---

BOX 8.1: CONNECT & RELATE

Fun Home

The award-winning musical *Fun Home* was based on a graphic novel about family tensions while growing up in a funeral home (aka "fun home"). As the protagonist tries to untangle family dynamics, the audience glimpses family scenes played out across the years, the effects parental behavior can have on children, and the ways in which there are secrets even in closeness. Along with the protagonist, the audience gains new insights into past behavior and grapples with decisions that adults made.

It is a powerful performance, and its content can encourage reconsideration of your earlier life and the effects of family relationships. What are the most crucial examples that you can describe of the way that your family of origin affected you?

These focal points (i.e., high and low in each of the two orientations) are used to derive four family types (Figure 8.3). The four types of families are (1) consensual, where a family is high in both, (2) protective, where a family is high in conformity and low in conversation, (3) pluralist, where a family is low in conformity and high in conversation, and (4) laissez-faire, where a family is low in both.

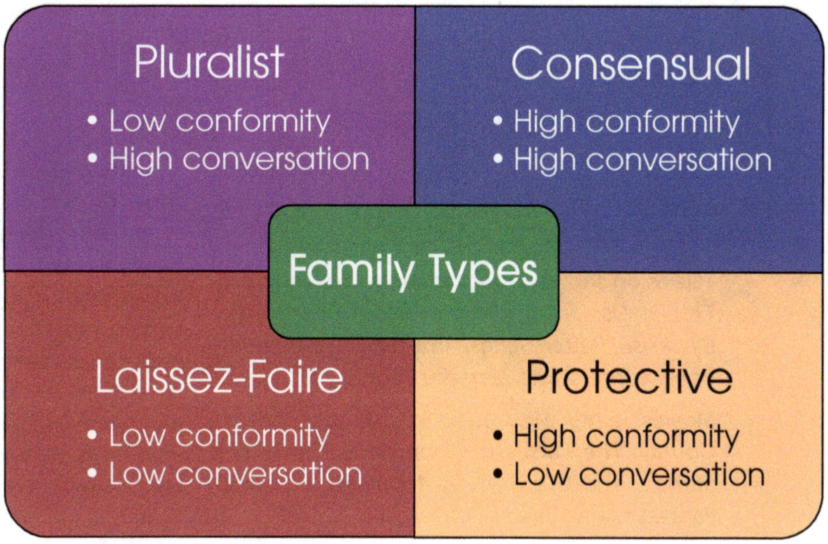

In **consensual families**, interest centers in both conversation and conformity. Children are encouraged to share their views and participate in making decisions, though parents make the final decisions, and may use their parents' worldview to evaluate information. Through discussion, there is hope for agreement.

In **protective families**, there is high interest in conformity but low interest in conversation. Family membership and adherence to family rules are prioritized over individuality and forming one's opinions independently. Little time is devoted to gathering input from children or explaining the reasons underlying decisions. Parents shoulder the responsibility, protect children from decisions, and expect agreement, and children may learn to accept the authority of others and to avoid engaging in conflict.

In **pluralist families**, priority is placed on conversation and not on conformity. Thus, there is considerable discussion of ideas and opinions, with family members making their own conclusions. In this context, children learn to develop their own positions and arguments to support their ideas. Independent thinking is valued and supported.

In **laissez-faire families**, little interest is placed on conversation or conformity. Thus, there tends to be less communication and conversational engagement as well as less emphasis on influencing others' thoughts or making decisions for them. Children likely have room for independence but also are not anchored by accountability to a family unit.

What do you see as the strengths and weaknesses of each type of family? Certainly, what works for one group may not work for another group. Although there may not be an absolute best approach, each approach is associated with particular outcomes (e.g., consider the examples and questions in Box 8.2). For example, children's communication skills, including comfort in social situations and strategies for addressing conflict, are influenced by these family dynamics.[5] Additionally, engaged, consistent, and positive parenting techniques are associated with adolescents having better psychological adjustment.[6]

FIGURE 8.3 Four family types based on orientations to conversation and conformity.

consensual families Interest centers in both conversation and conformity. Children are encouraged to share their views and participate in making decisions, though parents make the final decisions, and may use their parents' worldview to evaluate information

protective families High interest in conformity but low interest in conversation. Parents shoulder the responsibility, protect children from decisions, and expect agreement, and children may learn to accept the authority of others and to avoid engaging in conflict

pluralist families Priority is placed on conversation and not on conformity. There is considerable discussion of ideas and opinions with family members making their own conclusions. Independent thinking is valued and supported

laissez-faire families Little interest is placed on conversation or conformity. There tends to be less communication and conversational engagement as well as less emphasis on influencing others' thoughts or making decisions for them

---

BOX 8.2: CONNECT & RELATE

Reel and Recorded Families

Several popular television shows and films feature family relationships. The list of examples is long, but the following are a few to consider:

- Television shows: *Modern Family, Queen Sugar, This is Us, Fresh Off the Boat, Transparent, Black-ish, Better Things, Arrested Development,* Bob's Burgers, and *Jane the Virgin*

- Movies: *Little Miss Sunshine, August: Osage County, The Pursuit of Happyness, The Family Stone, The Farewell, Moonlight, Captain Fantastic,* and *The Hollars.*

Pick one of these options or select another TV show or film about a family or families. Then, answer the following questions about the one that you have selected. Of the four family types, which one(s) is illustrated in this work? Give specifics examples of behavior that led you to this conclusion. How is communication used to negotiate changes within the family? How would you describe the level of independence and interdependence present in the family relationships? What other observations do you have about family communication in this show or film?

Families have been undergoing considerable change in the past few years, and much of the research on families has focused on White, middle-class families. Thus, a broader understanding of today's families is needed.[7] For example, blended families (e.g., stepchildren), grandparent-headed families, and two-mom or two-dad families have become more common. As researchers widen their studies to include more diverse families, clearer understanding of family communication and family relationships and their impacts will emerge.

Although family relationships are generally seen as involuntary ones, some people are creating voluntary families—families of choice.[8] These chosen families blur traditional lines between friendships and families.

Friendships

Unlike family relationships, friendships are **voluntary relationships**.[9] Friendships are often shorter in duration than family relationships; usually form between peers, or people of approximately equal power (e.g., you and a classmate, rather than you and your school's president); and are informal.[10] Because friendships are voluntary, they are sometimes regarded as more vulnerable to change or dissolution than family relationships; however, because these are relationships developed by choice, they are often rewarding, highly valued ones.

Friendships can be classified into one of three types: associative, receptive, and reciprocal.[11] **Associative friendships** form due to an association, or physical connection, with another person. If you form a friendship with someone who lives beside you in an apartment complex, sits beside you in your communication course, or has a cubicle next to yours at work, the friendship might be considered

voluntary relationships
You choose your relational partner or partners

associative friendships
Form due to an association, or physical connection, with another person; might be closer to friendly relations than friendships

an associative one. This type of friendship is not characterized by a high level of closeness or intimacy, and such relationships might be closer to friendly relations than friendships.[12] Because there is not a high level of closeness, communication would be friendly, but would not likely involve deep self-disclosure. Think back to information from the previous chapter on relationship development. You might consider escalating an associative friendship to a deeper one based on filtering, communication, and your impressions. Alternatively, you may be comfortable with this type of friendship and engage in communication that maintains the current relationship. In this case, if the association ends (e.g., you no longer are neighbors, have class together, or are employed in the same workplace), then the friendship may dissolve.

Although friendships usually form among peers or equals, sometimes differences can be transcended. Friendships that form in spite of such differences are called **receptive friendships**. Although each person benefits from the relationship, tangible benefits may seem to flow more in one direction. For example, suppose that you develop a friendly relationship with one of the partners at the law firm where you work, and eventually, that relationship transitions into a friendship. Because you are hoping to attend law school after you complete your undergraduate degree, it is helpful to have a friend who can field your questions, make suggestions on improving your chances of acceptance, and provide overall guidance. In a social exchange analysis (Chapter 7), it might appear that you are accumulating most of the rewards in the relationship. However, your friend may also see the rewards as outweighing the costs. For example, the lawyer may enjoy being a mentor, recall positive memories based on the questions that you ask, gain new insights from learning about your perspective, and benefit from making a difference in your life. Thus, each of you may be satisfied with the relationship, despite not being peers.

In some cases, associative or receptive friendships are escalated and become closer. In **reciprocal friendships**, benefits from the relationship are shared equally. Of course, as noted earlier, relationships are not stable entities but instead experience changes; thus, there may be fluctuations in rewards and costs.[13] For example, you may enjoy talking with your friend about every stressful homework assignment and benefitting from social support; however, if your friend is trying to navigate a family member's challenging medical problems, you may forego discussing your homework and spend more time supporting your friend. Across time in the relationship, each friend would receive about the same level of benefits from the relationship—hence, the label comes from this reciprocity. Reciprocal friendships are the most time-consuming ones, but they are also the friendships that people desire and value the most, our idealized view of what a friendship should be and would entail. Key qualities in reciprocal friendships include trustworthiness, honesty, intimacy, freedom, shared activities, affection, compromise, and loyalty, and these qualities influence the communication of friends.[14]

receptive friendships Form despite differences in power; tangible benefits may seem to flow more in one direction

reciprocal friendships Form between equals; benefits from the relationship are shared equally

In addition to the more immediate benefits that friendships bring to your life, a number of studies document positive outcomes from friendships. For example, friendships contribute to your level of happiness, school and/or work performance, as well as overall physical and mental health.[15] These outcomes are certainly ones that help you treasure your friendships even more (consider, for example, the questions in Box 8.3).

Friendships are escalated and maintained through self-disclosure. As you meet and get to know people, you share information about yourself and learn about them. As relationships develop, the types of information disclosed deepens. For example, self-disclosure can be categorized as observations, thoughts, feelings, and needs.[16] These categories move from the most general, or surface-level, disclosures to deeper ones. By sharing observations, you disclose information about your experiences. For example, you might disclose that you are a transfer student. You might also share your thoughts about being a transfer student with a disclosure (e.g., "It was a hard adjustment the first semester but transferring here was such a good decision"). If your disclosure deepens to the level of sharing feelings, you would share your emotions, such as by saying, "I feel so fortunate and happy to be studying here!" The final category, needs, involves the deepest disclosure. Continuing with the example above, if disclosing a need, you might share that you need to make better grades to get into graduate school or need to get an internship but do not know how to start identifying possibilities. In the process of getting to know people and making decisions on potentially escalating a relationship, such as becoming better friends, we consider factors such as the potential risks and benefits of self-disclosure. The disclosures above about being a transfer student may not carry much risk, but what if the topic were your immigration status or divorce? Decisions about what, when, and how much to self-disclose influence friendship development as well as the development of other relationships.

BOX 8.3: CONNECT & REFLECT

Recognizing and Celebrating Friendship

In the United States, there are holidays to celebrate many relationships. These holidays include Mother's Day, Grandparent's Day, Valentine's Day, and wedding anniversaries. But there is not an official holiday to celebrate friendships. Some friends do celebrate their relationships, though, with activities such as Friendsgiving and Galentine's Day. Do you and your friends celebrate in these or other ways? Given the important role of friendship in our lives and its many positive outcomes, should we have a specific holiday to recognize and celebrate friendship?

Workplace

Your workplace is more than just a job or a paycheck. It is also a place where you connect with others and have opportunities to build relationships (Figure 8.4). As the work week has grown longer and 24/7 connectivity has emerged, workplace relationships have taken on increasing importance. Although you probably do not choose the people with whom you work, you can choose who you develop relationships with at work; thus, to some extent these are voluntary relationships. With the focus on getting tasks done and accomplishing overall objectives, sometimes relationships at work are overlooked; however, they have important outcomes, both for employees and the organization.

At the workplace, several types of relationships are formed, including relationships with your coworkers, supervisors, subordinates, and potentially external people connected to the organizations (e.g., customers, vendors, etc.). As relationships develop beyond the roles slated in an organizational chart, they may transcend the formality and hierarchy associated with many organizations. How do close friendships and romantic relationships influence the workplace? What if your friend becomes your boss? What if you become romantically involved with a coworker and you have a bad breakup? These and related questions (e.g., Box 8.4 explores considerations surrounding viewing workplaces as families) take on increasing importance, as people spend more time at work than in other contexts.

FIGURE 8.4 Work friends: How would you describe your friendships at work or school?

Workplace relationships evolve in different ways and into different forms. Some become mentoring relationships. Some involve bullying behavior. Some escalate to romance. Others may not involve romance per se but have overtones of that type of relationship, such as when people refer to their "work spouse."[17] And many develop into important friendships.

Because coworkers are essentially peers, at least organizationally, their friendships might often be classified as reciprocal in the typology listed above. These workplace-based friendships have both potential rewards and costs for individuals

---

BOX 8.4: ETHICAL QUESTIONS

Workplaces as Families

Many organizations like to describe the workplace as a family. This metaphor can convey a sense of belonging, loyalty, informality, and care, and it can reinforce allegiance to the organization. Of course, some organizations are family businesses. Otherwise, is this way of viewing an organization accurate and fair? Consider the following questions as you analyze the organization-as-family metaphor:

- When leaders say that an organization is like a family, what do they mean?

- In what ways might an organization be like a family?

- In what ways are organizations unlike families?

- What are the benefits of defining an organization as a family, and who benefits most?

- What are the downsides of seeing an organization as a family? Who is most at risk of potential harms?

- When leaders say that an organization is like a family, do workers believe them? Why, or why not?

- To what extent is labeling an organization a "family" accurate, fair, and just? Explain your answer.

and the organization. Past research has examined possible positive outcomes, such as task and social support; employee development; and increased cooperation, creativity, productivity, morale, and job satisfaction. Research also has examined possible negative outcomes, such as prioritizing social or emotional needs over tasks, role ambiguity, jealousy from other coworkers, and altered development within or across teams (e.g., silos, subgroup formation, etc.).[18]

As workplace friendships develop, individuals and organizations will continue to experience their upsides and downsides. Just as in other interpersonal relationships, inherent dialectical tensions also will need to be recognized and carefully managed, especially to try to achieve greater positive outcomes for individuals and workplaces.

Romantic

When some people hear the word "relationship," they think of a romantic relationship. For example, if your aunt asked, "Are you in a relationship?" you likely would not say, "Yes, beyond you and other family members, I have at least 10 close friends at school and a couple more at work." Instead, you would comment on whether or not you are romantically involved. Romantic relationships command a lot of interest from people, and most people seek out one or more romantic partnerships across their lifetimes. For many in the United States, romantic relationships are voluntary ones where one chooses a partner.

As you grow up, you learn about loving relationships. For example, you learn from relationships within your family as well as the ways that such relationships are portrayed in books, films, and similar works. This knowledge shapes your approach to forming and maintaining romantic relationships as well as your expectations for such relationships. Happy romantic relationships are often associated with a number of benefits,[19] such as better mental and physical health (Figure 8.5). For example, when people are in happy romantic relationships, they are less likely to experience depression or heart disease and are more likely to have strong immune systems.

A number of communication behaviors have been linked to happiness and satisfaction in romantic relationships. These behaviors include providing support, demonstrating liking and affection, engaging in playful communication and humor, and talking openly.[20] Such behaviors are sometimes called prosocial or prorelational ones, as they work to sustain the relationship and engage and support one's partner.[21] Other examples include providing reassurance, having a positive attitude, and handling conflicts productively. If less positive communication behaviors are frequently employed, the relational partners are less likely to be satisfied with or committed to the relationship. Examples of such

FIGURE 8.5 Celebrating a long-term relationship

behaviors include avoiding and withholding, inducing jealousy, and engaging in negative conflict behaviors.[22]

How a couple manages conflict has important individual and relational outcomes. Of course, when relational interdependence is high, conflict is inevitable. The important issue is not the presence of conflict but rather how such conflict is handled. In particular, conflict management is linked to health outcomes. When relational conflict is not dealt with productively, there may be negative impacts on mental health,[23] such as anxiety, depression, distress, hurt, and lingering anger, as well as physical health.[24] When conflict is managed well, stress may be reduced and understanding increased, resulting in increased closeness and satisfaction.[25] Unfortunately, once partners experience ongoing dissatisfaction, they are likely to experience greater conflict, which may perpetuate hurt and increase instances of negative conflict behavior.[26] This negative cycle may continue to damage the relationship. For these reasons, conflict has been frequently examined in the context of romantic relationships.

In Chapter 7, you considered general conflict styles. Examining conflict patterns, one well-known typology of conflict in romantic relationships discusses four types of couples.[27] Each type involves a different orientation to conflict that results in specific communication behavior. These behaviors affect overall relational satisfaction outcomes, and the types have been used to predict likelihood of divorce. The four types are validating, volatile, avoiding, and hostile, summarized in Table 8.1.

When engaging in a conflict, a **validating couple** uses positive communication behavior. Of the conflict styles discussed in Chapter 7, they are likely to employ compromising or collaborating techniques. They show concern for each other and search for areas of agreement and ways to establish common ground. They listen to each other, are calm, and, as the label for this type indicates, validate each other, demonstrating an interest in and valuing of their partner's thoughts and opinions. The overall goal is to work collaboratively to solve the conflict.

In contrast, a **volatile couple** does not remain calm—they have heated arguments, and their conversations may escalate quickly to conflicts. But just as quickly, they may resolve their differences, and they are skilled at making up after a disagreement. Volatile couples tend to engage in straightforward communication, openly sharing their feeling and opinions, and they may have frequent conflicts and enjoy disagreement. When in conflict, they may powerfully assert their position but simultaneously demonstrate positive relational behaviors, such as touch and smiling, that show affection for their partner. Their communication shows respect for their partner and they do not engage in disrespectful actions. One component of relational satisfaction is the ability to argue, resolve differences, and make up, and volatile couples have strengths in these areas.

The next type is the opposite of a volatile couple. **Avoiding couples** do just what the label indicates—they work to avoid conflicts. Because they strive to avoid conflict, their arguments are infrequent. Rather than the heated arguments that volatile couples have, avoiders have carefully conducted exchanges

validating couple Uses positive communication behavior, such as compromising or collaborating. They show concern for each other and search for areas of agreement and ways to establish common ground

volatile couple Has heated arguments, and conversations may escalate quickly to conflicts. When in conflict, they may powerfully assert their position but simultaneously show affection for their partner

avoiding couple Works to avoid conflicts. They sometimes believe that by giving issues time the conflict will decrease or disappear; thus, it is better to not confront it

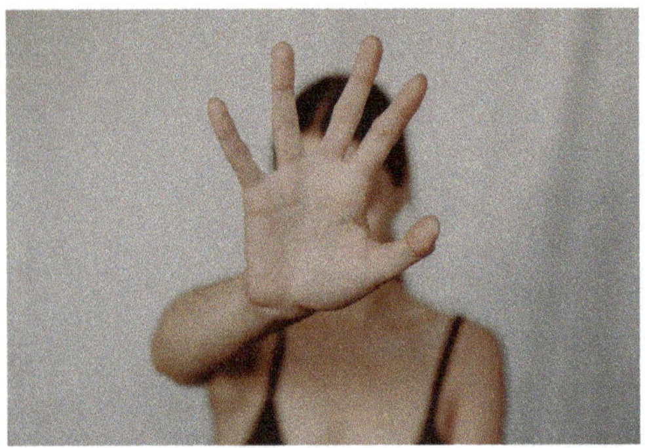

FIGURE 8.6 What are the positives and negatives of avoiding conflict?

hostile couple Has frequent arguments that use negative communication behavior. Disagreements are cutting, harmful, sarcastic, critical, and insulting—showing disdain for their partner. Defensiveness is high, and listening and an interest in the other's views are low

with minimal emotion. They sometimes believe that by giving issues time the conflict will decrease or disappear; thus, it is better to not confront it. Additionally, they may prefer to keep the focus on areas of agreement rather than assume what they regard as risks inherent in an argument (Figure 8.6).

Despite the differences across the three types of couples above, each can be satisfied with their relationship. Each type has strengths and weaknesses, and the relational partners may have differing preferences in negotiating differences. The fourth type, however, involves behavior that is less helpful, less healthy, and more likely to result in relationship dissolution (Table 8.1).

This fourth type is the **hostile couple**. These couples tend to have frequent arguments and use negative communication behavior in their arguments. In particular, the frequency of their negative behavior far outweighs the frequency of positive behavior. These couples have disagreements that are cutting, harmful, sarcastic, critical, and insulting—showing disdain for their partner. Defensiveness is high, and listening and an interest in the other's views are low. Given these behaviors, it is easy to see why satisfaction would be low and commitment would wane. Studies of marriages show that hostile couples are likely to divorce.

Your understanding of romantic relationships and your skills in conflict communication have important impacts on your relationships. Of course, beyond the relational context, there are outside influences. For example, the COVID-19 pandemic resulted in many couples spending considerable time together and having limited interactions with others. What do you think the influences of these shifts would be?

In the above sections, relationship contexts—family, friendship, workplace, and romantic—have been examined separately. Each context is broad and important enough to have courses taught about it, so consider taking one or more such courses in the future. It is important to note that the contexts can overlap. For example, a family business involves working with family members, so both workplace and family relationships would be included. If you start the business with

TABLE 8.1 Summary of Couple Types With Conflict Approach and Examples of Communication

COUPLE TYPE	CONFLICT APPROACH	EXAMPLE COMMUNICATION
Validating	Collaboration	Listen; show interest in fully understanding partner's points; demonstrate kindness; validate importance of partner and relationship
Volatile	Openness	Disagree with points but maintain positive relational orientation (e.g., show affection and respect)
Avoiding	Do not engage	Stress areas of agreement; ignore areas of disagreement
Hostile	Defensive	Defend, rather than listen; critical and insulting

Note: Based on Gottman's typology

BOX 8.5: LIVE IT LOCAL
Building and Strengthening Relationships

Consider the material in this chapter and the previous one. Identify a group or organization in your community that focuses on youth and relationships. It could be one that you are a member of or have worked with in the past or one where you would like to volunteer or work. Its focus on relationships could be about developing healthy personal relationships, such as dating relationships or friendships, or on cultivating mentoring relationships or work-based connections. Look at the website and any print materials of the group or organization. From these chapters, what recommendations do you have for expanding the discussion on building and strengthening relationships? Your strategies might center in communication to maintain relationships, to expand relational networks, to handle conflict in positive ways, or something else. What specifically should these youth understand and do? If they follow your suggestions, what outcomes should they anticipate?

your spouse, it would involve a romantic relationship too. If you also employ friends, then all four contexts might influence your communication simultaneously. Across time, relationships also evolve. As noted above, relationships are dynamic and thus undergo constant change. Across time, larger scale changes may take place. For example, parents who were caregivers to their children may later have their children as caregivers to them, and spouses may become former spouses but also friends.

Interpersonal Relationships Across the Life Span

In this and previous chapters, you have learned about the important roles that relationships play across contexts and time. You have also become familiar with the potential that they offer for important outcomes from relational satisfaction and happiness to better mental and physical health. The absence of close, healthy relationships also is associated with negative outcomes. For these reasons, interpersonal relationships are increasingly becoming recognized as a public health issue.[28] Given their importance, what ideas do you have for building and strengthening relationships (Box 8.5)?

Healthy interpersonal relationships directly contribute to overall health and life span. For example, people with close, positive relationships have significantly better survival rates, even in spite of factors such as age and other health conditions. Further, lacking such relationships is as harmful to one's health as other risk factors, such as smoking and obesity.[29] For many people, loneliness is a persistent experience. In the United States, more than 40% of people over the age of 60 report frequent loneliness.[30] Beyond the frequency, they often experience high levels of loneliness. The frequency and intensity contribute to worse health outcomes and increased risk of death. These examples are but a few of the reasons why relationships contribute directly to overall public health and need attention as a public health issue.[31]

FIGURE 8.7 We start building relationships early and continue throughout life.

Clearly, given the importance of interpersonal relationships across the life span, it is important to develop communication skills to cultivate and maintain such relationships.[32] The development of such skills begins in childhood, and the understandings of relationships that you develop then shape your future relationships (Figure 8.7). Studies indicate that one's family-of-origin experiences influence relationships throughout life. For example, a longitudinal study found that if one's childhood was spent as part of a loving, positive family, then the person had greater likelihood of more satisfying romantic relationships.[33] This effect extended into later life (e.g., people in their 80s). Thus, childhood experiences foster development of important interpersonal communication skills useful throughout one's life. Positive childhood relationships also appear to influence health across the life span, providing a buffer to physical health issues (e.g., chronic health conditions, such as heart disease).[34] In short, one's family of origin sets the stage for relational development and associated skills; in turn, if one has a family of procreation, experiences there shape the child's future relationships.

Also, in childhood, you begin your first friendship experiences. Despite different phases across your life, the goals for friendships generally remain the same across the life span—to have people to talk and share experiences with, to have fun with, and to depend upon (e.g., emotional and/or task support), and fulfilling these goals contributes to overall health, both physical and mental.[35] Your friendship network is likely to increase in size through the college years, which is an especially good time to connect with an array of people and build friendships. For many people, finishing college means starting a new career, perhaps in a new location, working to build that career, and possibly marriage and starting a family. After marriage, the amount of contact with friends tends to decrease.[36] However, the quality of your friendships is considerably more important than the quantity of them.[37] As you age, the importance of friendship may increase, and friendships are a key contributor to overall happiness and health, perhaps more than other types of relationships.[38] Some research has examined sex/gender differences in friendships, exploring topics such as the types of activities in which friends engage and level of relational intimacy. Although there are not sex differences in likelihood of having a romantic partner, women are more likely to have best friends than men are, and women may have smaller friendship circles but deeper friendships than men do.[39] How might such differences influence quality of life as one ages?

Several studies have examined friendships of people in later life.[40] For example, older adults report greater satisfaction with their friendships than do younger people, and as people age, they may take steps to nurture positive relationships and allow others that are less healthy or less pleasing to wither.[41] As the life course continues, experiences (e.g., passing of friends and family, decreases in mobility, etc.) may increase the likelihood of loneliness. Given that the U.S. population is an aging one and that more people in middle and later life are not married and do not have children, developing a better understanding of loneliness potential and social support needs is important.

Workplace relationships are important during your career years, and such relationships may be important sources of friendship, mentoring, and career advancement. After retirement, former workplace relationships may transition outside that context (Figure 8.8). For example, relationships in the community may be built or deepened. Additionally, through activities such as volunteering, relationships similar to workplace ones may develop. In some cases, people seek to preserve facets of their work identity or maintain work-based ties. For example, sometimes groups of former colleagues form breakfast or lunch clubs where they meet to socialize, catch up, and talk about the past.[42]

FIGURE 8.8 How is retirement likely to influence relationships?

Romantic relationships are important ones throughout adolescence and adulthood. Across phases of the life span, couples tend to value multiple elements of their relationship, including romance, other-orientation, and friendship.[43] With transitions in work relationships after retiring as well as the passing of friends and family members across time, later life may present increased time to connect with others, but also fewer opportunities to do so, which may contribute to loneliness. Being in a marriage can buffer some feelings of loneliness as well as offer companionship and social and task support. Several studies have shown that married persons, especially men, live longer and are healthier than single individuals.[44] Of course, if the marriage is a happy one, these benefits may be even greater. Some studies have shown changes to communication patterns, such as conflict management, across life phases that may facilitate increased satisfaction in later life (Figure 8.9).[45]

FIGURE 8.9 What communication strengths have you observed in older persons?

Across the life span, you will change, and your relationships will change. The communication within your family has changed since you were a toddler and will evolve further over time. In mid-life, many children become the caregivers that their parents were to them earlier. As your friends begin to have children and as their children grow up, friendships are likely to be altered. If you marry and/or have children, your friendships and other family relationships also will change. Sometimes parents are raising children and taking care of their own parents at the same time, referred to as the "sandwich generation." All of these changes highlight the dynamic, ever-evolving nature of interpersonal relationships (Figure 8.10).

FIGURE 8.10 What do you think fuels relational happiness over the life span?

CHAPTER REVIEW

- **Explain how context influences interpersonal relationships.**
  You form relationships across a variety of contexts. Each of those contexts influences your views about what a relationship should be like and the types of behavior involved (e.g., level of independence or interdependence). Some relationships are voluntary ones, such as choosing your friends, and others are involuntary, such as your family of origin.

- **Discuss key considerations for interpersonal relationships in different contexts: family, friendship, workplace, and romantic.**
  Across the contexts, the ability to manage tensions and conflict is critical in building and maintaining happy, satisfying relationships. It is also important to remember that relationships are dynamic—they constantly change and feelings ebb and flow.

- **Describe interpersonal relationships across the life span.**
  Interpersonal relationships are important in your life across contexts and time. Close, positive relationships have a number of benefits from social support and happiness to better mental and physical health. The absence of close, healthy relationships is associated with negative outcomes, such as poorer immune system functioning and shorter life span. Given the potential for positive and negative outcomes, it is important to develop communication skills to build and maintain relationships across your lifetime.

DISCUSS AND APPLY

1. Consider the following quote by Kendall Hailey about family: "The greatest gift of family life is to be intimately acquainted with people you might never even introduce yourself to, had life not done it for you." How can you relate this quote to the chapter's points on family relationships?

2. What influences do you believe that the COVID-19 pandemic had on family, friend, work, and romantic relationships? Try to list both positive and negative examples.

3. The sections above primarily deal with communication within relational contexts. How might factors outside that context influence relationships? For example, how might national politics, social justice movements, income, health, and similar factors impact relationships?

4. Is conflict with a friend or a romantic partner more challenging? Why?

5. Many workplaces have policies against certain types of relationships (e.g., nepotism, dating coworkers, romantic relationships with anyone you supervise). Should your workplace be able to determine the types of relationships that you can have? Explain your answer.

6. The term "conscious uncoupling" has been used to describe a way of ending romantic relationships. What does this approach convey about relational contexts?

7. The Spice Girls song "Wannabe"[46] contains this lyric: "If you wannabe my lover, you gotta get with my friends." How do one's friends influence romantic relationships?

8. Given the life phases that you have experienced so far, how have your family relationships changed? How have your friendships changed?

CREDITS

CHAPTER 9

Small Group Formation

The Kentucky Student Environmental Coalition (KSEC) has its roots in the first Power Shift Convergence organized by the Energy Action Coalition. The 2007 event in Washington, DC, brought together young people interested in tackling climate change issues in their own communities. There, students from Kentucky met and formed the KSEC. That same year, the newly minted KSEC held a meeting at Western Kentucky University and developed its mission statement and priorities. The group leveraged basic communication technology—the telephone—to bring together students from around the state interested in environmental issues. Three years later, the group organized a statewide day of action, advocating for the use of more sustainable resources on college campuses. More recently, the group supported Fossil Fuel Divestment Day and sent representatives to the latest Power Shift annual meeting. Using social media as well as in-person actions and educational events, the group has influenced state legislation and campus policies. What started with a few students who met at a conference has turned into a powerful student-led organization with more than 20 college and high school affiliates.[1]

This chapter describes the nature of small groups in a pervasive communication environment, identifies types of small groups you might join, discusses the phases of group development, highlights the roles members fulfill in a group, explains the requirements for leading a group, and offers strategies for building successful small groups.

Joining Small Groups in a Pervasive Communication Environment

Before mobile communication technologies, groups usually formed based on physical proximity—the people you sat next to in class, the neighbors on your block, the coworkers in your building. Groups still do form that way—that is how the KSEC got its start. But the internet offers many more opportunities for people to form and join groups that were not available before. Now individuals who have never met in-person join groups on Facebook, Twitter, Instagram, and similar social networking sites. In addition, social media platforms provide an avenue for groups to assemble quickly and distribute their messages to a wider audience.

For example, activist groups often post event details on Instagram and Twitter just hours before a protest is set to start (Figure 9.1). The groups then use social media to document the protests, sharing videos and photos viewed around the globe.

A pervasive communication environment and the ability to connect almost anytime with anyone from anyplace have altered the opportunities for joining groups, but the human needs groups fulfill have not changed. According to Abraham Maslow, human beings experience eight types of needs: physiological, safety, belonging-ness, esteem, cognitive, aesthetic, self-actualization, and transcendence.[2] These needs range from the most concrete, such as food, water, and shelter, to the most abstract, such as feeling self-fulfilled and helping others to achieve their

FIGURE 9.1 Small groups of local activists often use social media to organize and document large protests.

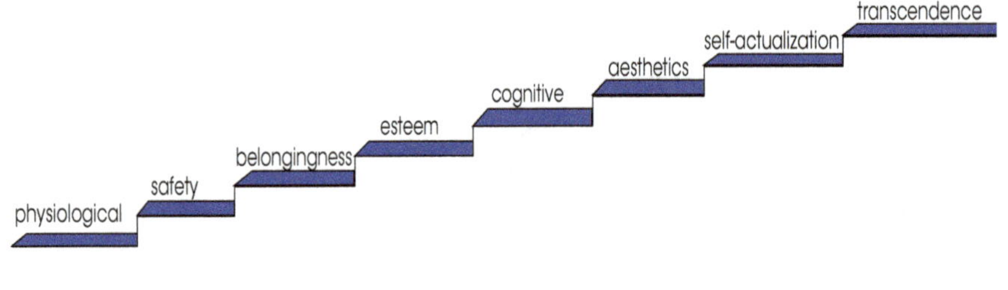

FIGURE 9.2 Maslow's hierarchy of needs represents a process from fulfilling basic needs to more abstract ones.

potential. Although often depicted in a pyramid, a ladder, as shown in Figure 9.2, better represents Maslow's idea of a gradual procession to transcendence—and possibly beyond.[3] Although you experience these needs internally, you fulfill them in interactions with other people.[4]

Meeting Physiological Needs

Physiological needs include what you must have to keep your body functioning, such as food, water, and sleep. Communities often designate small plots of land as "pea patches" or communal vegetable gardens. Those willing and able take on the various tasks of caring for the produce. In some U.S. cities and towns, groups have formed to oppose fluoridating public water systems, arguing the additive poses a health risk. Parents of children with sleep problems meet in online support groups to discuss ways to effectively help their kids.

physiological needs
What you must have to keep your body functioning, such as food, water, and sleep

Meeting Safety Needs

After addressing physical needs, you want to feel safe, secure, and free from harm. Neighborhood watch groups address these **safety needs**. The Map Your Neighborhood program brings together small groups of neighbors to make plans for how to keep each other safe in the event of a disaster.[5] If you are concerned about your personal safety, you might join a martial arts club that teaches self-defense or a Facebook group that discusses such safety issues. During the COVID-19 pandemic, two doctoral students at the University of Hawai'i at Mānoa brought together a group of volunteers to produce kits for cloth face masks that people could easily sew and wear.[6]

safety needs What you need to feel safe, secure, and free from harm

Meeting Belongingness Needs

Belongingness needs refer to the human desire to feel connected with others, accepted, and loved. A wide range of groups fulfills this need, such as clubs centered on an activity you enjoy (e.g., skateboarding, chess, bowling) and social organizations (e.g., fraternities and sororities). First-year college students often join campus-associated social media groups to help them integrate into and feel a part of their new community.[7] More informal groups—your family and friends—also contribute to your sense of belongingness.

belongingness needs
What you need to feel connected with others, accepted, and loved

Meeting Esteem Needs

Esteem needs center on the desire for others to recognize your achievements and view you as competent. Participating in professional organizations, campus groups, and student clubs allows you to demonstrate your skills and expertise. When others praise you for a job well done, such as effectively organizing a fundraiser for your group, they are addressing your esteem needs. Online and in-person, support groups meet esteem needs by giving members a sense of control over their lives.[8]

esteem needs Desires associated with having others recognize your achievements and competency

Meeting Cognitive Needs

You also join groups for **cognitive needs**—to gather information, explore your world, and better understand something. For example, if you are a public relations major, you might join your college's chapter of the Public Relations Student Society of America (PRSSA). The organization posts PR-related information on its various social media sites and sponsors an annual conference and a case study competition. You also join groups to fulfill cognitive needs when you know little about a topic but want to know more—that is why study groups are popular in courses students find especially challenging.[9]

cognitive needs Desires associated with gathering information, exploring the world, and better understanding something

Meeting Aesthetic Needs

Aesthetic needs include a desire for balance, order, and beauty (Figure 9.3). If you think your life focuses too much on school and work, you might join a hiking club or similar group to give yourself greater balance in your daily activities. Music and art appreciation groups satisfy human needs for beauty. For example, at the University of Music and Performing Arts in Graz, Austria, the Institute for Music Education started the Meet4Music Project to gather small groups from a wide range of community members who were interested in collaborative music experiences. Guided by trained facilitators, the groups participate in loosely structured improvisational sessions where they engage in musical activities. Research on the project found that by making music together participants developed an appreciation for the joint production of something artistic and creative. That is, they could enjoy both the music as something beautiful as well as the process of creating it.[10]

aesthetic needs Desires for balance, order, and beauty

FIGURE 9.3 People join local theater and musical groups to meet their aesthetic needs for beauty.

Meeting Self-Actualization Needs

Your desire to achieve your full potential reflects your **self-actualization needs**. Groups that give you a sense of self-fulfillment help you self-actualize. You may join reading groups, online discussion forums, or local clubs because they strengthen the sense of your abilities and self. For example, a study of a New York

self-actualization needs Desires associated with wanting to achieving your full potential

City synagogue that served a Modern Orthodox Jewish community found that younger members gravitated toward small group interactions outside the traditional communal services to experience greater authentic personal fulfillment.[11] Smaller, more intimate groups can provide the space for more satisfying one-on-one exchanges not possible in larger groups.

Meeting Transcendence Needs

transcendence needs
Desires associated with setting aside personal concerns and focusing on something greater than one's own self

When you help others reach their potential you fulfill your **transcendence needs**. That is, you rise above your personal concerns and focus on something greater than yourself. Altruism is one form of transcendence, where you take action not for yourself but because it benefits others. For example, in self-help groups individuals assist each other in understanding their common experiences.[12] A study of Syrian women who had been forced to emigrate to Turkey due to their own country's civil war found that those who joined a self-help group felt more empowered and experienced less depression through supporting each other.[13] Interestingly, volunteer firefighters cite two forms of altruism underlying their membership in their local fire department. The first is the desire to help people in their community. The second, which develops over time, is the gratification from supporting and mentoring other firefighters.[14] Some junior high and high schools use peer mentor programs, where older students tutor younger students in subjects such as reading and math. The tutors fulfill their transcendence needs, as students increase their skills. Research on teaching about sustainability using peer mentors found that the mentors developed a clearer and more comprehensive understanding of sustainability issues. More importantly, they felt more connected to their own lives, their peer mentor group, other people, places, and the natural world.[15]

The groups you join often meet more than one need. In addition, your motivation for staying with a group may differ from why you initially joined (Figure 9.4). For example, research on two online women's groups in India found that while members felt a sense of belonging to the groups through their online communication, meeting in-person at various venues such as cafes, parks, and other public spaces satisfied their needs for self-actualization.[16] That is, the online environment allowed them to form initial connections, but face-to-face contact offered a space for them to have more self-reflective conversations. New college graduates may join professional groups for belongness needs but later find their esteem, cognitive, and other needs are met through interactions with the group's members. Research on cycling groups has found that many cyclists join because they do not enjoy cycling alone (safety and belonging needs) and stay because the groups fulfill self-esteem and self-actualization needs.[17] And you join some groups because you have little choice, such as many work groups, yet you may

FIGURE 9.4 People often join small groups to meet multiple needs, and their motivations for staying in a group may change over time.

discover those groups meet your needs for recognition, understanding, balance, and the like. In Box 9.1, explore the reasons why people in your community join volunteer groups.

BOX 9.1: LIVE IT LOCAL
Volunteer Groups in Your Community

Volunteer groups provide a range of services to local communities, such as teaching classes for seniors, working with elementary school recreation programs, and assisting people who are unhoused. Research has found that college students join volunteer groups for a wide range of reasons, including a sense of connection (belongingness needs), increased competence (esteem needs), to learn about something (cognitive needs), and to improve others' lives (transcendence needs).[18] Contact two or three volunteer groups in your community and ask to meet with a few of the volunteers. Interview or meet informally with them to explore why they joined the group and why they continue to volunteer. What did you find out about joining volunteer groups that confirmed what you have learned in this chapter? What surprised you? How might joining a volunteer group meet your own needs?

Characteristics of Small Groups

What exactly is a small group? Are two people a group? What about 50 people? How small is small? When does a collection of individuals become a group? This section identifies the characteristics of a small group based on this definition: three or more interdependent people who recognize one another as unique individuals and communicate to achieve a perceived shared goal. Let's unpack and discuss each of these characteristics.

Three or More People

Dyads consist of two people interacting. When a third person joins, the dyad becomes a triad, or a small group, and the interaction dynamics change in important ways. First, because dyads have only two people, they cannot make decisions by voting or majority rule. Second, dyads allow for just two interaction possibilities: (1) Person A addresses Person B, or (2) Person B addresses Person A. For dyadic partners to interact, they must communicate with each other. But once a third person is added, there are nine possible configurations,[19] as shown in Table 9.1.

TABLE 9.1 Interaction Possibilities in a Triad

PERSON A'S INTERACTION POSSIBILITIES	PERSON B'S INTERACTION POSSIBILITIES	PERSON C'S INTERACTION POSSIBILITIES
Person A → Person B	Person B → Person A	Person C → Person A
Person A → Person C	Person B → Person C	Person C → Person B
Person A → Persons B and C	Person B → Persons A and C	Person C → Persons A and B

With each person added to the group, the interaction possibilities increase geometrically,[20] as shown in Table 9.2. When you think in terms of possible interactions, the complexity of small group communication becomes apparent. Similar to the interaction possibilities in a triad, the relationship opportunities among three people are more complex than between two people, who have only one possibility.[21] Even adding one person can change the group's dynamics. For example, research on a game-based mobile learning activity in a museum found that students reported higher levels of engagement with each other in four-person groups than those in five-person groups.[22]

TABLE 9.2 Geometric Progression of Communication Possibilities

NUMBER OF PEOPLE	NUMBER OF INTERACTION POSSIBILITIES
2	2
3	9
4	28
5	75
6	186
7	441
8	1,056

From: Bostrom, R. N. (1970). Patterns of communicative interaction in small groups. *Speech Monographs, 37,* 258.

The third way small groups differ from dyads is that small groups allow members to form coalitions, or temporary alliances between people. You have probably experienced this yourself when several group members join together to oppose others in the group. Coalitions may shift, depending on the nature of the task and topic under discussion. For example, you may find yourself agreeing with some group members on how to divide up the workload but disagreeing with the same people on setting deadlines.

Interdependent People

Interdependent individuals need each other to achieve their objectives. Without interdependence, people think in terms of themselves rather than the group.[23] The astronauts who work together on the International Space Station provide an example of interdependence (Figure 9.5). For Expedition 61, six astronauts from Italy, Russia, and the United States repaired the Alpha Magnetic Spectrometer, which detects cosmic rays and gives scientists insights into the universe's dark matter. While two astronauts spent more than 6 hours in a spacewalk painstakingly performing

FIGURE 9.5 NASA's Expedition 61 crew needed each other to complete repairs and conduct research aboard the international space station.

repairs on the AMS, their colleagues inside the station provided essential feedback and technical support. Team members successfully worked together to make the repairs, and engineers back on Earth successfully powered up the AMS so it could once again collect data on the mysteries of the universe.[24]

Recognizing One Another as Unique Individuals

Human relationships vary on a continuum from role-based to person-based. With **role-based relationships**, people communicate with each other in terms of the job or function they fulfill, such as instructor, student, salesperson, customer, server, or diner. Most of your daily interactions are role-based; you comment on the weather as the store clerk scans your items and you ask the driver where to transfer as you board the bus. **Person-based relationships**, such as friendships and kinships, involve recognizing the other as a distinct human being; you know the person's name, likes and dislikes, interests, skills, and inclinations.

Although some relationships are completely role-based and others completely person-based, many fall somewhere in between. For example, you may take the same bus to school or work every day and get to the know the bus driver—their name, favorite sports teams, upcoming vacation plans—so the relationship becomes more person-based. And you probably have some friends and family members you feel closer to than others. For instance, there is a big difference between the cousin you say a few words to at the occasional family gathering and the cousin who has acted as your college mentor.

The requirement that small group members recognize one another as unique individuals constrains a small group's size. Ideally, you want enough people to accomplish the group's objective—no more and no less. If you have too many group members, some will not have enough work to do. If the group is too small for the task, then group members will feel overburdened and overwhelmed. In addition, too few members result in a lack of multiple perspectives, ideas generated, and information sources.[25] A recent study with undergraduate student groups assigned to solve a hypothetical problem found that those with three or four members experienced higher levels of trust, commitment to the group, and awareness of each other than larger groups. In addition, students had more positive experiences in groups with fewer than seven members.[26]

Most researchers consider more than 20 people a large group in part because communication among participants becomes less personal. When a group's size exceeds 20, members tend not to share as much relevant information as those in smaller groups.[27] Getting to know other group members and interacting with them as unique individuals become more difficult as a group's size increases. In addition, not participating is easier when relationships are role-based, as often found in large groups, rather than person-based, which is more common in small groups. For example, a study of online Reddit communities found that as groups became larger, participation decreased and there was more turnover.

role-based relationships
Relationships in which people communicate with each other in terms of the job or function they fulfill

person-based relationships
Relationships in which you recognize others as distinct human beings

Rather than communities, the larger groups were more like crowds with a few members speaking to an audience.[28] Because smaller groups provide more opportunities for person-based communication, they tend to offer higher levels of satisfaction, engagement, and perceived effectiveness, as well as lower levels of conflict, than do larger groups.[29]

FIGURE 9.6 Small groups, especially those interested in social change, often evolve into large groups and organizations as more people become interested in the groups' goals.

Achieve a Perceived Shared Goal

For a group to exist, members need a reason to interact. A perceived shared goal provides that reason. But what if group members do not agree on a goal? Research has shown that *perceptions* of shared goals are more important than *having* the same goals.[30] Study participants were asked the degree to which they thought work group members shared the same goals. Groups with higher levels of perceived shared goals were more productive. When individuals think they share a goal, they are more willing to work together to achieve it. Participants also were asked to write out the goals of their work group. Interestingly, similarity in goal statements did not impact productivity. Box 9.2 encourages you to discuss ways to balance individual and group goals.

Many of the activist groups involved in the racial justice movement started as small local groups and evolved into large groups, with some becoming national and even international organizations as more people were drawn to the groups' goals (Figure 9.6). For example, the National Urban League, the oldest and largest

---

BOX 9.2: ETHICAL QUESTIONS

Openness and Small Group Goals

Open communication promotes cooperation in achieving a group's goals. Yet individual goals influence group work as well. How can you balance your own goals with the group's goals? Should you tell others of your goals for the group or keep quiet? What if your goals conflict with the group's goals? Communicating ethically in your group requires that you reflect on your own objectives and then explain them to the other members. Discussing each group member's goals allows the group to identify possible ways to integrate those goals within the group's purpose. In addition, group members may modify their own or the group's goals after a frank conversation. One way to encourage transparency and openness in balancing individual and group goals is to include goal discussion as an agenda item for some group meetings, particularly in early stages of the group's development or when new members join. Ask group members to bring a list of what they want to achieve in the group and the goals they hope the group will achieve. Discuss both individual and group goals, with special attention to determining how individual goals will help—or hinder—the group's goal attainment. Have you ever had such a discussion in a group? If so, what was the outcome? How might you bring up the topic of individual goals in future small group experiences?

civil rights organization in the United States, began as the Committee on Urban Conditions Among Negroes with a handful of people in 1910. Today, the organization has 90 affiliates in nearly three quarters of the nation's states.[31] More recently, three women—Patrisse Khan-Cullors, Alicia Garza, and Opal Tometico—founded Black Lives Matter in 2013. In addition to 17 chapters in the United States and Canada, Black Lives Matter has millions of online members and has spearheaded hundreds of protests against racial injustice in the United States and around the globe.[32]

Types of Small Groups

All groups are categorized along three dimensions: the extent to which they are social versus work related, voluntary versus involuntary, and spontaneous versus planned, as shown in Figure 9.7. Groups high on the social dimension form primarily for companionship and camaraderie. Group members may enjoy similar activities, such as playing pool or hiking. Social groups also may center on addressing personal or professional competencies, as with self-help groups and student clubs. Groups with an emphasis on work focus on accomplishing specific tasks, solving problems, and making decisions in the workplace. Committees, task forces, and project teams are examples of work groups.

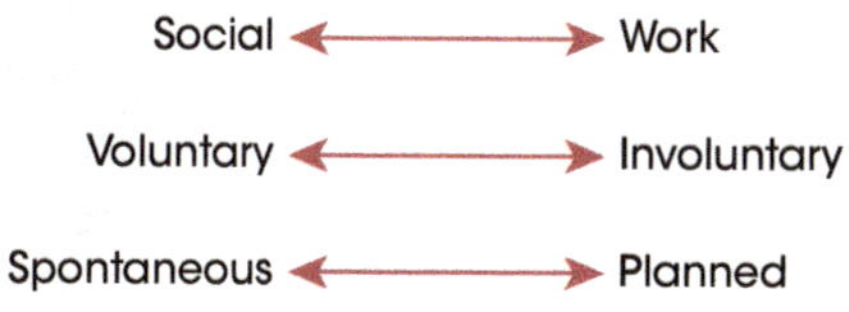

FIGURE 9.7 Groups can be categorized along these three dimensions.

Group membership also may be voluntary, the second dimension in Figure 9.7, as with study groups and community sports teams. One advantage of voluntary groups is members belong out of choice, so they bring a different kind of commitment. However, members of voluntary groups may be more apt to leave when conflict arises, they sense a lack of support or appreciation, or other life activities interfere.[33] When group membership is involuntary, members are required to join. Organizations may mandate employees to work in teams to complete tasks. Instructors often assign students to discussion and project groups.

Spontaneous groups, the third dimension in Figure 9.7, arise out of communication with others. Friendship groups generally fit this category, as do informal groups that develop in the workplace. When people interact with each other over time, bonds begin to grow, members start to think of themselves as a group, and they perceive a common goal, as with coworkers who meet every week for lunch or friends who get together for a yearly ski trip. In contrast, planned groups form for a purpose, such as completing a task or gathering to share similar experiences. Such groups may be voluntary or involuntary, social or work. But the group has a particular reason for existence, whether it is completing a project at work or learning more about gardening.

Your first experience with small groups is a social, involuntary, planned one—your family, also called the **first place group**. As a social group, families teach group members about how to interact with others. Along with socializing people into the larger society, parents, siblings, grandparents, aunts, uncles,

first place group
The family, which is a social, involuntary, planned group

cousins, and other family members provide important lessons in making decisions, solving problems, sharing feelings, listening, and accomplishing tasks as a group. Membership in families is involuntary because individuals do not get to choose their kin. Family groups are planned in that they serve many purposes, such as working together to provide food and shelter for members.

second place groups Work groups, which are planned and may be voluntary or involuntary

With the current emphasis on team-based organizing, **second place groups**, or work groups, are part of nearly every workplace. This chapter primarily is concerned with second place groups, although many of the concepts discussed apply to all types of groups. Second place groups focus on work and are planned but may be voluntary or involuntary. For example, an employee may be assigned to a project group and may volunteer to serve on a committee. Work groups generally serve problem-solving and decision-making functions. Quality circles assess an organization's performance and how well the organization functions. For instance, if a company noticed an increase in employee turnover, a small group of managers would convene to investigate the problem and offer solutions. Standing committees address tasks an organization faces regularly, such as recruiting new members and giving out performance rewards. Ad hoc committees form for a specified purpose and time. For example, if a company were considering whether to provide day care for employees' children, an ad hoc committee could be assigned to research the issue and present a report by a certain date. Once committee members completed their task, the committee would be disbanded.

third place groups Social groups, which may be voluntary or involuntary

Families prepare individuals for interacting in a range of **third place groups**, social groups such as school clubs, sports-related activities, and therapeutic groups. Although third place groups often are voluntary, as with book clubs and self-help groups, some members may be forced to join. For example, first-time drug offenders may be ordered to attend recovery self-help group meetings, and students may be required to join a community group as part of a service learning project. In the past, third place groups met only in-person. Now, many third place groups meet online, often using video conferencing software, such as Zoom, Join.me, Google Meet, and Skype. For example, during the COVID-19 pandemic, hundreds of Alcoholics Anonymous meetings were available online.[34]

Groups change over time, so a group may begin as a spontaneous social gathering of friends and morph into a neighborhood group that tutors kids after school. Alternately, members of a planned work group may start getting together after work and form a spontaneous social group that endures long after the work group no longer exists.

Small Group Phases

Small groups typically go through four phases or stages in their development: forming, storming, norming, and performing. As group members get to know each other, encounter various tasks and challenges, and achieve their objectives, how group members interact with each other changes. Although not every group progresses through these phases in the same way, the model provides a general framework for understanding how group members interact over time. Figure 9.8 summarizes the phases of small groups.

The **forming phase** occurs when group members first join together. In this phase, group members get to know each other and the problem or task they need to address. Group members typically feel uncertain about each other, their tasks, and the group's goals. Communication focuses on learning about group members and what the group needs to accomplish. Conversation usually is polite, with light humor and the avoidance of conflict. Whether meeting in-person or online, communication is tentative, revealed in awkward pauses and unintentional interruptions and talk-overs. During the forming phase, group members orient themselves to the task and develop initial ideas about how to interact with each other and ways to proceed with accomplishing their work.

Conflict provides the key characteristic of the **storming phase**. Now that group members have grown more accustomed to each other, they are more likely to express disagreement openly. Three common dialectics, or tensions, in small groups contribute to the issues groups face in the storming phase. The independence-interdependence dialectic reflects each group member's desire for both autonomy and connection. That is, you want to be part of the group, but to do that, you must give up some of your independence. Closely tied with the independence-interdependence dialectic is the group identity–self identity dialectic. Group members ask themselves, How am I the same as the others in the group? How am I different? The third dialectic, inclusion-exclusion, points to the tensions associated with group boundaries. Determining group membership may appear a simple matter, but group boundaries often are flexible and fluid. For example, who do you include in your family? What happens in blended families? How many generations are included? What about close friends? Are pets part of the family? In all small groups, members draw boundaries at some point, although not all group members will agree on the same ones. What you consider family may be different from what your parents, siblings, cousins, and others consider family. These tensions in small group interactions are inevitable and ordinary. But how group members manage conflict will impact their ability to work together to accomplish their goal. Successful conflict management,

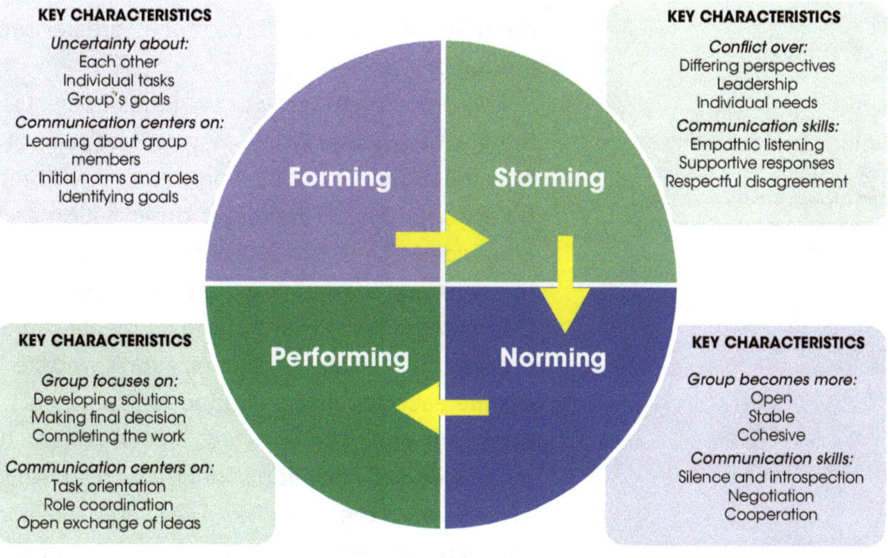

FIGURE 9.8 Small groups generally progress through these four phases, although the progression may not be linear as group members revisit earlier issues.

forming phase The first phase of group development when group members first join together

storming phase The second phase of group development when groups encounter conflict and tensions

discussed in Chapter 10, produces greater group cohesion and cooperation among group members.

As group members work out their conflicts in the storming phase, they enter the **norming phase** where they develop rules of interaction, or norms, that guide how the group functions. Some norms will be specific to the group; others will be consistent with the larger organization and culture in which the group is embedded. Some group norms are explicit in that group members talk about the rules and may even write them down. For example, the California Coastal Commission holds monthly public meetings that follow specific rules and procedures outlined on the organization's website.[35] One rule specifies the amount of time each speaker is allotted for comments. In most cases, however, group norms are implicit or unstated. Greeting rituals, turn-taking, and other interaction norms generally develop over time and are seldom discussed. Yet frank conversations about these implicit rules can make group functions more transparent and increase the likelihood that group members will work to adopt useful group norms.

Central to small groups is some sort of outcome or product, whether it is a report on campus climate, improving members' musical skills, caring for a sick patient, or reaching a verdict in a jury trial. In the **performing phase**, the group accomplishes the task it set out to do. How a group functions in the forming, storming, and norming phases directly impacts the group's ability to achieve its goal while maintaining productive relationships among members in the performing phase.

Over time, groups may return to earlier phases as new members join, other members leave, the group faces new tasks, and changes occur in the environment that affect the group (Figure 9.9). For example, during the COVID-19 pandemic, groups across the spectrum—from families to knitting groups to management teams—held real-time meetings via video conferencing. These groups revisited their interaction norms, revising them to meet the demands and constraints of online communication.

norming phase The third phase of group development when groups create norms of interaction that guide how the group functions

performing phase The fourth phase of group development when the group reaches goal attainment

FIGURE 9.9 Small groups may go back to previous development phases in response to internal and external changes, as when moving from in-person to online meetings.

Group Roles

Often, group roles emerge in group interaction, with members fulfilling responsibilities such as asking questions and giving information as the need arises. In other cases, group roles are assigned. For example, research on groups that meet exclusively online found that assigning group roles at the start of group formation improved the group's performance.[36]

Consider the various small groups to which you belong. What roles do you play in those groups? **Task roles** focus on moving the group toward its goal and accomplishing specific objectives or assignments that must be completed. Group members who provide relevant information, seek others' opinions, and elaborate on others' ideas fulfill task roles. **Maintenance roles** are concerned with facilitating constructive relationships among group members, as with managing conflict and encouraging everyone to participate. **Self-centered roles** interfere with the group reaching its goal and can lead to poor relationships in the group. Making inappropriate jokes, dominating the group's discussion, and refusing to participate are disruptive behaviors.[37] An analysis of communication behaviors in group meetings found that self-centered roles reduce group members' satisfaction with the meeting and short circuit the discussion of ideas.[38] Table 9.3 provides examples of the types of roles members may play in a group.

Group members often fulfill multiple roles, and the roles they fulfill may change over time as the group evolves. Recent research on workplace teams in four different countries found that job satisfaction increased with the number of roles a group member fulfilled.[39] Taking on multiple roles can give individuals a sense of self-worth and accomplishment as they contribute toward achieving the group's goal. As a group member, take on task and maintenance roles while avoiding self-centered ones. The most effective groups match group member competencies with group needs.[40] If you are someone who enjoys doing research, then the information giver role is a logical one for you. However, if the group already has plenty of people fulfilling that role, consider your other skills that might not be as evident, such as managing conflict (harmonizer) and identifying the relationships among ideas (coordinator).

Leading Small Groups

Small group leaders are those members who influence others through communication, moving the group forward toward its goal.[41] An individual may emerge as a group's leader, or someone may be appointed leader of a group. An elected or appointed group leader is a **designated leader**. Small groups may rotate the leader role among members either by design or by individual member's initiative. Leaderless groups often encounter more difficulties achieving their goals than groups with identifiable leaders. However, leaderless groups likely suffer from a lack of leadership rather than simply not having a leader. That is, groups may be able to function without leaders, but they cannot function without competent leadership.[42] Effective small group leadership motivates others to contribute, matches group members' expertise with the group's goals, and provides direction for the group.

TABLE 9.3 Group Member Roles

CATEGORY	TYPE	BRIEF DEFINITION	EXAMPLE
TASK ROLES	**Initiator**	Proposes new ideas; initiates discussion	"Consider the problem from the student's viewpoint rather than the teacher's perspective."
	Information seeker	Requests information and evidence	"What did you find in your research on mandatory advising for college students?"
	Information giver	Provides information and evidence	"A recent article in the *Chronicle of Higher Education* reported that cellphones are increasingly used for cheating in college classrooms."
	Opinion seeker	Asks others to share their opinions on the topic	"How do you feel about the city's new parking policy?"
	Opinion giver	Offers own opinion about topic	"I think our organization should require members to volunteer their time at least 4 hours a month."
	Elaborator	Extends on others' ideas	"To add to your idea, we could also give conference participants a list of local green businesses."
	Coordinator	Links ideas together	"Combining those two proposals might provide an even stronger rationale for the project."
	Orienter	Reviews group's progress and keeps discussion on track	"We've come up with several ideas to encourage more employee participation but recall that our charge is first to identify the current level of that participation."
	Evaluator	Assesses proposed ideas and solutions	"Based on the criteria we developed to evaluate each solution, the second one we discussed is the best."
	Energizer	Moves group's activities toward its goals	"There must be other possibilities. With all the talent in this room, I'm sure we can generate more creative solutions to this challenging problem."
	Procedural technician	Schedules meetings and acquires supplies	"Our video conference is noon Eastern Time. I'll email everyone the agenda and reports for discussion."
	Recorder	Records and archives group discussions	"I've downloaded the group's recorded video meeting and posted it to our secure server."
MAINTENANCE ROLES	**Encourager**	Praises others' work and contributions	"Thank you for finding those relevant articles for our discussion."
	Harmonizer	Manages conflict and reduces tensions among members	"Let's try to find some common ground here."
	Compromiser	Willing to change position if others will	"I'll complete my section of the report a day early if someone else will do the proofreading."
	Gatekeeper	Mindful of each person's communication time so no one dominates and no one is left out	"We want to make sure there's time for each person to contribute their thoughts on this issue."
	Process observer	Monitors and evaluates group processes and procedures	"The speakers list doesn't allow for free-flowing discussion."
	Follower	Listens and accepts others' ideas	"I agree with the way we've divided the workload in the group."

(*continued*)

TABLE 9.3 Group Member Roles *(continued)*

CATEGORY	TYPE	BRIEF DEFINITION	EXAMPLE
SELF-CENTERED ROLES	**Aggressor**	Uses personal attacks	"You're lazy and incompetent."
	Blocker	Opposes others' ideas without offering alternatives	"I won't support that proposal. There's no way it will work."
	Recognition seeker	Promotes self and own ideas	"When I was the president of my high school theater club, I earned the student service award my senior year."
	Self-confessor	Discloses personal information not relevant to the group	"I had this dream last night about surfing these massive waves in Australia."
	Joker	Use inappropriate humor and jokes	"I heard this great joke last night on one of the talk shows. You'll love it! Ha ha!"
	Dominator	Controls the group's conversation	"Oh, I know all about the company's new procedures for information technology security. First …"
	Help seeker	Appears helpless to avoid contributing	"Well, I've never used the library databases before. I'm just no good with figuring out how they work."
	Isolate	Does not participate in or contribute to the group	"I don't have anything to add."

Adapted from Kenneth D. Benne and Paul Sheats, Functional Roles of Group Members, *Journal of Social Issues*, vol. 3, no. 2, pp. 41-49. Copyright © 1948 by John Wiley & Sons, Inc.

Approaches to Leadership

Four approaches to leadership are especially relevant to small groups: democratic, transformational, participative, and distributed.[43] You likely are most familiar with **democratic leadership**, where group members vote on who will serve as leader. The leader then consults with the group on decisions, typically through some sort of discussion and voting process. Democratic leadership tends to engender confidence and trust in group members, as they have input in choosing the leader and decision making.[44] However, democratic leadership assumes group members are knowledgeable about the topics under consideration and are eager to participate in solving the problems the group faces.

Transformational leadership promotes change in the individual and group. Leaders inspire and motivate others to identify change that needs to occur and devise a plan to implement it. Transformational leaders offer themselves as role models for change, providing the group with a vision for what members can work on together and accomplish.[45] Research on transformational leadership has found that group leaders taking this approach support cooperative interaction norms and information sharing among group members, which lead to greater creativity and innovation in problem solving.[46] However, too much emphasis on change can overburden group members, resulting in undue stress and unnecessary work.[47] Thus, effective transformational leaders must balance innovation with stability in the group.

democratic leadership
Group members vote on who will serve as leader

transformational leadership The leader promotes change in the individual and group

participative leadership The leader shares power with group members, empowering them to take initiative in their work and in making decisions

distributed leadership Occurs when group members share leadership responsibilities, completing tasks and maintaining relationships when the need arises

In **participative leadership**, the leader shares power with group members, empowering them to take initiative in their work and in making decisions. Research on this approach to leadership in small groups has found an increase in the group's creativity, sharing of ideas, and perceptions that each group member's contributions matter. However, groups using participative leadership tend to be less efficient in coming to a final conclusion.[48]

With **distributed leadership**, group members share leadership responsibilities, completing tasks and maintaining relationships when the need arises (Figure 9.10). Rather than one person in the group taking charge of all task and relationship facilitation, all group members share in fulfilling leadership functions. Research on small group decision making and leadership supports this idea of distributed group leadership. Groups in which one person performs all the leadership functions often are not as successful as those in which leadership activities are shared.[49] In single-leader groups, members are concerned with giving everyone a chance to speak. In contrast, shared-leadership groups prioritize contemplation and reflection—listening to what others have to say. Although leaders typically emerge or are designated in a group, all group members are responsible for engaging in leadership behaviors. In this sense, leadership is distributed among group members.

How are the four approaches to leadership—democratic, transformational, participative, and distributed—enacted in small group communication? Find out for yourself by observing and talking with group leaders and members in the activity proposed in Box 9.3.

FIGURE 9.10 Distributed leadership encourages group members to contemplate and reflect on what others have to say.

Leading Community Groups

Different groups require different approaches to leadership. Find out how the leaders of groups in your community lead both from theirs and their followers' perspectives. First, identify a few small groups in your community, either where you live, in the city or town in which your campus is located, or on the campus itself. Contact the groups and arrange to observe one or two of their meetings. Then conduct a few informal interviews with the groups' leaders and some of the members. What leadership approach did you observe the groups' leaders using? How did they describe their leadership approach? How did group members describe their leaders' approaches? What similarities did you find across the interviews? What were some differences? Based on your observations and interviews, how effective were the leaders in leading their groups? How will you apply what you have learned to your own roles as a small group leader and follower?

Taking the Lead

While there is no magic formula that makes someone a good leader, there are leadership behaviors that are more likely to increase the group's success in carrying out its work and achieving its goal summarized in Table 9.4. First, effective leaders promote a sense of "us." The leader is instrumental in developing the group's identity, moving the group from a collection of individuals to an integrated unit.[50] For example, the leader may encourage members to come up with a group name or logo to identify the group.

TABLE 9.4 Examples of Effective Leadership Behaviors

BEHAVIOR	BRIEF DESCRIPTION	EXAMPLE
Promote a sense of "us"	Develop group identity	"We could not have made today's deadline for our report without everyone's efforts. Let's give ourselves a round of applause."
Communicate authentically	Honest, informed communication	"Additional survey data from the marketing team offers a credible counterpoint to our original plan. This information is not what I expected, but I've reviewed the team's methods and the research is sound."
Practice humility	Humble and confident, not arrogant and conceited	"I apologize for not sending out the meeting materials sooner. I realize I didn't give you much time to review the documents. I'll give you a quick summary so we can at least start our discussion."
Celebrate diverse perspectives	Encourage all group members' voices	"To be sure we get everyone's perspective on the design for the new program, I suggest we review it in three steps. First, I've set up an online document where all can make suggestions and comments. Second, we'll individually review those notations. Third, we'll have a synchronous online conversation to make final decisions."
Create and implement a vision	Focus the group on the future	"Let's go back to our initial vision. How does our current plan fit with where we envision the group in the future? Do we need to revisit the plan or our vision?"

Second, effective leaders communicate authentically in two ways. One, effective leaders communicate with group members from a place of honesty without hidden agendas, political motives, or personal goals, building trust with and among group members.[51] Two, especially in a pervasive communication environment where deepfakes and pseudo news proliferate on social media (Chapter 3), effective leaders provide accurate, credible, and reliable information relevant to the group's goals and objectives.

Third, effective leaders practice humility, indicating a sincere regard for others and expressing genuine interest in what others have to say. With humility comes humbleness, or an absence of arrogance. Humble leaders recognize that they do not—cannot—know everything. Humility invites other group members to participate and contribute. Leaders who practice humility are confident yet not conceited. For instance, effective leaders do not interrupt others, engage in bullying, or shift the blame for their own errors.[52]

Fourth, effective leaders celebrate diverse perspectives, recognizing and welcoming the many different worldviews and standpoints individuals bring to group processes, functions, and outcomes. Such leaders welcome all group members' voices and honor their lived experiences. Moreover, leaders identify points of intersection and difference in group members' perspectives, developing strategies for balancing competing ideas and managing conflict productively.[53] For example, rather than wait for a group meeting to discuss a draft report, the leader might set up a Google Doc on which all members can make suggestions and comments.

Fifth, effective leaders create and implement a vision. Practitioners and researchers agree that whatever approach a group takes to leadership, an effective leader presents a compelling vision for the group and facilitates the group's ability to enact that vision.[54] An effective vision statement is clear, concise, future-oriented, aspirational, actionable, and inclusive. Group members understand the group's direction, what actions will get them there, and how all group members will be a part of the effort. Group members often contribute to the group's vision, but it is up to the leader to guide that vision and assure its implementation. Use Box 9.4 to develop or revise a vision statement for a group to which you belong.

BOX 9.4: CONNECT & REFLECT

Envisioning a Vision

What makes a vision statement compelling? Find online examples from well-known organizations, such as Alphabet, Amazon, Ben & Jerry's, Feeding America, Harley-Davidson Motor Company, Habitat for Humanity, Instagram, St. Jude Children's Research Hospital, SpaceX, and Zappos. Which statements resonate with you? Why do you find those vision statements appealing?

While effective leaders communicate a clear and compelling vision for the group, members are part of shaping that vision and bringing it to life. With your group for this class or another group of which you are a member, develop, revisit, or revise the group's vision statement by first answering the following questions:[55]

1. Why does our group exist?

2. What is our group's primary goal? What do we hope to achieve?

3. How would we describe our group's culture to those outside the group? What words or phrases best reflect who we are as a group?

4. Why would others want to join our group?

Review your answers to those questions as you craft a single sentence that captures the essence of your group's vision. After you draft a vision statement, discuss these questions:

1. How inspiring is your vision statement? What makes it inspiring? To what extent does it motivate you to do your best?

2. How will you use your vision statement to guide your actions as a group?

3. How might your vision statement change in the future?

4. What have you learned about leadership and vision statements that you will use in future small groups?

Facilitating Teamwork and Discussion

Regardless of the approach group members choose to handle the question of leadership, there are group functions and processes that require leadership behaviors. A designated leader may fulfill some or all of these functions, or more likely, they will be distributed among group members. Two key areas of small group work that especially require effective leadership are facilitating teamwork and leading discussion.[56] The following strategies will help you and your group work together in a cooperative fashion:

- *Use "us" and "we" rather than "I" and "you"*. This simple shift in language promotes a we-not-me orientation in what you say and do.

- *Share all rewards with the group*. When you receive praise from those outside the group, give credit to all group members.

- *Keep disagreements focused on facts and issues, not personalities*. Engaging in name-calling and other derogatory communication is both unethical and counterproductive to group work. Arguments focused on individual people are destructive to the group. However, constructive conflict on issues may spark useful and creative thinking among group members and facilitate the critical thinking process.

- *Challenge any hidden agendas that hinder attainment of the group's goals*. Hidden agendas promote an individual's plans or objectives, which may not align with the group's goals. Hidden agendas contribute to a defensive communication climate and prevent productive group communication. Surface hidden agendas in a calm and constructive way to prevent later, damaging conflict in the group.

- *Recall that group work includes task and relationship dimensions*. Groups need both efficiency and satisfying interactions among group members. When discussions get particularly heated or serious, appropriate humor can relieve tensions and help group members achieve some perspective on the task at hand.

Whether meeting online or in-person, discussion leadership plays a key role in group decision making, problem solving, and the group's overall effectiveness.[57] Your group may want to assign group members to lead specific discussions, rotating the responsibility for this task. Or you may want to put one person in charge of leading all group discussions. Follow these procedures for effective discussion leadership:

- *Distribute the agenda in advance of the meeting*. Include key meeting elements, as with the time and place, purpose, and items for discussion. Review Figure 9.11 for a sample agenda.

- *Briefly review or explain the discussion's purpose*. Although group members should be familiar with why the group is having a discussion, make the purpose clear at the start.

- *Identify group procedures*. Group members must agree on the procedures to follow for solving problems and making decisions (Chapter 10).

Communication Club Executive Council Agenda
September 5, 2023, 3–5 pm
Communication Building, Room 305 and
Video Conference Meeting ID 12345678 Password 12345

Purpose: By the end of this meeting we will have identified three events the Club will sponsor this semester.

Item	Time	Outcome
1. Self-introductions	3–3:10 pm	Review group members' names and check attendance
2. Minutes from previous meeting (attached)	3:10–3:15 pm	Make needed revisions and approve minutes Identify tasks yet to be completed
3. Managing meeting tasks	3:15–3:30 pm	Develop list of meeting tasks (e.g., arrange for room, schedule video conference, take minutes) Assign/volunteer group members to each task
4. Report on last year's events (attached)	3:30–4:00 pm	Identify what worked and what didn't for the events the Club sponsored last year
5. Possible events for the Fall 2023 semester	4:00–4:45 pm	Identify one professional event, one philanthropic event, and one social event the Club will sponsor Determine potential dates for the events Assign one person to serve as lead coordinator for each event
6. Future meetings	4:45–5:00 pm	Identify topics/issues for this semesters' meetings

FIGURE 9.11 Distributing an agenda prior to a meeting serves as a reminder for group members and also encourages them to prepare for the meeting.

- *Balance completing tasks with maintaining effective interpersonal relationships.* Set aside meeting time for socializing so that task time can be used for accomplishing group work.

- *Assist the group in covering all items on the agenda.* Attend to how much time is spent on each agenda item. You may have to limit discussion on some items to allow time to discuss all items. However, it may be the case that the group simply has an overly ambitious agenda. In such a situation, the group must decide which items to discuss now and which to place on the agenda for the next meeting.

- *Listen carefully to what each group member has to say, especially to group members who do not speak often so you can encourage their participation.* Match the group's goal to your listening orientation (Chapter 4). Effective listening is a key leadership behavior and essential for making sound decisions.

- *Promote a supportive communication climate that encourages all group members to share their thoughts.* This does not require that everyone will speak for the exact same amount of time. However, no group member should dominate the conversation.

- *Keep disagreements focused on issues rather than people.* Do not let your ego get involved when group members disagree.

Good Leaders Need Good Followers

Leading implies following. Without followers, there are no leaders. Leaders depend on their followers to complete tasks, carry out group decisions, and support the group's goals. Good followers make for good leaders. Not everyone can lead all the time, and not every group member is interested in leading. As with leaders, group members vary in their effectiveness as followers, with some exhibiting competent followership that moves the group toward its goals and others who impede the group's processes.

Although typically construed in negative terms, such as submissive, passive, and mindless, competent followers do more than follow instructions. In fact, sometimes good followers do not do what they are told. For example, in 2009, nearly 20 students on San José State University's swim team alleged that the team's physician engaged in inappropriate touching during physical examinations. The university investigated and found no wrongdoing at the time. However, the swim coach did not drop the matter and compiled a report that he shared with individuals within and outside the university. Due to his efforts, 10 years later, the university reopened the investigation. When the university's athletic director ordered her deputy to discipline the swim coach for raising the allegations in public, he declined. In addition, the new investigation found the student-athletes' claims of sexual abuse were true.[58] Both the swim coach and deputy athletic director refused to go along with university leadership, demonstrating the courage needed for effective followership.

Followership refers to the activities of those who are subordinate to a group's leader. Recent research has identified the centrality of followership to effective group work. For example, a study of informational technology (IT) teams found that followers' skills and confidence were directly related to positive team outcomes and perceptions of leaders' abilities.[59] Research on organizational responses to crisis found that followers' contributions proved essential to achieving consensus on successful adaptation strategies. Interestingly, leaders' competing visions slowed teams in their progress to reach viable solutions, whereas followers were more likely to achieve an integration of their perspectives and agreement on the best way forward.[60] Not surprising, a study of ship repair teams found that followership voice—followers effectively expressed their viewpoints—contributed to leaders' proficiency, which in turn produced positive group outcomes.[61]

Followership encompasses two primary dimensions: active engagement and independent critical thinking.[62] Followers who are actively engaged contribute to the group's tasks, help out others when needed, complete their own work on time, and support the path the leader has charted for the group. Followers who are independent critical thinkers offer constructive feedback and comments, consider alternative solutions, and challenge faulty decision-making processes.

followership The activities of those who are subordinate to a group's leader

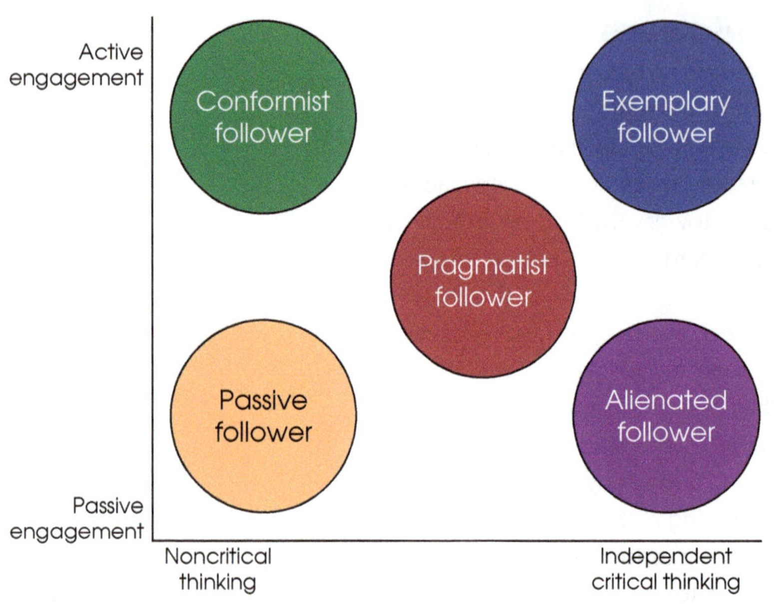

FIGURE 9.12 Followership behaviors vary in their productive contributions to the group and the group's leadership.

As Figure 9.12 shows, active engagement and applying critical thinking skills lead to followers who offer the greatest support for the group and the leader.

Research on followership suggests that those followers defined as exemplary are essential to establishing effective relationships between leaders and followers.[63] Importantly, group members do not simply decide to take on particular followership roles. Rather, those roles emerge out of group members' communication as they negotiate group processes and procedures. For example, in a group's initial interactions—the forming phase discussed previously—followers may express uncertainty, admit to a lack of knowledge, or seem hesitant in voicing their perspective.[64] In that way, followership roles are fluid and often change over time. Similarly, group members may move between leading and following. Research on interprofessional health care teams identified shifting from leadership and followership roles as central to connecting team members' expertise and delivering comprehensive patient care.[65]

In practice, effective followers exhibit several key qualities that provide support for the group's leadership and at the same time offer a check on leaders who may stray away from the group's goals and objectives.[66] These key qualities include:

- *Loyalty*: Effective followers are committed to the group's goals and the leader's vision, demonstrating their commitment by engaging with other group members and completing assigned tasks.

- *Adaptability*: Effective followers adjust to changes in the group and the larger environment in which the group functions by proactively monitoring activities within and outside the group.

- *Collaboration*: Effective followers work cooperatively with other group members, seeking solutions designed to incorporate all members' perspectives and acknowledging others' contributions.

- *Courage*: Effective followers take initiative, speak out against injustice, engage in independent thinking, and critically examine their own and others' motives.

- *Integrity*: Effective followers are honest, trustworthy, and caring in their communication and actions, interrogating their own ethical practices in their group work.

As these qualities suggest, competent followers are fully enmeshed in their small groups, exhibiting high levels of active engagement and independent critical thinking—these are the exemplary followers on which leaders depend. Table 9.5 offers examples of good followership behaviors.

TABLE 9.5 Examples of Effective Followership Behaviors

QUALITY	BRIEF DESCRIPTION	EXAMPLE
Loyalty	Committed to group and leader	"I believe in our mission and the research I'm reporting on today supports our long-term goals."
Adaptability	Proactively adjust to change	"The CEO just announced a new reorganization plan. What are some ways our group could respond to those changes?"
Collaboration	Seek integrative solutions	"Ramael and Aundria developed a spreadsheet that identifies similarities and differences in our team's three proposals to give us a better perspective of how the various ideas might fit together."
Courage	Take initiative, fight injustice	"While I agree online registration is one solution, it seems to me that this strategy privileges people with computer skills and reliable internet access and leaves out those who are less tech savvy or have limited internet access."
Integrity	Follow ethical practices	"I can't take credit for that idea. One of my professors mentioned it in class last week. I could interview her and find out more about it."

Building a Successful Small Group

As you likely have experienced, not all groups achieve their goals or produce a satisfactory outcome. Research on group success and failure identifies six aspects of small group work that will move the group in a positive direction: group cohesiveness, agreement on group roles, cooperative group processes, emotional intelligence, supportive communication climate, and social facilitation.[67]

Group Cohesiveness

Relationships among group members directly impact group performance. When group tensions in the storming phase remain unresolved and group members pursue their personal goals, the group is more likely to fail. In contrast, groups that productively navigate and manage the tensions associated with group work more often reach a successful outcome.[68]

A sense of unity, cohesiveness, and support generally characterize positive interpersonal relationships among group members. Cohesiveness serves a key function in uniting the group, encouraging group members to share information and resources.[69] More cohesive groups also exhibit increased engagement with the task, which results in more creative solutions and better task performance.[70] That is, group members who have a greater sense of togetherness view the work they have to do as more interesting and exhibit more commitment to a positive outcome. A recent study of online and in-person college student project groups found that the online groups were more cohesive, more satisfied with their group experience, and performed better on the project than the in-person groups. Interestingly, the online group members were more polite to each other and trusted each other more than the students in the in-person groups.[71] The cohesiveness that developed in the online groups facilitated positive interactions and a focus on the group's objectives. Another study that examined the use of a social network messaging app found that online student teams who used the app more often were more cohesive and performed better than the teams that used the app less often.[72]

Two types of cohesiveness occur in small groups: task cohesion and social cohesion. **Task cohesion** refers to good working relationships among group members, agreement on desired outcomes, and ways to achieve those outcomes. **Social cohesion** occurs when group members are attracted to the group and want to remain in it.[73] Generally, groups with higher levels of task and social cohesiveness are more satisfied with group outcomes, produce better quality outcomes, and enjoy greater productivity from individual group members.[74] Social cohesion is a key element in the group's viability. That is, groups with higher levels of social cohesion are more likely to stay together over time. However, very high levels of both types of cohesion can be too much of a good thing. High levels of social cohesion may be especially problematic, causing group members to waste time socializing rather than working, get distracted from the group's goal, and avoid giving constructive criticism. Too much task cohesion, although less of a problem in group performance, can lead group members to focus exclusively on the task and ignore interpersonal relationships, become overly critical of each other, and promote a defensive communication climate.

Agreement on Roles

A group's structure develops from the norms and roles created in the norming phase. Norms that stress a common goal and roles that match group members' expertise contribute to a group's success. Explicitly discussing the specific roles group members fulfill increases role clarity and reduces role ambiguity. **Role ambiguity** occurs when a group member is unsure of the responsibilities and behaviors associated with a role, how performance of the role is evaluated, and what will happen if the responsibilities are not met.[75] Say, for instance, you volunteer to serve as your group's recorder, but you are unclear on what you need to do. To reduce your role ambiguity, you ask the other members of the group, and they explain that they expect you to take notes during in-person meetings and archive all online group conversations. Persistent role ambiguity leads group members to lose confidence in their abilities to complete a task and reduces task cohesion among group members. However, initial role ambiguity can lead to group

---

task cohesion Involves good working relationships among group members and agreement on desired outcomes and ways to achieve those outcomes

social cohesion Occurs when group members are attracted to the group and want to remain in it

role ambiguity Occurs when a group member is unsure of the responsibilities and behaviors associated with a role, how performance of the role is evaluated, and what will happen if the responsibilities are not met members exploring the possibilities associated with a particular role and developing innovative ways to fulfill it.[76] For example, a group member acting as the gatekeeper could view the role as involving speaking and listening opportunities, rather than just focusing on speaking. Leaders also influence group members' degree of role ambiguity (Figure 9.13). When leaders are clear about the group's objectives and empower group members to achieve the group's goals, members generally experience less ambiguity about their roles.[77]

FIGURE 9.13 When group members feel empowered to make decisions, role ambiguity is reduced and the group is more productive.

Fulfilling the various task and maintenance roles in the group moves the group toward its goal. When group members have conflicting goals and roles are unfulfilled or poorly aligned, members will struggle to achieve their goal. In addition, self-centered roles and flagrant violations of the group's norms negatively impact the group's performance. Research on food preparation teams in a fast-food chain found that behaviors associated with self-centered roles, such as aggressor, joker, and dominator, caused all members of the teams—even those not engaging in the behaviors—to make more mistakes in completing their tasks.[78] Leaders can discourage poor team member behavior, and do so without resorting to punishment. For example, a study of three technology companies in China found that team leaders who encouraged members' passion for their work and developed positive relationships with team members reduced the likelihood that self-centered roles would surface in the group.[79] As another example, members of college student project teams who viewed their leaders as trustworthy and behaving with integrity were less likely to engage in disruptive and obstructive behaviors.[80]

Cooperative Group Processes

When group members cooperate rather than compete, they are more likely to achieve successful outcomes.[81] Cooperative group members share information and other resources, help each other when needed, welcome and acknowledge the value of differing perspectives, and view themselves as working toward the same goal (Figure 9.14). Openly discussing and identifying clear expectations for the group and group members promotes a more cooperative communication climate.[82]

FIGURE 9.14 Successful groups develop cooperative processes for working together.

Designing a productive reward system encourages group member cooperation.[83] Rewarding group members just for participating hinders the development of cooperative behaviors because the reward is not linked specifically with cooperation. Group members must be rewarded for working together, as with collaborating on a project proposal or cochairing a meeting. A reward may be a simple acknowledgment of the person's efforts or something more substantial, as with getting to select the group's next task to complete.

Perceptions of fairness also influence levels of group cooperation. Developing norms that reinforce equity in rewards encourages group members to communicate cooperatively. Although group members may contribute to the group's performance in different ways, all expect to be rewarded for their efforts. And when group members do not do their work, others expect those who fail to cooperate to endure some sort of equitable punishment or lack of rewards.[84] Punishment can be effective in sanctioning poor behavior and encouraging cooperation, although it must be used with care and not arbitrarily.[85] The most effective punishment systems for uncooperative behaviors involve all group members in the decision to punish.[86] That is, just as your group determines what and how cooperative behaviors should be rewarded, you also decide together on what and how uncooperative behaviors should be punished.

Emotional Intelligence

Emotional intelligence involves promoting intellectual and emotional growth through the application of emotional knowledge in appropriately identifying, expressing, and regulating emotions verbally and nonverbally (Chapter 3). When group members effectively manage emotions, they are more likely to engage in information-seeking behaviors essential to completing problem-solving tasks.[87] In the academic setting, emotional intelligence contributes to better assignment performance for student groups. For example, a study of science and technology master's student teams found that those with higher levels of emotional intelligence earned higher scores on their project reports.[88]

Emotional intelligence in small groups contexts is essential for appropriately using emotions in decision-making and problem-solving situations. Individuals with high emotional intelligence are better able to recognize emotions, apply them in thought processes, understand what an emotion means, and manage emotions more productively than individuals with low emotional intelligence. Those with high emotional intelligence display more confidence in their abilities to work in small groups and communicate more positively, openly, and agreeably with others.[89] For example, research on university library disaster response teams found emotional intelligence contributed to teamwork, effective communication, trust, responsiveness to change, and taking initiative.[90] Research on groups that meet solely online found that emotional intelligence promoted collaboration, accountability, active listening, and integrity.[91]

Group members who emerge as leaders typically exhibit higher levels of emotional intelligence than those not chosen as leaders. Moreover, leaders with higher levels of emotional intelligence are evaluated as more effective in both the task and relationship aspects of group work.[92] Emotional intelligence sensitizes leaders to others' motivations, making it easier for leaders to coordinate the group and improve performance. Groups in which leaders have higher emotional intelligence are more committed to the group and its goals.[93] When all group members have higher levels of emotional intelligence, the group is more likely to achieve its goal and attend more closely to group processes. That is, not only do team members get the work done but they also interact cooperatively and productively. Moreover, higher levels of group member emotional intelligence increase satisfaction with the group's communication and levels of support from group members as well as lead to more effective decision making.[94]

Some groups bring together participants with high levels of emotional intelligence. But for most groups, members have varying levels of emotional intelligence. However, group members can increase their emotional intelligence by becoming aware of and discussing their emotions, expressing empathy, acknowledging others' emotions, demonstrating sensitivity to their own and others' complex emotions, and helping other group members manage their emotions.[95]

FIGURE 9.15 A supportive communication climate encourages all group members to participate and contribute to the group.

Supportive Communication Climate

As group members interact with each other, they create a **communication climate**, or the tone of interaction among group members. A group's communication climate influences how comfortable members feel participating, the way they approach disagreement, how they view diverse perspectives, and the strategies they use to address power issues in the group.[96] When groups foster a supportive communication climate, the quality of their work improves (Figure 9.15).[97] Developing a supportive communication climate and equality in participation promotes goal attainment. For example, research on global online teams found that relationship-building messages exchanged early in each team's development facilitated a supportive communication climate and improved the team's performance.[98] In addition, groups with more supportive communication climates are more likely to embrace innovation and express overall satisfaction with the group.[99] Defensive communication, poor listening, and not sharing information often lead to a group's failure to reach its goal.

Trust is a key component in developing an effective communication climate in small groups. Box 9.5 asks you to consider ways to develop trust in your small groups.

communication climate The tone of interaction among group members

cognitive trust A logical reason to believe in other group members

affective trust The emotional bond and reciprocity of care and concern group members feel toward each other

---

BOX 9.5: CONNECT & REFLECT

Trust and Productivity

Trust among group members contributes to a supportive communication climate and makes groups more productive. Group trust requires both cognitive and affective trust. **Cognitive trust** refers to a logical reason to believe in other group members. For example, you trust the other members of your group to effectively complete the tasks they have taken on. **Affective trust** refers to the emotional bond and reciprocity of care and concern group members feel toward each other. Groups that report high levels of cooperation, express shared values, and feel a strong *esprit de corps* exhibit affective trust. When groups make a commitment to dialogue—a commitment to understanding what others think and feel—they also foster cognitive and affective trust. In turn, group trust improves a group's problem-solving and decision-making abilities, quality of tasks completed, and on-time performance. In addition, group trust reduces the mistakes group members make.[100] Consider the groups you belong to and how you might apply dialogue and trust to enhance productivity. To what extent do group members cognitively trust each other? To what extent do they affectively trust each other? How might you facilitate more trusting relationships in your groups?

Social Facilitation

People often respond to participating in groups in one of two ways: putting forth *more* of an effort than they would if working alone or putting forth *less* of an effort than if working alone. The first, **social facilitation**, results in groups outperforming individuals. That is, when a group and an individual attempt to complete the same task, the group does a better job because members work harder together than they would on their own. For example, research has found social facilitation positively impacts exercising—individuals generally exercise for a longer time and with greater effort in groups than alone.[101] The second, **social loafing**, results in individuals outperforming the group. When group members exhibit social loafing, they contribute less to the group's efforts than they would if they were working on their own. Social loafing has a direct negative impact on group performance, whereas social facilitation has a positive impact.[102] For example, research on student study groups at four universities in Taiwan found that higher levels of social loafing reduced overall group learning. That is, when group members contributed less, the group was less productive and learned less of the class material.[103] A study of teams using 3D virtual worlds to collaborate on projects found that increased social loafing resulted in decreased knowledge sharing.[104] Thus, when group members engage in social loafing, they keep key information to themselves rather than giving it to the group.

Groups can take specific steps to combat social loafing and promote social facilitation.[105] First, working collaboratively in the group's forming stage encourages all group members to contribute to group decisions, such as defining the group's goals and objectives. Developing an inclusive, supportive communication environment invites all group members to participate, builds cohesion, and reduces social loafing.[106] Second, research shows that mindfulness plays a key role in reducing social loafing.[107] As you learned in Chapter 4, with mindful communication you consciously process information in your interactions. When you are mindful, you are less likely to engage in social loafing because you are attending to the members of your group and considering how you might contribute. Third, linking specific group members with specific tasks, holding individuals accountable for completing tasks, and recognizing task completion minimize social loafing and maximize social facilitation. Discussing individual group member roles and desired outcomes reduces role ambiguity and opportunities for social loafing. Fourth, rewarding group members for a job well done encourages them to put forth greater effort. Identifying appropriate consequences for not completing tasks discourages social loafing. Fifth, careful documentation of all the group's work, including collaborative efforts as well as individual contributions, clearly identifies social facilitation and social loafing behaviors. Starting this practice in the group's forming phase allows the group to quickly detect instances of social loafing and work to change that behavior. Finally, providing constructive feedback to all group members on their performance helps them pinpoint areas of strength and areas needing improvement.

---

social facilitation Occurs when individuals put forth more effort in a group than on their own

social loafing Occurs when individuals put forth less effort in a group than on their own

A small group's effectiveness in the performing phase of group development depends on how well the members have handled the challenges presented in the forming, storming, and norming phases. Still, groups that move smoothly through those phases may still not achieve their goals. However, group cohesiveness, agreement on task and maintenance roles, cooperative group processes, emotional intelligence, supportive communication climate, and social facilitation increase the likelihood a group will succeed. Table 9.6 summarizes these factors and provides examples associated with group success and failure.

TABLE 9.6 Factors Contributing to Small Group Success and Failure

FACTOR	BRIEF DESCRIPTION	SUCCESS	FAILURE
Cohesiveness	Positive working relationships and agreement on groups' outcomes	Sense of group unity	Personal goals pursued over group goals, unresolved tensions
Group roles	Group member behaviors and responsibilities	Necessary task and maintenance roles are fulfilled, effective leadership	Unfulfilled or undefined roles, self-centered roles, poor leadership
Processes	Procedures and activities	Cooperative, efficient, organized, prepared	Uncooperative, inefficient, disorganized, unprepared
Emotional intelligence	Expressing feelings and motivations	Trust, respect, pride	Overconfidence, distrust, unenthusiastic
Climate	Communication tone and practices	Supportive communication, active listening, equal participation	Defensive communication, poor listening, information not shared
Social facilitation	Commitment to the group	Dedicated, engaged, informed	Disinterest, social loafing, unmotivated

CHAPTER REVIEW

- **Identify the reasons people join small groups.**

  Although a pervasive communication environment has provided you with new opportunities for joining small groups, the reasons for wanting to join have not changed. Joining small groups can fulfill eight human needs: physiological, safety, belongingness, esteem, cognitive, aesthetic, self-actualization, and transcendence.

- **Discuss the characteristics of small groups.**

  A small group is three or more interdependent people who recognize one another as unique individuals and communicate to achieve a perceived shared goal.

- **Identify and describe the types of small groups.**

  Individuals participate in a range of small groups in their everyday lives. Groups may be work-based or social-based, voluntary or involuntary, and planned or spontaneous. Family provides the first place group, preparing people to function in other social or third place groups, as well as work or second place groups. Individuals join some groups voluntarily, such as school clubs, and join other groups involuntarily, such as assigned work groups. Many third place groups form spontaneously; second place groups are typically planned.

■ **Recognize and discuss the phases of small group work.**
As they develop, groups progress through four primary phases: forming, storming, norming, and performing. Successfully navigating through each phase provides a foundation for achieving the group's goal.

■ **Explain the roles that group members may fulfill in a group.**
Group roles emerge as members interact with each other. Task roles focus on achieving the group's goal. Maintenance roles are concerned with facilitating productive relationships among group members. Disruptive or self-centered roles distract the group from its task and harm the group's interpersonal relationships.

■ **Describe the key components of leading small groups.**
Democratic, transformational, participative, and distributed leadership are the most relevant approaches to leading small groups. Applying a distributed approach to leadership encourages all group members to take responsibility for leading the group. Effective leadership behaviors include promoting a sense of "us," communicating authentically, practicing humility, celebrating diverse perspectives, and creating and implementing a vision. Facilitating teamwork and effectively leading group discussions are two essential aspects of small group leadership. Good leadership requires good followership. Effective followers exhibit loyalty, adaptability, collaboration, courage, and integrity.

■ **Identify the central features of building effective small groups.**
The chances of successful group outcomes are increased when groups are more cohesive, group members have agreement on task and maintenance roles, group processes are cooperative, group members display emotional intelligence, the group promotes a supportive communication climate, and group members adhere to social facilitation rather than social loafing.

DISCUSS AND APPLY

1. Identify two or three groups to which you belong. What were your motivations for joining those groups? What needs do those groups fulfill for you? To what extent have your reasons for staying with the group changed over time?

2. Consider a time when you were interacting with one other person and a third person joined the conversation. How did the interaction change with the addition of the third person?

3. Reflect on groups to which you belong. Are the relationships more role-based or person-based? How do you know?

4. How is your first place group—your family—a small group? How did your family prepare you to interact in second and third place groups?

5. Go to the 123test website and complete the Team Roles Test. What were the results of the test for you? Which roles did the test identify as best matching your personality? To what extent do you agree with the results? Disagree? What insights has the test given you into the roles you fulfill in a group?

6. Describe a current or past small group that had high task cohesion. What were the group's outcomes? How satisfied were you with the group?

7. Research suggests that too much social cohesion can cause problems in groups. Have you ever experienced this? What happened in the group?

8. Think about a small group experience in which the climate was characterized by supportive communication. How did you feel about interacting in that group? Now think about a small group experience in which the climate was characterized by defensive communication. How did you feel about interacting in that group? How do you think you might have changed the communication climate in that group from defensive to supportive?

9. Social loafing is a problem in many small groups. What have you done in the past to combat social loafing? How effective was that approach? How will you work to avoid social loafing in future groups?

CREDITS

Learning Outcomes

- Describe how to promote creativity in small groups.

- Describe the eight-step approach to problem solving in small groups.

- Explain how groupthink hinders effective decision making in groups.

- Define critical thinking and explain its role in the decision-making process.

- Identify and describe the primary approaches to decision making in small groups.

- Discuss the role of power in group problem solving and decision making.

- Identify the types of and approaches to conflict in small groups.

CHAPTER 10

Small Group Problem Solving and Decision Making

When the COVID-19 pandemic first struck in 2019, one key to understanding and disrupting the virus was testing both those who were symptomatic and asymptomatic to better identify how and when the virus traveled throughout the human population. Enter computational geneticist Dr. Paradis Sabeti and a team of about 20 researchers at her laboratory with the Broad Institute of MIT and Harvard. Dr. Sabeti's team had extensive experience with virus outbreaks from their work diagnosing and identifying the source of the Ebola virus in West Africa. In early 2020, she and her team turned their attention to using genome sequencing to better test and develop a vaccine for the novel coronavirus. In addition, her team's innovative work formed the foundation for Sentinel, a system of diagnostic tools that the team hopes will serve as a pandemic early-warning system allowing for a more proactive response to viral outbreaks in the future.[1]

People often associate innovative discoveries with individuals, such as Marie Curie and the discovery of radioactive materials, Tim Berners-Lee and the creation of the World Wide Web, and Randice-Lisa Altschul and the first disposable cell phone. As with stopping a pandemic, however, brilliant solutions to complex problems typically arise from people working together in a group. For example, Altschul, a toy designer, brought on a team of experts to execute her abstract idea for an inexpensive disposable cell phone.[2]

Not all groups face challenging and unique problems. But all groups must solve problems, make decisions, and manage conflict as they complete tasks and achieve their goals. As you know from your own experience, groups vary in their success at goal attainment and managing relationships among group members. You likely have participated in competent and productive groups, dysfunctional and ineffective groups, and groups somewhere in between those two extremes. In this chapter you will learn about different ways to solve problems, make decisions, leverage technology, and develop skills in overcoming some of the more common challenges that confront small groups.

Small Group Problem Solving in a Pervasive Communication Environment

There is no doubt that employers expect job candidates to solve problems in groups. A recent survey of employers by the National Association of Colleges and Employers revealed problem solving and teamwork as the top two skills recruiters look for on an applicant's resume.[3] A pervasive communication environment gives group members the ability to tap into global networks of information and people, greatly enhancing the potential for creative solutions to complex problems. A LinkedIn Learning survey of nearly 3,000 managers, 1,700 learning and development (L&D) professionals, and 2,000 workplace learners in 18 different countries identified creativity as the most in-demand soft skill organizations seek in their employees.[4] Early views of creativity focused on the idea of an individual developing something completely and totally new. More recent definitions recognize the importance of multiple contributors in the creative process.[5] As a social process, **creativity** involves communicating with others to develop alternative, novel ideas and solutions to problems.[6] All approaches to problem solving depend on a group's ability to think creatively.

creativity Communicating with others to develop alternative, novel ideas and solutions to problems

FIGURE 10.1 Creativity is a collective effort.

Facilitating Creativity in Your Group

Creating a climate for creativity in your group provides the foundation for generating new ideas. Six group practices contribute to a productive communication climate in which group members can unleash their creative powers: recognize creativity as a collective effort, provide support for innovation, create a shared vision, establish cooperative goals, develop a supportive communication climate, and take a task orientation.[7]

Recognize Creativity as a Collective Effort

In recognizing creativity as a collective effort, group members understand the unique contributions and creative roles each person brings to the group (Figure 10.1). Sponsors, for instance, promote new ideas inside the group, keeping interest up and attention on them. Shapers, in contrast, map out details and extend the initial ideas. In addition, individuals outside the group also may contribute to the group's creative activities. For example, sounding boards are people not connected with an idea who bring new and fresh perspectives on it. Specialists have in-depth knowledge in particular aspects of an idea, adding their expertise to its development.[8] Experts are especially essential to group creativity, bringing in new information and combining existing information in new ways.[9] Box 10.1 offers practical advice for developing an ethical community of practice for creativity.

community of practice
Occurs when group members share ways of working and learning together that help them strive toward common goals

---

BOX 10.1: ETHICAL QUESTIONS

Developing an Ethical Community of Practice for Creativity

When working in small groups, developing an ethical community of practice based on cooperative participation can enhance the creative process, discourage displays of competition, and promote an ethical approach to creativity. A **community of practice** emerges when group members share ways of working and learning together that help them strive toward common goals. Ethical practices include explicitly sharing knowledge and information, actively including all group members in discussions, applying active listening skills, and welcoming disagreement.[10] These practices help groups avoid unethical behavior, such as stealing others' ideas, dismissing alternative perspectives, and disenfranchising specific group members, while at the same time encouraging everyone to participate and maximize the group's creative potential. How might you develop these practices in your group to promote innovation while at the same time communicating with respect and fairness?

Support Innovation

Groups support innovation when members are motivated to try new things, rewarded for new ideas, and encouraged to develop entrepreneurial skills. Group members actively seek alternative solutions, encouraging each other to improve on ideas presented. Support for innovation also requires tolerating mistakes, viewing them as a normal part of innovation, expressing a willingness to challenge traditional ideas, and finding ways to learn from those mistakes.[11] For example, a study of college students in learning communities found that failure produced closer interpersonal relationships, greater creative collaboration among group members, and ultimately more innovative solutions than when groups faced minimal challenges.[12]

Create a Shared Vision

When group members have a shared vision, they have a mutual mental image about the group's future or the group's task, and members rely on this vision for structuring their actions. For example, a college task force on developing new online tools for the campus career center may develop a shared vision of an interactive tool students can use to engage in mock employment interviews. Embracing this shared vision helps focus the group's creative ideas and provides a purpose for the group's innovative activities. In general, promoting a shared vision increases a group's creativity.[13]

Develop Cooperative Goals

Developing cooperative goals encourages group members to perform well because they view themselves as interdependent, relying on each other to reach the group's objective (Figure 10.2). In a cooperative climate, group members easily share information and help one another. In contrast, groups with competitive goals pursue their own interests regardless of the others' needs and the group's goal. Competition leads group members to withhold information and view others' contributions with suspicion.[14]

FIGURE 10.2 Cooperative goals promote creativity in small groups.

Facilitate a Supportive Communication Climate

Facilitating a supportive communication climate encourages group members to attend to others' ideas and opinions, consider how those in other groups might view an issue or problem, and give and accept constructive criticism. Supportive communication, discussed in Chapter 4, focuses on description, a problem orientation, spontaneity, empathy, provisionalism, and equality. Communicating in supportive ways contributes to a group climate in which members feel safe to voice unusual and novel ideas. Simple acknowledgments of other group members' contributions, as with thanking them for their ideas and sharing their perspectives, increases the level of a group's creativity.[15] In contract, disrespectful and especially aggressive communication negatively affects the group's ability to generate creative ideas.[16]

Exhibit a Task Orientation

Groups exhibiting a task orientation link ideas to the group's objective, agree on standards of excellence for the group's performance, and demonstrate a concern for high-quality decisions and products. Although you might think a task orientation would inhibit creativity, focusing on the task serves to channel the group's creativity. Group members waste creative efforts when they fail to provide structure for their activities. Creativity, attention to detail, and efficiency can coexist in group interactions.[17] When group members view their creative ideas as contributing to the task, they evaluate themselves and the group as more productive, and they are more satisfied with the group.

Creativity Techniques for Your Group

Once your group has developed a climate for creativity, specific techniques encourage members to approach problems in creative ways. Your group has many approaches available to facilitate creativity. The ones discussed below and summarized in Table 10.1 serve as a starting point for all groups, whether you are meeting in-person or online.[18]

TABLE 10.1 Creativity Techniques for Small Groups

TECHNIQUE	BRIEF DESCRIPTION	USEFUL FOR ...
Brainstorming	Idea generation without evaluation	Developing, combining, and building on ideas
Lotus blossom technique	New ideas radiate out from initial one	Identifying relationships among ideas
Nominal group technique	Ideas generated individually and voted on	Contentious issues or to manage dominant group members
Six thinking hats	Different hats associated with different ways of thinking	Explicitly considering a problem or topic from a variety of viewpoints
Synectics	Multiple word association techniques	Especially difficult problems

Brainstorming

The basic rules of brainstorming are:

- Generate as many ideas as possible.

- Record all ideas.

- Do not dismiss any idea as too silly or outrageous.

- Combine and build on ideas.

- Do not critique or evaluate any ideas.

Often, the first rule of brainstorming, generating ideas, receives the most attention. However, coming up with novel ways to combine and build on ideas can lead to key innovations, as you will find out in Box 10.2. Groups generally find refraining from criticism the most difficult rule to follow. Yet it is the most essential because criticism can halt the flow of ideas. Later you will evaluate all the suggestions proposed, but not during brainstorming.

BOX 10.2: CONNECT & REFLECT

Connecting Ideas With "Yes-And"

Adopting a yes-and mindset in which group members recognize others' contributions and then seek to form a bridge between them provides one method for making connections between seemingly disparate ideas.[19] Group members ask questions such as "How might the popularity of smartphones be used to help students learn math?" or "If junior high students read the Harry Potter books, then why don't they read their textbooks?" Or a group member simply may ask, "And if we connect those two ideas, what might be the outcome?" Committing to a yes-and mindset requires actively listening to the others in the group, reflecting on their ideas, expressing a curiosity about how those ideas might be connected, and phrasing questions that open up the possibilities for innovative idea combinations. How might you try out the yes-and mindset in your group?

Lotus Blossom Technique

Like a lotus blossom, this technique begins with a central theme from which ideas radiate out. The central theme may be the problem the group wants to solve or an initial idea to solve it. The process works this way:

1. Write down the central idea.

2. Brainstorm for eight ideas associated with it.

3. Take each of those ideas and write them down, brainstorming for eight more ideas based on each original one.

Now you have gone from one lotus blossom to nine. Although you could continue to generate additional blossoms from all the ideas you have generated, at this point it usually is best to identify the ideas that are the most original and useful and develop them as the centers of more lotus blossoms.

Nominal Group Technique

Unlike most creative-thinking strategies for small groups, the nominal group technique calls for minimal communication among group members. This technique is especially useful for contentious issues or situations in which one or two members dominate the group. This technique follows four steps:

1. Group members generate ideas individually.
2. Ideas are recorded as each member reports on their list of ideas.
3. Group members go through the compiled list and raise questions about specific items as needed. Ideas are not evaluated, however.
4. Group members vote to choose the top idea.

If the group has generated a long list of ideas, the group might need several rounds of voting, during which a leader gradually narrows down the number of choices, to choose the top one.

Six Thinking Hats

With this technique, each member wears a different hat, and each hat is associated with a different way of thinking, as shown in Table 10.2. You can make real hats or use tags to identify which person is wearing which of the following hats:

- *White*: uses purely logical thinking, focusing on facts and figures, but avoiding judgments
- *Red*: uses intuitive thinking and expressing emotions
- *Black*: critically identifies the reasons why an idea or solution *will not* work
- *Yellow*: favorably identifies the reasons why an idea or solution *will* work
- *Green*: concentrates on provocative ideas
- *Blue*: focuses on the process the group is using to generate ideas, such as determining whether the group needs more black hat or yellow hat thinking

TABLE 10.2 Six Thinking Hats

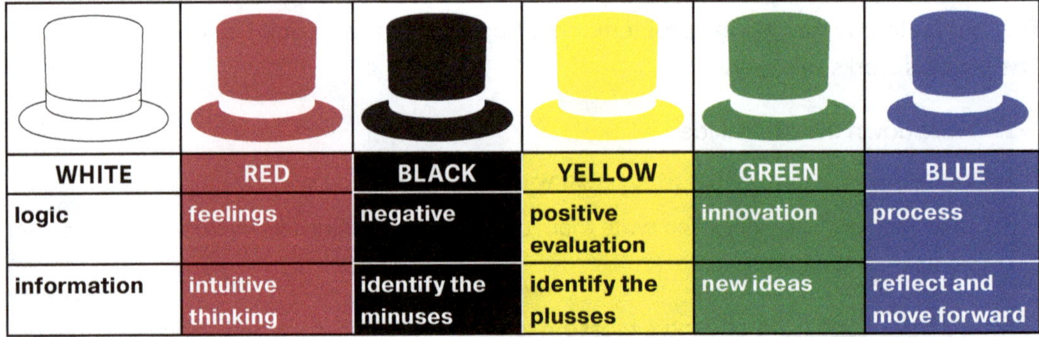

WHITE	RED	BLACK	YELLOW	GREEN	BLUE
logic	feelings	negative	positive evaluation	innovation	process
information	intuitive thinking	identify the minuses	identify the plusses	new ideas	reflect and move forward

Groups may use these hats in two ways. First, all group members might take on a single hat, focusing on one type of thinking. Alternately, each group member may choose or be assigned a hat, contributing in ways aligned with the hat's way of thinking. Group members may also exchange hats to give them fresh perspectives on the ideas under discussion. Group members do not need to meet in-person for this system to work. The six thinking hats technique can be applied easily to online meetings or phone conferences.

Synectics

This technique focuses on words and word associations to generate ideas. The group must identify one person to serve as leader, at least for this part of the discussion, because the leader decides when a word association approach is not working and the group should try a different one. After the group has identified the problem and gathered, selected, and integrated relevant information, members can use one of these techniques to brainstorm for wild and wacky ideas:

- *Analogies and metaphors*: The leader asks group members to consider what the problem is like, as with "Building a new events center on campus is like creating a sand castle when the tide is coming in."

- *Association*: The leader provides the group a word and then asks them to form associations with it in one of several ways. For example, the leader might ask group members to free associate based on contrast (opposition), similarity, or contiguity (ideas or objects in proximity to each other).

- *Trigger words*: The leader asks the group to think of the problem in terms of a word, such as *subtract, animate, parody, contradict, distort, isolate, fantasize*.

These word exercises help group members to use language in thinking about the problem and solutions in new ways. After group members feel they have exhausted all their ideas, they can move on to the evaluation stage and identify the best solution.

There are many other techniques available to your group to facilitate creative solutions to problems. Integrating such techniques with a systematic approach to problem solving will increase your group's chances of developing the best solutions.

An Eight-Step Approach to Problem Solving

Many small group communication theorists and researchers argue that small groups need an orderly procedure to follow in problem-solving processes so that personal biases, individual goals, and outside influence are less likely to lessen the quality of the group's work. The eight-step problem-solving model will help your group approach problems in a creative, yet structured, way. Eight core steps, summarized in Table 10.3, form the basis of this approach: constructing the problem, gathering information, selecting information, integrating information, generating ideas, evaluating ideas, planning the implementation, and monitoring the implementation.[20]

TABLE 10.3 Eight-Step Problem-Solving Model

CORE STEP	OUTCOME
1. Constructing the problem	Clear statement of the problem
2. Gathering information	Understanding of past solutions
3. Selecting information	Array of relevant information
4. Integrating information	List of conclusions
5. Generating ideas	Broad range of alternatives
6. Evaluating ideas	Rank-ordered list of alternatives
7. Planning the implementation	Strategy to put idea into practice
8. Monitoring the implementation	Evaluation of idea's effectiveness

Constructing the Problem

What problem are you trying to solve? Determining the problem or issue the group faces provides the basis for developing solutions. A university's Fall Welcome Committee composed of students and faculty, for instance, wanted ideas for encouraging students to participate in a variety of activities at the start of the academic year. Initially, committee members identified the problem as low student attendance at the activities. Upon closer examination, however, members realized the issue was student motivation; students needed a reason to attend. Thus, the group constructed the problem as "What will motivate students to participate in these fall activities?" Without a good understanding of the root problem, group members may come up with plenty of original ideas that have little or no utility.

Gathering Information

When your group encounters a problem, your impulse may be to start generating ideas. But creative problem solving requires originality, so you need to know what others have done in the past before you can offer a new alternative. Once you have constructed the problem, gather information on it so you better understand the solutions that have been tried previously. In the university Fall Welcome Committee example, members examined past practices at its own and other campuses to identify both successful and unsuccessful strategies to encourage student participation.

Selecting Information

In this step, groups go through the information they have gathered and identify what is most relevant to the problem, what provides the most insight into the problem, and what seems to hold the greatest potential for new avenues of thought. Along these lines, the Fall Welcome Committee discarded strategies designed for activities that were not planned for that year. Overall, in selecting information, groups want to avoid both information overload and information underload—too much or too little information (discussed in Chapter 15).

Integrating Information

In this step, as they would when solving a puzzle (Figure 10.3), group members review the information and figure out ways the various strands might fit together. Again returning to the example of the Fall Welcome Committee, the members identified three groups of strategies for motivating students to attend events: prizes for competitions, free food, and a T-shirt giveaway.

Generating Ideas

Generating ideas is the part of problem solving with which you are probably most familiar. Any of the creativity techniques, such as brainstorming and six thinking hats, will help your group come up with innovative ideas. The process is most effective once the group has a solid foundation of information from which to consider alternatives. Thus, after reviewing the problem statement and the information they had gathered, the Fall Welcome Committee realized they were focusing solely on the events as sources for motivating students' attendance. This narrow focus limited the alternatives, so the committee turned to other motivational sources, such as parents, peers, and faculty, which in turn generated a list of potential ways those groups might motivate students to attend the events.

FIGURE 10.3 When you integrate information in the eight-step problem-solving model, you determine how the various bits of information you have fit together, like the pieces of a puzzle.

Evaluating Ideas

In this step, group members return to the problem they identified at the beginning of the process. Next, they evaluate ideas on several criteria, such as creativity, practicality, and feasibility. Is the idea innovative? Does it solve the original problem? Can the group implement it? To return to the Fall Welcome Committee, identifying assignments related to the events that faculty could integrate in their classes was a new idea that had not been tried in previous years. Committee members agreed this solution could increase event attendance. And to test the feasibility of their idea, committee members generated a few assignments relevant to several different courses.

Planning the Implementation

A group may have the best idea imaginable, but without a plan to implement it, no one will ever know about it. The faculty and students on the Fall Welcome

Committee developed a list of possible course assignments, some applicable to all events and others relevant to specific events. Then they developed a webpage listing each event with links to relevant assignments. Distributing the information was easy: The group simply emailed the link to all campus faculty along with a message explaining the Fall Welcome program and the purpose of the assignments.

Monitoring the Implementation

In this part of the problem-solving process, the group receives feedback on the implementation of the idea. Did the idea work as planned? Does the group need to make adjustments? In the case of the Fall Welcome Committee, members canvassed the faculty to find out how many used the suggested assignments and asked how those assignments might be improved. The group also collected data on student attendance at the events to find how why students decided to attend, whether they liked the events they attended, and whether attendance was linked to the assignments (Figure 10.4).

Although the eight core steps of the model are presented here in a linear fashion, groups often go back and forth among them. For example, when gathering information, a group may discover that the problem requires further clarification. Or group members may evaluate the ideas proposed and then need to generate additional ones. Still, each part is essential in the successful implementation of a solution to a problem.

FIGURE 10.4 In the Fall Welcome Committee example, the group checked student attendance at events to find out which ones were the most popular.

Decision Making in a Pervasive Communication Environment

critical thinking Purposefully interpreting, analyzing, and evaluating information to solve problems and make decisions

The access to a wide range of information sources and people a pervasive communication environment provides does not guarantee that groups will make the best decisions. Although researchers have identified many reasons why groups make poor choices, failing to apply effective critical thinking skills is always in the top of the list. **Critical thinking** involves purposefully interpreting, analyzing, and evaluating information to solve problems and make decisions. In addition, critical thinkers strive to gather a wide range of evidence, welcome differing perspectives, consider a wide range of options, and apply what they have learned to address novel problems.[21] **Groupthink**, when group members agree on a course of action or solution to a problem without fully discussing the advantages and disadvantages of the range of alternatives, occurs in the absence of critical thinking.[22] Without full and open discussion of all options the group might consider, groupthink often leads to disastrous results because group members fail to identify the most effective solution.[23]

Groupthink can have serious consequences. For example, an analysis of media reports on groups protesting quarantine restrictions and vaccinations during the COVID-19 pandemic revealed these groups exhibited the characteristics of groupthink, such as high cohesion, isolation from other groups, and pressures on group members to conform with group norms. The researchers suggested that a public health campaign to encourage these groups to follow quarantine and vaccination guidelines would work only if the fundamentals of groupthink were addressed. Messages would need to focus on reaching out to groups in positive and welcoming ways, reducing group members' feelings of isolation, and developing strategies to modify group norms.[24]

Avoiding Groupthink and Promoting Critical Thinking

How do you know if your group is experiencing groupthink? Examine your group for the symptoms listed below:[25]

- *Promotional leadership*: In the early stages of decision making, the group's leader strongly endorses one solution or idea over all others.

- *In-group bias*: Group members view themselves positively and those outside the group negatively, leading to the perception that information from inside the group is superior to information from outside the group.

- *Pressure to conform*: Disagreement with the majority of group members is discouraged and seeking agreement from all group members is encouraged.

- *Hidden profiles*: Individual group members fail to reveal information that is contrary to the group's prevailing view.

Research shows that facilitating critical thinking is one of the most successful ways to reduce the chances your group will experience groupthink and increase the likelihood of making sound decisions.[26] For example, if a group member says, "Most of the students on this campus cheat on their exams," other group members would ask, "What is the basis for your assertion?" "When you say 'most,' how many students do you mean?" and "What is the source of your information?" Asking these and similar questions encourages group members to carefully examine and critique information so the group is more likely to make good decisions.

groupthink A process in which group members agree on a course of action or solution to a problem without fully discussing the advantages and disadvantages of the range of alternatives

Eight strategies, summarized in Table 10.4, will encourage you and the members of your group to avoid groupthink and actively engage in critical thinking: make clear and relevant statements, clarify group goals, contribute new information, link ideas together, provide supporting evidence, offer critical assessments, identify practical utility, welcome a wide range of viewpoints, and frame ideas within a broader perspective.[27]

TABLE 10.4 **Strategies for Facilitating Critical Thinking in Your Group**

STRATEGY	APPLICATION
Make clear and relevant statements	Keep group members on track by asking them to explain ambiguous statements
Clarify group goals	Remind the group of the goal of the discussion
Contribute new information	Invite group members to offer their ideas
Link ideas together	Show how different ideas are related
Provide supporting evidence	Include data to support your claims
Identify the practical utility	Demonstrate how an idea might be put into practice
Welcome a wide range of viewpoints	Encourage group members to engage in active listening
Frame ideas within a broader perspective	Consider the impact of the group's proposed solution

Clear and relevant statements keep the group focused and on topic. There are two ways you can contribute in this area: (1) Strive for clarity and relevancy in your own communication, and (2) encourage other group members to do the same. Avoid inserting irrelevant comments or tangential information that distracts the group from its primary discussion. To help others with clarity and relevance, ask questions to bring the discussion back to the original topic. For example, if a group member makes a comment that appears unrelated to the group's discussion, you might ask, "I'm not quite sure how that's related to what we've been discussing. Would you help make the link for me?" The question is polite, yet moves the group back on track.

To clarify the group's goal for a particular collaborative effort, determine in advance the desired outcome from the interaction and the steps that will help the group complete its task. As group members are working together, remind them of the reason they are together and what they hope to accomplish. This structure sharpens group members' focus on critiquing the information they gather and the arguments they present.

Contributing new ideas, solutions, and information to the group promotes critical thinking by broadening the foundation of information that members may use to evaluate claims. In addition to offering your own new ideas, motivate others to do the same by asking for their input.

When you link ideas together, you put your reasoning skills to work. Consider how the various pieces of information, premises, and conclusions your group has discussed fit together. Linking ideas together provides new information for the group to consider. For example, in discussing what do to for a holiday event, members of a student organization may have determined the budget, identified the kinds of food and beverages they plan to serve, and listed whom they will invite. Then one discussant might say, "Our budget, food requirements, and invitation list suggest we need a venue close to campus or easily accessible on public transit, with a comprehensive yet inexpensive menu." In linking together the various bits of information, the group member has given the group something new to consider.

Effective critical thinking requires high-quality and sufficient information. As you present arguments to your group, provide evidence to support your claims and encourage others to do the same. Use the criteria for evaluating information—reliability, validity, currency, and relevancy—to provide your group with high-quality information.

Demonstrating how the group might put an idea into practice is how you identify its practical utility. Suppose you are on a task force charged with fostering community among employees in your organization. In discussing possible activities that might interest workers, one group member suggests organizing reading groups. You consider how the company might implement this idea, such as rotating the committee members who choose the book and forming another committee to organize the book discussions. Critical thinkers place ideas in context and offer practical ways to implement them.

Research confirms the value of welcoming a wide range of viewpoints in small groups. Groups with high levels of diversity in areas such as ethnic background, socioeconomic status, dis/ability, age, gender, and other demographic characteristics consistently produce higher quality decisions, better and more creative solutions to problems, and demonstrate more effective critical thinking skills than homogeneous groups.[28] Welcoming a wide range of viewpoints increases the likelihood that group members will examine and question closely the arguments presented.

Expressing interest in others' perspectives requires a commitment to dialogic listening. As discussed in Chapter 4, dialogic listening involves listening *for* culture and listening *to* culture as well as detecting both similarities and differences among group members. In addition, dialogic listening encourages you to listen for inequities—whose ideas does the group readily receive and whose ideas does the group ignore? Welcoming a wide range of perspectives means encouraging all group members to participate before reaching a conclusion (Figure 10.5). Apply promoting an inclusive communication climate in your group in Box 10.3.

FIGURE 10.5 Dialogic listening encourages group members to welcome a wide range of perspectives.

Framing ideas within a broader perspective involves placing the group's discussion in a larger context. For example, when groups make decisions or implement solutions to a problem, those actions will impact others outside the group. Considering factors outside the group's immediate environment gives the

---

BOX 10.3: CONNECT & REFLECT

Achieving Inclusive Excellence

Developing a critical thinking disposition in a small group requires expressing interest in others' perspectives, even when you are unfamiliar with or disagree with them. Rejecting the advantages of group diversity assures a fast track to groupthink. Inclusion goes beyond merely accepting difference in your group. When group members promote inclusion, they support cultural diversity, value interdependence, and implement egalitarian group practices.[29] The ethical small group communicator helps the group achieve inclusive excellence—welcoming, sharing, and valuing the experiences of all group members. Consider a group to which you currently belong. What have group members done to welcome and value the experiences of all group members? How might group members help the group better achieve inclusive excellence? What can you do to promote egalitarian group practices in your group?

---

group a wider viewpoint on the ideas under discussion. When group members fail to consider the larger implications of their ideas or focus on only a narrow aspect of the topic, they ignore important information essential to effective critical thinking.

Thinking critically provides a key foundation for effective decision making in small groups. Group members have a range of strategies they can apply to facilitate critical thinking in their groups across all modes of small group communication. An analysis of 16 studies on online classroom discussion, for instance, found that participants demonstrated effective critical thinking skills along multiple dimensions, such as providing critical assessments, linking ideas, and offering new information and solutions. The results suggest that online discussion allows students to reflect on their own and others' contributions, facilitating the application of critical thinking skills.[30] When you interact in your small groups, consider the ways you can apply the key strategies of critical thinking both online and in-person, such as reflecting on what others have said before adding your contributions.

Approaches to Decision Making in Small Groups

Small groups have several options in how they make decisions. Choosing a strategy depends on the task at hand, group member preferences, and constraints on decision making, such as time and resources. Summarized in Table 10.5, this section discusses six common approaches: majority rule, autocratic rule, consultative rule, expert rule, unanimity, and consensus.

TABLE 10.5 Approaches to Decision Making

APPROACH	BRIEF DESCRIPTION	ADVANTAGES	DISADVANTAGES
Majority rule	Group members vote to determine top choice	Fair and decisive	Minority voices not heard, lack of critical thinking
Autocratic rule	The leader makes the decision	Make decisions quickly, useful for routine tasks	Group members may feel disenfranchised, decision may not be the best one
Consultative rule	The leader consults with group members before making a decision	Input from group, encourages group cooperation	Group members may not feel invested in final decision
Expert rule	Group members rely on an expert to make a decision	Decision based on expertise related to subject	May not consider other aspects of the problem
Unanimity	All group members agree or no one objects	Brings group members together	True unanimity difficult to achieve, lack of objection does not equal support
Consensus	The solution is acceptable to all group members	Signals support for decision, creates community	Complex and time consuming

Majority Rule

Majority rule, the alternative with the most votes wins, is perhaps the most common and widely used decision-making process in the United States. This is the way elections are decided and is the default decision-making process in guides like *Robert's Rules of Order*.[31] The advantages of majority rule are that it is both fair and decisive. That is, all votes are equal, and there is a clear winner. However, majority rule can create problems when the votes are tied. The group then must determine a course of action, as with voting again, flipping a coin, or some other alternative. In addition, this form of decision making can infringe on minority rights. For example, in research on gender and decision-making models, women are less likely to voice disagreement when their gender is in the minority, although both genders are more likely to voice disagreement when they are in the majority.[32] When groups use majority rule to make decisions, those in the minority may feel hesitant to speak up, so not all perspectives are heard. In addition, majority rule can increase participants' negative views of conflict, particularly for those in the minority.[33] Thus, group members may go along with the majority opinion even if they disagree with it to avoid conflict. Majority rule also can lead to group members quickly converging on one alternative and reaching agreement too hastily, resulting in groupthink and stifling critical thinking.[34]

majority rule Making a decision whereby the alternative with the most votes wins

Autocratic Rule

In **autocratic rule**, the decision rests entirely with the leader. This approach can be effective in small groups when an immediate decision is needed. For instance,

autocratic rule A type of decision making where the leader alone makes the decision for the group

a paramedic team responding to a highway traffic accident may have little time for discussion. The team's leader must decide without delay what course of action to take and direct the team to implement it. In addition, there may be minor decisions that are routinely left up to a group's leader to make, such as how to format a document or which refreshments to order for an in-person meeting. Also, when group members are more uncertain of themselves and the group, they tend to prefer having a leader make decisions.[35] Still, when leaders make decisions without input from the group, group members often feel disenfranchised and demotivated, the decisions may not be the most effective ones, and the group's goals may not be achieved.[36] In most cases, group members prefer a more collaborative approach to decision making.

Consultative Rule

consultative rule The group's leader makes the decision after discussing it with the group's members

Consultative rule in small groups involves the leader discussing a decision with the other group members but making the final decision. This type of decision making can increase group members' commitment to and satisfaction with the group.[37] In addition, consultative rule can produce better decisions and more cooperative relationships in the group than autocratic rule. However, participatory decision making that involves all group members in making the final decision produces more reasoned decisions than a consultative approach.[38]

expert rule Occurs when groups rely on a professional in a specified topic area to make a decision for the group

Expert Rule

Expert rule occurs when groups rely on a professional in a specified topic area to make a decision for the group (Figure 10.6). Someone within a homeowners association, for instance, may be a retired civil engineer and therefore knows how drainage systems work and connect to the municipal system. They would be a natural person to turn to when the group faces a water problem. However, although an expert may be a vital resource in a decision-making process, that expert information is only one part of the process. Due to highly specialized knowledge, the expert position is easy to intentionally or unintentionally abuse for personal gain or to serve a particular bias. In the homeowners association case, financial constraints or concerns about landscaping also are considerations in a decision on drainage work.

FIGURE 10.6 Having an expert make a decision can be effective when group members know little about the specific topic or issue.

Unanimity

Unanimity occurs when all group members are in complete agreement about a decision. When combined with voting, a unanimous decision is one in which all group members vote the same way. *Robert's Rules* also describes unanimity as occurring when no objections are raised. However, abstention in a decision process does not necessarily represent a strong signal of unanimous support. For example, the Security Council in the United Nations often makes binding decisions, such as trade sanctions, with some members abstaining. The decision is made, but it is clear that not all members supported it enough to vote for it. However, they did not object enough to oppose it. In addition, if group members feel pressured to come to a unanimous decision, some participants may be hesitant to express differing viewpoints.[39] The decision may become a product of groupthink rather than thoughtful deliberation. Interestingly, a study on online groups discussing climate change found that unanimity exerted less of an influence to conform in those groups than in face-to-face discussions.[40]

> **unanimity** A decision in which there is complete agreement among all group members

Consensus

Consensus involves reaching a solution acceptable to all group members. Consensus decision making can be complex and slow, yet good for weighing all aspects of a problem and its potential solution. Activist groups, such as those associated with the racial justice movement and other protest groups, often apply a consensus model in their decision making. For example, research on German antinuclear groups collaborating on possible collective action found participants achieved consensus by first agreeing on a vaguely worded proposal and then gradually moving toward more specific ideas before coming to an agreement on a final plan of action.[41]

> **consensus** A decision-making process where the solution is what is acceptable to all members

Similarly, research has found that a multistage process in which group members periodically assess their progress increases the likelihood that the group will achieve true consensus.[42] Group members begin by explaining their individual preferences, identifying areas of agreement and disagreement, and determining what is needed for consensus. As discussion continues and ideas are offered, group members return to their individual preferences and note where they might make adjustments based on the group's discussion. The group again assesses areas of agreement and disagreement as well as their progress toward consensus and considers what adjustments they might make as a group to reach a desired outcome. Viewing consensus as a process of discussion-assessment-adjustment recognizes that both the group as a collective and individual group members may change their views based on additional information. Thus, consensus done well builds support for the decision and helps to create community and solidarity.[43]

The choices you make for decision making should be based on the situation and composition of each group. For example, consensus is particularly useful for complex issues, but can be a problem if there are tight time constraints. In Box 10.4, you will explore how groups in your community make decisions.

BOX 10.4: LIVE IT LOCAL
Decision Making in Local Government Groups

City and county government groups often must make difficult decisions due to competing interests, local regulations, and limited resources. City councils and county boards of supervisors provide useful examples of decision making that directly impacts the everyday lives of local citizens. Attend or watch online several meetings of your local city council or county board of supervisors. What approach or approaches do the group members take to decision making? How effective is the group in making decisions? What factors seem to influence the group's decision-making processes? Are there ways you think the group could improve its decision making? If so, what suggestions do you have? What have you learned about decision making in your community?

Power in Decision Making and Problem Solving

power A force that gives one communicator the ability to influence others to take an action that would not be taken without that motivating force

empowerment A cooperative form of power that seeks to enhance the capabilities and the influence of group members

Small groups are sites of influence attempts and the exercise of power. **Power** is a force that gives one communicator the ability to influence others to take an action that would not be taken without that motivating force. Power is not a static property of an individual, group, organization, or the environment. Rather, power operates in the in-between spaces in contexts and relationships.[44] Because power functions within specific contexts and relationships, who exercises power and when may change over time and with tasks and topic areas.

Empowerment is a cooperative form of power that distributes power among group members, enhancing their capabilities and influence (Figure 10.7). Empowerment practices in small groups increase group productivity and effectiveness, strengthen group members' commitment to the team, encourage greater group member creativity, and facilitate better interpersonal relationships among group members.[45]

As a group member, you earn the ability to exercise power by becoming an expert on the group's task, gathering relevant information, developing alliances, rewarding others for good work, and demonstrating active involvement in the group. For example, you may suggest that your group's final report be divided among the members, with each person responsible for a different section. All group members become experts on their topic areas, developing a basis for influence in the group. Members of your group might also form alliances or coalitions as a way to pool their resources.

FIGURE 10.7 When group members are empowered, they share power cooperatively within the group.

Empowerment does not mean that all group members have the same amount of power in the group at all times. Power distributions typically change and mutate as group members interact over time. Thus, Jithen might exert more power during the first meeting, Gabriela during the second meeting, and Farsheed during the third meeting. Even within meetings the group's power dynamics will not remain static. However, central to empowerment is that all group members' voices are heard in decision making and problem solving. Identify how power is distributed in your group in Box 10.5.

BOX 10.5: CONNECT & REFLECT

Locating Power in Your Group

How do you know who exerts power in your group? Small group culture provides one place to identify how power is distributed in your group (Figure 10.8). Culture is both a tool in the exercise of power and a power resource in that rituals, practices, stories, language, and other aspects of culture can be used to a group member's own advantage. By investigating your small group's culture, you can uncover processes that typically work behind the scenes in the group's interactions. Thus, you can view what you take for granted in small groups in a new way by examining what is usually unexamined. Discuss the questions that follow to discover the hidden aspects of power in your group:

FIGURE 10.8 Examining your group's culture can give you insight into how power is enacted in your group.

1. *Vocabulary*: What words are used to describe various group members and their roles? Who controls the language the group uses? How does the group define itself?

2. *Practices*: How are decisions made in the group? Do coalitions gather before group meetings to plan their strategies? Are meetings sites of hidden agendas and self-promotion? To what degree do group members have autonomy in completing tasks and making decisions?

3. *Stories*: Who tells stories? Who are the winners in the stories? Who are the losers? Who makes decisions in the stories? Who carries out those decisions? Who gives out rewards/punishments? Who receives those rewards/punishments?

4. *Rituals*: Does the meeting not start until the group's leader arrives, or can anyone begin the meeting? What are some of the group's rituals? How did they develop? Who participates in rituals? Who does not?

In identifying how power is evident in your group's culture, go beyond the surface to the values underlying cultural elements. Stories, practices, vocabularies, and rituals legitimize power structures, leading group members to accept such structures without question. In analyzing group culture, group members should ask, "Who is privileged in our group?" and "Who is left out of decision-making and problem-solving processes?" Then group members should ask, "How can we more equitably distribute power in our group?" That is, group members should work together to empower themselves. How can practices serve to empower group members? What rituals will encourage all group members to exert influence in the group? What might be a more appropriate way to talk about your group?

Sources of Power

Power in groups stems from several sources, summarized in Table 10.6. Expert power arises from knowledge and experience in a particular area that is relevant to the group's task. For example, if your group is researching a topic and you know quite a bit about it, you have expert power within the group. Expert power also can come from a particular ability to get things done that benefit the group, such as running a meeting efficiently.

TABLE 10.6 Sources of Power

TYPE	BRIEF DEFINITION	EXAMPLE
Expert	Knowledge or skill related to topic	Renowned researcher in specific area
Interpersonal network	Relationships with others	Friendship with someone who has resources the group needs
Reward/Punishment	Give benefits or enforce costs	Express praise
Legitimate	Assigned to a position	Top level of organizational hierarchy
Referent	Affiliation with others	Belonging to a prestigious organization
Avoidance	Appear to avoid exercising power	Displaying disruptive behaviors

Interpersonal network power comes from a member's relationships with others or access to information that can bring needed resources to the group. Anyone can have unique and useful networks or connections. For example, having a good friend in information technology (IT) may be the key to getting faster help with your group's technical problems.

The power of reward/punishment is the ability to control members by giving praise or material benefits or taking them away or otherwise harming them. This type of power can be emotional, as in sarcastic wit, or behavioral, as in getting someone fired. Often, minor rewards can be effective, such as praising a group member for giving an excellent presentation. Minor slights, such as not copying some group members on an important email or text because they missed a key meeting, can serve as a form of punishment.

Legitimate power is assigned to a position in a group, organization, society, or culture, such as the power associated with the president of a student organization. The position may not be one assigned within the group. For example, a retired police officer in a neighborhood group may command higher respect and influence due to the status in society typically associated with police officers. Referent power is based on a member's affiliation with respected groups or people. Someone with an advanced degree or influential contacts can exercise referent power. Unlike interpersonal network power, these relationships or affiliations may or may not lead to benefits to the group or its goals.

Finally, with avoidance power, defiant and deviant members appear to avoid exercising power in the group, but actually seek to influence others' behaviors.

Inappropriate comments, tardiness, and other disruptive behaviors are ways to control a group.

Power and Culture

The notion of power, what constitutes power, and how to address power imbalances in small groups varies across cultures. **Power distance** refers to how those with less power view power inequalities. Low power distance cultures value minimizing differences between members, sharing power, participation, and democratic leadership. Countries such as the United States, Netherlands, and Israel have laws and norms that reinforce this perspective. In low power distance cultures, people in low power positions do not accept unequal power distribution in organizations and other institutions. It is not unusual in these cultures for workers to challenge supervisors or socialize with them.

High power distance cultures have a relatively strong emphasis on maintaining power distances, and those in lower levels accept inequality in power distribution. Norms, laws, and values that maintain distance between authorities and the population guide countries such as Mexico, Singapore, and South Korea.[46] Still, views of power distribution change over time and are not necessarily uniform within any one culture. However, power distance does influence group member participation. Research has found that, generally, group members from low power distance cultures are more willing to speak out than those from high power distance cultures.[47]

In a world increasingly shaped by globalization, the chances that groups will have members from many cultures makes it important to acknowledge and address cultural differences in conceptualizations of power. As you learned earlier in this chapter, the synergy of diverse backgrounds and experiences offers a critical resource for successfully completing group tasks. You build an affinity with the members of your small group, and the more you work with them, the more likely you are to share power.

power distance How those with less power view power inequalities

Managing Conflict in Small Groups

Conflict occurs when individuals are interdependent and perceive that they have incompatible goals. All groups encounter conflict. How group members manage it influences how effectively the group achieves its goal. While some conflicts can spur on the group to consider innovative and creative ideas, other conflicts can prevent group members from engaging in productive conversations and accomplishing their tasks. This final section of the chapter identifies different types of conflicts groups may encounter and strategies for managing those conflicts.

conflict Occurs when individuals are interdependent and perceive that they have incompatible goals

Types of Small Group Conflicts

Summarized in Table 10.7, six types of conflicts emerge in small groups: identity, content, value, relational, meta, and serial.[48] Although discussed separately, some overlap among the types of conflict can occur. Still, recognizing the different types of conflicts can help your group best determine how to manage them.

TABLE 10.7 Types of Small Group Conflicts

TYPE	BRIEF DEFINITION	EXAMPLE
Identity	Group members reject image of self	You view yourself as highly competent, and other group members comment negatively on your work
Content	Disagreement about issue or topic	Group members' ideas diverge on the next step to take in a project
Value	Differing evaluations of something	Group members disagree on the effectiveness of the proposed solution
Relational	Discord stemming from the group's interpersonal relationships	Some group members want the group to have more social relationships, while other group members want strictly professional relationships
Meta	Disagreement over how the group manages conflict	You want the group to take a more cooperative approach to conflict, while other group members take a more competitive approach
Serial	Recurring unresolved conflict	Group members continually argue about how to structure their meetings without reaching an agreement

Identity conflicts occur when the self-image a group member projects is partially or completely rejected by other group members. Recall from Chapter 3 that your identity arises from your interactions with others and their support or lack of support for the self you present. Suppose you are a few minutes late for a group meeting, although usually you are a punctual person. A conflict might arise if other group members suggest you are undependable or uncommitted to the team because that is not how you view yourself.

Content conflicts center on a specific issue or topic and are common in small groups. Group members may clash over their views on current events, essential life skills, or the substance of the task they have been assigned to complete. These conflicts may be about public issues that are widely discussed outside the group, such as news stories, or private issues pertaining just to the group, such as the structure of group meetings. Conflicts over issues only tangentially related to the group can be left unresolved if they do not directly impact the group's ability to function. However, conflicts related to topics within the group can cause lasting damage if not addressed effectively.

When group members engage in value conflicts, they disagree about their evaluation of the worth, quality, or condition of something. In value conflicts, group members may hold differing views on whether something is right or wrong, good or bad, strong or weak, or imaginative or dull. For instance, in choosing a topic for a group presentation, members may hold strong and opposing views on how interesting or unique the topic might be for the audience. When managed productively, these conflicts can lead to new and innovative ideas, as with choosing an especially novel topic for a presentation that the audience will find especially intriguing.

Relational conflicts are about the relationships between and among group members. Conflicts may arise from differences in how group members define their relationships with each other or how they define the group's relationship to those outside the group. For example, you might consider the people in your work group professional colleagues, but others in the group might think of the group members as friends. A disagreement could occur when group members reveal personal information and pressure you to do the same. Relational conflicts are common in the earlier phases of group formation (Chapter 9) but also can surface in later phases, especially when original members leave and new members join.

When group members argue about how they are engaging in conflict, that is a metaconflict, or a conflict about how to do conflict. For example, some group members may feel comfortable with heated arguments, while others would prefer to take the volume down a notch or avoid the conflict altogether. When approached constructively, metaconflicts can help group members identify useful strategies for managing conflict and better prepare the group for future identity, content, value, and relational conflict.

Finally, serial conflicts are recurring conflicts that the group does not resolve. Any of the five previous conflict types also could be a serial conflict if the group returns to it again and again. For example, group members may not reach an agreement on the value of something or their methods for conflict communication. When left unresolved, serial conflicts can produce an undercurrent of animosity in the group and interfere with the group achieving its goal.

Approaches to Conflict Management

There are five approaches to conflict that apply to small groups: avoidance, competition, accommodation, compromise, and collaboration.[49] Summarized in Figure 10.9, the approaches vary according to group members' assertiveness and cooperativeness.[50] Choosing which approach to use depends on many factors, including the type of conflict, type of group, and relationships among group members.

With an avoidance approach to conflict, group members sidestep or evade the conflict altogether. Avoiding a conflict reflects both a lack of assertiveness and cooperation. When a disagreement is about something that is not all that important and will have little impact on the group's processes and outcomes, avoidance may be an appropriate

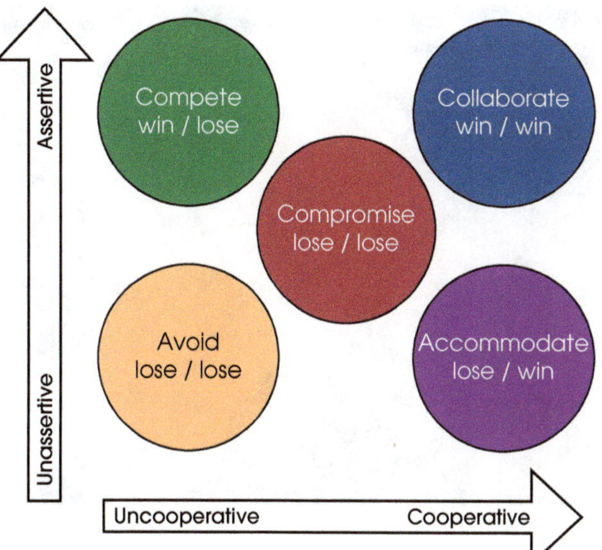

FIGURE 10.9 Approaches to managing conflict in small groups

strategy. However, avoidance is considered a lose/lose conflict strategy because the issue never is addressed, so group members do not have the opportunity to fully discuss their viewpoints.

A competitive approach to conflict is low on cooperation and high on assertiveness to the point of aggressiveness. With such an approach, conflict is viewed as a contest in which some group members win while others lose. When the stakes are high, as with the completion of a project, a competitive approach can be appropriate. For example, if one group member insists on a course of action that will sabotage the work others have completed, then not cooperating with that group member and charting a path that keeps the group on track are reasonable responses.

When group members use accommodation in conflict situations, they cooperate with others while not expressing their own wishes. If a disagreement addresses something about which you do not have particularly strong feelings or the topic is not all that important to you, then accommodation is a good course of action. For instance, if some group members want to use Zoom for a meeting, others would rather use Google Meet, and you do not have a preference, you could just go along with whatever the majority decides. While accommodation can be an appropriate approach to conflict in some instances, in others it can keep the group from learning about all group members' perspectives, lead to groupthink, and result in poor problem solving and decision making.

Compromise involves finding a middle ground between competing ideas. When group member compromise, each side gives up some of what they want. In that sense, everyone loses and no one wins. For example, if some group members think everyone should interview five students for their project on social media use and other group members think three would be sufficient, deciding on four interviews each would be a compromise. This approach to conflict can prove a useful way to settle differences but might hinder the group's ability to find more creative or innovative solutions. In addition, because no one is satisfied completely with the outcome, everyone is somewhat unhappy.

Collaboration involves high degrees of cooperation and assertiveness. Especially complex problems require taking a collaborative approach to conflict to encourage group members to consider all possible perspectives and solutions (Figure 10.10). Collaboration often is effective in managing serial conflicts, which require fresh viewpoints that might offer a path out of recurring disagreements. For instance, if group members consistently argue over how to distribute limited resources, a collaborative

FIGURE 10.10 When group members use collaboration to manage conflict, they find innovative solutions to problems.

approach to the conflict might explore additional resource avenues. With avoidance, competition, accommodation, and compromise, innovative solutions likely would not be considered.

Conflict is an inherent part of working in groups. Identifying the type of conflict and the best way to manage it will facilitate group members' working relationships and achieving the group's goals.

CHAPTER REVIEW

- **Describe how to promote creativity in small groups.**
  You can facilitate creativity in your group by recognizing creativity as a collective effort, providing support for innovation, creating a shared vision, establishing cooperative goals, developing supportive communication, and taking a task orientation. Concrete techniques to encourage creativity in small groups include brainstorming, lotus blossom technique, nominal group technique, six thinking hats, and synectics.

- **Describe the eight-step approach to problem solving in small groups.**
  The components include constructing the problem, gathering information, selecting information, integrating information, generating ideas, evaluating ideas, planning the implementation, and monitoring the implementation. Group members may go back and forth between components but completing each is essential to innovative problem solving.

- **Explain how groupthink hinders effective decision making in groups.**
  Groupthink occurs in the absence of critical thinking. Four group behaviors indicate when groupthink is at work: promotional leadership, in-group bias, pressure to conform, and hidden profiles.

- **Define critical thinking and explain its role in the decision-making process.**
  Critical thinking involves purposefully interpreting, analyzing, and evaluating information to solve problems and make decisions. Facilitating critical thinking in the small group decision-making process involves making clear and relevant statements, clarifying group goals, contributing new information, linking ideas together, providing supporting evidence, offering critical assessment, identifying the practical utility, welcoming a wide range of viewpoints, and framing ideas within a broader perspective.

- **Identify and describe the primary approaches to decision making in small groups.**
  With majority rule, group members vote and the alternative with the most vote wins. In autocratic and consultative rule, the leader makes decisions. Expert rule is when decisions are dependent on a specific expertise associated with the problem. Unanimity involves all group members agreeing on an alternative or no one objecting. Consensus is making a choice acceptable to all group members.

- **Discuss the role of power in group problem solving and decision making.**
  Power in groups comes from a number of sources, although no one is ever powerless in a group. Power within a group is often communicated by behavior and is culturally bound.

What is considered dominant or submissive in one culture is not considered the same way in another—this is a complicating factor in a globalizing world.

■ **Identify the types of and approaches to conflict found in small groups.**
Groups engage in identity, content, value, relational, meta, and serial conflicts. Group members use avoidance, accommodation, compromise, competition, and collaboration in managing those conflicts.

DISCUSS AND APPLY

1. Think of an example when you have experienced creativity in a small group. How did group members encourage creativity? How did you feel participating in the creative process?

2. Try out the eight-step approach to problem solving in a small group setting. How well did it work? What have you learned about small group problem solving from trying out this approach?

3. Think about a small group decision in which you participated. What decision-making approach did the group use? How effective was it?

4. Have you ever experienced or observed groupthink occurring in a group? What behaviors indicated groupthink was at work? What will you do in the future as a group member to avoid groupthink?

5. Think of an example in which you have been empowered in a group. How did that affect group members' interactions?

6. Reflect on a time in which you encountered a serious conflict in a group. What type (or types) of conflict was it? How did you respond? How did the group manage the conflict? How satisfied were group members with the outcome?

CREDITS

Public Speaking

CHAPTER 11

Developing Your Speech Purpose and Topic

For her last speech as First Lady of the United States, Michelle Obama spoke at a White House ceremony honoring high school counselors. After thanking the honorees and other participants, she addressed her message to the young people in attendance:

> Do not ever let anyone make you feel like you don't matter, or like you don't have a place in our American story—because you do. And you have a right to be exactly who you are. But I also want to be very clear: This right isn't just handed to you. No, this right has to be earned every single day. You cannot take your freedoms for granted. Just like generations who have come before you, you have to do your part to preserve and protect those freedoms. And that starts right now, when you're young.

Obama went on to inspire her audience with the stories of high school students and counselors as well as those from her own family. Then she called her audience to action:

> I want our young people to know that they matter, that they belong. So don't be afraid—you hear me, young people? Don't be afraid. Be focused. Be determined. Be hopeful. Be empowered. Empower yourselves with a good education, then get out there and use that education to build a country worthy of your boundless promise. Lead by example with hope, never fear. And know that I will be with you, rooting for you and working to support you for the rest of my life.[1]

When you give a speech, you always have a purpose in mind. In this case, Michelle Obama wanted to convince the audience that young people must pursue an education and then give back to their communities. Later in the speech, she used the high school counselors as examples of people making a difference and encouraged the students in attendance to chart their own path to personal and community success. Making her purpose clear and discussing her points in ways that engaged her audience helped her achieve her goal (Figure 11.1).

This chapter begins by identifying how public communication has both changed and stayed the same in a pervasive communication environment. Then you will learn how to develop your speech purpose, topic, and working outline so that you can reach your objective when you give a speech.

FIGURE 11.1 When you make the purpose of your speech clear, as Michelle Obama did in her speech to high school counselors, you are more likely to achieve your goal.

Public Communication in a Pervasive Communication Environment

Have you heard of these speeches?

- "How Schools Kill Creativity"
- "The Best Stats You've Ever Seen"
- "Averting the Climate Crisis"
- "Simplicity Sells"
- "Greening the Ghetto"
- "Why We Do What We Do"

If you have, then you are familiar with TED (Technology, Design, and Education) Talks, the website that features an international list of speakers discussing topics that range from activism to youth. These six speeches, posted online in June 2006, were the first TED Talks. Within 3 months, a million people had watched those videos.[2] Today, millions of people from around the world watch TED Talks every day. TED Talks provide just one example of how a pervasive communication environment has extended the influence of a speech from one audience to many audiences.[3]

You might think that in these times of tweets and texts public speaking skills have lost their importance, but that is not the case. In fact, skills associated with public speaking, such as finding and evaluating information, organizing that information, and clearly presenting ideas, are among the top characteristics employers want in job candidates.[4] In addition, a survey of midlevel and senior managers across several industries in the United States found executives ranked ethics, critical thinking, and communication (writing and speaking) in the top 10 most important nontechnical or soft skills essential to success in the workplace.[5] Moreover, a LinkedIn Learning survey of 1,700 learning and development (L&D) professionals found that communication was one of the highest priority soft skills they want to build in employees.[6] As has been true for decades, if not centuries, developing your public speaking skills is crucial for your personal, academic, and professional success.

Changes in Public Communication

A pervasive communication environment has changed public speaking in three main areas: ethics, information access and evaluation, and audience diversity. In the area of ethics, the internet simply makes it easier to plagiarize. **Plagiarism**, derived from the Latin word for kidnapper, *plagarius*, is the act of taking someone's work and presenting it as your own.[7] In a survey of more than 63,000 undergraduate students in the United States, 36% reported they had copied material from an internet source without citing it, 14% said they had made up a false bibliography, and 7% admitted turning in work someone else had done. Another survey of more than 71,000 undergraduates found that 40% admitted to cheating on written assignments.[8] You must take extra care in keeping track of the sources of your information and then cite them in your speeches. In addition, while you might be tempted to use a speech you find online, the consequences for doing so can range from embarrassment when you cannot answer questions about the topic to a more severe penalty for academic dishonesty. Consider the cases of school administrators who plagiarized parts of their speeches in Box 11.1.

The ease with which information can be distributed online requires diligence in evaluating sources and the information they provide. As discussed in more depth in Chapter 12, examine the author's credibility and determine if their

BOX 11.1: ETHICAL QUESTIONS

Penalties for Plagiarism

The two high school seniors thought the commencement speech for their graduation sounded familiar—and it was. Newtown, Massachusetts, School Superintendent Daniel Fleishman used parts of a speech the state's governor had presented just days before. Fleishman admitted there were a few similarities, but noted they constituted a small portion of his speech. The penalty? A fine of one week's wages, or about $5,000.[9] In contrast, a week before the Newton incident, Brenda Hodges, another Massachusetts school superintendent, was accused of plagiarizing part of her commencement speech. She denied the charge, stating that she got some ideas from another speaker, but did not copy anything word for word. However, she did admit that she should have credited her sources.[10] The penalty? She was forced to resign. What do you think? What should be the punishment when a school official plagiarizes? How does the degree of plagiarism—borrowing a few ideas, copying a few lines, using the entire speech—influence the penalty? What are the penalties for plagiarism at your school?

perspective is unbiased and fair. Relying on the online databases available from your campus library is one way to assure that the information sources you use are credible and reliable (Figure 11.2).

A pervasive communication environment has broadened the scope of who may view a speech. Audiences often are more culturally diverse. New communication technologies allow speakers and audience members to be some distance apart—even in different countries. And speeches that in the past were presented only for an in-person audience now may be recorded and viewed online by many others. For example, when former president Barack Obama gave a eulogy for civil rights leader and U.S. Representative John Lewis, millions of people around the world viewed the speech live during the funeral and then millions more watched the video posted on YouTube, C-SPAN, and many news sites.

FIGURE 11.2 Online library databases are excellent sources for credible information.

Enduring Traditions in Public Communication

Although the ubiquitous nature of communication has changed some aspects of public speaking, the essential process has remained the same for centuries. Scholars in ancient Rome identified five elements, or canons, of public

TABLE 11.1 **Five Canons of Rhetoric**

CANON	LATIN NAME	BRIEF DEFINITION	DEMONSTRATED WHEN THE SPEAKER …
Invention	*Inventio*	Creating the ideas and arguments for a speech	Presents novel information and sound arguments.
Arrangement	*Dispositio*	Organizing the points of a speech	Orders ideas in a pattern the audience can follow easily.
Style	*Elocutio*	Choosing language to convey the speaker's message	Uses words that spark the audience's interest and imagination.
Delivery	*Actio*	Using the voice and body in presenting a speech to an audience	Engages the audience with a lively voice and meaningful gestures.
Memory	*Memoria*	Relying on memory to recall what the speaker planned to say as well as depth of knowledge on a topic	Transitions smoothly from one point to the next and answers audience questions clearly and completely.

invention One of the five canons of rhetoric, invention is developing the ideas and arguments for a speech

arrangement One of the five canons of rhetoric, arrangement is the order in which the speaker organizes the ideas in a speech

style One of the five canons of rhetoric, style refers to the language the speaker uses in a speech

delivery One of the five canons of rhetoric, delivery involves using the voice and body to present a speech to an audience

memory One of the five canons of rhetoric, memory is the speaker's depth of knowledge about a topic and ability to remember what to say when giving a speech

speaking summarized in Table 11.1 that remain applicable today.[11] **Invention** is developing your ideas and arguments for your speech. The element's Latin word *inventio*, or discovery and inquiry, suggests that speakers actively research their speech topics.[12] **Arrangement** is putting your ideas in some sort of order that fits with the speech purpose and information found during invention. The Roman's called this element *dispositio*, or organization, which referred to the overall structure of a speech—introduction, body, and conclusion—as well as the pattern of organization for the main points. **Style** reflects the language you use to convey your message. From the Latin *elocutio*, to speak, effective word choice promotes both clarity and audience interest. **Delivery** is using your voice and body to present your speech. The element's Latin term *actio*, or activity, underscores the importance of a speaker's physical voice and movement in presenting a speech. Finally, **memory** refers to recalling what you planned to say. *Memoria*, or memory, encompasses both remembering your speech and having a deep knowledge of your topic, knowing much more about it than you present in the speech itself.

The remainder of this chapter gets you started on invention by having you consider your speech's purpose, topic, and initial main ideas.

Determining the General Purpose of Your Speech

You probably have heard people say, "That's just rhetoric" or "Those are only words," but speeches and words have the power to make things happen. A well-crafted speech, such as Michelle Obama's address to school counselors and students, can change listeners' attitudes or call them to action. Or a speaker may want audience members to know more about a topic or view a topic from its more humorous side. That is, speakers have reasons for giving speeches—they want to accomplish something. The **general purpose** is the overall goal a speaker has for a speech. The speaker may seek to inform, persuade, or entertain the audience. Although each general purpose may involve elements of the other, as with a speech to inform that includes some humor, an effective speech will focus on a single general purpose.

When you give a **speech to inform**, discussed in Chapter 15, you are concerned with the audience learning more about your topic. Speeches to inform typically address five different categories: objects and places, people and other living things, processes, events, and ideas and concepts. You use informative speaking every day, such as explaining how something works or describing someplace you have visited.

In a **speech to persuade**, discussed in Chapter 16, you want to influence your audience members' beliefs, values, attitudes, or behaviors. With speeches to persuade, you must first get your audience to accept your position on a topic and then convince them to internalize it. For instance, say you want your audience to sign a petition for a community garden project on campus. Your audience must think such a project is a good idea before they will be willing to sign the petition.

Speeches to entertain encourage audience members to enjoy themselves as they consider the lighter side of topics. Special occasions, such as celebrations and tributes, often invoke speeches to entertain. These speeches are meant to amuse and charm audience members. For example, if you were giving an after-dinner speech, you might talk about a few comical situations from a recent vacation or highlight some humorous incidents you experienced in your first job.

Once you have determined the general purpose for your speech, you can consider the topics you might want to talk about.

general purpose The speaker's overall goal for the speech

speech to inform A speech in which the speaker wants the audience to learn more about a topic

speech to persuade A speech in which the speaker wants to influence audience members' beliefs, values, attitudes, or behaviors

speech to entertain A speech in which the speaker encourages audience members to enjoy themselves as they consider the lighter side of a topic

Brainstorming for Speech Topics

In some speaking situations, you will not choose your topic, such as when you are giving a report at work. But in other cases, you will choose the topic you are going to talk about, as with speaking at a neighborhood meeting organized to identify local concerns.

Choosing a topic in today's world can be easier than in the past because you have so many resources to help you. If you start thinking about the speech you are going to give well in advance of the occasion, you will have more time to consider the range of topics from which you can choose. Initially, you want to brainstorm for as many potential speech topics as you can. Recall from Chapter 10 the rules for brainstorming:

- Generate as many ideas as possible.
- Record all ideas.
- Do not dismiss any idea as too silly or outrageous.
- Combine and build on ideas.
- Do not critique or evaluate any ideas.

Develop a list of at least 15 topics you might speak about. Ask yourself these questions to start the brainstorming process:

- What topics do I feel passionate about?
- What topics do I argue or debate about with my friends?
- What is trending on Twitter, Facebook, Instagram, TikTok, and other social media?
- What current events are dominating the news?
- What unique experiences have I had in my life?
- What are my favorite activities or hobbies?

Check newspapers, magazines, TV and radio news shows, and other media sources for ideas. Box 11.2 provides suggestions for identifying issues of current

---

BOX 11.2: LIVE IT LOCAL

Topics of Interest in Your Community

What are the issues people are talking about in your community? How can you find out? Here are some strategies:

- Attend a meeting of your local city council or county board of supervisors.
- Attend a local neighborhood group meeting.
- Sign up for electronic newsletters from your local representative, such as your city's mayor or district representative.
- Sign up for a neighborhood listserv or discussion group.
- Join your neighborhood's Facebook group.

- Read the letters to the editor in your local newspaper for 1 week.
- Read the editorials printed in your local newspaper for 1 week.

How are these issues similar to or different from you own? For ones that are similar, how might you get more involved and become part of the conversation? For those that are different, how might you make your voice heard? For instance, consider writing a letter to the editor of your local newspaper or speaking up at a neighborhood meeting.

TABLE 11.2 Speech Topic Websites

WEBSITE	BRIEF DESCRIPTION	USEFUL FOR …
Buzzle Speech Topics	Hundreds of topics as well as tips for topic development	Developing topic ideas for specific types of speeches, such as informative and persuasive
Debatabase	More than 1,000 topics for debate from the International Debate Education Association	Clarifying the specific points and counterpoints associated with a topic
ProCon.org	Nonpartisan information about the pros and cons on 50 controversial issues	Identifying the key differences in positions on important topics
Public Agenda Reports and Resources	Nonpartisan information on public issues, such as education, the economy, and immigration	Determining the current points of contention and agreement on public policy topics
Speech Topics Generator	Randomly generates up to 10 speech topics at a time	Coming up with novel approaches to topics (click on "expand" for each topic)
Wikipedia	Online, free, open-access encyclopedia	Browsing through categories and subcategories on a wide range of topics

interest that are close to home. Table 11.2 includes a sampling of online resources that can help you generate ideas for speech topics.

Mind mapping also can help you brainstorm for possible topics. With **mind mapping** you diagram your ideas so you can visualize how they are related to each other.[13] You begin your mind map with one idea and then start branching off from it, drawing lines to show how those ideas link to each other. Figure 11.3 provides an example for the general topic idea "music." Lines show how ideas are linked together and how they are related to each other. *Floating topics* are related to the original idea, but not directly connected to the current structure. At this stage in the brainstorming process, you want to use mind mapping to generate ideas.[14] So add whatever comes to mind, even if it does not exactly fit with your topic. As shown in Figure 11.3, the topic "music" led to other possible topic areas, such as reality television shows and the Salem witchcraft trials.

mind mapping A way to diagram ideas to visually represent how they are related

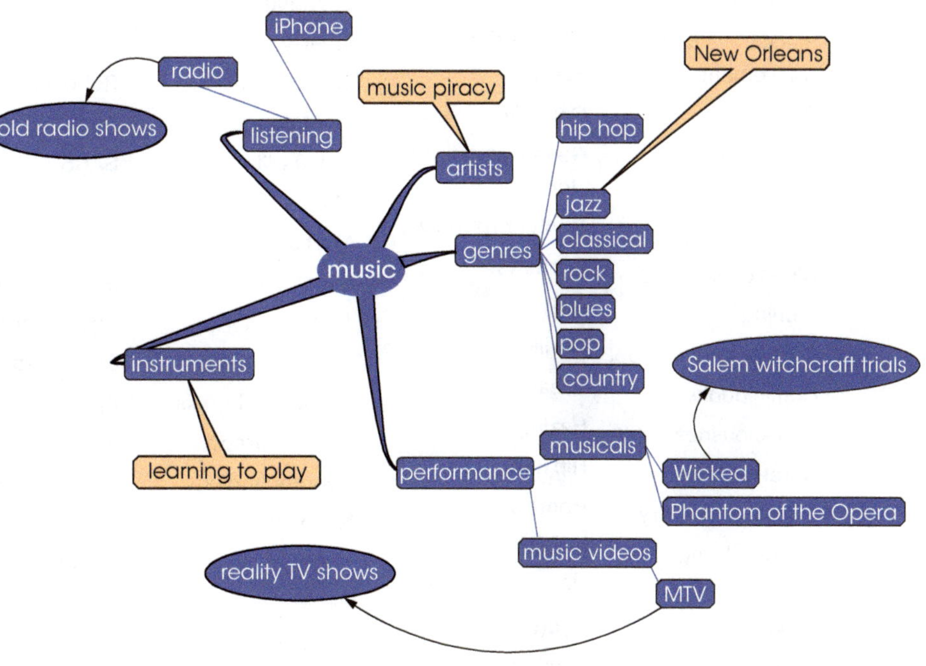

FIGURE 11.3 Completing a mind map on a general topic, such as music, can lead you to more specific topics or other topics altogether.

For additional speech topic ideas, Table 11.3 lists ones our students have suggested during many class brainstorming sessions. Refer to these topics as you create your own list or mind map.

Use whatever techniques work for you in coming up with a robust list of topics. After developing that list, you are ready to evaluate and then select the best topics for your speeches.

TABLE 11.3 Possible Speech Topics

Acupuncture	Eagles	Jazz	Organ donation	Time management
Agoraphobia	Earthquakes	Jellyfish	Origami	Tornados
Aliens	Ebola	Job sharing	Owls	Tsunamis
Amusement parks	Ecotourism	Juggling	Pandas	Ukulele
Antibiotics	Electric cars	Juries	Paralympics	Unicorns
Artificial intelligence	Emotional intelligence	Kaleidoscope	Poetry	United Nations
Astrology		Kangaroos	Prisons	Urban homesteading
Badminton	Fast food	Karaoke	Quakers	Vampires
Ballroom dancing	Firefighting	Kayaking	Quantum mechanics	Ventriloquism
Batteries	Fireworks	Killer bees	Racial justice	Video games
Bicycle helmets	Flamingos	Labor unions	Rain forests	Volcanoes
Black holes	Fly fishing	Learning styles	Rainbows	Water desalinization
Black Lives Matter movement	Food insecurity	Liberty	Reality TV shows	Wind power
	Fossils	Lie detectors	Robotics	Women in the military
Bullying	Gambling	Lightning	Rock climbing	World War II heroes
Charter schools	Genetically modified food	Living wills	SARS	Wristband fitness trackers
Childhood obesity		Magic	Sea stars	
Chocolate	Geothermal power	Martial arts	Semiconductors	Xanadu
Clean coal	Go-karts	Mediation	Serious games	Xeroradiography
Cloning	Golf	Meditation	Solar energy	Xylographs
College student debt	Gorillas	Minimum wage	Space exploration	Yellowware
Comic books	Great Lakes	Multiple sclerosis	Spiders	Yerba maté
Consciousness	Hazing	Music therapy	Steampunk	Yodeling
Coral reefs	Hip hop	Nanotechnology	Stem cell research	Yoga
Cosmetic surgery	Home schooling	National parks	Stress	Zebras
Cyberbullying	Honey bees	Nebulas	Student-athlete union movement	Zen
Cycling	Hurricanes	Nuclear energy		Zeppelins
Dieting	Hydroponics	Nutrition	Surfing	Zinc
Dinosaurs	Identity theft	Occupational therapy	Sustainable agriculture	Zombies
DNA evidence	Illiteracy	Ocean exploration		Zoos
Dolphins	Investing in the stock market	Online dating	Teamwork	Zootechnics
Dragons		Online games	Telemedicine	
Drought	IQ test	Online privacy	Third-hand smoke	

Evaluating and Selecting Speech Topics

Competent communicators demonstrate effectiveness in reaching their goals and appropriateness in adapting to norms and contexts (Chapter 1). In choosing a speech topic, competent communicators consider their own interests, the audience, available resources on the topic, and the occasion. Answer the questions that follow about each topic to determine the best topics for your speeches:

- *Your own interests and knowledge*: How interested am I in the topic? What do I know about the topic? Is this a topic I would feel comfortable talking about with an audience?

- *The audience*: What does my audience already know about the topic? How much does my audience care about the topic? Can I spark my audience's interest in the topic?

- *Available resources*: Where can I get information on the topic? How readily available is information on the topic? Where can I find resources on the topic? Whom could I contact who would know about the topic?

- *The occasion*: What is the purpose of the occasion and my speech? How appropriate is the topic for the speaking occasion? If there are other speakers scheduled to present, how well will my topic fit with theirs? Can I cover the topic in the time allotted for the speech?

Let's go through an example for a classroom speech with a topic from Table 11.3. Suppose it is about halfway through the term. You have exams and papers, as well as a speech due in your communication class. In addition, your boss has changed your work schedule. Stress seems like the perfect topic for a speech. You have experienced it and you are interested in it. Your audience is your classmates, so they probably are experiencing midsemester stress too. The campus student health center sponsors workshops on managing stress, which will provide one place to start for resources. In analyzing the occasion, you cannot be sure of what your classmates will talk about, but the topic seems appropriate for the time in the semester. Also, you think you can cover at least three points about stress in the time limit your instructor has assigned to the speech. In this case, stress seems like an ideal topic. As shown in Figure 11.4, the ideal topic is at the intersection of the four criteria: The topic interests you, it interests the audience, you can find plenty of information on it, and it is appropriate to the occasion.

Not all topics will meet all four criteria for choosing a good topic. Does that mean you should not choose them? Not necessarily. For instance, suppose you want to give a speech on volcanoes, one of the topics listed in Table 11.3. You are especially interested in the topic because you visited Popocatépetl Volcano near

FIGURE 11.4 An ideal topic choice fits with your and your audience's interests and knowledge, is relevant to the occasion, and has sufficient information available.

FIGURE 11.5 If you can get your audience interested in a topic that meets all the criteria for a good topic, then it becomes a possibility.

Mexico City a few years ago and know it erupted quite recently. There are many resources on volcanoes, and the topic fits with the informative speaking assignment you have in your class. But as shown in Figure 11.5, you are not sure your audience will care about volcanoes. In this case, you could pique your audience's interest in several ways, such as starting your speech with a captivating image of the Popocatépetl Volcano taken by NASA after the eruption or telling a brief anecdote about your visit to the volcano (Figure 11.6). Volcanoes may not seem like an ideal topic at first, but it is one that you could move into the ideal topic category. However, if the topic is one that you are fairly certain your audience will not find interesting, choose a different one. You want a topic that will excite—not bore—your audience.[15]

FIGURE 11.6 A captivating image, like this one from NASA taken days after Popocatépetl Volcano in Mexico erupted, can instantly capture your audience members' attention and pique their interest in your speech topic.

As you review the list of topics in Table 11.3, mark the ones you find most interesting and cross off the ones that do not interest you. Research shows that your interest is the most important factor in a choosing a successful topic.[16] If you are not passionate about the topic, your audience will not be either. Choose topics the excite you—ones that you want to talk about with your classmates, friends, and others. As Figure 11.7 shows, a topic in which you are disinterested is a poor choice, even if you can find a great deal of information on the topic, your audience is interested in it, and the topic is relevant to the occasion.

You may identify a topic in which you and your audience are interested and you can find information on it, but it does not fit with the occasion, as

FIGURE 11.7 If you are not interested in a topic or know little about it, it is not a good speech topic choice.

shown in Figure 11.8. For instance, the time allotted is too brief to cover the topic, or the topic would be inappropriate for the situation. Topics such as immigration, capital punishment, and legalizing drugs are ones that draw intense debate, but tend not to lend themselves to the occasion of a classroom speech. Because people have such strong feelings about these topics, a short speech does not provide the format for the needed in-depth discussion. When a topic is inappropriate for the occasion, cross it off your list.

Some topics may be of great interest to you and your audience as well as relevant to the occasion. However, if you cannot locate a wide range of resources that provide new information on the topic for your audience, that will not be a good topic for you to choose, as shown in Figure 11.9. For instance, say you wanted to talk about your campus's recent decision to ban smoking. Although the topic is relevant to your classmates and the occasion, there likely is not much new information you can provide.

Finally, if a topic does not meet more than one criterion, choose a different one. For example, if your audience likely will not be interested and you cannot find much information on a topic—even if it is one you feel passionate about—then do not choose it. Selecting an excellent topic is the first step toward giving an excellent speech. If you choose your topic with care and thoughtfulness, developing your speech will be all that much easier. Box 11.3 suggests collaborative strategies for choosing the best topics for your speeches.

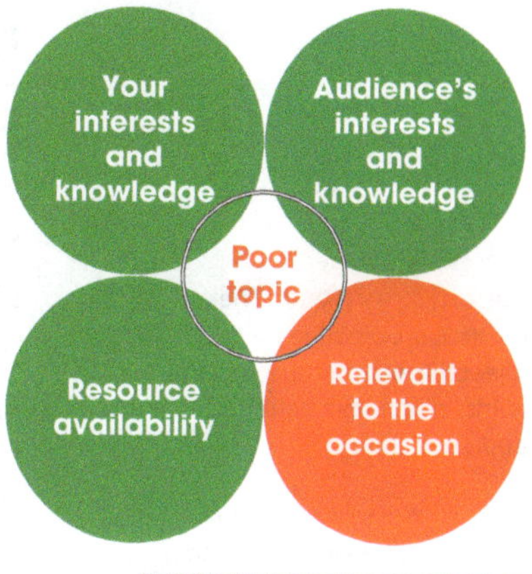

FIGURE 11.8 A topic must fit with the speaking occasion to be a good choice for your speech.

FIGURE 11.9 If you cannot find information on a topic—even if it meets the other three criteria—it is not a good choice for your speech.

BOX 11.3: CONNECT & REFLECT

Crowdsourcing Speech Topics

You have heard of crowdsourcing for all sorts of activities, from launching a small business to helping the survivors of a natural disaster. You can use this approach to evaluate your speech topics as well, getting feedback and ideas from a wide range of people, including friends, classmates, and family. You might take an informal approach, posting a brief explanation of the speaking occasion or assignment and your topic ideas on your social media accounts, asking for comments. Or you might set up a short online survey on Google Forms, SurveyMonkey, or Typeform with a description of the speaking context and topics you are considering and send it out for feedback. Either way, you can use the power of crowdsourcing to assess how others respond to your topic and even get suggestions for approaching, researching, and developing your final topic choice. Try it out and then consider the effectiveness of this strategy for evaluating speech topics.

Phrasing the Specific Purpose and Thesis

specific purpose The particular objective for a speech topic that the speaker wants to achieve with the audience

thesis A single declarative sentence that expresses the essence of a speech

Now that you have chosen your topic, the next step is to link your topic with the general purpose to phrase the specific purpose. The **specific purpose** is the particular objective for your topic that you want to achieve with your audience by the end of your speech. The specific purpose merges your topic with the general purpose to clearly state your reason for giving the speech.[17] The **thesis**, or central idea, expresses the essence of your speech in one declarative sentence.[18] The general purpose, specific purpose, and thesis work together to form the foundation of your speech and will guide you in researching your topic and developing and organizing your ideas.[19]

Informative Speeches

When you give a speech to inform, you want your audience to learn more about a topic. Here are three sample topics with the general purpose, specific purpose, and thesis.

EXAMPLE 11.1

Topic: Art Deco

Title: "The Meaning of Art Deco"

General purpose: To inform

Specific purpose: To inform my audience about the beliefs Art Deco represented

Thesis: Art Deco represented beliefs in the power of modern technology, social progress, and efficiency.

FIGURE 11.10 An informative speech on the history of roller coasters would trace their development from the past to the present.

EXAMPLE 11.2

Topic: Roller Coasters

Title: "Roller Coasters: From Hills of Ice to Steel Dragons"

General purpose: To inform

Specific purpose: To inform my audience about past and current roller coasters

Thesis: Roller coasters have evolved over the past 4 centuries from tracks built on icy gentle hills to wooden tracks near the beach and more recently to steel mountains that drop riders from dizzying heights. (Figure 11.10)

EXAMPLE 11.3

Topic: Color of Change

Title: "Color of Change and Racial Justice"

General purpose: To inform

Specific purpose: To inform my audience about the racial justice activities of the nonprofit organization Color of Change

Thesis: Color of Change responds to racial injustice in six areas: criminal justice, media justice, voting rights, technology justice, White nationalism, and economic justice.

Persuasive Speeches

Speaking persuasively involves attempting to convince audience members to change their attitudes, beliefs, values, or behaviors. When your goal is to persuade, you go beyond simply educating your audience about a topic to actively encouraging them to think differently about it and possibly take some sort of action. For instance, if your topic were bamboo, you might want to convince your audience to switch from buying cotton clothing to buying bamboo clothing, as bamboo is more sustainable to grow. Below are three additional examples of speeches to persuade.

EXAMPLE 11.4

Topic: Silence

Title: "Practice Silence Every Day"

General purpose: To persuade

Specific purpose: To persuade my audience to spend 10 minutes every day in silence

Thesis: Practicing silence improves your health, makes you happier, and increases your productivity.

EXAMPLE 11.5

Topic: High-Speed Rail

Title: "Building High-Speed Rail in the United States"

General purpose: To persuade

Specific purpose: To convince my audience that they should support building more high-speed rail lines

Thesis: More high-speed rail lines should be built in the United States because they reduce pollution, boost economic development, and cut commute times.

EXAMPLE 11.6

Topic: Genetic Information

Title: "Protect Your DNA"

General purpose: To persuade

Specific purpose: To persuade my audience that only government agencies and research institutions should have access to the results of individuals' DNA tests

Thesis: DNA test results are essential for science and health-related government agencies and nonprofit research institutions, but should not be available to pharmaceutical companies, advertisers, and for-profit health care organizations.

Entertaining Speeches

With entertaining speeches, your overall goal is to amuse your audience. You want your audience laughing and smiling, rather than thinking deeply about a topic. For instance, rather than an informative speech on the history of roller coasters, you might delight your audience with humorous stories about your experiences riding roller coasters. Below are three additional examples of entertaining speeches.

FIGURE 11.11 An entertaining speech on sand art might include humorous tales of work as a sand artist.

EXAMPLE 11.7

Topic: Sand Art

Title: "The Ebb and Flow of Art in the Sand"

General purpose: To entertain

Specific purpose: To entertain my audience with stories from my experience one summer as a sand artist for a Florida beach resort

Thesis: The two ingredients in sand art—sand and water—make for some unexpected outcomes. (Figure 11.11)

EXAMPLE 11.8

Topic: Waiting Tables

Title: "My Favorite—and Not So Favorite—Restaurant Customers"

General purpose: To entertain

Specific purpose: To entertain my audience with humorous descriptions of customers at a high-end restaurant

Thesis: There are five types of customers at high-end restaurants: the Star, the Camper, the Joker, the Critic, and the Gem.

EXAMPLE 11.9

Topic: Choosing a Pet

Title: "When the Right Pet for You Is a Pet Rock"

General purpose: To entertain

Specific purpose: To entertain my audience with anecdotes about my attempts to find the right pet

Thesis: After an exhaustive search in which I met pets from parrots to poodles, I finally found the right pet for me—a pet rock.

Although each example above features a different topic, the same topic may produce speeches that are quite different based on the general purpose, specific purpose, and thesis. Table 11.4 compares the specific purpose and thesis statements for speeches to inform, persuade, and entertain on the same topic, swamps.

TABLE 11.4 Comparison of Specific Purpose and Thesis Statements for One Topic

TOPIC	TITLE	GENERAL PURPOSE	SPECIFIC PURPOSE	THESIS
SWAMPS	"Swamp Animals"	To inform	To inform my audience about the types of animals that live in a swamp	Many common mammals, amphibians, and reptiles live in swamps.
	"Save Our Swamps"	To persuade	To persuade my audience that swamps must be protected from draining and development	Swamps must be saved because they provide flood protection, purify water, and offer a safe habitat for many of the plants and animals that people rely on for food.
	"It's Alive in a Swamp!"	To entertain	To entertain my audience with descriptions of the offbeat creatures that live in swamps	Snake birds, flying foxes, and skinks are just a few of the friendly creatures I've met in a swamp.

Whatever topic you select, creating the specific purpose and thesis will start you on a clear path to further developing and researching your topic.

Developing the Working Outline for Your Speech

As you build your speech, you will create three outlines: the working outline, the complete-sentence outline, and the presentation outline. The **working outline** reflects your initial planning for your speech and includes the main points and

working outline The outline that reflects the initial planning for a speech and includes the main points and possible subpoints using key phrases

	Type	Brief Description	Helps you ...	In chapter ...
This chapter →	**Working**	Draft outline with possible points and subpoints	Organize initial ideas and identify information sources.	11: Developing your speech topic and purpose
	Complete-sentence	Full-sentence outline with all parts of speech and bibliography	Fully develop each point and demonstrate how ideas link together.	13: Supporting, organizing and outlining your ideas
	Presentation	Speaking outline that distills speech to essential points	Identify key words for practicing and presenting your speech.	14: Delivering your speech

FIGURE 11.12 This chart compares the working, complete-sentence, and presentation outlines.

possible subpoints using key phrases. You revise this outline as you research and prepare your speech. When you are ready to move into the phase of finalizing your ideas, you will develop a *complete-sentence outline* written in complete sentences that includes all the aspects of your speech, such as the thesis, main points, transitions, and bibliography. The *presentation outline* is a brief outline with key words and phrases that you will use when you give your speech. Figure 11.12 explains the functions the three outlines serve in the speechmaking process.

In developing your working outline, you will notice if your topic is too narrow or too broad. If your topic is too narrow, you will not have a sufficient number of points to talk about or you will lack enough information for the time limit requirements. If your topic is too broad, you will have too many points and too much information for the time limit.

Let's take one topic introduced earlier, stress, and go through the process of creating the working outline for a 5–7 minute informative speech that a student could give in class.

Planning Your Speech With the Working Outline

You begin the working outline by recording the topic and general purpose:

Topic: Stress

General purpose: To inform

Next, consider what aspects of stress you want to talk about with your audience. Do not evaluate at this point; brainstorm for subtopics associated with stress. Just like mind mapping can help you brainstorm for possible topics, it can help you expand or narrow your topic as needed and consider differing perspectives on it.[20] Figure 11.13 provides an example of mind mapping for the general topic of stress.

The mind map suggests several avenues for developing the topic: how to manage stress, its causes, its effects, and different types of stress. To narrow down your topic, you consider:

- Your interests in the various aspects of the topic

- What your audience knows about the topic and their interest in it

- Available resources

- The occasion, including the time limit

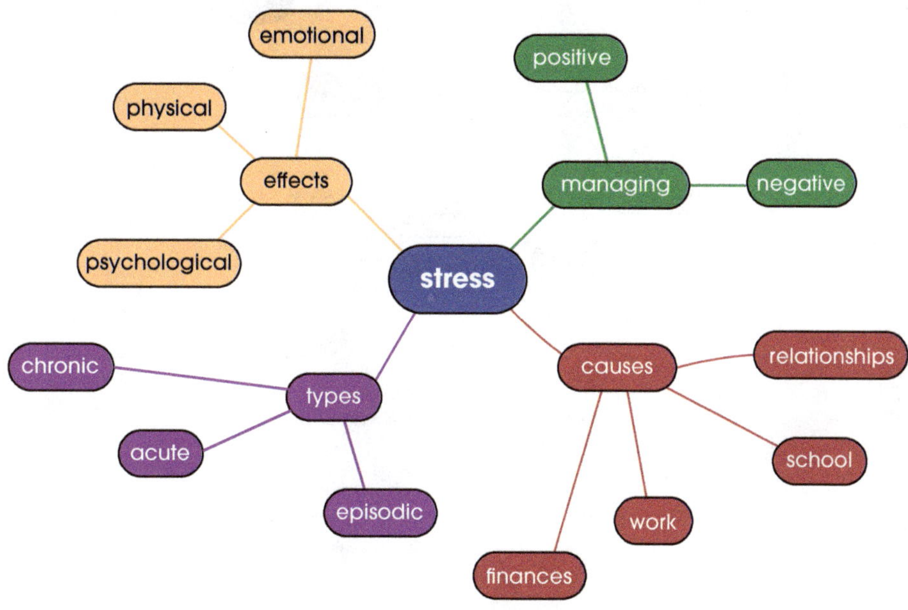

FIGURE 11.13 An initial mind map for "stress" as the speech topic.

Although you have some general knowledge about stress, you would like to learn more about the types of stress and think your audience would as well. A quick internet search revealed plenty of information on this aspect of the topic from credible courses. And you think there is sufficient information for the time limit.

Now you can add the specific purpose, thesis, and main points to your working outline:

Topic: Stress

Title: "The Different Types of Stress"

General purpose: To inform

Specific purpose: To inform my audience about the different types of stress

Thesis: There are three types of stress: acute stress, episodic stress, and chronic stress.

Main points:

    I. Acute stress

    II. Episodic stress

    III. Chronic stress

Your working outline is starting to take shape. But you want to have a better idea of what you likely will cover within each main point. Return to your mind map and develop the area associated with the different types of stress, as suggested in Figure 11.14.

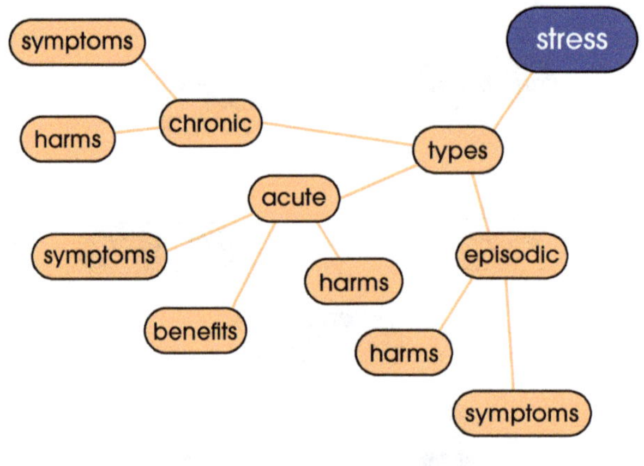

FIGURE 11.14 A mind map of the different types of stress.

Now the working outline can include subpoints for each main point.

Topic: Stress

Title: "The Different Types of Stress"

General purpose: To inform

Specific purpose: To inform my audience about the different types of stress

Thesis: There are three types of stress: acute stress, episodic stress, and chronic stress.

Main points:

    I. Acute stress

        A. Symptoms

        B. Benefits

        C. Harms

    II. Episodic stress

        A. Symptoms

        B. Harms

    III. Chronic stress

        A. Symptoms

        B. Harms

The mind mapping exercises and working outline provide clear guidance about what materials you need to gather to give your speech.[21] As you research your speech, refer to the working outline to keep you on track. Also, revise the working outline as needed. For instance, you may find that the three types of stress have important differences in what causes them, so you would expand your working outline to include "causes" as a subpoint for each main point. Box 11.4 gets you started on your own mind map for your speech topic.

BOX 11.4: CONNECT & REFLECT

Mind Map Your Speech Topic

Mind maps can help you in each step of your speech, from choosing your topic, to developing your points, to noting research relevant to your ideas. Using a mind map of your speech along with your working outline will give you a visual representation of how all the parts of your speech fit together.

Free versions of mind mapping applications are available for laptops, smartphones, and tablets. Or you can draw your mind map on a piece of paper, your laptop, or tablet. Whichever method you use, follow these five basic steps:[22]

1. Start with a single word or phrase.

2. Add topics or ideas that are related to that word or phrase, drawing lines or branches to show the connection to your main idea.

3. Add subtopics to the topics, drawing a line to connect each subtopic to the topic.

4. If you think of something that apparently is not related to anything on your mind map, add a *floating topic* and draw an arrow or dashed line from the word or phrase that made you think of the new idea.

5. If a topic or subtopic calls out a particularly interesting idea to you, indicate that with a *call out topic* identified in a different color.

You can use images in place of or in addition to the words in your mind map. Make your mind map colorful so you can visualize how your ideas are linked together. Now try it out with your own speech.

CHAPTER REVIEW

■ **Discuss the ways in which a pervasive communication environment has changed public communication.**
Today's speakers must be more vigilant in avoiding plagiarism, information is more widely available but must be evaluated more carefully, and audiences are more diverse.

■ **Identify and recognize examples of the three speech purposes.**
A speech will have one general purpose: to inform, to persuade, or to entertain. Successful speeches focus on one general topic.

■ **Describe how to brainstorm for topic ideas.**
Brainstorming involves generating and recording as many ideas as possible, not dismissing any ideas, combining and building on ideas, and avoiding any evaluation or critique.

- **Discuss how to evaluate and select a topic for a speech.**
  Evaluate topics based on four criteria: the speaker's interests and knowledge, the audience's interests and knowledge, the availability of resources, and the requirements of the occasion.

- **Identify examples of appropriately phrased specific purposes and thesis statements for speeches.**
  The specific purpose merges the speech topic with the general purpose to identify the speaker's overall objective for the speech. The thesis statement captures the essence of the speech.

- **Create a working outline for a speech.**
  The working outline includes the speech topic, title, general purpose, specific purpose, thesis, and main points. This outline guides you in developing your ideas as you research your speech.

DISCUSS AND APPLY

Review this student speech on U.S. chess grandmaster Irina Krush and answer the questions at the end.

IRINA KRUSH CRUSHES CHESS

What games were you playing when you were 5 years old? I was playing Candy Land, Go Fish, and Tic-Tac-Toe. I played T-ball, tag, and hopscotch. There was the occasional video game, like Animal Crossing and Mario Kart. And of course, I played imaginary games I made up with my friends. However, I was not playing chess. Were you? I hadn't even figured out checkers at that age. Yet Irina Krush learned how to play chess at age 5 and a year later was beating players her father's age and older. Irina Krush may not be a household name to you, but in the universe of chess she is. I'm going to introduce you to Irina Krush, who today at 36 is a chess grandmaster as well as a renowned chess teacher. I'll start with her early years and trace her path to her current place in the chess world as both player and mentor.

Born in Odessa, Ukraine, Christmas Eve day in 1983, Krush moved with her parents to the United States when she was 5 years old, where they lived in Brooklyn. She first started learning about chess on the flight from her birth country to America. When she and her family arrived in their new home, Krush's father gave her more formal lessons in the game. She was such a fast learner that at age 6, she entered her first tournament. This was at the Manhattan Chess Club in New York, where she played against adults and ended up winning $20.

The following year, when she was 7, she traveled to Poland to play at the world chess tournament for girls younger than 10. She earned an award for youngest player. Just 2 years later, at age 9, Krush beat her first chess master. At such a young age, and not a master yet herself based on the requirements established by the Fédération Internationale des Échecs (or the International Chess Federation), that was quite a feat.

In 1995, the top two contenders in the U.S. Women's Chess Championship withdrew from the event, allowing Krush to participate. As the *Los Angeles Times* reported, at 11 years old, she became "the youngest American ever to compete in the national championship." Krush didn't win the championship that year, but in her final round, she played one of the ultimate winners to a draw. In an interview later that year with a writer for the teen magazine *Girls' Life*, Krush explained why she doesn't feel any pressure when she plays: "I play because I love the game. I don't play for my dad or for my coach. It's not the best feeling to lose, but I get over it. I know I'll have to lose my share of games in order to learn more moves." As Krush realized at just 11 years old, the most important lessons come from failing.

What made Krush stand out in the world of chess was not just her age but also her gender. For example, a September 1998 article in *The Salt Lake Tribune* titled "Prowess of Women Chess Warriors" reported on three young women, one of them Irina Krush, who recently had out-performed male players. For instance, Krush, who was only 14 years old at the time, played a 68-year-old grandmaster to a draw. In 1998, she was the top American player under 18—female or male. And in the top 50 players under 21 in the United States that year, Krush was the only female on the list. That also was the year Krush won the U.S. Women's Chess Championship for her first time, the youngest ever in the history of the tournament.

Krush attended Edward R. Murrow High School in Brooklyn, New York. The school is noted for its arts programs, although its chess team produced several chess grandmasters and international masters in addition to Krush, including Alex Lenderman, Anna Hahn, and Salvijus Bercys. In an article for *The New York Times*, Eliot Weiss, a math teacher at the school and the team's coach, had this to say about Krush: "Coaching Irina is like coaching Wayne Gretzky. How do you coach Wayne Gretzky?" And not content just to play chess, Krush also played basketball and swam competitively in high school.

In 1999, world chess champion Garry Kasparov played a game of chess against the world on the internet. Today you might shrug your shoulders and think, "So what?" But 20 plus years ago the internet still was young; the first free web browser for the public had come out just 6 years earlier. Although Kasparov eventually won, Krush was one of four analysts on the world team. These ana-lysts suggested moves and then the more than 50,000 people in 75 countries who were part of the world team voted. The *Calgary Herald* reported, "Since the 10th move, when Krush proposed using Black's queen to open a surprise position on the board—the apparently novel move which threw the game wide open and put Kasparov under pressure—her suggestions have almost invariably been accepted." The game lasted 62 moves over 4 months, in large part due to Krush's exceptional chess strategies. However, the event was not without controversy. Krush encoun-tered technical difficulties, and in the 58th move, her suggestion was not relayed to the team. She ended up resigning shortly before the competition ended.

In addition to winning the U.S. Women's Chess Championship at 14, she won 6 more times, in 2007, 2010, 2012, 2013, 2014, and most recently in 2015. And in 2013, at age 29, she became a grandmaster, the only female American chess player to earn that title. Interestingly, while the lower levels of chess competition are separated into women's and men's groups, that's not true of the most elite levels. In an NPR interview, Krush had this to say about the division: "I just don't see the point having these separate women's titles. I'm not sure what they indicate. Women can play with men—they do play with men now. They can earn the same titles as men."

Last spring Krush was in the news for something completely unrelated to chess—she survived an especially serious case of the novel coronavirus. One of the early known cases of the virus, Krush initially spent 2 days in the hospital after her diagnosis. She was sent home to recuperate, yet it would take 2 more trips to the emergency room after experiencing intense pain and tremendous difficulty breathing before she was on the road to recovery. She relied on her chess philosophy to cope with the illness. As she said in an interview, "It's not just a disease. It's a life trial. Chess players know what it's like to be in a bad position, to suffer. I realized it was going to be a long game, with no easy victory."

Even while suffering from the virus, she continued to play chess as competitions moved online. She shared first place with another player in an online women's chess tournament called Isolated Queens II, organized to give players a chance to keep their skills sharp during the pandemic. In addition, she played in the first Online Nations Cup, where the Chinese team barely squeaked by Krush's U.S. team in the Superfinal match. In discussing Krush's performance, the team's captain told a *New York Times* reporter, "Without her, we never would have finished second or made it to the playoffs."

Today, Krush teaches more than she competes. She offers a class at the Marshall Chess Club in New York City, one of the oldest chess clubs in the United States and where she one won her first tournament. In the summer, she teaches a 3-week chess camp for kids in New York City called the True Grandmaster Competition Prep Camp. She's highly committed to promoting the next generation of chess players, especially girls and young women. While she enjoys teaching, her latest and highly successful experiences with online chess tournaments have her taking a renewed interest in competitive chess.

As you've learned today, chess grandmaster Irina Krush accomplished many firsts in the world of chess. She won her first tournament at age 6. A year later, she was the youngest participant in a world youth chess championship. At 11, she was the youngest American ever to compete in the national championship. At 14, she was the only female among the top 50 chess players under 21 years old in the United States. Finally, she was the first, and still only, American female grandmaster. Now she has come full circle, teaching where she won her first chess tournament. Who knew that $20 prize would be the start of an incredibly impressive career.

References

Goldman, A. (1996, March 31). Master of the game. *Girls' Life*, *2*, 14. https://search.proquest.com.libaccess. sjlibrary.org/docview/219954072?accountid=10361

Krush, I. (2020, May 28). In *Wikipedia*. https://en.wikipedia.org/wiki/Irina_Krush

Lyman, S. (1998, November 29). Teen chess phenom krushes her opponents. *The Salt Lake Tribune*. https:// search.proquest.com.libaccess.sjlibrary.org/docview/288866062?accountid=10361

Macklin, K. (1999, January 17). Chess is this teenager's life (except for a certain boy). *New York Times*. https://nytimes.com/1999/01/17/nyregion/neighborhood-report-midwood-chess-is-this-teenagers-life-except-for.html

Peters, J. (1995, November 26). Chess; National news [home edition]. *Los Angeles Times*. https://search. proquest.com.libaccess.sjlibrary.org/docview/293114112?accountid=10361

Phillips, S. (2010, August 15). *A gender divide in the ultimate sport of the mind*. National Public Radio. Available https://npr.org/templates/story/story.php?storyId=129214019/?storyId=129214019

Prowess of women chess warriors. (1998, Sep 13). *The Salt Lake Tribune*. https://search.proquest.com.lib-access.sjlibrary.org/docview/288833353?accountid=10361

Waldstein, D. (2020, May 27). A chess prodigy's return to health brings cheer to the game. *The New York Times*. https://nyti.ms/2TEKXOL

Wastell, D. (1999, September 12). Kasparov fears krushing defeat: 15-year-old girl gives champ cause to pause. *Calgary Herald*. https://search.proquest.com.libaccess.sjlibrary.org/docview/244788204?accountid=10361

Whitworth, D. (1999, October 21). We were robbed, say world chess team. *The Times*. https://search.pro-quest.com.libaccess.sjlibrary.org/docview/318179326?accountid=10361

QUESTIONS FOR DISCUSSION

1. How appropriate was the student's topic choice? How well does it meet the criteria for choosing a topic?

2. What is the general purpose of the speech?

3. How would you phrase the specific purpose for the speech?

4. How successful do you think the student was in achieving the specific purpose? Give examples to support your answer.

5. What is the thesis of the speech?

6. Identify the main points for the speech.

7. Overall, how would you evaluate the speech? What did the speaker do well? What would you suggest the speaker improve on?

8. What have you learned that you will apply in your own speeches?

CREDITS

Researching Your Speech

Learning Outcomes

- Discuss the ways in which a pervasive communication environment has changed how speakers research their speeches.

- Identify internet resources for researching your speech.

- Identify library resources for researching your speech.

- Explain how to use interviews to gather information for your speech.

- Differentiate between credible and untrustworthy information sources.

- Discuss ways to avoid plagiarism when researching your speech.

In her presentation at the American Public Health Association's webinar "Racism: The Ultimate Underlying Condition," Dr. Amani M. Allen, executive associate dean at the University of California, Berkeley's School of Public Health, discussed systemic racism's physical effects with the more than 7,000 people attending the presentation. Using effects of the COVID-19 pandemic as an example, she told the audience:

> We see that for example with COVID-19, where Black Americans are overrepresented in COVID-19 deaths. And we see the same pattern across the majority [of] states where data on race have been collected and reported. On average, non-Hispanic Blacks have a rate of COVID-19 deaths approximately four-and-a-half times that of non-Hispanic Whites. While Latinos have a rate approximately 3.5 times that of non-Hispanic Whites. We also see it unfolding with Indigenous Americans. Non-Hispanic American Indian or Alaska Natives have a rate approximately five times that of non-Hispanic

Whites in some places. And while Latinos are dying from COVID-19 at a rate similar to their share of the population, they're dying at rates above their population share in many states.[1]

Framing her topic within current events and current factual data made the presentation more relevant to the audience and underscored Dr. Allen's knowledge of the topic. She went on to address additional evidence of systemic racism:

A study led by one of my former doctoral students, Dr. Marilyn Thomas, now a post-doc at UCSF [University of California, San Francisco], used data from the *Washington Post* to examine the risk of being shot and killed by police by armed status, being armed versus being unarmed for Black and White men. ... And this analysis that I'm going to share with you used data from 2015 through 2018.

Dr. Allen then presented the data, using graphs and charts to explain the findings. Agreeing with activists in the racial justice movement, she concluded:

And so, we can take a few things away from this analysis. First, ... Black men are more likely to be shot and killed by police when they are unarmed compared to White men. Second, the disparity in armed status is exacerbated among Black males perceived as mentally ill, and among those over the age of 44. Some have suggested that focusing on policing policies may help reduce the heightened risk of being shot and killed while unarmed among Black versus White men. But in another study also led by Dr. Thomas, she found that while several policies helped reduce police killing rates for White men, those indicated here in bold [refers to digital slide], such as having a mission statement, and using videos, such as body and vehicle cams, none of the policies showed a significant impact on reducing the killing rate for Black men.

Dr. Allen's message is persuasive in part because of her position—she is an expert on race and socioeconomic health disparities. But she also did her research, including facts to support her arguments. For instance, she reported on data that demonstrated COVID-19 has affected Black Americans disproportionately, stating, "On average, non-Hispanic Blacks have a rate of COVID-19 deaths approximately four-and-a-half times that of non-Hispanic Whites." In addition, she referred both to her own research as well as the findings of other researchers, further underscoring her extensive knowledge of the topic (Figure 12.1).

FIGURE 12.1 In her webinar, Dr. Amani M. Allen argued that systemic racism in the United States is reflected in the disproportionate COVID-19 death rates and police killings for Black people.

Whether you are giving an informative, persuasive, or entertaining speech, you must be the expert on your topic. You become the expert by gathering information—facts and data compiled about your topic—from a wide range of sources. In this chapter, you will learn how to research your topic and find the best information for your speech.

Researching Your Speech Topic in a Pervasive Communication Environment

Before the internet, searching for information was a job for academics or research specialists, such as librarians. Most information was available only in university libraries, from expensive or difficult to access databases, or paper documents that existed in a few locations. Today, nearly everyone engages in some form of researching, whether to learn more about a current event, understand how something works, plan a trip, or locate information for a class project.

The easy access to information a pervasive communication environment provides has brought with it an increase in pseudo or fake news, as you learned in Chapter 3. Because communicators can share information from nearly anywhere, anytime, with anyone, producing and distributing deliberately false information has become much easier. Typically posted on social media, pseudo news appears to be true and travels faster and farther than actual news.[2] Interestingly, few people are taken in by fake news. In a recent study of about 15,000 people over 3 years, researchers found most people get their news from television and their direct encounters with fake news online are quite low.[3] However, a review of research on fake news found that the majority of Americans learn about such false stories when they are covered in the mainstream media as journalists seek to correct misinformation. Unfortunately, this can result in a wider dissemination of pseudo news and a greater likelihood that some people will believe the false story rather than the correction.[4] In addition, a Pew Research Center study found that 50% of Americans believe fake news is a greater problem than violent crime, climate change, racism, illegal immigration, terrorism, and sexism. Survey respondents felt fake news has a negative impact on how confident people are in government, how confident they are in each other, and the ability of political leaders to do their work.[5]

Sorting out fact from fiction and reporting on the information you have found to others requires a new type of literacy, **digital literacy**, or "the ability to use information and communication technologies to find, evaluate, create, and communicate information, requiring both cognitive and technical skills."[6] Digital

digital literacy The ability to use information and communication technologies to find, evaluate, create, and communicate information, requiring both cognitive and technical skills

literacy goes beyond typing a few words into a search engine and clicking on the links on the first page of results. You must engage a complex array of skills to gather, assess, and present information for your speeches. This section examines the myths associated with digital literacy and identifies ways for you to improve your digital literacy.

Digital Literacy Myths

There are four primary myths associated with digital literacy.[7] Identifying these myths underscores the importance of digital literacy for all communicators across all contexts.

Myth 1: Digital Literacy Is the Same as Information Literacy

information literacy
Recognizing when information is needed and having the ability to locate, evaluate, and use effectively the needed information

According to the American Library Association, you demonstrate **information literacy** when you "recognize when information is needed and have the ability to locate, evaluate, and use effectively the needed information."[8] While information literacy is part of digital literacy, today's pervasive communication environment requires much more than basic information-gathering and assessment skills.[9] Thus, the digitally literate person uses diverse technologies to find and evaluate information, understands and applies ethical standards in using that information, effectively communicates what they have learned to others, and integrates these skills into all aspects of their lives.[10] For example, you might use several search engines and library databases to locate texts and images related to your speech, evaluate what you find, and then present that information orally and visually to both an in-person and online audience. Digital literacy is essential in every step of the speechmaking process.

Myth 2: Digital Natives Are Digitally Literate

Just because you have grown up with smartphones, Instagram, and Google does not mean you have a high degree of digital literacy (Figure 12.2). Although you may feel comfortable searching for information online, you may not have the training and skills essential for effectively locating, assessing, and presenting that information. As Bernd W. Becker, associate librarian at San José State University, observes:

FIGURE 12.2 Even people who have lived all their lives in a pervasive communication environment—so-called digital natives—need to work on their digital literacy skills.

The reality is that everyone struggles with digital literacy. Younger generations might have the technical skills, but lack the refined cognitive skills to find, evaluate, create, and communicate. Older generations might have the cognitive skills, but lack the refined technical skills to find, evaluate, create, and communicate. No one is born digitally literate, and no one's age will determine their digital literacy skills.[11]

Everyone needs to work on their digital literacy.

Myth 3: Digital Literacy Is Only for Academic Work

In a pervasive communication environment, digital literacy is more than simply something you need for your schoolwork. Digital literacy is foundational to your everyday life. Whatever your career path, digital literacy skills are essential to success in your professional life. For instance, research has found that individuals with better digital literacy skills are more entrepreneurial.[12] Digital literacy increases worker productivity, as organization members are better able to efficiently search for, evaluate, and present relevant information using the latest digital tools.[13] Lack of digital literacy skills can hurt an individual's employability and job prospects.[14] Digital literacy also impacts your willingness to engage with your community, current events, and political issues. Research has shown that higher level digital literacy skills are associated with more effective information-seeking and analysis, a greater interest in local and national news, and an increased engagement in politics.[15] Finally, although digital literacy skills will not necessarily improve your health, they will improve your ability to find and evaluate health-related information.[16]

Myth 4: Digital Literacy Skills Do Not Change

As with everything in a pervasive communication environment, digital literacy skills change as the technology changes. Strategies that worked 5 years ago might not be applicable today, as online resources develop, evolve, or even cease to exist. For example, in the early days of the internet, users depended much more on web directories to find information because search engines were quite limited in their capabilities. Now search engines are much more sophisticated, offering suggestions based on your original search query. In addition, innovators regularly add new search tools, each with their own features and characteristics.

Developing Your Digital Literacy Skills

Developing your digital literacy skills is a lifelong process. You have acquired some skills in the digital world as you have gone about your everyday life in a pervasive communication environment. But to truly improve your digital literacy, you must focus on sharpening those skills. The strategies that follow are applicable to researching your speech as well as other aspects of your life.

Fight Fake News

The power of pseudo or fake news rests in its ability to entice people to believe it. Sensational stories, ideas that confirm your own perspective, and flashy images are all ways that fake news gets noticed and then spread via social media. However, you can stop fake news in its tracks by following the guidelines outlined by the International Federation of Library Associations and Institutions in Figure 12.3.

FIGURE 12.3 The International Federation of Library Associations and Institutions suggests that you can spot fake news by checking the source, author, date, references, and your biases; reading beyond the headlines; finding out if it is satire; and asking the experts.

Apply Your Critical Thinking Skills

When you apply your critical thinking skills, you carefully analyze and evaluate information so the conclusions you reach are based on sound information and good reasons.[17] Critical thinkers distinguish between arguments and opinions. As you will learn in Chapter 16, arguments are based on reasoning and evidence. For instance, consider the following two examples about climate change.

EXAMPLE 1

Back in the '70s, they were warning us that we were about to enter an "ice age" and there was speculation about spreading coal dust on the glaciers so they would start melting. Then the rhetoric changed; we were about to cook under the blazing sun of "warming" and my young grandson told a neighbor not to burn his pile of junk as he would be contributing to the hole in the ozone—he'd been told that in school. One of our daughters was taught that by a certain year all the trees on the planet would be gone. She believed that until she went up in a plane and saw nothing BUT trees; the grand pronouncements, made with such gravity, were so much blather. Well, these predictions failed, so it has been renamed "global climate change," which covers either event, warming or cooling. ... Now, come to find out, our planet has been cooling since 1998. The world temperature is 1.08 degrees cooler than it was in the late '90s. This may come as a shock to many, but there is really nothing to this "global warming" hoax. Turns out, this has all been an attempt to keep the money coming to those who are collecting funds for various "anti-climate change," entities who think they are helping to save our planet from an invented "disaster."[18]

EXAMPLE 2

Scientists attribute the global warming trend observed since the mid-20th century to the human expansion of the "greenhouse effect"—warming that results when the atmosphere traps heat radiating from Earth toward space. ... On Earth, human activities are changing the natural greenhouse. Over the last century the burning of fossil fuels like coal and oil has increased the concentration of atmospheric carbon dioxide (CO_2). This happens because the coal or oil burning process combines carbon with oxygen in the air to make CO_2. ... In its Fifth Assessment Report, the Intergovernmental Panel on Climate Change, a group of 1,300 independent scientific experts from countries all over the world under the auspices of the United Nations, concluded there's a more than 95 percent probability that human activities over the past 50 years have warmed our planet.[19]

Which one is an argument? Which one is an opinion? In applying your critical thinking skills, you likely chose Example 2 as the argument because the author provided scientific evidence and information from a credible source to support the conclusion that human activities are contributing

FIGURE 12.4 Based on scientific research and evidence, experts worldwide agree that human activity is the major cause of global climate change.

to global climate change (Figure 12.4). In Example 1, the author presents an opinion without any sound evidence. But, you may ask, what about the statement "The world temperature is 1.08 degrees cooler than it was in the late '90s"? A quick online search reveals that tracking the trend of the Earth's temperature requires a long-term assessment. For instance, NASA Jet Propulsion Laboratory science writer Alan Buis reports in his blog post "Nope, Earth Isn't Cooling" that over the long term—since 1880—the average global temperature has increased 2 degrees Fahrenheit.[20] Critical thinkers always question and analyze the information they find. Whether searching for or presenting information, your critical thinking skills are essential to developing your digital literacy.

Seek Information From a Variety of Sources

Defaulting to a Google search is easy. It also demonstrates poor digital literacy skills. As you will learn in this chapter, there are many different metasearch engines, search engines, and specialized search engines you can use to find information online. In addition, your library has hundreds of databases available for you to access. This chapter will introduce you to some of them.

Go Outside Your Technology Comfort Zone

Both for locating and presenting information, avoid defaulting to the technology you already know. Try a new library database to research your speech topic. Consider how images and sound files might make the presentation of your topic clearer for your audience. As you learn new technology skills, you will enhance your digital literacy and expand both your information-seeking and information-presenting options.

Become an Active Prosumer

As you learned in Chapter 1, immersion in a pervasive communication environment means you are both a producer and consumer of information. Students typically think of themselves as information consumers, viewing videos, reading text messages, listening to podcasts. Yet you also produce information, often in digital form, as when uploading a video to YouTube, sending a text, and leaving a voicemail message. As an active prosumer, you attend more closely both to the digital information you produce and consume, practicing what you are learning in this and other classes to expand your digital communication repertoire. Complete Box 12.1 as a first step in becoming a more active prosumer.

BOX 12.1: CONNECT & REFLECT

Trying Out Your Digital Literacy Skills

Digital literacy is part of every step of the speechmaking process. Apply what you have learned about digital literacy in this activity.

1. Choose a topic from the list below or another current topic in which you are interested.

   - 19th Amendment to the U.S. Constitution
   - AI (artificial intelligence)
   - antiracism
   - astronomy
   - coal mining
   - cyberbullying
   - derechoes (inland hurricanes)
   - elephant seals
   - esports
   - greenhouse gas
   - home schooling
   - identity theft
   - learning styles
   - murder bees (Asian giant hornets)
   - urban gardening

2. Search online for information about your topic. Find at least five text (written) sources and five images or videos.

3. Evaluate the sources. How comprehensive are they? How much did you learn about the topic from each source?

4. Using the textual and image sources you found, create a 2-minute presentation on your topic using four digital slides.

5. Share your presentation with your classmates or friends who might be interested in the topic. Ask for their feedback on your presentation.

How effective were your skills in locating information on your topic? Did you encounter any fake or pseudo news? If so, how useful were the guidelines in Figure 12.3 from the International Federation of Library Associations and Institutions in spotting it? How carefully did you evaluate the information? How well did you integrate your digital slides with your oral presentation? What have you learned that you will use in future speeches?

The sheer volume of information available online and from other sources poses a challenge for even the most skilled researcher in today's pervasive communication environment. Successfully researching your speech topic requires developing critical skills for searching the web, locating information from traditional library sources, interviewing experts, creating clear criteria for evaluating information, and applying strategies for avoiding plagiarism.

Internet Resources

If you are like most people, when you search for information online you go to Google, which has captured 86% of the global search market.[21] But Google is just one search engine, and

FIGURE 12.5 Search for both relevant text and image sources when researching a topic.

keyword A word used to find information when using search software, such as an internet search engine or library database

search engines are just one option for searching the internet. Metasearch engines, specialized search engines, and web directories provide additional options for locating the best research for your speech topics.

Keywords drive most searches. A **keyword** is a word used to find information when using search software, such as an internet search engine or library database. The advantage of a keyword search is that you do not have to know exactly what you are looking for, which can lead to unexpected sources. For example, if you searched using the keyword "wildfires," you likely will find information on current, historical, local, national, and worldwide wildfires as well as text, image, and video sources (Figure 12.5). The disadvantage is that language ambiguity and other factors can cause you to miss sources or become overwhelmed with the number of sources a search returns. For instance, searching Google with the keyword "wildfires" produces nearly 2 billion results. Here are some tips for generating a variety of useful keywords in your searches:

- *Use the thesaurus.* Thesaurus websites will give you a wider range of synonyms than most word processing programs.

- *Use alternate keywords suggested by search tools.* Many search engines and databases offer additional keywords to improve your search. Try out these keywords to narrow and expand your search.

- *Use sources such as encyclopedias, dictionaries, and library catalogues.* These sources provide overviews of topics and can provide additional keywords.

- *List and track your keywords.* Keep track of your keywords so you can add to them, note dead ends, and concentrate on keywords that give you the best results.

Metasearch Engines

metasearch engine A search engine that that uses several search engines at the same time

A **metasearch engine** uses several search engines at the same time. The most useful aspect of a metasearch engine is how it organizes and categorizes information based on a set of frames, or ways of structuring the information. Carrot[2] groups the results of your search into clusters that you can view three ways: list, treemap, and piechart. Let's say you search for "opioid crisis." In addition to the list of links you would expect from any search engine, you also can choose to display the results as treemap or piechart. When you click on an area or cluster in the treemap or piechart, the results are shown the right side of the page. The clusters allow you to explore specific aspects of a topic and the treemap and piechart offer useful visual representations of your topic (Figures 12.6 and 12.7). Table 12.1 lists several popular metasearch engines and their features.

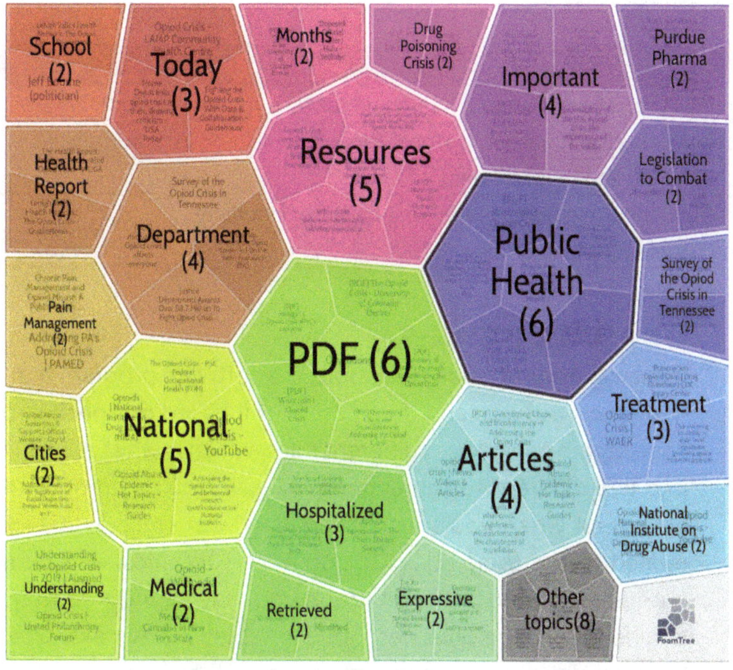

FIGURE 12.6 Semantic map from the results of the "Racial Justice Movement" search on swisscows.com.

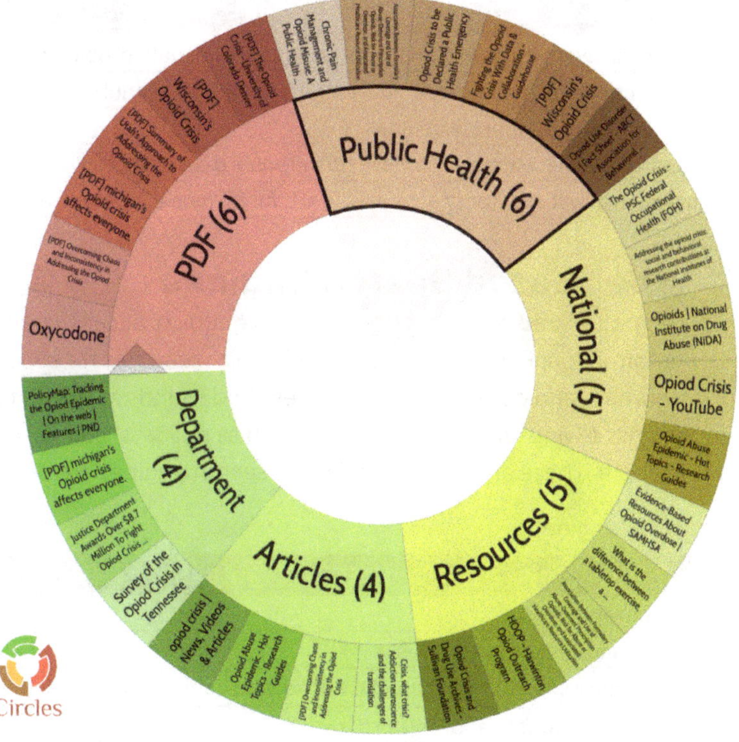

FIGURE 12.7 Semantic map from the results of the "Racial Justice Movement" search after clicking on "Movement"

TABLE 12.1 Example Metasearch Engines

NAME	FEATURE
Dogpile	Lists suggestions based on your search under "Are you looking for?"
Ecosia	Searches Wikipedia, Amazon, YouTube, Bing, Google, and Ecosia Travel
info.com	Searches White and Yellow Pages
Search Encrypt	Erases local browsing history after 30 minutes of inactivity
Yippy	Groups results by topics or "clouds"

Search Engines

search engine Software program that searches for information on the web and then compiles it based on specific rules or parameters

A **search engine**, such as Google, is a software program that searches for information on the web and then compiles it based on specific rules or parameters. Each search engine uses different rules, which is why typing in the same keywords into two different search engines will give you different results. Search engines are useful when you start researching your topic to get general information related to the topic, ideas for possible subtopics, and people who might be authorities on the topic. Although you may be inclined to use only one search engine, try out the ones described in Table 12.2. For example, the search engine WolframAlpha searches in a query format, answering questions about people, math problems, history, health, culture, education, and geography.

TABLE 12.2 Example Search Engines

NAME	FEATURE
Bing	Gives instant answers for definitions, sports scores, and math problems
DuckDuckGo	Use ! (Bang) command for site-specific searches, such as !youtube
Qwant	Lists news results in a separate column
Startpage.com	Private search engine, anonymous view, no personal data collected
WolframAlpha	Get answers to questions about astronomy, art, math, music, physics, and other topics

Specialized Search Engines

specialized search engine A search engine designed to access specific types of information

A **specialized search engine** is designed to access specific kinds of online information. For example, Google Scholar searches for scholarly work in books, journals, and conference paper sites. FindSounds searches for online audio files across a wide range of categories, such as holidays, games, nature, mayhem, and people. These and more specialized search engines are described in Table 12.3.

TABLE 12.3 Example Specialized Search Engines

NAME	SPECIALTY AREAS
Atomic Lyrics	Artists, songs, and lyrics
FindSounds	Sound files, from animals to vehicles
Google Scholar	Scholarly literature in academic journals, books, and conference proceedings
IconArchive	Icons grouped by category and keyword
PDF Search Engine	PDF, CHM, DOC, RTF, and TXT files
PubMed	Citations in biomedicine
Sweet Search	Educational information from sites such as the Smithsonian, PBS, and the Library of Congress

Web Directories

A **web directory** is an index or database of websites that professionals and experts review and categorize by topics and subtopics. Although less common than they were when the internet was in its early stages, web directories offer curated content both in general categories and specialized areas. Web directories can be helpful in finding higher quality and more specific information than general search engines. For instance, the staff at the Smithsonian Libraries created the Library and Archival Exhibitions on the Web database that includes thousands of museum, library, and nonprofit historical society online exhibitions. You can find books, posters, photographs, sound and video recordings, engravings, and other archival material on a host of topics. JoeAnt is a volunteer-run indexing site where editors collect and organize information. Several web directory examples are listed in Table 12.4.

web directory An index of websites that professionals and experts review and categorize by topics and subtopics

TABLE 12.4 **Example Web Directories**

NAME	PRIMARY SEARCH AREA(S)
Best of the Web	Businesses, blogs, and forums worldwide
Blogarama	Blogs by topic area
Directory of Open Access Journals	Scholarly articles from over 130 countries
GoGuides	Relevant websites and images in 24 categories, from arts to vacations
HotFrog	Local businesses in the United States
JoeAnt	Websites and blogs by subject
Library and Archival Exhibitions on the Web	Library, museum, and historical society online exhibitions worldwide
The WWW Virtual Library	General information on a wide range of topics

Library Resources

No other institution is more dedicated to finding, gathering, and providing information on such a wide variety of topics as the library. Librarians are research and information-gathering specialists, trained to assist you in finding the information you need. Learning about the full range of materials and expertise available at your campus library will benefit you in researching your topic and for all future research projects (Figure 12.8). Box 12.2 suggests how to access this critical resource, your campus librarian.

Books

Books generally go through a rigorous review process. Authors must be experts on their topics for their books to get published. The strength of books lies in the historical information they provide or the topics they cover that are not especially time sensitive. Because of the time delay in

FIGURE 12.8 Your campus library offers many resources for researching your speech, including books, periodicals, government publications, and reference materials.

BOX 12.2: LIVE IT LOCAL

Meet Your Campus Librarian

Do you know the name of a librarian at your campus library? If not, now is the time to introduce yourself. Librarians are experts in general searching, researching, finding materials, and using search technology. Some also have areas of expertise, such as art, reference materials, or politics. You can spend hours trying to locate information that a librarian can find in moments. It is now easier than ever to get individual help from librarians, as nearly all college libraries offer live chat, call-in, or email options for their patrons. You also can make an appointment and get one-on-one extended research assistance at your campus library. Bring your list of questions and description of your class assignment. You will save time and learn about the techniques for using information resources.

writing, editing, and publishing, books are not the best source for current events. For example, if you are trying to find the latest trends in how people use social media, a book would not be your best bet. However, you might be able to find books about media use in the past, such as people's use of television and radio, which would give your social media topic some perspective.

Use your library's online catalog to search for books by keyword, title, or author. Most libraries allow users to check out books online and pick them up later. Many books now are digitized in e-book or audiobook format, so you can review them immediately.

Periodicals

Periodicals, which include newspapers, magazines, and journals, provide more current information than books do. The print version of a newspaper is more current than a magazine or journal, although newspapers and magazines typically update their websites throughout the day with the most recent news. Magazines usually have more in-depth coverage of a specific topic because they are published less often than newspapers.

If you are searching for information from a specific newspaper or magazine, use the search feature on the news outlet's website. More often, you will want to search multiple news sources simultaneously. In those cases, use an electronic library database, such as ProQuest Newsstand or Gale OneFile News. These databases offer several options for advanced searching, such as date range, subject heading, and source type. Special tools can assist you further with your search. For example, ProQuest Newsstand includes search forms customized for specific subject areas, such as the arts, health and medicine, and science and technology. Gale OneFile News features Topic Finder that helps you identify additional relevant keywords for your search, visualizing suggestions in tile or wheel format.

Academic journals publish scholarly research on a wide range of topics. If you were researching a speech on suburban sprawl, for instance, you might want to look for articles in journals such as *Geography and Urban Environment*, *Health and Space*, *City & Community*, and *Geoscape*. Using library databases, such as Academic Search Complete, Business Source Complete, Communication and Mass Media Complete, Factiva, Ingenta, JSTOR, ScienceDirect, Social Sciences Full Text, and

Web of Science, you can search multiple scholarly journals from a single database. Some databases cover general topics, as with Academic Search Complete. Others are more specialized, such as JSTOR, which includes older issues of journals in the humanities, literature, language, music, and social sciences. Table 12.5 introduces you to several library databases and when you would want to use them.

TABLE 12.5 Example Library Databases

DATABASE	BRIEF DESCRIPTION	USE TO SEARCH …	FEATURE
Academic Search Complete	Full-text scholarly articles in wide range of disciplines	Academic journals	Receive notifications about new publications on a topic
ARTstor	Images in art, architecture, the humanities, and social sciences	Image reference materials	Tools to organize images into groups
Civil Rights Digital Library	Digital video archive of the U.S. civil rights movement.	Videos on the civil rights movement	Browse by events, places, people, or topics
Communication and Mass Media Complete	Full text articles for more than 500 journals in communication	Academic journals	Profiles of more than 5,500 authors
CQ Researcher	Reports on current topics written by journalists	Reference materials	Issue tracker that groups together related reports
Encyclopaedia Britannica	Articles on thousands of topics	Reference materials	Research tools such as compare countries and world atlas
Factiva	National and international business news	News organizations	Query Genius to maximize your search
Gale OneFile News	Full text newspaper and magazine articles, and TV and radio show transcripts, in the United States and worldwide	News organizations	Topic Finder for additional keyword ideas
JSTOR	Full text articles for more than 2,000 journals in the sciences, humanities, and social sciences	Academic journals	Search by discipline
Latino Literature	Poetry, plays, and fiction by Latinx writers	Poetry, short stories, novels, and plays	Search for authors by nationality or ethnicity
North American Indian Thought and Culture	Biographical and autobiographical documents on North American Indigenous peoples	First-person accounts, including oral histories, diaries, and letters	Browse by historical events, people, content types, or place discussed
Pop Culture Universe	Articles on all topics related to popular culture	Reference materials	Search specific time periods, such as 1910s: The Age of Opulence and 2010s: Social Revolution
Slavery & Anti-Slavery: A Transnational Archive	Books, serials, manuscripts, and Supreme Court rulings on the slave trade, abolition, and emancipation	Reference materials, websites, and collections	Featured images and documents
ProQuest Newsstand	Full text national and international news	News organizations	Customized subject search areas
Social Sciences Full Text	Index of journals and conference proceedings in the social sciences	Academic journals	Set preferences for your search

Government Publications

The U.S. government produces massive amounts of information, including transcripts of Senate hearings, geological surveys, monographs, handbooks, and pamphlets. The *Catalog of United States Government Publications* indexes more than 500,000 publications from the three branches of the U.S. government. The search results include links to documents available online or library locations for documents in paper format. Another government publications database, *HeinOnline Government, Politics, and Law Collection*, indexes worldwide documents dating back to the 1700s and may be available in your library. In addition to the regular search function, you can browse by category, such as International Treaties and Agreements, Organization; Periodicals; and U.S. Federal; or specific database, as with U.S. Presidential Impeachment Library; Gun Regulation and Legislation in America; and COVID-19 in America: Response, Issues, and Law.

Reference Materials

Some reference materials, such as maps, atlases, and historic print materials, can be accessed only in your library's building. In addition, libraries often have special collections that people have donated over the years. These collections may include materials that exist nowhere else. Find these materials through your library's online catalog and then go to the library to view them.

Many reference materials that once were available only in your library now have been digitized. Several library databases can help you locate these materials. American Government includes maps, images, and documents and audio-video recordings produced by the U.S. government. ARTstor includes more than 2.5 million images in the arts, architecture, humanities, and sciences. Many encyclopedias are available online through library databases, such as *Encyclopaedia Britannica, Encyclopaedia of Judaism, Garland Encyclopedia of World Music,* and *Encyclopedia of Popular Music.* Table 12.5 (page 279) includes sample databases that feature reference materials.

FIGURE 12.9 Interviewing an expert on your topic gives you insights and information not available from other sources.

Interview Sources

Interviews can provide information not available from other sources and personalize your topic (Figure 12.9). For instance, if you are presenting an informative speech on participating in a community garden, interviewing someone who engages in that activity can give you firsthand information about it. Interviewing an expert and incorporating that information into your speech can help gain your audience's interest as well as enhance your credibility. Follow the steps below for a successful interview.[22]

Preparing for the Interview

1. *Identify possible interviewees.* Choose experts on your topic who can give depth and insight that would interest your audience.

2. *Contact interviewees.* Explain the purpose of the interview and arrange a time to meet in-person, via the phone, or in a video chat.

3. *Do your research.* Have some basic knowledge of your topic and interviewee.

4. *Develop your questions.* Construct an interview guide with a list of least 10–15 questions related to your topic. Then review each question for relevancy and clarity. Make sure the interviewee will understand what you are asking and be able to answer your question. Figure 12.10 provides a sample interview guide.

Interviewee: Ashley Smith

Organization: Ashley's Apps

Interview purpose: To find out the strategies to become a successful entrepreneur

Questions

1. The profile of you in the local business magazine stated that Ashley's App is your third start up. Tell me about the first business you started.
2. What motivated you to start that business?
3. How did you raise the funding for your first start up?
4. What did you find the most challenging in starting your first company?
5. What surprised you the most about starting your first company?
6. The magazine article mentioned that your most-recent business is Ashley's Apps. What inspired you to start this new company?
7. On the Ashley's Apps website, you state that these apps are "designed by girls and women for girls and women." Why did you think it was important to have smartphone and tablet applications that girls and women design?
8. What have you found the most rewarding about starting your latest company, Ashley's Apps?
9. Say I wanted to start my own business. What are the essential skills I would need to start my company?
10. What could students do now to help prepare them to start their own business?
11. Which college courses would you say are the most helpful in preparing to start your own business?
12. What else should students know about starting their own company?

FIGURE 12.10 In this sample interview guide, the student has prepared for an interview with a local entrepreneur.

Conducting the Interview

1. *Open the interview.* Give a friendly greeting and establish rapport with your interviewee. Remind the interviewee of the interview's purpose, preview the topics you will cover, and ask for permission to record the interview.

2. *Ask your questions but be flexible.* As you ask your questions, listen care-fully to your interviewee's responses. Follow up with additional questions if a topic area seems especially useful or interesting. Skip questions if the interviewee has already covered that topic.

3. *Take notes.* Recordings sometimes fail. Notes also will help you highlight key insights from the interview and ask any needed follow-up questions.

4. *Close the interview.* End the interview with a brief recap of the main points you got from the interview. Ask the interviewee if you may contact them with any additional questions you have. Thank the interviewee for participating.

Integrating Interview Information Into Your Speech

1. *Explain why your interviewee is an expert.* Give your interviewee's title, organizational affiliation, or experience with the topic so audience members know that the person is well-qualified to talk on the subject.

2. *Use information from the interview to confirm or contrast with other sources.* If the interviewee provides a different perspective on your topic than what you found from other sources, include it to give a more well-rounded view of your topic. If your interviewee agrees with other sources on an important point, note that as well.

3. *Use a few direct quotes from the interview.* Integrating one or two key statements the interviewee made puts a human face on the topic. Too many quotes will bore your audience, but a few statements that truly stand out will get your listeners' attention.

Conducting an informational interview with an expert on your topic often provides you with additional depth and perspective missing from other sources. However, you must follow ethical standards when preparing for and conducting the interview as well as using the information in your speech. Complete the activity in Box 12.3 to explore the many facets of ethical communication in interviews.

Evaluating Information Sources

source credibility The trustworthiness, soundness, and believability of an information source

As you research your topic, you evaluate the information you gather. A pervasive communication environment provides an overwhelming number of potential sources of information with varying degrees of quality. You want to have credible sources for your speeches. **Source credibility** refers to the trustworthiness, soundness, and believability of an information source. Determining a source's credibility involves assessing four areas: the author, the publisher, the process, and the information.

BOX 12.3: ETHICAL QUESTIONS

Following an Ethical Code of Conduct in Interviews

When you interview someone for your speech topic, in many ways you are acting as an investigative journalist: You want to know more about something, and the person you are interviewing is the best source. Journalists are bound by ethical codes that cover how they conduct and then report on interviews. For example, the Society of Professional Journalists lists four principles in its code of ethics (Figure 12.11):

1. Seek Truth and Report It

2. Minimize Harm

3. Act Independently

4. Be Accountable and Transparent[23]

   View the full text on the organization's website. How might you apply these principles when conducting informational interviews for your speeches? What ethical issues might you face in such interviews? How will you handle them?

FIGURE 12.11 Applying the society of professional journalists' ethical code will guide you in ethically conducting and using interviews in your speeches.

Evaluating the Author

Identify the author of the information source. Is the person an expert on the topic? What biases might the author have related to the topic? For instance, a scientist in biotechnology likely views genetically modified foods more favorably than an organic farmer. Also, if the author belongs to a professional organization, find out if it has a code of ethics or conduct. Although not a guarantee, this indicates compliance with a certain standard of professionalism in the information source. If the author has some sort of certification or license to practice, that suggests training or expertise the author can document.

If the author is an organization or institution, also consider what biases it might have. You would expect a local chamber of commerce to have a different view of business taxes than a local city council. The type of organization—for-profit, nonprofit, educational, political, governmental—influences the stance it will take on a topic. All sources have biases; your job is to determine what those biases are and how they might influence the information you have gathered from those sources.

Evaluating the Publisher

Identify who published the information. If you cannot determine who published something, as with a webpage or pamphlet, you should question the credibility of the source. When the publisher is known, such as a national newspaper or well-respected professional organization, consider the oversight used to ensure that certain standards or levels of quality are maintained. At its best, oversight is implemented by an editorial review board or by experts who have no personal stake or interest in the products they oversee or review. For example, most scholarly journals use a review process in which the reviewers do not know the author of the work.

Also consider the laws or regulations that are in place to constrain or hold the author or publisher accountable for the quality of the work. For example, libel and slander laws provide legal restraints on newspapers, television, magazines, and other media organizations that constrain media outlets from making claims they cannot support.

Evaluating the Process

Consider how the information was produced. Did someone just post it on a website? Who reviewed the information before it was made available? How open to public scrutiny is the collection of data, analysis, or editing of the publication? The more transparent these processes, the more likely the author or publisher is working in an ethical fashion.

Evaluating the Information

Finally, closely examine the information presented. How current is the information? Even with historical information, new evidence surfaces that changes what people have accepted as fact. For instance, with recent archeological discoveries, scientists know much more about dinosaurs today than they did even 10 years ago.[24]

Pay particular attention to faulty information or misinformation. Information may appear unacceptably biased or suspect for a variety of reasons that raise warning signs. If information seems suspicious to you, scrutinize it in more detail. Table 12.6 lists the major warning signs that information may be faulty.

TABLE 12.6 Faulty Information Warning Signs

TYPE	DESCRIPTION
Major omissions	Leaves out contradictory information, fails to include analysis from other sources, or fails to explain or challenge conflicts that exist in the field
Factual errors	Incorrect references to people, places, organizations, dates, or well-established facts
Misrepresentation of others' work	Incorrect or missing references to what others have written or produced
Failure to identify sources, affiliations, funding, or author	Does not identify possible conflicts of interest based on sources of information, professional or other affiliations, funding for work, or who the author is
Value-laden language or hyperbole	Uses name-calling, personal attacks, inflammatory language
Decontextualized information	Presents information out of the context in which it was created, such as leaving out when events happened, when information was written/created, or how events or information relate to the argument at hand

Staying on alert for these faulty information warning signs and applying the Credibility Analysis Checklist in Table 12.7 will help you find sound information for your speech.

TABLE 12.7 Credibility Analysis Checklist

ASSESSMENT AREA	QUESTIONS TO CONSIDER
Author	Who is the author? What are their affiliations? What are the person's credentials, experience, and reputation? How is the author an expert on the topic? What type of organization created this information? Is the organization for-profit, not-for-profit, educational, political, governmental? How does that impact the quality of the information?
Publisher	Who is the publisher? What systems ensure the accuracy and quality of the information? Are they reputational, legal, regulatory, commercial, or some other system?
Process	How was the information gathered? What was the decision making or editorial process?
Information	How current is the information? How complete is the information? Are any of the faulty information warning signs present?

Avoiding Plagiarism When Researching Your Speech

Plagiarism is taking someone's work and presenting it as your own (Chapter 11). Plagiarism is not new, but the internet and other digital media have made plagiarizing much easier. The improvement in search technology and access to library databases provides you with quick access to information on scores of topics. The tremendous amount of information available and the ability to easily copy and paste material makes cheating as easy as a few mouse clicks. The link between public and private information can appear blurry when so much is available so readily for free. Still, students who plagiarize miss out on the learning that occurs in completing an assignment. Moreover, students who cheat will not have the knowledge and experience to succeed outside of school. In addition, colleges and universities apply strict penalties for plagiarism, which can range from a failing grade in the class to expulsion from the school.

How Could It Be So (Copy)Wrong When It Feels So (Copy)Right?

Each year there are 126.7 billion viewings of illegally downloaded U.S. television shows. The movie industry alone loses more than $40 billion in yearly revenues from private digital content. Of the online bandwidth used today, 24% is devoted to pirated movies, television shows, songs, books, and video games.[25] In spite of U.S. and international laws prohibiting the downloading of copyrighted material, this illegal practice continues to increase.[26]

FIGURE 12.12 The U.S. Constitution is the basis for copyright law in America.

copyright A legal framework designed to protect the rights of creators over their works

fair use doctrine A legal principle that allows the limited use of an author's work if attribution is given to the source

Copyright refers to the legal framework designed to give authors and creators control over the works they produce. The basis for copyright or intellectual property laws in the United States is Article I, Section 8, of the United States Constitution (Figure 12.12):

> The Congress shall have power ... to promote the progress of science and useful arts, by securing for limited times to authors and inventors the exclusive right to their respective writings and discoveries.[27]

Copyright laws, including the Digital Millennium Copyright Act (DMCA) of 1998 and international treaties and organizations such as the World Intellectual Property Organization (WIPO), ensure that copyright protections are global. These laws protect original published and unpublished works authors develop from others using those works. According to the DMCA, anything created in digital form—from the CNN website to your friend's Facebook page—is copyrighted automatically. However, in the United States, the **fair use doctrine** allows you to use limited portions of an author's work if you attribute the source of the information.

Ethics and Citing Sources

Ethical researching and speaking involve giving credit when credit it due. Some students believe that citing others' work means they are not thinking for themselves. To the contrary, citations tell your audience that you have done your research on a topic. In many ways, research and speaking are like a long conversation. You have your own ideas and analysis, but they are more convincing when you consider others' viewpoints. Reading and citing others' work is like listening to what they have to say.

Plagiarism ranges on a continuum from accidental to deliberate. Some of the more obvious forms of plagiarism, such as copying someone else's speech or buying a speech from an online vendor, are clearly deliberate. However, it is also unethical to copy sections of a work, such as a paragraph or a few sentences, without noting the source. Patchwork plagiarism or remix, where you compile information from several sources without citing them, is the most common form of plagiarism among college students.[28] Less obvious, but also considered plagiarism, is failing to give credit for another's work that inspired your own. For instance, if you read a magazine article on Belgian dance music and then decided to give an informative speech on it, you would need to include that article in your reference list. Table 12.8 identifies 10 different types of plagiarism found in college students' work.[29]

TABLE 12.8 Types of Plagiarism

TYPE	BRIEF DEFINITION	APPLICATION TO PUBLIC SPEAKING
Clone	Complete copy of someone else's work	You give a speech you found online.
Ctrl-C	Partial copy of someone else's work	You give part of a speech you found online but add your own introduction and conclusion.
Find-Replace	Change only a few words of some else's work	You give a speech you found online, using a thesaurus to replace some of the words the source used.
Remix or Patchwork	Paraphrase several sources without attribution	You use several sources for your speech but fail to cite them during the speech or in your outline.
Recycle	Reuse your own work	You give a speech you gave in a previous class.
Hybrid	Cite some sources and not others	You neglect to cite all your sources when you give your speech.
Mashup	Copy work from several sources	You find several speeches online about your topic and use parts of each one for your speech without citing your sources.
404 error	Incomplete or inaccurate citations	You cite your sources in your outline, but the entries are wrong or missing information.
Aggregator	Citations are correct, but no new information is included	You repeat what others have said without adding any new analysis or your own interpretation.
Retweet	Citations are correct, but format and style are copied	You cover the same or similar main points as another speaker or writer on a topic in much the same fashion.

Adapted from Turnitin. (2012). *White paper: The plagiarism spectrum* (p. 4).

You do not need citations for all information. For example, you do not need to cite your own experiences, observations, and analysis. There is also common knowledge, or general information that is well known. If you find the same basic information undocumented in multiple sources and it is something that your audience likely will know already, then it is common knowledge. For example, statements such as "New York City's nickname is the Big Apple" and "On average, college graduates have greater earning power than high school graduates" reflect information that is commonly known. When you are referring to commonly known information, you do not need to cite a source.

In the *Complete Guide to Referencing and Avoiding Plagiarism*, author Colin Neville argues you must include the source for any of the following:

- A table, figure, photo, or similar image

- A theory, process, or model that someone developed

- Word-for-word quotes or definitions

- Paraphrases of sentences or paragraphs[30]

Neville's final word of advice: When in doubt, cite your sources.

Strategies for Avoiding Plagiarism

You can take steps to avoid plagiarism. Here are specific strategies you can use as you research your speech topics:

FIGURE 12.13 Developing a clear note-taking system is one strategy for avoiding plagiarism when researching your speech.

- *Develop a clear note-taking system.* Before you start taking notes on a source, record the complete citation. Then as you take notes, highlight or bracket any direct quotations and include the page number. In addition, include your own thoughts about what you have read or heard, using underlining or parentheses to separate out your ideas from those of the source's author (Figure 12.13).

- *Develop sound paraphrasing techniques.* Paraphrasing requires more than just changing a few words. Read or listen to the original source, consider what it means to you, and then write about it in your own words.

- *Develop the art of identifying and citing key quotes.* Use quotations sparingly and choose ones that are memorable and will resonate with your audience.

Learning how to effectively summarize and paraphrase material, integrate quotations, and take notes will help you avoid plagiarism. In Box 12.4, you will create procedures you can use when you research your speeches.

BOX 12.4: CONNECT & REFLECT

Developing a Source Citing System

People who get caught plagiarizing often cite poor or sloppy note-taking as the source of the problem. But even unintentional plagiarism is wrong. As a class or in a small group, do the following:

1. *Devise a note-taking system.* Create a practical, easy note-taking system that clearly differentiates between what is copied from the source and what are your own thoughts on the topic.

2. *Practice paraphrasing.* Take a few paragraphs from this book (or another book) and have each person write out the main ideas in their own words. Then compare what you wrote. Which ones do the best job of paraphrasing the material? How can you improve your paraphrasing?

3. *Identify key quotes.* Find a few key quotes from this book or another source. Compare the quotes the students in your group or class chose. What makes for a great quote? How should you cite a quote in your speech?

4. *Discuss takeaways.* What have you learned from completing this activity that you will use as you research your speech topics?

CHAPTER REVIEW

■ **Discuss the ways in which a pervasive communication environment has changed how speakers research their speeches.**

A pervasive communication environment requires the development of digital literacy skills to find, evaluate, and communicate relevant information. Developing digital literacy skills involves fighting fake news, applying critical thinking skills, going outside your technology comfort zone, and becoming an active prosumer of digital information.

■ **Identify internet resources for researching your speech.**

Internet resources include metasearch engines, search engines, and web directories. Effective use of keywords is critical in locating relevant information.

■ **Identify library resources for researching your speech.**

Library resources include books, periodicals, government publications, and reference materials. You search for these resources online using the library catalogue and databases. Your librarian is an essential ally in helping you find the materials you need for your speech.

■ **Explain how to use interviews to gather information for your speech.**

Interview experts on your topic to gather information not available from internet and library sources. Prepare for the interview by developing a list of questions; conduct the interview, taking good notes and preferably recording it; and then integrate key pieces of information from the interview into the speech.

■ **Differentiate between credible and untrustworthy information sources.**

Evaluate the author, publisher, process of producing the information, and the information itself to distinguish between credible and untrustworthy information sources.

■ **Discuss ways to avoid plagiarism when researching your speech.**

Developing a clear note-taking system, sound paraphrasing techniques, and the art of identifying and citing key quotes are strategies to avoid plagiarism.

DISCUSS AND APPLY

1. How digitally literate are you? What ideas do you have to improve your digital literacy skills?

2. When you have to find information about a topic, how do you begin? Where do you usually look for information?

3. What search engine do you usually rely on? Why that one? Try out one of the other search engines suggested in this chapter. What do you like about this search engine? What do you not like?

4. Now that you have read this chapter, how might you change your search strategies? What suggestion for finding information might you try out?

5. Identify a topic you are thinking about using for a speech in this class. Now develop a research strategy. Where will you start? What other information sources do you think will help you find the best information?

6. Try out at least three different library databases. What did you search for in each one? How easy or difficult was each database to use? What did you learn about each database that you did not already know?

7. Reflect on your own regular sources of information, such as television shows, magazines, and websites. Why do you choose those sources? What kinds of biases might these sources have?

8. Evaluate one information source, such as a webpage, news story, or journal article, based on the four evaluation areas: author, publisher, process, and information. How credible is the information?

9. Describe your note-taking system when you are researching a topic for this class or another one. What steps do you take when recording your notes to avoid plagiarism?

CREDITS

CHAPTER 13

Supporting, Organizing, and Outlining Your Ideas

Learning Outcomes

- Identify the types of supporting materials used in public speaking.

- Recognize the common patterns of organizing ideas in a speech.

- Discuss and apply the three principles of outlining.

- Demonstrate how to develop a complete-sentence outline.

- Identify effective and ineffective speech introductions and conclusions.

In the initial months of the COVID-19 pandemic, governors and mayors around the country ordered residents to stay at home. In Illinois, University of Chicago Medicine's Dr. Emily Landon, the medical director of antimicrobial steward-ship and infection control, explained the rationale for the governor's order at a press conference.[1] Early in her speech, Dr. Landon said: "Despite doing our best to prepare for a respiratory virus pandemic, we now find ourselves facing a brand new virus with too little information, not enough personal protective equipment, changing protocols every single day, and no second chances." After she thanked all the health care providers who have been on front lines in combatting the pandemic, she went on to say: "But the real problem is not the 80% who will get over this in a week. It's the 20% of patients, the older, those that are immunocompromised, those that have other medical problems who are going to need a bit more support." Then she explained the importance of social distancing in stopping the spread of the virus and what happened in places that did not put stay-at-home orders into practice. She stressed the need to shelter in place, even if it felt like an overreaction, saying: "In short, without taking drastic measures, the healthy and optimistic among us will doom the vulnerable. ... These extreme restrictions may seem, in the end, a little anticlimactic. Because it's really hard to feel like you're saving the world when you're watching Netflix from your couch." Finally, she made clear what she expected from her audience, closing with: "Public health and hospitals have been working hard for a long time. And now, it's your turn to do your part, a huge sacrifice to make, but a sacrifice that can make thousands of differences, maybe even a difference in your family, too."

Dr. Landon talked directly to the people of Illinois, explaining in clear language and a step-by-step way the rationale underlying the stay-at-home orders. The information she provided and how she organized her ideas were central to convincing her audience. First, she identified the problem—despite the health care community's best efforts, the virus was spreading. Then she presented the solution—Illinois residents had to sacrifice and follow the stay-at-home orders.

In this chapter, you will learn how to support your ideas, choose effective patterns of organization for your speeches, and how to outline what you plan to say.

Supporting the Ideas for Your Speech

When you research your speech (Chapter 12), you gather information to support the main ideas you want to convey to your audience. You have five types of supporting materials at your disposal for your speeches: facts and statistics, examples, definitions, testimonies, and narratives.

Facts and Statistics

fact Something that is known to be true based on experience and observation

A **fact** is something that is known to be true based on experience and observation. For example, some facts about the United States are that the national bird is the bald eagle, the highest geographic point is Mount McKinley in Alaska, the oldest college is Harvard, and the capital is Washington, DC. These are all observations about the United States that can be verified. Most of the facts that you accept to be true are based on others' experiences rather than your own. When a group of four scientists report in a well-respected scholarly journal that more than 268,000 tons of plastic are floating in the ocean,[2] you accept that as a fact even though you did not make the observation yourself. Instead, you rely on the scientists' explanation of their methods that their procedures produced accurate results. As another example, a student's persuasive speech on joining a student organization included facts, such as the number and types of such organizations, the kinds of events the organizations sponsored, and leadership opportunities in the organizations, to demonstrate the positive aspects of campus involvement. The student referred to the sources of the facts, which gave the audience confidence that the facts were accurate.

A **statistic** is a numerical representation of observations, often involving a comparison of variables. When pollsters report on the results of a survey, they are presenting statistics based on how people responded to the survey questions. For example, a Pew Research Center survey found that in the early months of the COVID-19 pandemic a majority of Americans (51%) were following news about it very closely and more than one third (39%) were following pandemic-related news fairly closely. Importantly, nearly half (48%) of those surveyed reported they had viewed or read made-up news about the novel coronavirus.[3] The statistics provided insight into how and what people were learning about the virus (Figure 13.1).

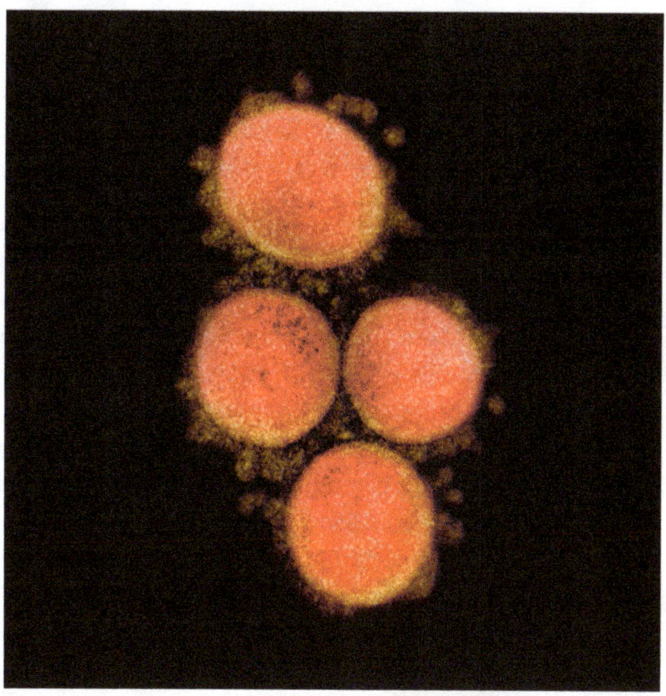

FIGURE 13.1 The Pew Research Center's survey on American's consumption of COVID-19 news provided statistics about how and what kinds of information respondents sought out.

Speakers rely on statistics to show the size of a problem or compare differences between groups. U.S. National Transportation Safety Board Chair Deborah A. P. Hersman used statistics to support her argument that strategies are needed to encourage women who are in STEM (science, technology, engineering, and math) professions to stay in their fields, stating:

> The vision of 85 years ago led to today's reality: according to the latest census, among young adults age 25–29, 58% of those with advanced degrees are women. But women graduates in STEM fields are another story. A recent study found that in the U.S., Brazil, China, and India, roughly one-third of women graduates in science, engineering, and technology feel "stalled." They say that within a year, they are likely to leave not only their jobs, but their whole field. In the U.S., 86% of women in those fields say they lack mentors.[4]

statistic A numerical representation of observations, often involving a comparison of variables

Hersman demonstrated that although more women are choosing to earn their degrees in STEM fields, those numbers are not translating into an increase of women in the STEM workforce.

Facts and statistics can promote agreement because they are verifiable. In addition, they can identify trends, such as increases in the number of students choosing a particular major and decreases in the popularity of a video game. For example, in an April 2020 press briefing on the COVID-19 pandemic, Michigan Governor Gretchen Whitmer used facts and statistics to report on her state's current situation:

> And now, for our update on cases. As of today's date, we have 32,000 positive cases. That's 576 new cases as well as sadly 2,469 deaths, 77 new deaths since yesterday. Statewide the number of patients who are in the hospital with a suspected or confirmed COVID-19 has declined from a high point about 10 days ago. In fact, we've seen a 15% reduction in the last 10 days.[5]

The number of cases and deaths underscored the magnitude of the problem, yet the reduction of hospitalizations suggested the state was making progress in combatting the disease.

While facts and statistics provide key support for your arguments, they can be distorted or taken out of context. For example, a 50% increase in bicycle thefts on your campus in 1 year sounds alarming. However, if there were only two thefts last year, a 50% increase would be just three thefts this year. In addition, facts and statistics can gloss over important anomalies and differences. For instance, if the average grade in a class is a C, you might assume that most students earned a grade close to a C. Alternately, there might be an even distribution of grades from A to F. But it could be that half the students earned As and half earned Fs, resulting in the C average for the class. Thus, the average itself is not a good indication of student performance in the class. Explore how to make sure you are using facts and statistics in ethical way in Box 13.1.

BOX 13.1: ETHICAL QUESTIONS

Just the Facts

From the bicycle theft example and your own experience, you know that understanding facts requires putting them in context. The first of the National Communication Association's principles introduced in Chapter 1 offers guidance in using facts and statistics in your speeches: "We advocate truthfulness, accuracy, honesty, and reason as essential to the integrity of communication."[6] For example, in arguing against racial injustice in the United States, you might state the fact that more White people than Black people are killed by police each year. While that is true, it is not the whole story. To provide a more complete picture, you would need to acknowledge that Black people are killed by police at a higher rate—twice as high—as White people.[7] Because there are more White people than Black people in the United States, accurately reporting how many in each group are killed by police each year requires you to cite the proportion of such deaths for each group, not the raw numbers. How can you check your facts for your speech? In addition to doing your own careful research, several nonpartisan websites regularly complete fact checking on current topics. FactCheck.org, sponsored by the Annenberg School for Communication, focuses on checking facts cited by politicians and political groups. Snopes fact checks urban legends. Politifact, developed by the Poynter Institute, investigates information presented in mainstream and social media. Our.News offers a mobile app that identifies the accuracy of online news stories. Explore these and other fact-checking resources. How can they help you follow NCA's first ethical principle in your speeches? How might you use a fact-checking site in your research? What other strategies might you use to fulfill the principles of an ethical communicator in public speaking?

Examples

An **example** is something that is used to represent an entire group. When Dr. Susan Mboya, president of the Coca-Cola Africa Foundation, gave a speech on women's empowerment at Meredith College, she used the examples of her grandmother, her mother, and herself to demonstrate that empowerment across time and generations has changed (Figure 13.2). In giving the example of her grandmother, Dr. Mboya said, "For my grandmother, empowerment meant marrying the man who she chose to marry, rather than a man whom she was forced to marry. A man who was educated enough to treat her as an equal, and to open her eyes to new worlds, which he did."[8] She then went on to describe her mother's successful career as well as her own achievements. By comparing the three examples, Mboya demonstrated how all three women were empowered, but in different ways.

Appropriately chosen examples, like the ones Dr. Mboya used, make a topic less abstract for an audience and give listeners something they can visualize or more easily understand. However, for an example to be effective, it must share enough characteristics with the other members of the group to serve as a valid representative of that group. Also, an example may oversimplify a complex topic and may not generalize to the group it is supposed to represent.

example Something that is used to represent an entire group or class of things

FIGURE 13.2 In her speech on empowerment, Dr. Susan Mboya used examples to support her argument.

Definitions

A **definition** provides the meaning of a word or phrase. Speakers may offer straightforward definitions, as with providing the dictionary or denotative meaning of a word. Or they may include their own experiences and emotions associated with the word, or the connotative meaning. In addition, speakers may define a word in terms of what it is not. For example, in her acceptance speech at the National Book Awards, author Ursula K. Le Guin defined writing a book in this way:

definition The meaning of a word or phrase

> We need writers who know the difference between production of a market commodity and the practice of an art. Developing written material to suit sales strategies in order to maximise corporate profit and advertising revenue is not the same thing as responsible book publishing or authorship. ... Books aren't just commodities.[9]

Le Guin defined both what writing a book is—an art—and what it is not—producing a commodity. She used this definition to forcefully argue that writers must not let publishers and sales departments dictate what authors write.

Speakers use definitions to clarify how they are using a term and to identify the parameters or boundaries of a concept or problem. However, definitions may be poorly constructed or artificially constrain the boundaries of a concept.

For example, early definitions of communication equated communication with all human behavior, making it difficult to identify what was communication and what was not. Other definitions privileged speaking over listening, leading to simplistic models of human communication.

Testimonies

testimony Information from an expert or authority on a topic

When speakers include information from an expert or authority on a topic, they are using **testimony** to support their ideas. Snapchat's CEO Evan Spiegel used testimony in his keynote address at LA Hacks when arguing that new technologies constrain individual's self-expression:

> Technology has perpetuated the myth of the transparent glass house and created a culture that values popular opinion over critical thought. ... And increasingly, we live in a time when, as [National Constitution Center President and CEO Jeffrey] Rosen describes, "intimate personal information originally disclosed to our friends and colleagues may be exposed to—and misinterpreted by—a less understanding audience."
>
> Every time we express ourselves, we do so with the understanding that things we say might become permanently and publicly known. We are encouraged to express ourselves in ways that are accepted by the largest possible audience. We lose our individuality in favor of popular acceptance.[10]

Quoting Rosen, a legal scholar and commentator likely well known to his audience, enhanced Spiegel's credibility on the topic and provided support for his view on self-expression.

Effective testimony can strengthen your ethos, or credibility, as a speaker and highlight the research you have done on your topic. In using testimony, you must be sensitive to possible biases the expert may have and recognize that the expert's viewpoint is one of many.

Narratives

narrative A story or anecdote

A **narrative** is a story or anecdote that speakers tell to illustrate a point. Narratives can provide an effective way to help your audience learn more about a topic. Research on scientific topics, for instance, has found that when speakers tell a story related to the topic, audience members are more likely to understand the underlying scientific principles and find the speaker more persuasive.[11] In addition, when audience members identify with a character in a story, they are more likely to get emotionally involved with the topic and more likely to support a positive outcome for the character, even when the character has behaved poorly.[12]

Because humans are natural storytellers and story listeners, narratives can prove a powerful tool to engage your audience. However, telling more stories will not prove necessarily more informative or persuasive. Stories must dramatize a topic, so one vivid story is more engaging than telling several uneventful stories.[13] In her Nobel Peace Prize lecture, Malala Yousafzai told this story to underscore the need to support education for girls worldwide (Figure 13.3):

> One of my very good school friends, the same age as me, who had always been a bold and confident girl, dreamed of becoming a doctor. But her dream remained a dream. At the age of 12, she was forced to get married. And then soon she had a son, she had a child when she herself was still a child—only 14. I know that she could have been a very good doctor.
>
> But she couldn't ... because she was a girl.
>
> Her story is why I dedicate the Nobel Peace Prize money to the Malala Fund, to help give girls quality education, everywhere, anywhere in the world and to raise their voices.[14]

Yousafzai could have skipped the story and simply stated she was using the funds from winning the Nobel Peace Prize to help educate girls. But the story she told helped her audience better visualize the costs of not educating girls and identify with the topic, making her reasons that much more convincing. Although narratives are a powerful way to develop your ideas, they present only one perspective on a topic. In addition, effective stories must be relevant and appropriate to the subject. Box 13.2 considers the power of stories in public speaking and the ethical implications of embellishing them.

FIGURE 13.3 When she gave her Nobel Peace Prize acceptance speech, Malala Yousafzai used stories to persuade her audience of the importance of educating girls around the world.

BOX 13.2: ETHICAL QUESTIONS

Truth or Fiction?

There is no doubt that audiences find stories appealing. Listeners often identify with the characters in a story and develop an emotional attachment to them. Stories make issues more real for audience members. But what happens when the details in stories are exaggerated? For example, study participants who viewed a film about a political candidate that included obviously exaggerated stories developed a more positive view of the candidate if they simply watched for enjoyment. However, those who listened for facts were not swayed by the embellished narratives. Still, the exaggerations did not have a negative effect on attitudes toward the candidate for either the appreciative or critical listeners.[15] These findings suggest that speakers who stretch the truth when telling a story likely will be viewed more favorably, at least by some audience members. Even though the outcome might be positive for the speaker, what are the ethical issues associated with exaggerating what actually happened in a story? Should speakers use the strategy of an embellished narrative if it will help them win over their audience? What would you do?

Successful public speakers integrate multiple types of supporting materials into their speeches to develop their ideas. Table 13.1 summarizes the types of supporting materials and their strengths and weaknesses.

TABLE 13.1 **Types of Supporting Materials**

TYPE	DEFINITION	STRENGTHS	WEAKNESSES
Facts and statistics	Verifiable observations and numerical data	• Foster agreement • Identify trends	• Can be distorted and taken out of context • Ignore anomalies
Examples	Representations of a larger class or group	• Visualize topic • Reduce topic abstractness	• Oversimplify issues • Not generalizable
Definitions	Meanings of words and phrases	• Delineate concepts • Specify parameters of a problem	• Incorrect or poor • Artificially constrain the topic
Testimony	Information from experts	• Authority on topic • Expert knowledge	• Bias • Limited viewpoint
Narratives	Stories and anecdotes	• Audience identification • Visualize topic	• Present only one perspective • Inappropriate or irrelevant

Organizing the Ideas for Your Speech

Every speech has four parts: the introduction, the body, the conclusion, and transitions, as shown in Figure 13.4. The introduction provides orientation to your speech and gets the audience ready to listen. The body is the main substance of the speech and where you develop your points. The conclusion wraps up your speech and leaves your audience with a feeling of closure. You use transitions to tie the parts of your speech together. Organizing the ideas for your speech assists you in identifying what information to include and what information to exclude from your research.

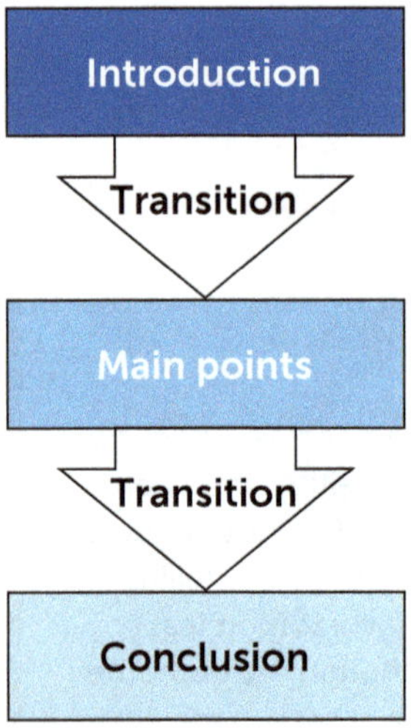

Applying Patterns of Organization

Organizing your ideas requires that you order your points in a way that supports the specific purpose for your speech. Use your working outline (Chapter 11), which includes the general purpose, specific purpose, thesis, and main points for your speech, to determine the best way to organize your speech. Useful patterns of organization usually align with the general purpose of a speech.

Patterns of Organizations for Speeches to Inform

In a speech to inform, you want your audience to learn more about a topic. Four patterns of organization work well with informative speaking: topical, chrono-logical, narrative, and spatial.

FIGURE 13.4 There are four main parts to every speech: introduction, main points, conclusion, and transitions.

With a **topical pattern of organization**, you arrange your points based on the main themes or elements associated with a topic. If you planned to give a speech on local organizations supporting the racial justice movement, you could organize it by the four main areas in a topical pattern.

topical pattern of organization An organizational pattern in which the main points in the body of a speech are arranged based on major themes or elements associated with a topic

Topic: Local racial justice organizations

General purpose: To inform

Specific purpose: To inform my audience about four local organizations active in the racial justice movement

Thesis: Four organizations near our campus that are active in the racial justice movement are Showing Up for Racial Justice Bay Area, Black Organizing Project, East Oakland Collective, and Urban Peace Movement.

Main points:

I.   Showing Up for Racial Justice Bay Area is the local chapter of a national organization that mobilizes White people to advocate for racial justice.

II. Black Organizing Project is a grassroots Black-member led community organization based in Oakland that works to improve the local school system and develop a new generation of leaders.

III. East Oakland Collective is a community group that promotes local civic engagement, economic opportunities, and more comprehensive solutions for the underserved.

IV. Urban Peace Movement is a youth-serving organization that empowers young people through cultural, policy, and systemic change.

A **chronological pattern of organization** traces a topic over time, such as the history of a subject or the steps needed to complete something. A speech on the history of jazz would use a chronological pattern of organization (Figure 13.5).

chronological pattern of organization A pattern of organization in which the main points in the body of a speech are arranged based on when they occurred in time

Topic: Jazz

General purpose: To inform

Specific purpose: To inform my audience about how jazz music has changed over time

Thesis: Changes in jazz coincided with events in U.S. history, including Prohibition, the Great Depression, and World War II.

FIGURE 13.5 A chronological pattern of organization would work well for an informative speech about the history of jazz.

narrative pattern of organization A pattern of organizing the main points in the body of a speech to tell the story of a topic as a series of dramatic events

Main points:

I.   Jazz saw its popularity soar in the speakeasies of Prohibition.

II.   Big bands and their swing music form of jazz dominated the music scene during the Great Depression.

III. During World War II swing music fell out of favor and was replaced with another type of jazz, bebop.

A chronological pattern of organization highlights the history of a topic or the order in which events did or should occur (Figure 13.5). The pattern is flexible, in that you may start at the end and work your way back to the beginning, as with showing a completed sculpture and then explaining the steps it took to make it. Box 13.3 asks you to delve into the history of your community and develop a chronology of key events.

The **narrative pattern of organization** tells the story of a topic as a series of dramatic events. Although similar to the chronological pattern of organization, the narrative pattern frames the topic as a single story or a group of stories. More than recounting the order in which events happened, the narrative pattern of organization incorporates the basic elements of storytelling: the setting, the characters, the conflict, and the resolution. This pattern of organization fits well with speeches about objects and things, people, and events. For instance, applying a narrative pattern of organization to a speech on the history of jazz takes it in a different direction than the previous chronological pattern example.

BOX 13.3: LIVE IT LOCAL

Tracing the Chronology of Your Community

How much do you know about the history of the community in which you live? Now is an excellent time to find out! Using the research skills you developed in Chapter 12, discover the major events that have shaped your community into what it is today. Access your library's databases to review newspaper articles from decades past and interview people who have lived in the community for a long time. Develop a timeline that traces the key turning points in your community's history, applying a chronological pattern of organization. Then consider speaking about these events to audiences who might be interested, such as elementary classes and community organizations.

Topic: Jazz

General purpose: To inform

Specific purpose: To inform my audience about the origins of jazz

Thesis: Three musicians claim to be the father of jazz: Buddy "King" Bolden, Ferdinand "Jelly Roll" Morton, and Nick "Joe Blade" LaRocca.

Main points:

I. Jazz was born in New Orleans, but there is discord about the parentage of this music genre.

II. Travel back to the New Orleans music scene in 1895 and hear Buddy "King" Bolden and his band's fresh and improvisational style.

III. Jump ahead 20 years and tap your foot to "Jelly Roll Blues" by Ferdinand "Jelly Roll" Morton, one of the first jazz compositions ever published.

IV. Wind up your gramophone in 1917 and listen to the first jazz recording, "Livery Stable Blues," produced by the Nick "Joe Blade" LaRocca's band.

V. King, Jelly Roll, and Joe Blade—three fathers of jazz.

In this example, the speaker suggests a conflict in the origins of jazz, sets the scene in New Orleans, introduces the three main characters, and offers a resolution to the problem. Presenting the topic this way offers a more personal view of jazz and adds drama to its history, with the three characters driving the story (Figure 13.6). As with the chronological pattern of organization, the narrative structure allows for flexibility. For instance, the speaker might introduce the characters all at once or at different points in the speech, offer the resolution before identifying the conflict, or gradually reveal elements of the scene. Whatever the order in which the speaker arranges the story, drama represents the central feature of the narrative pattern of organization.

When you arrange your ideas using a **spatial pattern of organization**, you order them in relationship to where they are in physical proximity to each other. An informative speech explaining the physical layout of your campus, for example, might use a spatial pattern of organization that starts at one end of the campus and moves progressively to the other. A spatial organization pattern also works well to identify geographical locations, as with a speech on botanical gardens in your state.

FIGURE 13.6 A narrative pattern of organization offers another way to present an informative speech on the history of jazz. In this example, the speech highlights the characters in the story, as with Ferdinand "Jelly Roll" Morton, third from left above.

spatial pattern of organization An organizational pattern in which the main points in the body of a speech are arranged based on where they are in physical proximity to each other

Topic: Botanical Gardens

General purpose: To inform

Specific purpose: To inform my audience about botanical gardens in our state

Thesis: Each quadrant of our state had a botanical garden: Metropolitan in the Southeast, Lakeview in the Northeast, Wintergreen in the Northwest, and Rosewood in the Southwest.

Main points:

I. Metropolitan Botanical Gardens is in the southeast part of the state.

II. Lakeview Botanical Gardens is in the northeast part of the state.

III. Wintergreen Botanical Gardens is in the northwest part of the state.

IV. Rosewood Botanical Gardens is in the southwest part of the state.

cause-effect pattern of organization A pattern of organization in which the main points in the body of a speech are arranged in a way that attempts to show that one event or series of events led to a particular outcome or set of outcomes

FIGURE 13.7 Using the cause-effect pattern of organization, Dr. Neil deGrasse Tyson argued this photograph of Earth changed how people thought about the planet.

Patterns of Organizations for Speeches to Persuade

A successful persuasive speech involves reinforcing or changing audience members' attitudes, values, beliefs, or behaviors. Three organizational patterns are used for persuasive speeches: cause-effect, problem-solution, and Monroe's motivated sequence. The **cause-effect pattern of organization** attempts to show that one event or series of events led to a particular outcome or set of outcomes. For example, in his commencement speech at Rice University, astrophysicist and author Neil deGrasse Tyson used this pattern of organization when he argued that the 1968 photograph an Apollo 8 astronaut took of Earth (shown in Figure 13.7), later titled *Earthrise*, led to fundamental changes in how people viewed the planet and human activity. He noted that in the 5 years after the photograph was publicized—on the first Earth Day—the federal government created the Environmental Protection Agency, and Congress passed the Clear Air and Clean Water Acts. He also linked the founding of Doctors Without Borders to the photograph, asking, "Where did that come from? Did anyone before that photo think of earth as a place without borders? No!" Then he went on to argue, "Something changed about us after the publication of that photo. It was a cultural response to our presence in space. ... We went to the moon to explore it, but in fact, we discovered earth for the first time." Thus, he concluded, the quest to explore is one taken on by a culture or group, not private enterprise. He later suggested that the momentum for exploration had waned but encouraged the Rice University graduates from all majors to pursue a culture of exploration and innovation.[16] Tyson demonstrated that the photograph of Earth taken during the Apollo 8 mission fundamentally changed how people thought about the planet.

A cause-effect pattern also is appropriate for a speech on declining bee populations in which the speaker might want to show the negative effects of human activities on bees worldwide.

Topic: Decline of Bees

General purpose: To persuade

Specific purpose: To persuade my audience that human activities are the primary cause of the global decline of bees

Thesis: Human activities, including the use of pesticides and herbicides, are killing millions of bees around the world.

Main points:

I.   Pesticides meant to kill problem insects kill bees as well.

II.   Herbicides designed to kill weeds contain chemicals that also kill bees.

III.   The number of bees worldwide has declined dramatically since the 1990s.

When you use the cause-effect pattern of organization, you must be sure that what you identify as the cause really does lead to the effect. Some speakers confuse correlation with causation. **Correlation** is a statistical term that indicates when two variables are related to each other, but that statistic alone does not show causality. For instance, research demonstrates that increased speech anxiety is associated with an increased heart rate.[17] But does one cause the other? That is unclear. Speakers may feel their heart rate increase and then feel more nervous about the speech, or they may experience anxiety, which increases their heart rate. Similarly, early research on sustainable consumption, such as shopping and driving less often, concluded that training people in mindfulness increased sustainable activities. However, a recent analysis found that once variables such as individuals' feelings of connection with nature were factored in, mindfulness training had no effect. Mindfulness still may play a role in sustainable consumption, but the relationship likely is not direct.[18] In Box 13.4, consider the issues that might arise when something identified as a cause really is not one.

In the **problem-solution pattern of organization**, the speaker establishes that a problem exists and then suggests how to solve it. In defining the problem, the way to solve it should point naturally to the solution the speaker suggests. Dr. Landon used this pattern in her speech quoted at the start of this chapter. First she identified a problem: "But the real problem is not the

correlation A statistical term that indicates when two variables are related to each other

problem-solution pattern of organization A pattern of organizing the main points in the body of a speech to show that a problem exists and then suggest how to solve it

BOX 13.4: ETHICAL QUESTIONS

Correlation vs. Causation

The human brain naturally seeks out patterns (Chapter 3). That's why people so easily—and wrongly—assume that where there is a correlation between two variables, one causes the other. Ciarán Gilligan-Lee, senior researcher at University College London and Babylon Health, offers this example:

> Data from seaside towns tells us that the more ice creams are sold on a day, the more bathers are attacked by sharks. Does this mean that ice cream vendors should be shut down in the interests of public safety? Probably not. A more sensible conclusion is that the two trends are likely to be consequences of an underlying third factor: more people on the beach. In that case, the rise in ice cream sales and shark attacks would both be caused by the rise in beachgoers, but only correlated to each other.[19]

The correlation lacked context. Without considering the context for the relationship—more people on the beach in the summer—the statistic was nonsensical.

Interpreting correlations incorrectly can prove dangerous, especially when the relationship is spurious, or not genuine (Figure 13.8). Tyler Vigen, project leader at Boston Consulting Group, developed Spurious Correlations, a website dedicated to demonstrating the problems with using correlation to establish causation or even a relationship between two unrelated variables. For instance, he shows that over a 10-year period, there is a .985 correlation (1.0 is a perfect correlation) between the total revenue generated by U.S. arcades and the number of computer science doctorates awarded in the country. As an even sillier example, he found a .952 correlation between the number of people in the United States who drowned after falling out of fishing boat and the Kentucky marriage rate.[20] What can you do to avoid the causation equals correlation trap? How can you tell if a correlation is spurious? As an ethical speaker, how should you report research that is based on correlation? As an ethical listener, how might you question a speaker's claim of causality?

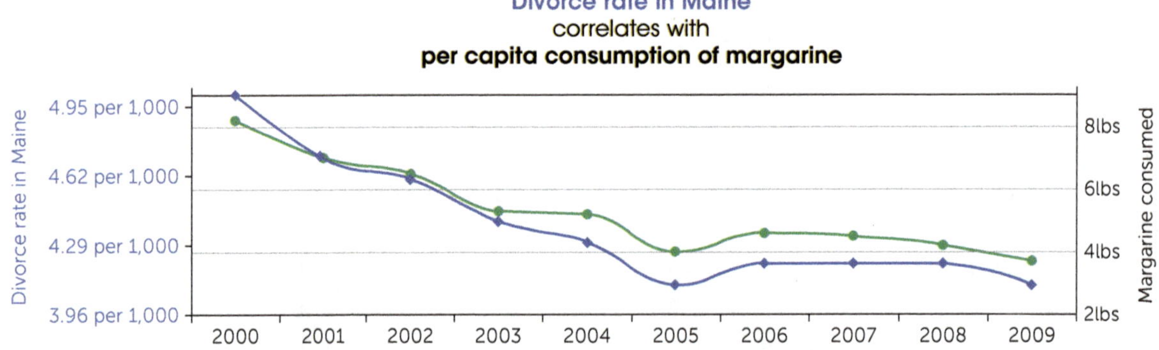

FIGURE 13.8 This silly and spurious correlation between the divorce rate in Maine and United States per capita consumption of margarine serves as a reminder of the limitations of the correlation statistic. *From*: Spurious Correlations website by Tyler Vigen.

80% who will get over this in a week. It's the 20% of patients, the older, those that are immunocompromised, those that have other medical problems who are going to need a bit more support."[21] Then she told her audience about the solution: "In short, without taking drastic measures, the healthy and optimistic among us will doom the vulnerable." By clearly establishing the problem that the virus was deadly for some people, she prepared her audience for the solution of "taking drastic measures," including staying at home and practicing social distancing.

The problem-solution pattern also works well in a speech designed to convince audience members to use clothes exchanges rather than buy new clothes.

Topic: Clothes Swapping

General purpose: To persuade

Specific purpose: To persuade my audience to use clothes exchanges rather than purchase new clothes

Thesis: Clothes swapping is the answer to the billions of dollars wasted each year on new clothes.

Main points:

    I. Producing new clothing has huge costs in water, fuel, and materials.

    II. People throw out more than half their body weight in clothes each year.

    III. Most clothes items are worn fewer than 12 times a year.

    IV. With clothes swapping, you get to enjoy new clothes at a fraction of the monetary and planetary costs.

Monroe's motivated sequence includes five steps specifically designed to motivate the audience to take action.[22] Speakers follow these steps in order when organizing their main points:

Step 1. *Attention*: Relate the topic to the audience.

Step 2. *Need*: Show there is a problem.

Step 3. *Satisfaction*: Provide the solution to the problem.

Step 4. *Visualization:* Demonstrate how the audience will benefit from implementing the solution or experience harm if the solution is not implemented.

Step 5. *Action:* Explain the action the audience members must take to implement the solution.

Monroe's motivated sequence A pattern of organizing the main points of a speech into five steps: attention, need, satisfaction, visualization, and action

Persuasive speeches in which you want the listeners to change their behaviors or take some sort of action are especially suited to the motivated sequence. Say, for instance, you wanted your audience to learn to play a musical instrument. You might use the motivated sequence to encourage them to try it out.

Topic: Musical Instrument

General purpose: To persuade

Specific purpose: To persuade my audience to learn to play a musical instrument

Thesis: Learning to play a musical instrument is good for relaxation and exercising your brain.

Main Points:

I.   Attention: Learning how to play a musical instrument can save your life.

II. Need: In this digital always-plugged-in world, people don't take enough time to relax while at the same time keep their minds active.

III. Satisfaction: Learning how to play a musical instrument engages your brain and helps you relax.

IV. Visualization: Imagine yourself relaxing in your living room strumming on a ukulele and just listening to the music.

V.   Action: Here is a list of the classes available at the local community college you can take to learn how to play a musical instrument.

In choosing a pattern of organization, consider the steps you will take your audience through to achieve the goal of your speech. Table 13.2 summarizes the different patterns of organization and gives a brief example for each one.

Outlining Principles: Parallelism, Coordination, and Subordination

parallelism The principle of outlining that refers to using a consistent numbering system and consistent language for your main and supporting points

Once you have identified the pattern of organization you plan to use for your speech, you can organize your ideas and research into a more detailed outline. Outlining is based on three principles: parallelism, coordination, and subordination. **Parallelism** refers to using a consistent numbering system and consistent

TABLE 13.2 Patterns of Organization

PATTERN	BRIEF DEFINITION	USED FOR SPEECHES TO …	USE WHEN YOU WANT TO …	EXAMPLE
Topical	Elements of a topic	**Inform**	Highlight the different aspects of a topic.	Common birds of prey are hawks, eagles, falcons, and owls.
Chronological	Sequence in which something occurs over time	**Inform**	Explain the history of a subject or the steps needed to do something.	Follow these seven steps to start your own business. …
Narrative	Tell a story or stories	**Inform**	Add drama or mystery to the topic.	Grammy-winner Billie Elish's story started when she uploaded her song "Ocean Eyes" to a music-sharing website.
Spatial	Physical proximity of the parts of a topic	**Inform**	Show how parts of the topic are related to each other in space.	My tour of U.S. baseball stadiums began with Fenway Park in the east and ended with Dodger Stadium in the west.
Cause-effect	Particular actions led to particular outcomes	**Persuade**	Demonstrate that an action or event produced a particular result.	Fast food has led to the obesity epidemic in the United States.
Problem-solution	Describe a problem and how to solve it	**Persuade**	Promote a particular solution to a problem.	Drought conditions in California will only worsen so we must build a desalination plant.
Monroe's motivated sequence	Five steps designed to motivate the audience to take action	**Persuade**	Get the audience to take a specific action.	Exercising your memory will give you lifelong learning skills.

language for your main and supporting points. Generally, points on outlines are numbered in this way:

I.   Main Point

   A. Supporting Point

      1. Sub-supporting point

         a. Sub-sub-supporting point

Use similar wording for each level for consistency as well. Let's return to the working outline on stress developed in Chapter 11 and expand the ideas for each main point.

Topic: Stress

Title: "The Different Types of Stress"

General purpose: To inform

Specific purpose: To inform my audience about the different types of stress

Thesis: There are three types of stress: acute stress, episodic stress, and chronic stress.

Main points:

  I. Acute stress
      A. Symptoms
      B. Benefits
      C. Harms

  II. Episodic stress
      A. Symptoms
      B. Harms

  III. Chronic stress
      A. Symptoms
      B. Harms

Using the principle of parallelism, a Roman numeral identifies each main point and the language used for each point is similar.

I.   Acute stress is the most common, stemming from the everyday demands of life.

II.   Episodic stress is acute stress that happens frequently.

III. Chronic stress is the stress individuals feel when they have no hope of an awful situation changing.

coordination The principle of outlining that refers to how the points of your speech are related to each other, with each point having the same level or degree of importance

Coordination refers to how the points of your speech are related to each other; each point should have the same level or degree of importance, especially the speech's main points. In the speech on stress example, each main point is of equal importance, as each identifies one type of stress. Compare the two outlines in Table 13.3 and identify the ways in which one demonstrates coordination among points and the other does not.

TABLE 13.3 Comparison of Outlines With and Without Coordination

WITH COORDINATION OF POINTS	WITHOUT COORDINATION OF POINTS
I. Acute stress is the most common, stemming from the everyday demands of life.	I. Acute stress and episodic stress are related in that episodic stress is when acute stress happens a lot.
II. Episodic stress is acute stress that happens frequently.	II. Some stress is good for you, as it can serve as a motivator.
III. Chronic stress is the stress individuals feel when they have no hope of an awful situation changing.	III. Chronic stress is really bad for you and can have long-term mental and physical health consequences.

Subordination refers to the hierarchical ordering of points, from large to small or general to specific. The first supporting point for the first main point in the speech on stress example reflects this ordering:

I. Acute stress is the most common stemming from the everyday demands of life.

   A. The symptoms of acute stress are emotional and physical.

      1. Emotional symptoms include feeling angry, anxious, and depressed.

      2. Physical symptoms include headaches, heartburn, upset stomach, sweaty palms, and dizziness.

subordination The principle of outlining that refers to the hierarchical ordering of points, from large to small or general to specific

Developing the Complete-Sentence Outline

You already developed your working outline as you researched your speech. Now you can use that outline as a basis for expanding on and organizing your ideas in the complete-sentence outline following the outlining principles of parallelism, coordination, and subordination. Figure 13.9 compares the three types of outlines you use in the speechmaking process.

Type	Brief Description	Helps you …	In chapter …
Working	Draft outline with possible points and subpoints	Organize initial ideas and identify information sources.	11: Developing your speech topic and purpose
Complete-sentence	Full-sentence outline with all parts of speech and bibliography	Fully develop each point and demonstrate how ideas link together.	13: Supporting, organizing and outlining your ideas
Presentation	Speaking outline that distills speech to essential points	Identify key words for practicing and presenting your speech.	14: Delivering your speech

FIGURE 13.9 This chart compares the working, complete-sentence, and presentation outlines.

The **complete-sentence outline** includes all the parts of your speech—the introduction, the main points, the conclusion, and transitions—along with your speech purpose and thesis, and your sources of information. This outline helps you more clearly delineate the structure for your speech and the order in which you will present your ideas.[23] The complete-sentence outline provides you with a comprehensive map of your speech, as well as the foundations that guided your research and the research itself. The more detailed your complete-sentence outline, the clearer the picture you will have about your ideas and the order in which you plan to present them.

Creating Strong Main Points

The working outline that you sketched out as you identified your speech purpose, topic, and initial ideas serves as the guide for developing the main points of your speech (Chapter 11). As you integrate the supporting materials you uncovered in your research (Chapter 12), your working outline will evolve into a more complete roadmap for your speech, as you add to refine your ideas.

Strong main points are clear, coherent, and comparable. First, when main points are clear, the audience easily identifies the topic, the support you are providing for your ideas, and how you want them to respond. Table 13.4 compares clear and unclear main points on the topic, famous beach boardwalks in the United States.

TABLE 13.4 Comparison of Clear and Unclear Main Points

Topic: Famous U.S. Beach Boardwalks

General purpose: To inform

Specific purpose: To inform my audience about famous beach boardwalks in the United States

Thesis: Although there are many beach boardwalks in the United States, four stand out: Coney Island, Atlantic City, Kemah, and Santa Cruz.

CLEAR MAIN POINTS	UNCLEAR MAIN POINTS
I. The Coney Island Boardwalk is in New York.	I. Guess which boardwalk is called Riegelmann Boardwalk.
II. The Atlantic City Boardwalk is in New Jersey.	II. I'm sure you know which boardwalk is the Boardwalk property in Monopoly.
III. The Kemah Boardwalk is in Texas.	III. There's one boardwalk that's on Galveston Bay.
IV. The Santa Cruz Beach Boardwalk is in California.	IV. Cocoanut Grove is part of the board-walk that's on the West Coast.

In the example, the clear main points in Table 13.4 directly relate to the specific purpose and thesis of the speech. In addition, audience members get a definite sense of what the speaker will discuss for each main point. The unclear main points leave the audience wondering which boardwalk the speaker will discuss next and the overall purpose of the speech.

Second, strong main points are coherent in that they fit with the topic, purpose, thesis, and each other. As you develop your main points, consider their relevance to what you hope to achieve with your speech. Are you staying on topic with your main points? Do your main points flow naturally from your purpose and thesis? Are your main points internally consistent? Table 13.5 compares coherent and incoherent main points for a persuasive speech on the tiny-house movement.

TABLE 13.5 Comparison of Coherent and Incoherent Main Points

Topic: The Tiny House Movement

General purpose: To persuade

Specific purpose: To persuade my audience to support the tiny house movement

Thesis: You should support the tiny house movement because these houses are sustainable, affordable, and community-oriented.

COHERENT MAIN POINTS	INCOHERENT MAIN POINTS
I. Tiny houses are sustainable because they use recycled and renewable resources.	I. I volunteer for an organization that builds tiny houses.
II. Tiny houses are affordable because they are inexpensive to build and situate.	II. I have met many people who live in tiny houses, and their experiences are positive.
III. Tiny houses are community-oriented because they are often placed in clusters.	III. The idea of tiny houses is at least 50 years old.
	IV. We all should support the tiny house movement.

The coherent main points in the tiny-house movement example illustrated in Table 13.5 are related to the topic, the purpose, the thesis, and each other. In that way, the main points are connected logically and the audience can understand how they work together to support the thesis. The incoherent main points are about the topic, tiny houses, but appear disjointed and unrelated to each other, failing to reinforce the speaker's central idea.

Third, strong main points are comparable, or similar in scope and importance. Think of your main points as breaking up your topic into proportions that reflect their contributions to achieving your specific purpose. In informative speeches, this translates into roughly an equal amount of emphasis and time on each point. In persuasive speeches, achieving balanced main points rests on the pattern of organization you apply. For example, if you use a problem-solution organizational pattern, you likely will spend about an equal amount of time on the problem and the solution. If you use Monroe's motivated sequence, the attention step will be quite short, whereas later steps will require fuller development. Table 13.6 shows the differences between comparable and noncomparable main points for an informative speech on the Liberty Bell.

TABLE 13.6. Comparison of Comparable and Noncomparable Main Points

Topic: The Liberty Bell

General purpose: To inform

Specific purpose: To inform my audience on the Liberty Bell's travels

Thesis: Although its hometown is Philadelphia, Pennsylvania, the Liberty Bell has traveled across America.

COMPARABLE MAIN POINTS	NONCOMPARABLE MAIN POINTS
I. In 1885, the Liberty Bell traveled to the World Cotton Centennial in New Orleans.	I. The Liberty Bell traveled to two cities in 1885: New Orleans, Louisiana, and Biloxi, Mississippi.
II. In 1893, the Liberty Bell traveled to the World Columbian Exposition in Chicago.	II. The Liberty Bell traveled back to Philadelphia in 1898.
III. In 1915, the Liberty Bell traveled to the Panama-Pacific International Exposition in San Francisco.	III. The Liberty Bell took a long train ride in 1914 and 1915 to reach the exposition in San Francisco, stopping in many places that I will tell you about.

The comparable main points for the Liberty Bell speech in Table 13.6 each address a specific trip the bell took, so are balanced in the attention they receive. In contrast, the noncomparable points are dissimilar, with the first point covering two cities the bell visited, the second a trip back to its home, and the third refers to the stops the bell made during a trip. With the noncomparable points, the speech will be unbalanced, with the speaker likely talking only briefly about the second point and at great length about the third point.

Connecting Your Ideas With Transitions

Transitions show your audience how your ideas are related and connect the three main parts of your speech, the introduction, body, and conclusion. The first transition occurs after the introduction when you signal to your audience that you are moving on to your first main point. A few sample transitions are:

- "I'll begin by explaining [*introduce first main point*] ..."

- "Let's start with [*introduce first main point*] ..."

- "Now that I've given you a preview, I'll first tell you about [*introduce first main point*] ..."

Transitions within the body of your speech link together the main points for your audience. These transitions tell the audience how the points are related and when you have finished one main point and are starting the next. Examples of effective transitions for the speech on the Liberty Bell include:

- "Eight years after the Liberty Bell traveled to New Orleans, it took a train ride to the World Columbian Exposition in Chicago." [*transition between first and second main points*]

- "While the Liberty Bell's trip to New Orleans wowed thousands of people, the bell's trip to the World Columbian Exposition in Chicago brought in an even larger audience." *[transition between first and second main points]*

- "In addition to trips to the East Coast and Midwest, the Liberty Bell traveled to San Francisco on the West Coast." *[transition between second and third main points]*

The last transition in your speech provides a bridge between your final main point and the conclusion. This transition indicates to your audience that your speech is winding down and you're nearly finished. There are several ways to transition to your speech closing as with:

- "To summarize, the key points I've covered are *[review main points]* ..."

- "In closing, I urge you to *[state action expected from audience]* because *[review main points]* ..."

- "As I have argued here today *[review main points]* ..."

Transitions provide guidance for your audience and demonstrate how your ideas and the parts of your speech work together as a complete unit. Table 13.7 summarizes the common types of transitions you might use in your speeches.

TABLE 13.7 Common Types of Transitions in Speeches

TRANSITION TYPE	WORDS/PHRASES	EXAMPLE
Time	After; Afterward; Before; During; Meanwhile, Now; Recently	Before the harmful health consequences of lead were known, nearly all houses in the United States were painted with lead-based paint.
Addition	Also; Equally important; Further; In Addition; Moreover	Moreover, taking a summer class will get students that much closer to graduation.
Contrast	Conversely; However; In contrast; Whereas; Yet	Yet most renters are unaware of these free services.
Similarity	In the same way; Correspondingly; Equally; Likewise; Similarly	Likewise, few small towns have comprehensive natural disaster response plans.
Consequence	Because; Consequently; For this reason; Therefore	For this reason, bright green is used on signs to get people's attention.
Order	First; Second; Third; And; Then; Last	Last, I'll explain how to follow up on a job interview.
Summary/ closing	Finally; In closing; In summary; To review	In summary, there are three general categories of tea.

The complete-sentence outline reflects the culmination of your fully developed speech: topic, general purpose, specific purpose, thesis, main points, transitions, introduction, conclusion, and references. Example 13.1 shows the final complete-sentence outline for the speech on stress. Review and assess that outline in detail in Box 13.5.

EXAMPLE 13.1

Topic: Stress

Title: "The Different Types of Stress"

General purpose: To inform

Specific purpose: To inform my audience about the different types of stress

Thesis: There are three types of stress: Acute stress, episodic stress, and chronic stress.

Introduction:

I. Research reported in *The Washington Post* and *Nature* has identified a mental-health crisis in the United States as people struggle with emotional stress due to the impacts of the COVID-19 pandemic.

II. Students experience stressful situations such as current events, midterms, changes at work, and class assignments like this speech.

III. Up to 90% of doctor visits in the United States are for problems that are stress-related.

IV. My research revealed that there are three types of stress: acute stress, episodic stress, and chronic stress.

Transition and preview: Today I will tell you about the symptoms and harms of each one as well as some surprising benefits.

Main points:

I. Acute stress is the most common, the American Psychological Association reports, stemming from the everyday demands of life.
   A. The symptoms of acute stress are emotional and physical.
      1. Emotional symptoms include feeling angry, anxious, and depressed.
      2. Physical symptoms include headaches, heartburn, upset stomach, sweaty palms, and dizziness.
   B. Research has found that some acute stress can be beneficial.
      1. The stress you feel about an uncertain situation, such as giving a speech, can motivate you to prepare for it.
      2. The stress you feel about encountering something new, such as trying something you've never done before, can help you recall what happened, as researchers recently reported in the medical journal *Neurobiology of Learning and Memory*.
      3. The stress associated with something that might hurt you, such as a car hurtling toward you as you cross the street, can help you take evasive action.
      4. The stress you feel when engaged in a challenging activity, such as a competitive sport, can help you focus more and enhance your performance.

    C. Too much acute stress can have psychological and physical harms.

        1. Psychological harms include overestimating your abilities and emotional distress.

        2. Physical harms you may experience are high blood pressure and migraine headaches.

Transition: Acute stress is manageable and can even be beneficial, unlike episodic and chronic stress.

  II. Episodic stress is acute stress that happens frequently.

    A. Like acute stress, the symptoms of episodic stress are emotional and physical, but are more extreme.

        1. Emotional symptoms include pessimism, hostility, insecurity, and aggressiveness.

        2. The physical symptoms are more severe than with acute stress, such as heart disease and hypertension.

    B. Episodic stress is psychologically and physically harmful.

        1. Psychologically, people who experience episodic stress often find their lives chaotic and view the world as a scary, turbulent place.

        2. Physically, episodic stress negatively impacts the body's immune system, leading to many health problems.

Transition: Episodic stress can become part of your life is you get accustomed to it, but it still pales in comparison to the seriousness of chronic stress.

  III. Chronic stress is the stress individuals feel when they have no hope of an awful situation changing.

    A. As with acute and episodic stress, the symptoms of chronic stress are emotional and physical, but are even more alarming.

        1. Emotionally, symptoms of chronic stress include negativity, little to no sense of humor, frequent sadness, panic, and anxiety.

        2. Physically, the symptoms of chronic stress are fatigue, over eating or loss of appetite, increased use of alcohol or drugs, and difficulty sleeping.

    B. As Dr. Pierce Howard states in his book *Stress: The Owner's Manual*, that chronic stress is extremely harmful both psychologically and physically.

        1. In the psychological realm, chronic stress can lead to depression, insomnia, addictions, and even suicide.

        2. Physically, chronic stress can cause strokes, kidney disease, cancer, heart disease, and other major life-threatening health problems.

Transition: Of the three kinds of stress, chronic stress clearly is the most serious and problematic to your mental and physical health.

Conclusion:

I.   There are three kinds of stress: acute, episodic, and chronic.

II.   Most stress is harmful, but some acute stress can be beneficial.

III.   Not everyone experiences stress the same way.

IV.   Knowing about the symptoms, benefits, and harms of stress should give you some insight into how you respond to stressful situations.

References

Abbott, A. (2021, February 3). COVID's mental-health toll: How scientists are tracking a surge in depression. *Nature*. Available from nature.com/articles/d41586-021-00175-z

American Psychological Association. (2019, November). *Stress in America 2019*. Available from apa.org/topics/stress

Ehrenfeld, T. (2018, December 7). The three types of stress. *Psychology Today*. Available from psychologytoday.com/us/blog/open-gently/201812/the-three-types-stress

Howard, P. (2014). *Stress: The owner's manual*. William Morrow.

Sazma, M., Mccullough, A., Shields, G., & Yonelinas, A. (2019). Using acute stress to improve episodic memory: The critical role of contextual binding.*Neurobiology of Learning and Memory*, 158, 1-8.

Sewart, A., Zbozinek, T., Hammen, C., Zinbarg, R., Mineka, S., & Craske, M. (2019). Positive affect as a buffer between chronic stress and symptom severity of emotional disorders. *Clinical Psychological Science*, 7(5), 914-927.

Wan, W. (2020, May 4). The coronavirus pandemic is pushing America into a mental-health crisis. *The Washington Post*. Available from washingtonpost.com/health/2020/05/04/mental-health-coronavirus/

BOX 13.5: CONNECT & REFLECT

Analyze That Outline

Review the complete-sentence outline in Example 13.1, answering the questions that follow.

1.   How well does the outline follow the principle of parallelism? Give an example.

2.   How well does the outline follow the principle of coordination? Give an example.

3.   How well does the outline follow the principle of subordination? Give an example.

4.   In addition to the list of references, how does the outline show the speaker's research?

5.   How clear are the speaker's main ideas?

6.   How effective are the transitions?

7.   Overall, how would you evaluate this outline? What suggestions do you have for the speaker to improve the outline?

8.   What have you learned from reviewing this outline that you can use in your own speeches?

Beginning and Ending Your Speech

Although you probably have been noting rough ideas for the introduction and conclusion of your speech as you created your outline, you cannot fully develop them until you have completed the body of your speech. The introduction of your speech sets the stage for the main points you will cover, and the conclusion reinforces those main points.

Creating the Introduction

As shown in Figure 13.10, you want to accomplish five tasks in the speech introduction: Get your audience's attention, give your audience a reason to listen, establish your credibility, reveal the topic and purpose, and preview your main points.

Here is the introduction for the speech on stress:

> Research reported in *The Washington Post* and *Nature* has identified a mental-health crisis in the United States as people struggle with emotional stress due to the long-term impacts of the COVID-19 pandemic. [*attention getter*] It's also midterms. And your boss has cut your hours at work. Your phone battery won't hold a charge. And you have to give a speech in your communication class. Feeling stressed just imagining those events? [*reason to listen*] I know I am. And we're not alone. For instance, studies on stress have found that up to 90% of doctor visits in the United States are for problems that are stress-related. [*credibility*] I decided to find out more about stress so I could better understand when and why people get stressed. [*topic and purpose*] My research revealed that there are three types of stress: acute stress, episodic stress, and chronic stress. [*preview of main points*]

These elements are part of every effective speech introduction. Together, they prepare your audience to listen to your ideas and help you achieve the specific purpose for your speech.

Creating the Conclusion

In the speech conclusion, you also have five tasks: Signal that the speech is ending, summarize your main points, reinforce your specific purpose, provide closure to the speech, and leave your audience with a memorable message, as shown in Figure 13.11.

Introduction
Get audience's attention
Give audience reason to listen
Establish credibility
Reveal topic and purpose
Preview main points

FIGURE 13.10 You need to complete these five tasks in the introduction to your speech.

Conclusion
Signal ending of speech
Summarize main points
Reinforce purpose
Make message memorable
Provide closure

FIGURE 13.11 You need to complete these five tasks in the conclusion to your speech.

The speech on stress concluded this way:

> In summary, [*signals ending of speech*] there are three kinds of stress: acute, epi-
> sodic, and chronic. Most stress is harmful, but some acute stress can be beneficial.
> [*summary of main points*] I know that the anxiousness I felt about this speech
> motivated me to prepare and practice for it. Not everyone experiences stress the
> same way. But knowing about the symptoms, benefits, and harms of stress gives
> you some insight into how you respond to stressful situations. [*reinforce purpose*]
> Phew! I feel less stressed now. I hope you do, too. [*memorable message*] Thank
> you. [*provides closure*]

A well-done conclusion for your speech does not guarantee you will achieve your purpose,
but it will make it more likely that your audience will view you and what you have to say more
favorably. Try out developing introductions and conclusions for five different public speaking
scenarios in Box 13.6.

BOX 13.6: CONNECT & REFLECT

Beginning at the Beginning and Ending at the End

Speakers often find introductions and conclusions the most challenging part of developing their speech. For each speaking scenario below, identify three main points you would cover and then develop a brief introduction and conclusion.

Scenario 1: You are speaking to sixth graders about everyday things they can do to reduce how much energy they use.

Scenario 2: You are speaking to small business owners in your local community about why they should create more internships for college students.

Scenario 3: You are speaking to parents at your college's first-year orientation about how they can help their children adjust to college life.

Scenario 4: You are speaking to your local city council about the possibility of starting a student exchange program with the city's sister city in Japan.

Scenario 5: You are speaking to the members of a local landscapers association about how to convince clients to plant more native plants.

Which scenarios were easier to develop an introduction and conclusion? Which were more difficult? What introduction and conclusion strategies might you try out in your own speeches?

CHAPTER REVIEW

- **Identify the types of supporting materials used in public speaking.**
  Facts and statistics, examples, definitions, testimonies, and narratives are the five types
  of supporting materials speakers use in developing their ideas.

- **Recognize the common patterns of organizing ideas in a speech.**
Speeches to inform typically use a topic, chronological, or spatial pattern of organization. Speeches to persuade usually use a cause-effect, problem-solution, or motivated sequence pattern of organization.

- **Discuss and apply the three principles of outlining.**
Outlines follow the principles of parallelism, a consistent numbering system and consistent language; coordination, each point having the same level or degree of importance; and subordination, a hierarchical ordering of points.

- **Demonstrate how to develop a complete-sentence outline.**
A complete-sentence outline includes all the parts of a speech—introduction, body, conclusion, and transitions—written in full-sentence form, along with the speech purpose, thesis, and references.

- **Identify effective and ineffective speech introductions and conclusions.**
An effective introduction for a speech gets the audience's attention, gives the audience a reason to listen, establishes the speaker's credibility on the topic, reveals the topic and purpose, and previews the main points. An effective conclusion for a speech signals the ending of the speech, summarizes the main points the speaker discussed, reinforces the purpose, makes the message memorable, and provides closure.

DISCUSS AND APPLY

Review this complete-sentence outline a student developed for a persuasive speech on learning to code and answer the questions at the end.

SPEECH OUTLINE FOR ANALYSIS

Title: "Learning to Code: It's for Everyone"

General purpose: To persuade

Specific purpose: To persuade my audience that they should learn to code

Thesis: Learning to code improves you complex problem-solving and critical thinking skills, both fundamental to success in today's workplace.

Introduction:

I. Twenty-five years ago, Apple founder Steve Jobs said in an interview, "Everybody in this country should learn how to program a computer, should learn a computer language. Because it teaches you how to think."

II. Since 2013, there's been a movement to teach computer coding skills more widely.

III. Learning to code addresses gaps that many people have in complex problem-solving skills and critical thinking.

IV.	In a recent *Forbes* article, nearly all 15 tech specialists agreed that basic coding skills are useful for everyone in any field.

V.	I completed a free coding course through FutureLearn, Video Game Design and Development: Introduction to Game Programming, where I improved my problem-solving and critical thinking skills while creating a mobile game.

Transition and preview: Few people have the complex problem-solving and critical thinking skills essential in today's technology-centered world; learning to code helps you develop those skills.

Main points:

I.	Complex problem solving requires you to respond to a dynamic situation in which the goals and methods for achieving the goals are ambiguous.

   A.	A study of young adults in South Korea, Germany, Singapore, Japan, Canada, Estonia, the United Kingdom, the United States, and Australia found that those in the United States had the poorest complex problem-solving skills.

   B.	Complex problem solving is the top skill employers seek in job applicants in today.

   C.	A recent study across 13 countries and various organizations, job types, and job levels found that effective complex problem-solving skills predicted career success.

Transition: In addition to lacking the complex problem-solving skills essential in the workplace, many U.S. students lack critical thinking skills.

II.	Critical thinking requires you to interpret, analyze, evaluate, and draw reasoned inferences about your observations.

   A.	In the book *Academically Adrift*, researchers Richard Arum and Josipa Roksa report that students' critical thinking skills do not improve in college.

   B.	Critical thinking is ranked third among the top 10 skills recruiters currently seek in job applicants.

   C.	Employers view critical thinking skills as essential for professional success, personal improvement, and the larger society.

Transition: Complex problem-solving and critical thinking skills are fundamental to career success, yet many students are deficient in those areas.

III.	Research shows learning to code improves your complex problem-solving and critical thinking skills.

   A.	A recent review of 10 studies found that students who learned to code also improved their problem-solving and critical thinking skills.

   1.	I found that just the one course in learning to code helped me think more critically and better approach complex problems.

2. Feedback from playtesting the game forced me to review the code and figure out why some things weren't working as I'd planned.

B. Learning to code is part of computer literacy, or the ability to effectively use computers and related technology.

    1. Although computing has been called the literacy of the 21st century, few college students arrive on campus with solid computing skills.

    2. Some schools, such as Broward College, Goodwin College, Central New Mexico Community College, and Eastern Washington University, have a computer literacy requirement for all undergraduate students.

C. LinkedIn Learning notes that computing is among the top 10 "hard skills" employers are looking for today.

    1. The jobs of the future will require you to be computer literate.

    2. A basic knowledge of computer coding will set you apart from other job candidates in your field.

Transition: Now that you know how learning to code will improve your complex problem-solving and critical thinking skills, as well as help you get a job, I'll tell you about how you can learn to code.

IV. There are three options for learning to code: courses here on campus, free online courses, and fee-based online courses.

A. Our Computer Science Department offers basic coding courses for non-majors, such as Python Programming for Nonmajors and Introduction to Computing.

B. Several online platforms offer free courses in coding.

    1. FutureLearn, where I took my coding course, offers courses such as Computer Programming for Everyone, Learn to Code for the Web, Functional Programming in Erlang, and Programming for Everybody: Getting Started With Python, all taught by experts and university professors.

    2. Teacher and writer Phil Mendez recommends freeCodeCamp, a nonprofit organization that has certificates, courses, and tutorials in JavaScript, Python, Linux, and other computer languages.

    3. Coursera offers courses such as C for Everyone: Programming Fundamentals, Python for Everybody, Introduction to Computer Programming, and Programming With JavaScript, HTML and CSS.

    4. Codeacademy has courses grouped into paths, such as Code Foundations.

    5. edX offers a free online course, Introduction to Computer Science, taught through Harvard University.

    C. There are many fee-based options for learning to code.

        1. Coursera, Codeacademy, edX, and other online venues also offer certificates and other options for a fee.

        2. Several universities offer online coding courses, such as Oregon State University's School of Electrical Engineering and Computer Science, Purdue University's Global program, and Grand Canyon University.

Conclusion:

    I. Most people, including college students, need improvement in their complex problem-solving and critical thinking skills.

    II. Learning to code not only improves those skills but also increases your computer literacy.

    III. You can learn the basics or even advanced coding by taking courses from the Computer Science Department on our campus, free online learning organizations, or fee-based online programs.

    IV. As Steve Jobs said, when you learn to code, you learn to think.

References

Arum, R., & Roksa, J. (2011). *Academically adrift: Limited learning on college campuses.* University of Chicago Press.

Cummins, P., Yamashita, T., Millar, R., & Sahoo, S. (2019). Problem-solving skills of the U.S. workforce and preparedness for job automation. *Adult Learning, 30*(3), 111–120.

Dörner, D., & Funke, J. (2017). Complex problem solving: What it is and what it is not. *Frontiers in Psychology, 8*, 1153. https://doi.org/10.3389/fpsyg.2017.01153

Forbes Human Resources Council. (2020, November 8). The top 10 skills recruiters are looking for in 2020. *Forbes.* https://www.forbes.com/sites/forbeshumanresourcescouncil/2020/11/09/the-top-10-skills-recruiters-are-looking-for-in-2021

Forbes Technology Council. (2020, March 20). Should everyone learn to code? 15 tech pros weigh in on why or why not. *Forbes.* https://forbes.com/sites/forbestechcouncil/2020/03/20/should-everyone-learn-to-code-15-tech-pros-weigh-in-on-why-or-why-not

Haber, J. (2020). *Critical thinking.* MIT Press.

Lafee, S. (2017). Coding: The new 21st-century literacy? Computer science gains a foothold in K-12 schooling, though qualified instructors and gender imbalance remain widespread challenges. *School Administrator, 74*(5), 38+.

Mainert, J., Niepel, C., Murphy, K., & Greiff, R. (2019). The incremental contribution of complex problem-solving skills to the prediction of job level, job complexity, and salary. *Journal of Business and Psychology, 34*(6), 825–845.

Mendez, P. (2020, May 7). How I taught myself code. *Medium.* https://medium.com/@mendez_phil/how-i-taught-myself-code-10685a463507

Pate, D. (2020, January 13). The skills companies need most in 2020—and how to learn them. *LinkedIn: The Learning Blog.* https://learning.linkedin.com/blog/top-skills/the-skills-companies-need-most-in-2020and-how-to-learn-them

Popat, S., & Starkey, L. (2019). Learning to code or coding to learn? A systematic review. *Computers & Education, 128*, 365–376.

Sen, P. (Director). (2012). *The lost interview* [Film]. Magnolia Pictures. https://magpictures.com/stevejobsthelostinterview/

Vee, A. (2017). *Coding literacy: How computer programming is changing writing.* The MIT Press.

QUESTIONS FOR DISCUSSION

1. What types of supporting materials does the speaker use to develop the main points? Give examples.

2. How effective were the supporting materials? What additional supporting materials might the student have used? Give examples for your answers.

3. What pattern of organization did the student use? How appropriate was that pattern of organization for the speech?

4. What pattern of organization might the student use for an informative speech on learning to code? What information would stay the same in the speech? What information would you delete?

5. How effective was the pattern of organization the student chose in motivating her audience to learn to code?

6. To what extent did the student follow the principles of outlining: parallelism, coordination, and subordination?

7. To what extent did the student include all the elements of an introduction? Give examples.

8. To what extent did the student include all the elements of a conclusion? Give examples.

9. Overall, how would you evaluate this speech? What did the student do well? What would you suggest the student could improve on?

10. What have you learned that you can apply in supporting, organizing, and outlining your ideas for your own speeches?

CREDITS

Delivering Your Speech

When artist, record producer, and philanthropist Missy Elliott received an honorary doctorate from the Berklee College of Music, she began her acceptance speech by crying and saying, "Let me just soak it in." This might seem like an inauspicious way to start a speech, but it earned her a standing ovation. She then went on to tell the graduates:

> There will be ups and downs—prepare for that. But don't give up. And I know you may hear that all that time. But if I had given up a long time ago, I wouldn't be standing here today. So you have come too far to quit. ... As long as you are breathing, it is never too late. Because people will tell you you're too old. People will tell you it will never work. Don't believe that because I'm standing here today and I'm quite sure these people (gestures to the students, faculty and administrators on the stage) can say the same thing. And you all can say this because look at you. You have graduated. ... And I want to say I love you all and I thank you all so much. I'm

so sorry that I'm a crybaby. But I'm about to wear this cap and gown in the shower, on my next video—everywhere I go. ... Thank you all so much.[1]

Elliott's authenticity, sincerity, and passion came through clearly as she delivered her speech (Figure 14.1). She smiled. She looked at her audience. She gestured. She used her voice to emphasize her points. She wiped a few tears from her eyes. And her audience responded. They clapped. They laughed. They cried.

Since humans first interacted with each other, *how* they communicated was as important as *what* they communicated. Ancient Roman orators and scholars identified how the different components in the public speaking process work together to present a coherent whole (Chapter 11). Four of the five canons of rhetoric—invention, arrangement, style, and memory—provide the background, or foundation, for your speech. Delivery is the foreground, when the speaker and audience come together and you and your ideas take center stage.[2]

This chapter explains how to deliver a speech with confidence, choose a delivery method, adapt to a presentation format, integrate presentation media, prepare the presentation outline, practice your speech, and, finally, present your speech to your audience.

FIGURE 14.1 When singer, songwriter, hip-hop artist, and public speaker Missy Elliott gave a speech at the Berklee College of Music, her authentic delivery resonated with her audience.

Delivering Your Speech With Confidence

You never will give a perfect speech. No one gives a perfect speech. Even Missy Elliott's speech was not perfect. She had a few bumpy transitions and missed words. Yet it was a strong, engaging, and inspiring speech because she connected with her audience in that moment. Your goal in delivering your speech is to give an excellent speech—not a perfect speech, but one for which you have prepared and practiced.

Like most people, you likely feel nervous about giving a speech. But are you more afraid of public speaking than death? You probably have heard that Americans fear speaking in public more than they fear dying, but that statement is not all that accurate. The original study conducted in the early 1970s simply asked respondents to check off their most-common fears, not rank-order them. Thus, public speaking was a common fear because people more often find themselves in situations where they need to speak in front of a group than they find themselves in death-defying situations. A recent study of college students found that when they were asked to rank-order their fears, death topped the list, followed by speaking before a group, financial problems, heights, and loneliness.[3] Thus, while public speaking is scary, death is scarier.

Still, there is no doubt giving a speech to an audience is nerve-wracking and speakers have real physical symptoms, including fluttering stomach, sweaty palms, dry mouth, increased heart rate, and shaking hands.[4] Students in particular are concerned with practical matters, such as how the audience will respond, disfluencies, negative evaluations, poor vocal delivery, and unexpected events.[5] However, **speech anxiety**, or the fear of speaking before an audience, is manageable. You have already made an excellent start in building your confidence by taking this class. Research shows that students who successfully complete a basic communication course such as this one reduce their speech anxiety.[6] The remainder of this section offers strategies for managing your speech anxiety and increasing your confidence, summarized in Table 14.1.

speech anxiety The fear of speaking before an audience

TABLE 14.1 Building Your Confidence in Speech Delivery

	ACTION	OUTCOME
Before your speech	Choose a topic you care about.	Motivated to talk about topic.
	Do your research.	Well-informed on the topic.
	Know your audience.	Understand audience's knowledge, attitudes, and values.
	Curate your presentation media.	Simple yet powerful visual and audio materials.
	Practice your speech.	More confidence, less anxiety.
Speech day	Get prepared.	Unhurried and calm.
	Channel your nervous energy.	Appropriately energetic presentation.
	Begin with confidence.	Appear confident and informed.
	Take an audience-centered approach.	Adapt to audience and promote understanding.
The future	Compliment yourself.	Recognize strengths.
	Identify challenges.	Determine ways to do better.
	Focus forward.	Create action plan for giving future speeches.

Building Your Confidence Before Your Speech

Managing your speech anxiety starts long before the day you give your speech. The following five strategies will build your confidence and lower your speech anxiety.

Choose a Topic You Care About

The first step in building your confidence is choosing a topic in which you are interested and comfortable talking about with others. In his book *Speak With Courage*, Professor Martin McDermott refers to this as following your bliss. He advises speakers that when selecting a topic, "come up with things you want to share so badly that—despite any fears about public speaking—you *have* to say them."[7] What do you feel passionate about? What topics do you want to share with others? What sorts of things interest you? What do you want to learn more about? Recall from Chapter 11 the first criterion in evaluating a topic is you—your interests and knowledge. When you are enthusiastic about a topic, it is easier to talk about, you will feel more confident, and your audience will perceive that confidence.

Do Your Research

Knowing your topic well increases your confidence in discussing it (Figure 14.2). Chapter 12 provides you with a comprehensive guide for researching your topic. Having an extensive knowledge base will give you flexibility in your presentation if you need to add or delete information. In addition, you will be prepared to answer audience questions with authority.

Know Your Audience

Learn all you can about your audience before your speech—their knowledge, attitudes, and values associated with your topic. Researching your audience allows you to tailor your speech to their interests and concerns, making them more receptive to your ideas. When you know your audience, you are better prepared for how they might respond to your speech, which increases your confidence.[8]

FIGURE 14.2 Doing your research so you know your topic well will increase your confidence in your public speaking abilities.

Curate Your Presentation Media

Presentation media are audio and visual materials that support the points or arguments of your speech. Although you might be tempted to include a lot of text on digital slides or other presentation media, do not do it. You will find yourself relying on, and even reading, your slides. Research on using digital slides, such as PowerPoint and Keynote, in presentations has found that speakers are *more* nervous when their slides include many lines of text and *less* nervous with fewer words on their slides.[9] In addition, your audience will experience information overload if you bombard them with text-heavy visual materials. Images combined with minimal text will increase the audience's comprehension of your message, likely leading to more positive feedback from your audience, and leave you feeling more confident.[10]

presentation media Audio and visual materials that support the points or arguments of a speech

FIGURE 14.3 Getting an early start on practicing your speech increases your confidence.

Practice Your Speech

Practice is one of the best ways to lower your speech anxiety and improve your confidence. Author, speech coach, and former news anchor Carmine Gallo puts it this way: "Relentless preparation is the single best way to overcome stage fright: know what you're going to say, when you're going to say it, and how you're going to say it."[11] Several studies have found that students who practice their speeches not only feel more confident but they appear more confident to the audience as well.[12] Moreover, the research is clear: The earlier students start practicing their speech and the more time they spend rehearsing, the more confident the audience rates them (Figure 14.3).[13] As you practice your speech, visualize the setting and your audience. Imagining yourself giving your speech and getting positive feedback from your listeners will increase your confidence and lower your anxiety.[14]

Building Your Confidence on Speech Day

Building your confidence continues on the day of the speech itself. Expect to feel some nervousness, but the following four strategies will help you manage that anxiety.

Get Prepared

Preparation on the day of your speech will reduce your nervousness and increase your confidence. Have all your materials for your speech ready, including your notes and presentation media. Arrive early so you are not rushed. Check to be sure any technology you are using is functional. Avoid making any last-minute changes to your speech. Feel confident in all your planning for this presentation.

Channel Your Nervous Energy

Channel your nervous feelings into a more energetic, but not frenetic, presentation. Research has found that speakers who engage in negative self-talk about public speaking feel less confident, whereas those who engage in positive self-talk feel more confident.[15] Relabel your anxiety as energy, enthusiasm, and a desire to share what you know with your audience.

Begin With Confidence

Portray confidence right from the start. If you think of yourself as a confident speaker, the audience will perceive you as a confident speaker.[16] Take a breath before you start speaking—there is no rush to begin. Make sure your audience is paying attention, make eye contact, smile, and then give your introduction. Once you have successfully presented the opening to your speech, you will find that your speech anxiety will lessen and your confidence will increase.[17]

Take an Audience-Centered Approach

Although the spotlight in public speaking is on the speaker, turn that attention back to the audience. Instead of thinking about yourself, think about the audience. As you are giving your speech, monitor their responses. Do your listeners seem to understand what you are saying? Are they following your points? Do they seem interested in your topic? Do you need to make any adjustments to increase audience understanding? When you take an audience-centered approach, you will be less concerned about yourself and more concerned about them (Figure 14.4).[18] And because you know your topic well, you will feel more confident talking with your audience about it.

FIGURE 14.4 Keeping your focus on the audience increases your confidence as you give your speech.

Building Your Confidence for the Future

Once you have given your speech, you might be tempted to think, "There, that's done." But you are not quite finished. Reflecting on your speaking performance will help you build your confidence for future speeches. When you take the mindset that you learn from all your experiences—both those in which you did well and those in which things did not go quite as expected—you can develop a plan for further improving your confidence.[19] Use the following strategies to continue building your confidence in your public speaking skills.

Compliment Yourself

Review your speech performance in your mind. What parts of your speech received positive responses? Think back to your audience and the listeners who nodded in agreement and smiled. How did you feel at those points in your speech? How did you show your confidence? Note all the aspects of your speech that went well and plan to use those strategies for your next speech.

Identify Challenges

Consider parts of your performance that did not go as you had hoped, but do not dwell on them or berate yourself. Ruminating on mistakes or negative feelings will not build your confidence and is not a productive way to improve your speaking skills.[20] Instead, simply acknowledge areas for improvement and develop a plan to address them. In addition, praise yourself for managing any missteps that occurred. For instance, if you accidentally skipped one of your digital slides and then went back to it, saying something like, "Before we move on, let me tell you about ..." or "I seem to have gotten ahead of myself," congratulate yourself for calmly fixing the problem.

Focus Forward

You have given your speech. Now consider what you will do the same and what you will do differently for future speeches—because there will be future speeches. Even once you leave this class, you will give speeches. Develop a plan for how you will approach the next speech, including how you will choose your topic, organize your ideas, engage your audience, and practice. Try out your plan for building your confidence in Box 14.1.

BOX 14.1: LIVE IT LOCAL

Implementing an Action Plan

One of the best strategies for building your confidence in public speaking is giving presentations outside the classroom. Research venues in your community for giving speeches, including local schools, organizations, and neighborhood groups. Take the plan you developed for increasing your confidence and apply it to speeches in your community. The more experience you get successfully giving speeches to a variety of audiences across a variety of situations, the better you will feel about your skills as a public speaker. Every time you give a speech is an opportunity to build your confidence.

Choosing a Delivery Method

Delivery methods vary in preparation, rehearsal, and presentation outcome. As you learned in Chapter 1, competent communicators adapt to norms and contexts in achieving their goals. As a competent public speaker, the method you choose must fit the audience and occasion. The four methods of delivering a speech—manuscript, memorized, impromptu, and extemporaneous—are discussed in this section and summarized in Table 14.2.

TABLE 14.2 Delivery Methods

METHOD	THE SPEAKER	ADVANTAGES	DISADVANTAGES	USEFUL FOR …
Manuscript	Reads a document written in advance	Careful selection of each word; precisely timed	Relies on written rather than oral language; inflexibility; lack of eye contact with audience	News reports and critical announcements
Memorized	Recites a text from memory	Present exact message without notes; free to make eye contact and gesture	Can appear mechanical and impersonal; requires extensive practice	Short speeches and those presented many times
Impromptu	Talks with little or no preparation	Responsive to audience and occasion; spontaneous and authentic	Disjointed ideas; no outside research; unpolished delivery	Answering audience questions during Q&A and speaking in informal contexts
Extemporaneous	Presents a planned, researched, and rehearsed speech with brief notes	Deep knowledge of topic; appear spontaneous yet polished; adapt to audience responses	Time commitment; requires some memorization; can cause speech anxiety	Oral classroom presentations, workplace reports, and creative proposals

Giving a Speech From a Manuscript

In **manuscript speaking**, you present a speech from a document you have prepared in advance. Speaking from a manuscript has several advantages. First, you identify ahead of time exactly what information you will include and what you will leave out. That is especially important in situations where the message you convey must be particularly precise, clear, and accurate. For example, when reporting the news, TV news anchors speak from a manuscript so they can provide their audience with correct information in the allotted time. Politicians typically use a manuscript, displayed on a teleprompter, for major speeches to stay on message and avoid gaffes. Second, you can carefully craft each word you want to use in your speech. If a misspoken word could lead to a catastrophic misunderstanding, as with a spokesperson announcing the terms of a new peace accord between two nations, then speaking from a manuscript is the best delivery mode.

However, manuscript speaking has several disadvantages. First, in writing out word-for-word what you plan to say, you cannot make any changes based on feedback from your audience. If your audience appears puzzled or confused, you keep reading your speech rather than adding in an explanation that might address their confusion. Second, written and oral language are not the same. Written language is more complex, with longer words and sentences. Oral language is more informal and conversational. Thus, the audience may have a difficult time following a manuscript speech, with its more sophisticated vocabulary and grammatical structure. Third, when speakers read a manuscript, they make infrequent eye contact with the audience, leaving the audience feeling ignored or neglected. While the appearance of little eye contact can be overcome with a teleprompter, effectively using such a cuing device requires extensive practice to avoid sounding robotic.

manuscript speaking
Giving a speech from a document written in advance

Giving a Speech From a Memorized Text

memorized speaking
Writing out a speech word-for-word in advance, committing it memory, and reciting it to the audience

Memorized speaking involves writing out your speech word-for-word in advance like manuscript speaking, but then committing to and reciting the text from memory. Speakers in high school and college speech competitions, which require giving the same speech many times, usually memorize their speeches. Short speeches for special occasions, such as wedding toasts and speaker introductions, often are memorized. Committing a speech to memory frees the speaker from relying on a manuscript or notes. In addition, the speaker can make eye contact with the audience, easily gesture, and move about as needed. As with actors in a play, speakers who give memorized speeches may commit both words and actions to memory. That is, rather than just memorizing the words of a speech, the speaker may memorize specific gestures, pauses, and vocal intonations.

Memorized speeches have their drawbacks, however. The clearest disadvantage occurs when speakers forget their lines. Because they have memorized the speech with each word in a specific order, a forgotten word or phrase can lead to an awkward pause or, worse, skipping entire sections or even abruptly ending the speech. As with manuscript speaking, memorized speaking's inflexibility prevents the speaker from adapting to audience feedback. Also, the speech may sound rote, as if the speaker is just saying the words without really meaning them. Finally, memorizing a speech, especially a lengthy one, takes time and many practice sessions.

Giving an Impromptu Speech

impromptu speaking Creating a speech on the spot with little or no planning

You have given many impromptu speeches, as with explaining directions to a friend or telling a story at a family event. **Impromptu speaking** involves creating a speech on the spot with little or no planning. Sometimes speakers are caught completely unaware, as with an instructor unexpectedly calling on a student in class. Other times speakers may have thought about what they are going to say, as with preparing for a job interview, but still respond with a short impromptu speech to an unanticipated question.

Because impromptu speeches are unplanned, audiences view them as genuine and authentic (Figure 14.5). Impromptu speeches arise directly from the moment and with the audience, giving these speeches the spontaneity memorized and manuscript speeches lack. However, impromptu speeches often lack structure and appear disjointed, making them difficult for the audience to follow. In addition, without advanced preparation,

FIGURE 14.5 With impromptu speaking, as at a protest or demonstration, the speaker responds to events unfolding in the moment and seems natural and sincere.

impromptu speeches rely on the speaker's knowledge base and do not include any outside research. Finally, because the speaker has not rehearsed beforehand, the delivery may be uneven and awkward. Still, you can use the advantages of impromptu speaking to overcome some of the disadvantages, as you will find out in Box 14.2.

Giving an Extemporaneous Speech

Most of the presentations you will give for this class will involve **extemporaneous speaking** in which you present your researched, organized, and practiced speech in a conversational style with minimal notes to jog your memory (Figure 14.6). With extemporaneous speaking, you know your topic and speech well, so you can talk with your audience about it, making adjustments as needed. Extemporaneous speaking appears spontaneous yet is carefully prepared, structured, and rehearsed. Thus, the speaker's delivery is polished while also conversational and interactive with the audience. Because extemporaneous speeches are well-researched, audiences view speakers as knowledgeable and credible. In addition, effective extemporaneous speeches have clear main points and are easy for the audience to follow. When speakers give an extemporaneous speech, they monitor audience feedback and adapt their speech as needed. For instance, if audience members seem to disagree with an idea, the speaker might present additional supporting information. Finally, extemporaneous speakers can make last-minute adjustments during the speech, as with referring to a previous speaker, a current event, or the presentation's setting. For all these reasons, this chapter primarily focuses on strategies for delivering a speech extemporaneously.

To the audience, extemporaneous speaking seems effortless and natural because the speaker is engaging and easily discusses the topics. However, one disadvantage of extemporaneous speaking is that it requires extensive planning and preparation. Effective extemporaneous speakers must thoroughly research their topic, analyze their audience, develop their ideas, and practice, practice, practice. Extemporaneous speaking requires a substantial investment of time for speakers to give a successful presentation. In addition, extemporaneous speakers must commit some parts of the speech to memory, as with the main points, to avoid an overreliance on notes. Like a memorized speech, forgetting a main point can confuse the audience. Finally, extemporaneous speaking can produce speech anxiety, which is why building your confidence, discussed in the previous section, is essential to this delivery method.

extemporaneous speaking Giving a researched, organized, and practiced speech in a conversational style

FIGURE 14.6 In extemporaneous speaking, you plan, prepare, and practice your speech, giving an excellent version adapted to your audience.

BOX 14.2: CONNECT & REFLECT

Practicing for Impromptu Speeches

While you cannot practice an impromptu speech, you can get yourself ready to give impromptu speeches. That is, by practicing impromptu speaking, you increase your expertise in this delivery method as well as your comfort level and confidence. Complete the activity below at least three times, either on your own or with a small group of classmates, following these instructions:

1. Pick a number between 1 and 10. Count down the topics listed to the right until you reach that number.

2. Set a timer for 2 minutes. Quickly jot down key words and phrases for the items in the template below, taking about 20 seconds for each one.

   a. Identify your first thoughts about the topic.

   b. Consider your own experience with the topic.

   c. Organize your ideas.

   d. Think of how you want to begin your speech.

   e. Think of how you want to end your speech.

3. After your time is up, use the brief notes you have written to speak on the topic for 1 minute.

4. Repeat at least two more times, using the same number you picked and starting with the first topic you chose.

5. Now reflect on your experience.

   a. Which topics were easier to talk about? What made them easy?

   b. Which topics were more challenging to talk about? Why were they more challenging?

   c. How did the template help you organize your ideas and thoughts?

   d. What did you do well?

   e. What areas of impromptu speaking do you need to work on?

   f. What did you learn that you can apply to future impromptu speaking situations?

IMPROMPTU SPEECH TOPICS

- Reducing consumerism
- How to be a better listener
- What you learn from traveling
- The importance of art
- Why I chose this school
- My best summer vacation
- Play isn't just for kids
- The future of self-driving cars
- What racial justice means to me
- Why nations have holidays
- My ideal job after graduation
- How to give the perfect gift
- The value of a college education
- My favorite time of the year
- How to be a good friend
- My approach to ethics
- The value of compassion
- Learning from failure
- How public education contributes to democracy
- The role of music in life
- Why all eligible voters should vote
- The most important lessons from my first job
- The meaning of love
- Defining politeness
- My favorite book or movie
- The qualities of successful people
- My role model
- What to do on a rainy day
- How to plan an event
- My three predictions for the future
- What it means to be wise
- The influence of advertising
- My favorite sport
- The person I'd most like to meet

Adapting to the Presentation Format

For thousands of years public speaking was equated with giving a speech in-person to an audience. A speech was an event that took place at a certain time in a certain place with a speaker and listeners. Although that format for public speaking still holds true today, the pervasive communication environment has untethered the "event" of a speech from time and space. That is, you can give a speech to an in-person audience that is livestreamed to others around the world and then recorded for even more people to view. You may skip the in-person audience altogether and simply livestream in the moment or record your speech for others to view later. The pervasive communication environment—the ability to connect anywhere, anytime, with anyone—has exponentially expanded the reach and scope of the public speaker.

Adapting to In-Person Presentations

You likely are familiar with in-person public speaking, where the speaker and audience are in the same physical location. Although throughout history this is the most common presentation format, speaking in-person still varies from a small conference room to a huge outdoor arena (Figure 14.7). Whatever the setting for your speech, become familiar with it as soon as possible. Find out the presentation media capabilities. Are there an LCD project and screen? Will you need to bring your own laptop or tablet? Is there a podium with a microphone? Outside of a regular classroom or small conference room, you will need a microphone to project your voice—you do not want to shout your speech.

No matter how large or small the space, your audience will expect you to connect with them. As you rehearse your speech, work on making eye contact with your imaginary audience, smiling, and gesturing. Whether it is a 50,000-seat stadium or a 20-person community meeting room, you are having a conversation with your listeners. You personalize your speech, referring to the audience, setting, and occasion. You make your audience feel like you are talking *with* them, not *at* them.

FIGURE 14.7 An auditorium, large lecture hall, standard classroom, and small conference room each have different constraints for giving a speech.

Adapting to Mediated Presentations

The technology for video conferencing dates back to the late 1800s, although businesses did not start using it until the 1970s.[21] Now, with many people working from home at least part time, video chat and conferencing are routine with platforms such as Google Meet, FaceTime, and Zoom. Podcasts are commonplace. Millions of people regularly upload videos to YouTube.

Mediated presentations take many forms, from streaming a webcast for a live remote audience to recording a speech and making it available for people to view later. Not surprisingly, assuring that all the technical aspects of your presentation are working is essential. Go through several trial runs before the day and time of your speech or when you plan to post it. Check that the audio is clear and sufficiently loud and the video looks as you want it to look. For example, you want an uncluttered background so the audience focuses on you rather than what is behind you. Resist the temptation to use a favorite photograph as a virtual background as it will distract from your presentation. Practice sharing your computer screen and moving through your digital slides. If your audience is live, as with a real-time webinar, have a colleague or friend manage audience questions so you can answer them either at the end of your presentation or during breaks between points.

Whatever the format for your mediated presentation, remain audience-centered, always considering the best way to deliver your message to your audience. Clearly articulate your words, use a moderate rate of speech, and keep a conversational tone.

Adapting to Blended In-Person and Mediated Presentations

There may be some cases where your speech is both in-person and mediated. That is, you have multiple audience members, some of whom are in the setting with you and others who are viewing or listening to your speech either in real time or sometime in the future. For instance, college presidents now routinely give their fall welcome speeches to in-person audiences, livestream the events for remote audiences, and record and upload the speeches to the colleges' websites for anyone with internet access to view at a later time. TED Talks offer another example, with speakers presenting to live audiences and the recordings uploaded to the TED website.

Who is the audience in a speaking situation that is both in-person and mediated? Should the speaker try to target anyone who might view the speech in that moment and in the future? That approach is not practical. If the speaker aims for too broad an audience, no one will be satisfied. Thus, the speaker must address the known audience, whether in-person or online. View a few TED Talks. You will notice that speakers engage the in-person audience because speakers have no way of knowing the future audiences for their speech. However, if you are giving a presentation at work to some coworkers in the room with you and others attending via video conference, then you can deliver your speech to both groups (Figure 14.8).

FIGURE 14.8 With a pervasive communication environment, presentations that blend in-person and mediated channels have become more common.

Integrating Your Presentation Media

The presentation media you include with your speech—or whether you include any presentation media at all—depend on the presentation format, the occasion, your speech content, and your audience. For some speaking situations, presentation media are required, such as a webinar. In other cases, audiences expect presentation media, as with a project leader's report to top management. Finally, for some speaking situations, especially special occasion speeches such as toasts and nominations, presentation media typically are not expected or needed. Ask yourself the questions in Figure 14.9 to help you decide if you should include presentation media in your speech.

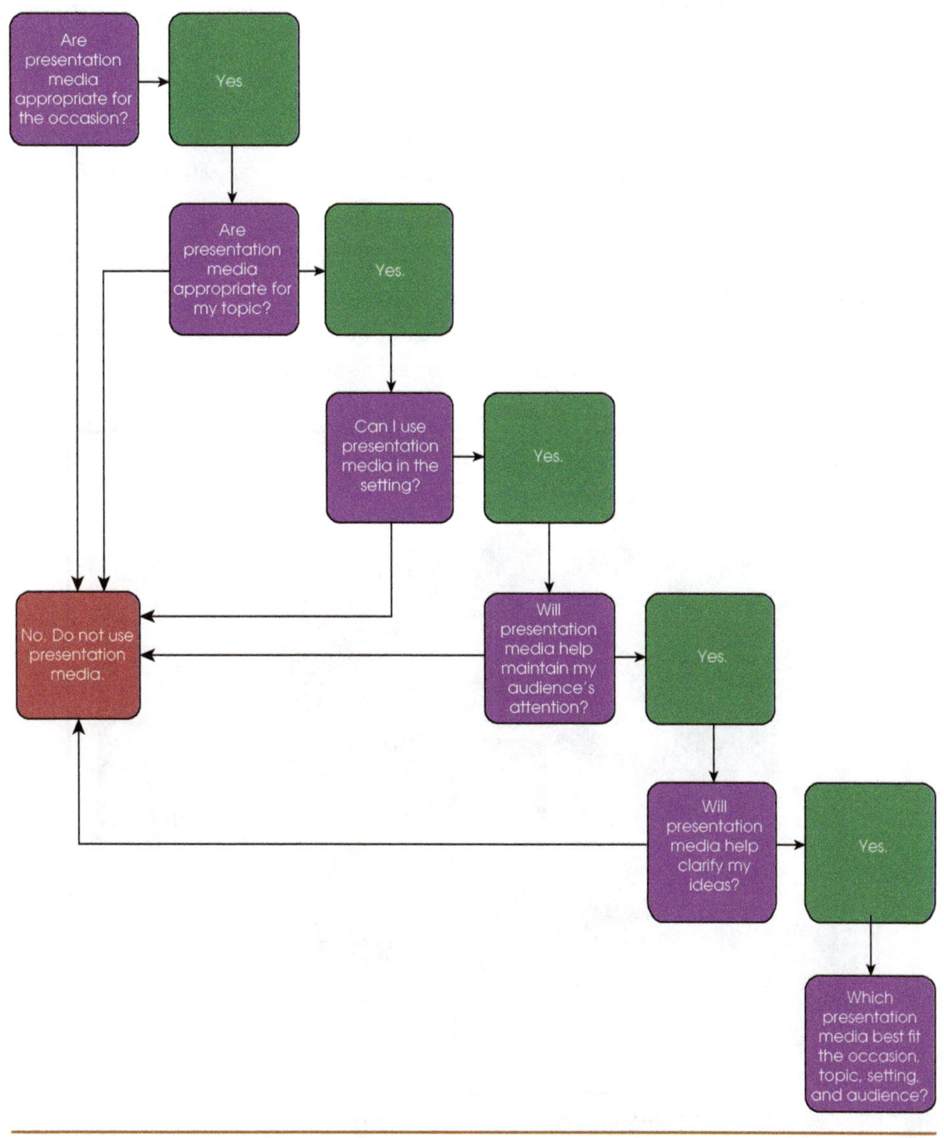

FIGURE 14.9 Ask the questions in this flow chart to determine when to use presentation media in your speech.

If you answer yes to the first five questions, then identify which presentation media best fit the occasion, setting, topic, and audience. You have many presentation media choices, such as whiteboards, smartboards, flipcharts, handouts, posters, objects, and, of course, digital slides with embedded images, text, audio, and video. Whichever presentation media you decide on, apply the following guidelines:

- *Use presentation media to support your content, not take the place of it*. Although well-chosen images can quickly demonstrate an idea or point, your words are the primary vehicle for conveying your message.

- *Keep presentation media simple.* Audience members should be able to understand the idea you are trying to get a across in a few seconds. For instance, if you want to include a video clip, show just enough to make your point. Similarly, too many images on a digital slide will lead your audience to wonder how to interpret them, as in Figure 14.10.

- *Minimize text.* Capture the essence of what you want your audience to know in meaningful words, phrases, and short quotes. Digital slides, handouts, and posters with large blocks of texts or long bulleted lists overwhelm your audience with too much information, as in Figure 14.11.

- *Create for clarity.* Clear and uncluttered presentation media are easy for your audience to understand. For audio files, check that the sound quality is crisp and the volume is loud enough for all audience members to hear. For visual materials, use large, screen-friendly fonts, sharp contrast, and vivid images, as shown in Figure 14.12.

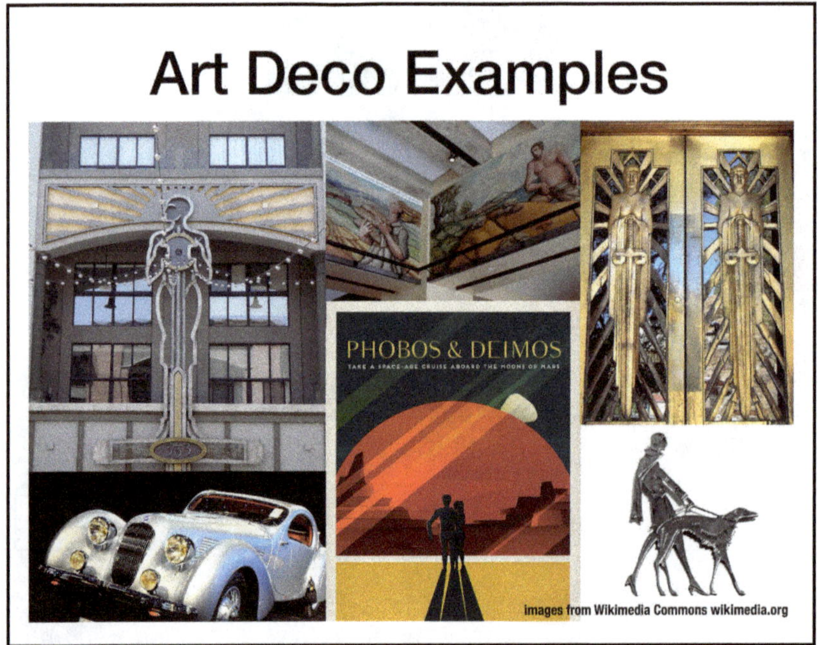

FIGURE 14.10 Too many images on one digital slide, poster, or handout overwhelm the audience.

The Roots of Art Deco

Art Deco is a visual art reflected in posters, architecture, paintings, and sculptures. Everyday things such as radios, automobiles, teapots, and clothing also incorporated Art Deco themes. One of the most famous examples of Art Deco is the Chrysler Building in Chicago.

Many artistic theories and styles influenced the look that became Art Deco.

- **Constructivism** started in Russia and was a geometric and angular art and architectural style.
- **Orphism,** which traces its history to Cubism, started in France in the early 1900s with the name from the Greek myth of Orpheus, who was a prophet and poet.
- **Futurism** started in Italy and focused on movement and technology across the arts and had a distinctive modern feel.
- **Functionalism** arose after World War II and when applied to architecture was concerned with all aspects of a building having a purpose.
- **Modernism** was experimental art that in part was in response to the Industrial Revolution and technological advances of that time.

FIGURE 14.11 Your audience will experience information overload with too much text.

FIGURE 14.12 This example has just the right amount of information for the audience.

- *Apply universal design principles.* **Universal design** means creating materials and environments that people can access regardless of ability.[22] Design your presentation media with all your audience in mind. If you play an audio file, have a transcript for Deaf audience members and those with hearing loss. If you show a short video clip, provide a description to set the scene for audience members who are blind or vision-impaired. If you have handouts, make a digital version available before or after your presentation for people who use a screen reader. Always consider the accessibility of your presentation media for everyone in your audience.

- *Be prepared to give your speech without your presentation media.* Even with the most thorough preparation, the unexpected can happen. The LCD projector light might be burned out. The promised flipchart might be absent. The speakers for the audio you intended to play might be on the fritz. Whatever the problem, be ready to rely on yourself and your words alone to convey your message.

Effective presentation media can gain your audience's attention, clarify a point, reinforce an idea, demonstrate a complex process, and enhance the audience's memory of your speech. In contrast, poor presentation media confuse the audience, distract from you and your main points, and make your speech memorable in negative ways. Choose your presentation media with care and develop them to support and not replace what you have to say.

Citing Your Sources

When you deliver your speech, your audience does not have access to your reference list. Therefore, you must tell your audience about the sources you used to develop your speech. Reading the list of all your sources at the end of your speech would be cumbersome. Instead, refer to your sources throughout your speech. **Oral citations** are acknowledgments of the sources for supporting materials. That is, you give a shortened reference to the source, but you do not state the full citation. When you acknowledge your sources, you demonstrate sound ethical judgment. In addition, you signal to your audience that you have researched your speech, which enhances your credibility. The following offer examples of orally citing your sources in your speech:

- "The Pew Center Research recently reported that a majority of Americans are concerned about how their personal data are being gathered, tracked, and used by government institutions and private corporations."

- "Video games can increase your social interactions and enhance your well-being. That's what researchers in Ireland found and reported in the journal *Perspectives on Psychological Science.*"

- "In a blog post on Columbia University's Earth Institute website, climate expert Renee Cho offers 35 easy ways to lower your carbon footprint. The first one is interrogating the food you buy and eat."

- "Last week I interviewed computer science professor Sue Jones about blockchain, bitcoin, and cryptocurrencies. What she had to tell me was fascinating, and I'm here to share it with you."

In addition to citing the sources for your speech content, you also must cite your sources for any video or audio materials you use. For example, if you play a short audio file from a podcast, briefly introduce the podcast title and speaker's name. For images on a digital slide, handout, or poster, include the source on the visual aid itself, as shown in Figure 14.13. Consider the ethical issues associated with citing sources in a public speaking class in Box 14.3.

FIGURE 14.13 Include the sources for your audio and visual materials as well as the text sources in your speech.

BOX 14.3: ETHICS 2.0

Source Citing vs. Plagiarizing

Professor Suzanne LaGrande reported on the blogging site Medium that one of her public speaking students plagiarized a TED Talk.[23] Shortly into the student's speech, she recognized it as a TED Talk she had shown many times in other classes. Later, when she met with the student, he made three claims:

- He did not plagiarize the entire speech.

- He cited his source.

- The instructor was treating him unfairly.

After the discussion, rather than failing the student for the speech or the class, she allowed him to give another speech, this time addressing an ethical issue. She first suggested plagiarism as the topic, but he declined, so they agreed on copyright infringement in music. He gave the speech, although he did not really address the ethical issues involved. From this experience, LaGrande suggests these reasons why people plagiarize:

- People plagiarize because they can and everyone else is doing it.

- People are confused about how to properly attribute their sources.

- No one can honestly say that their ideas are entirely their own.

- As with lying, people are overtly against plagiarism, but in practice endorse it.

- As long as someone is willing to pay for it, anything can be bought.

- Plagiarism in the "real world" is widely accepted.

What do you think? Did the student plagiarize the first speech? Is plagiarism as common as LaGrande argues? Consider these and other ethical issues in the questions that follow:

1. Was the student's first speech a case of plagiarism? LaGrande reported that 90% of the speech was from the TED Talk. Did citing his source mean he did not plagiarize the speech? What would you say to the student?

2. What do you think of LaGrande allowing the student to give a replacement speech?

3. Review and discuss each of the reasons LaGrande offers for why people plagiarize. What do you think of her list? What other reasons can you add?

4. What would you say to a friend or classmate whom you know plagiarized a speech or other assignment?

5. What have you learned about plagiarism that you can apply to your own communicative experiences?

Preparing Your Presentation Outline

You complete three outlines for every speech: working, complete sentence, and presentation, summarized in Figure 14.14. The last outline, the **presentation outline**, is a brief outline with key words and phrases that you will use when you give your speech. Developed from your complete-sentence outline (Chapter 13), the presentation outline acts as a memory device to remind you of what you plan to say and in what order. In addition, by having just key words on your presentation outline, you can give a more engaging speech and interact more directly with your audience.

The first step in developing your presentation outline is identifying **key words**—terms and phrases that represent the essence of an idea—that correspond with the points you plan to make. Use your complete-sentence outline as your reference point, noting the main points of your speech as well as essential supporting materials. Your presentation outline is a work in progress. You will revise it as you practice your speech and the speech day becomes closer. In the end, your presentation outline reflects the most basic structure of your speech and serves as a helpful tool in giving a conversational and extemporaneous talk, as shown in the example in Figure 14.15.

presentation outline
A brief outline with key words and phrases used when giving a speech

key words Terms and phrases that represent the essence of an idea

Type	Brief Description	Helps you ...	In chapter ...
Working	Draft outline with possible points and subpoints	Organize initial ideas and identify information sources.	11: Developing your speech topic and purpose
Complete-sentence	Full-sentence outline with all parts of speech and bibliography	Fully develop each point and demonstrate how ideas link together.	13: Supporting, organizing and outlining your ideas
Presentation	Speaking outline that distills speech to essential points	Identify key words for practicing and presenting your speech.	14: Delivering your speech

FIGURE 14.14 You develop a working, complete-sentence, and presentation outline for each speech.

Title: The First Amendment of the U.S. Constitution

1. Introduction

    a. [show digital slide of First Amendment text]

    b. adopted December 15, 1791

    c. 45 words, 5 freedoms

    d. freedom of religion, of speech, of the press, of assembly, to petition government

2. Main Points

    a. No state establishment of religion and free exercise of religion

        i. what it means—no national religion; free to worship, practice, observe religion

        ii. application—multiple places to worship

    b. Freedom of speech

        i. what it means—no censorship or legal sanctions

        ii. application—our campus free speech policy

    c. Freedom of the press

        i. what it means—disseminate ideas via media without censorship

        ii. application—our campus newspaper

    d. Freedom of assembly

        i. what it means—meet publicly, share ideas

        ii. application—last year's peaceful campus protest

    e. Petition government for redress of grievances

        i. what it means—make complaint without reprisal

        ii. application—We the People website

3. Conclusion

    a. [show digital slide of First Amendment text]

    b. 200+ years, 45 words, 5 freedoms

    c. history in the now

FIGURE 14.15 This presentation outline includes just the key words and prompts the speaker needs to give an engaging and conversational extemporaneous speech.

Practicing Your Speech

deliberative practice
A process in which you identify goals, get feedback on your performance, and always explore ways to improve

You know the old saying "Practice makes perfect." However, if you keep making the same mistakes when you practice something, such as singing a song in the wrong key, no amount of practice will make your performance any better. Instead, engage in **deliberative practice** where you identify your goals, get feedback on your performance, and always explore ways to improve.[24] As you learned earlier in this chapter, practicing your speech will help you develop strategies to channel your nervousness and increase your confidence. Follow the guidelines below to apply deliberative practice in rehearsing your speech:

1. List your goals for presenting your speech, keeping in mind your specific purpose and your audience. Recall that there is no perfect speech, but you can give an excellent speech.

2. Using your presentation outline, talk through your speech. Have your complete-sentence outline nearby in case you need to refer to it. Do this several times until you are comfortable with your presentation outline.

3. Using your presentation outline, time your speech as you talk through it to be sure you are within the time limit. If your speech is too long or too short, make any necessary revisions.

4. Transfer your revised presentation outline to notecards, small tablet, or phone. Go through your speech again until you are comfortable with your notes.

5. Using your presentation outline, present your speech in front of friends or family who can give you constructive feedback. Identify which parts of the speech are working well and which are not.

6. Revise your presentation outline, making adjustments based on the feedback you received as well as your own thoughts on how well you are achieving your specific purpose.

7. Rehearse and time your speech as if your audience were there, including any presentation media you plan to use. Note any changes you want to make and develop a plan for making those changes (Figure 14.16).

8. Using your phone or other video camera, record yourself presenting your speech as you plan to present it on speech day. View the recording, noting the areas you want to improve. This is a key step. Research on college students in a public speaking class found that students who video recorded and reflected on their speeches both improved their presentation skills and reduced their speech anxiety.[25]

9. Develop a plan for improving your delivery and record your speech again, implementing your improvements. Review the recording with friends,

FIGURE 14.16 Part of deliberative practice is talking through your speech, timing it, and making adjustments.

family, coworkers, classmates, or others who can give you constructive feedback. Note their suggestions and incorporate them into your plan for improvement. Ask yourself, "How can I better achieve my specific purpose for this speech?"

10. Continue rehearsing your speech until you feel at ease presenting it. Recall that you will experience some nervousness when you give your speech, but knowing your speech well will reduce that anxiety.

With deliberative practice, you take a reflective approach to practicing your speech. You focus on your goals, gather feedback from others, and identify ways to improve. This process of practicing will help you give a more confident, conversational, and engaging speech.

Presenting Your Speech

It is speech day! You have arrived early with all your materials. You have rehearsed using deliberative practice. Now is your opportunity to share with your audience the information you have gathered and organized. Your primary vehicles for giving your excellent speech are your voice and physical presence. This section offers guidelines for giving an extemporaneous speech, summarized in Figure 14.17, that you may adapt to your own abilities.

Adapting to Your Abilities

The U.S. Centers for Disease Control and Prevention reports that 61 million American adults, or 1 in every 4, live with some sort of disability. Of those, about 14% experience restricted mobility, almost 11% have restricted cognitive and memory functions, 6% are deaf or hard of hearing, and nearly 5% are blind or have serious vision loss.[26] Moreover, an estimated 48 million Americans have some degree of hearing loss, including 20% of teenagers.[27] Nearly 27 million adults in the United States report having trouble seeing even with corrective lenses.[28] And for those who identify as nondisabled, several scholars note that the ableist identity is fragile and temporary.[29] That is, injury, illness, and age transform the temporarily abled into people with disabilities.

While the notion of disability is fraught with identity, power, and linguistic issues, ignoring the constraints speakers face forces them into an unworkable

Adapt Your Delivery to Your Own Abilities

Use your voice.
- Project your voice so your audience easily hears you.
- Clearly articulate and correctly pronounce your words.
- Vary your speaking rate, pitch, and volume.
- Avoid verbal fillers and lengthy pauses.

Establish your physical presence.
- Dress professionally.
- Face your audience and make eye contact.
- Use purposeful movement and gestures.
- Close the distance between you and your audience.

Respond to your audience.
- Use the bucket method to prepare for questions.
- Answer questions completely and honestly.
- Interact respectfully with difficult audience members.

FIGURE 14.17 Follow these guidelines for delivering your speech.

delivery model. Thus, this discussion of delivery offers guidelines for the more predominant practices in presenting a speech. However, as the person who best knows your abilities, adapt these practices to you (Figure 14.18). For instance, if you use a wheelchair for mobility, determine in advance how you might want to move—or not move—when you are in front of your audience. If you have cognitive or information-processing difficulties, try out one of the many smartphone and tablet apps for recording and displaying notes. As one strategy, you might find that setting up your notes like a mind map or flow chart makes it easier for you to recall what you planned to say in your speech. If you have hearing loss and are comfortable speaking, then do so. However, if you are more comfortable communicating in American Sign Language (ASL), then give your speech in ASL and rely on an interpreter for your audience. Finally, if you have vision loss, consider your options for notes and presentation media. For example, you may want to have an assistant, such as a classmate, cue up your digital slides. Whatever accommodations you need, discuss them in advance with your instructor or, in the case of other public speaking situations, the person coordinating the event.

FIGURE 14.18 Adapt the best practices for delivering your speech to your own abilities.

Using Your Voice

In Chapter 6 you learned about paralanguage, nonverbal communication that involves the acoustics of speech—or how you say what you say. How you use your voice influences your audience's ability to understand you and your credibility.[30] Your voice must reach your entire audience, your words must be clear, and your vocal quality must vary in rate, pitch, and volume. In addition, you want to avoid lengthy pauses and verbal fillers.

Although some people naturally have a loud voice, most do not. **Vocal volume** is the loudness or softness of your voice. When you give your speech, you want the audience to hear you easily. This means you need to speak more loudly than in everyday conversations. To project your voice, take deep breaths from your diaphragm, maintain an upright posture, and relax your shoulders.[31] With this method, your breath will help carry your voice. If a microphone is available, do not hesitate to use it.

vocal volume The loudness or softness of a speaker's voice

Vocal clarity refers to how you articulate and pronounce your words. When you articulate your words, you enunciate each syllable so the words are clear and you avoid running words together. For example, you say, "Today, I'm going to explain phishing techniques that internet scammers use" rather than, "Today, I'm gonna 'splain phishin' techniques that innernet scammersuse." You also want to pronounce words correctly. Use an online dictionary if you are unsure of a word's correct pronunciation.

vocal clarity How clearly speakers articulate and how correctly they pronounce their words

monotone Little variety in a speaker's rate, pitch, and volume

vocal quality The rate, pitch, and volume variety of a speaker's voice

In any speaking situation, avoid speaking in **monotone**, where there is little variety in vocal rate, pitch, and volume. Instead, use your voice to emphasize your points, show interest in your topic, and highlight especially interesting ideas. **Vocal quality** includes rate, or how fast you speak, pitch, or how high and low your voice goes, and volume variety, or how much your voice varies in loudness and softness. Your rate, pitch, and volume should work together to convey your message to your audience and help you achieve your specific purpose. Your voice reveals how you feel about something and can bring up similar emotions in your audience. For example, you can speak a little more quickly to suggest excitement and a little more slowly to suggest calmness. Try out varying your vocal rate, pitch, and volume as part of using deliberative practice when rehearsing your speech. Box 14.4 offers tips on getting your voice ready to deliver your speech and relaxing your voice afterward.

BOX 14.4: CONNECT & REFLECT

Warm Up and Warm Down Your Voice

The best singers, actors, and speakers warm up their voices before their performances and warm down their voices afterwards. In *The Voice Exercise Book*, Jeannette Nelson gives several strategies for both.[32] Try out some of her suggestions. Start with warming up your voice:

- Practice deep breathing by taking in a deep breath and feeling it spread throughout your body. Let the breath out slowly, saying the letter "s." Breath in and out this way six times, then the seventh time, let your breath out as slowly as you can.

- Expand your resonance by humming, sending the sound from the top of your head to the bottom of your feet and especially into your chest and back.

- Open your voice by yawning or stifling a yawn and saying "ha" from low notes to high ones, like you are getting ready to sing a song.

- Articulate your words by trying some tongue twisters, such as "Green glass globes glow greenly."

Practice part of your speech or just talk aloud for a few minutes. Now try warming down your voice:

- Release the tension in your body by shrugging your shoulders and then letting them drop several times.

FIGURE 14.19 One strategy for warming down your voice is yawning or stifling a yawn.

- Relieve facial tension by yawning or stifling a yawn (Figure 14.19).

- Release your voice by softly and gently humming through the range or pitch of your voice.

- Hydrate by drinking room temperature water.

How do you feel after completing the exercises? What improvements did you notice in your voice after warming up? How did warming down help you relax? How might you use these exercises when you practice and on speech day?

Although people generally do not notice long pauses and verbal fillers such as "ah" and "um" in everyday conversations, disfluencies stand out when speaking in public. **Vocal fluency** refers to speaking with natural pauses and avoiding long hesitations and verbal fillers. Listeners typically evaluate speakers who use frequent pauses, whether silent or vocalized, more negatively, viewing them as less credible, knowledgeable, and prepared.[33] However, research shows that speakers who spend more time practicing encounter fewer disfluencies in their speeches.[34] As you rehearse your speech using deliberative practice, one of your goals is to reduce long pauses, ahs, ums, and other disfluencies until your speaking voice is assured and fluent.

vocal fluency Speaking with natural pauses and avoiding long hesitations and verbal fillers

Establishing Your Physical Presence

Along with your voice, you use your body to present your speech. Recall the types of nonverbal communication discussed in Chapter 6. Many of these come into play when you deliver your speech: artifacts (dress), face and eyes (facial expressions and eye contact), kinesics (gestures and movement), and proxemics (distance from your audience). Although *your* focus is on your audience, your audience's focus is on you and your message. You want your nonverbal messages to enhance your speech content, not distract from it.

Effectively establishing your physical presence begins with wearing the appropriate attire for the occasion, your audience, and your topic. For classroom speeches, dress more formally than you ordinarily would for school. For speeches outside the classroom, it is generally best to choose standard business attire. When you dress professionally, you signal to your audience you are prepared for the event, increasing your credibility even before you take the stage.

Before you start speaking, take a moment to look at your audience and smile. This eye contact and facial expression help the audience feel a connection with you and build your confidence. Throughout your speech, use your face to express emotions that parallel the verbal content of your speech, as with a frown for sadness and a neutral expression for a particularly serious point. Keep your eyes trained on your audience, glancing only occasionally at your notes. Your ability to truly look at your audience reflects the deliberative practice you completed for your speech. If you are projecting images on a screen, as with an LCD projector, avoid the temptation to look at the screen.

Whenever you are talking, your body should be oriented toward the audience. Making eye contact with your listeners helps keep your body facing them. If you use a whiteboard or chalkboard, either write on it before your speech to avoid turning your back to the audience or stop talking while you are writing. Make all your movement purposeful. For example, moving around the front of the room can help all your listeners feel included, but pacing from one side to the other will make them anxious. If you keep your shoulders relaxed and your arms comfortably at your side, you can gesture as you would in a conversation with friends.

FIGURE 14.20 Leaving the podium behind brings you closer to your audience.

Gestures should appear natural and spontaneous, not forced or contrived. Your gestures should convey your confidence and expertise.

In a classroom setting, you are typically just a few feet from the front row of your audience. This physical proximity can promote feelings of closeness and identification between student speakers and listeners. Even in more formal settings where you are speaking from a stage, you can close the gap between yourself and the audience by stepping out from behind the podium. This allows you to move more freely as you speak and removes a physical barrier between you the audience, as shown in Figure 14.20.

Just as there is no perfect speech, there is no perfect way to present your speech. Your delivery will depend on the occasion, the setting, the topic, the audience, and, most of all, you. As you practice and give speeches, you will become more accustomed to the public speaking context and more comfortable engaging with your audience.

Responding to Your Audience

Your audience is your partner in your presentation, working with you to make this event—your speech—happen. When you make every step of creating your message audience-centered, you are prepared to respond to your audience as you deliver your speech and during any question-and-answer session.

Answering Questions

Many speaking situations allow for audience questions, usually after the speaker has completed the presentation. Preparing for audience questions is part of planning for your speech. In fact, communication researchers have found that effectively answering questions from the audience is at least as important as delivery in the audience's evaluation of the speaker.[35] Gallo suggests using a "bucket method" to prepare for audience questions.[36] First, list the questions you think the audience might ask. Second, group them into categories, or buckets. Third, develop an answer for each category or bucket, but not each question. This way, you focus your preparation on the general areas audience members might want to know more about, rather than specific questions they might ask.

When an audience member asks a question, make eye contact with them and listen to the entire question. If you cannot hear or do not understand any part of the question, ask the audience member to repeat it. Restate the question if not all listeners heard it. Then, if you know the answer, clearly and concisely give it, addressing the entire audience. If you do not know the answer, avoid trying to bluff your way through a response. Instead, indicate you do not have that information and offer to research the answer. When speakers are prepared, they find that the question-and-answer session after a speech is a low-stress opportunity to engage in a stimulating conversation with the audience.

Handling Difficult Audience Members

Hostile or rude audience members are rare in a college classroom. However, listeners may react negatively or with strong emotions to some topics, such as gun control, climate change, immigration, and capital punishment. When this happens, above all else, remain calm. Your goal is to de-escalate rather than escalate the situation. Listen to what the person has to say. Respond respectfully and empathically, acknowledging the other person's perspective. Keep the focus on a discussion of ideas, not people or personalities. If your de-escalation attempts are not effective, suggest talking with the person after your speech, and then move on to the next question.

CHAPTER REVIEW

■ **Explain the strategies for delivering a speech with confidence.**
Building your confidence before speech day involves choosing a topic you care about, doing your research, understanding your audience, curating your presentation media, and practicing your speech. On speech day, get prepared, channel your nervous energy, begin with confidence, and take an audience-centered approach. After the speech, compliment yourself, identify challenges, and focus on the future.

■ **Identify and describe the four speech delivery methods.**
The speaker reads a document written in advance with the manuscript delivery method. With the memorized method, the speaker recites a text from memory. Impromptu speeches are given with little or no preparation. Extemporaneous speeches are planned, researched, and rehearsed.

■ **Identify the three presentation formats.**
Speeches may be delivered in-person, in a mediated format, or a blend between the two.

■ **Discuss how to integrate appropriate presentation media in your speech.**
The guidelines for using presentation media are (a) use presentation media to support your content not take the place of it, (b) keep presentation media simple, (c) minimize text, (d) create for clarity, (e) apply universal design principles, and (f) be prepared to give your speech without your presentation media.

■ **Explain how to accurately cite your sources when you deliver a speech.**
Use oral citations to acknowledge the sources for text, video, and audio content in your speech. Use visual citations to acknowledge sources for images.

■ **Prepare a presentation outline for a speech presented extemporaneously.**
Developed from the complete-sentence outline, the presentation outline includes the basic elements of a speech that remind speakers of what they plan to say.

■ **Demonstrate an effective process for practicing a speech.**
Deliberative practice involves identifying goals, getting feedback on performance, and exploring ways to improve. When applied to practicing a speech, speakers progress through several iterations of rehearsing, reviewing, and revising their speeches.

■ **Effectively present an extemporaneous speech.**
Effective extemporaneous speakers adapt their delivery to their abilities, use their voice, establish their physical presence, and respond to the audience.

DISCUSS AND APPLY

For the speech below, you will act as a consultant, identifying how the speaker can develop a presentation outline, what presentation media the speaker might include, and key aspects of delivery that will best convey the speaker's message to the audience. Assume the speaker is a college student giving an in-person or real-time (synchronous) online classroom speech.

SPEECH OUTLINE FOR ANALYSIS

Topic: Fun

Title: "There's More Fun in Fun Than You Might Think"

General purpose: To inform

Specific purpose: To inform my audience about the four types of fun

Thesis: There are four types of fun: hard fun, easy fun, serious fun, and people fun.

Introduction

    I.   What does "fun" mean to you?

    II. Game designer Nicole Lazzaro has identified four types of fun: hard fun, easy fun, serious fun, and people fun.

    III. Another game designer, Jane McGonigal, argues that when life is more fun, people are more engaged, committed, and motivated.

    IV. I played a game on my phone as I rode the light rail into school today, which was fun and lowered my stress about this speech.

    V. Having fun feels good.

Transition: Today, we're going on an adventure to find out about the four different types of fun.

Main points

I.   The first type is hard fun, which involves achieving a goal and triumphing over adversity.

    A.   With hard fun, you overcome some sort of challenge or you solve a puzzle.

    B.   Hard fun relies on strategy.

    C.   People play to win.

        1.   For example, research has found that people who play e-sports, such as NBA2K, the basketball video game, enjoy the games for the achievements, like winning championships.

        2.   Outside of games, people experience hard fun when they achieve an important goal, like running their first marathon or winning a debate tournament.

Transition: There's more to fun than winning.

II.   The second type is easy fun, which is immersive and adventurous, grounded in curiosity and surprise.

    A.   With easy fun, there's some sort of mystery you want to find out more about.

    B.   Easy fun relies on inquisitiveness and wonder.

    C.   People play to explore.

        1.   For example, research on Pokémon Go has found that many people play the game because they enjoy exploring public spaces.

        2.   You've probably experienced easy fun when you traveled to a new place, as I did in my study abroad class in Belize.

Transition: While it's fun to go on adventures, it's also fun to learn about things.

III.   The third type is serious fun, which is about exercising the mind and providing an escape from the drudgery of everyday life.

    A.   With serious fun, a boring task is made more interesting or intriguing.

    B.   Serious fun relies on excitement.

    C.   People play to get better at something.

        1.   Educational games, like those designed to teach about math or grammar concepts, make learning more exciting.

        2.   You can apply serious fun to accomplishing everyday task, like Jane McGonigal did when she created *Chore Wars* to make completing household chores less tedious.

    IV.   *Transition*: The other types of fun often involve the last type.

        A.   Finally, people fun arises from the joy of interacting and bonding with others.

        B.   With people fun, it's all about being with friends, whether the activity is cooperative or competitive.

        C.   People fun relies on our social needs and desire to connect to others.

        D.   Media studies professor Ian Bogost argues that generosity and selflessness in spirit lead to fun.

            1.   People play to be around other people.

            2.   I look forward to the once-a-month game night my friends and I have because I like hanging out with them.

            3.   If you think about it, much of your social life is about people fun, getting together with friends and family just because you enjoy being with them.

                a.   People fun was especially important in helping people cope with staying at home during the COVID-19 pandemic.

                b.   My aunt's bridge group moved its weekly card game online, using an app that worked on their tablets.

                c.   My friends and I went from playing Dungeons and Dragons the old-fashioned way—in-person—to playing online.

Conclusion

    I.   The four different kinds of fun—hard fun, easy fun, serious fun, and people fun—aren't mutually exclusive, so there's overlap among them.

    II.   The categories remind us that fun isn't just for kids and is a central part of being human.

    III.   Play theorist, author, and game developer Bernie Dekoven said in an interview, "Playing playfully is all about fun. That's how you know you're doing it right."

    IV.   Hard fun, easy fun, serious fun, and people fun—who knew there was so much fun in the world?

References

Bogost, I. (2016). *Play anything: The pleasure of limits, the uses of boredom, and the secret of games.* Basic Books.

Crawford, G., Muriel, D., & Conway, S. (2019). A feel for the game: Exploring gaming "experience" through the case of sports-themed video games. *Convergence: The International Journal of Research into New Media Technologies, 25*(5–6), 937–952.

Deep fun and the theater of games: An interview with Bernie Dekoven. (2015). *American Journal of Play, 7*(2), 137–154.

Koster, R. (2013). *A theory of fun for game design*, 2nd ed. O'Reilly Media.

Lazzaro, N. (2004). Why we play games: Four keys to more emotion without story. White paper for XEODesign available at xeodesign.com/the-4-keys-to-fun

Lazzaro, N. (2009). Understanding emotion. In C. Bateman (Ed.), *Beyond game design: Nine steps toward creating better videogames* (pp. 3–48). Cengage.

McGonigal, J. (2011). *Reality is broken: Why games make us better and how they can change the world.* Penguin.

Potts, R., & Yee, L. (2019). Pokémon Go-ing or staying: Exploring the effect of age and gender on augmented reality game player experiences in public spaces. *Journal of Urban Design, 24*(6), 878–895.

QUESTIONS FOR DISCUSSION

1. Which presentation format would you recommend for this speech—manuscript, memorized, impromptu, or extemporaneous? Why is your choice the best one?

2. Review Figure 14.9. Are presentation media appropriate for this speech?

3. What presentation media would you suggest? Give specific examples of presentation media you would include.

4. Develop a presentation outline for the speech.

5. Provide specific suggestions for how the speaker might cite the sources for the speech.

6. How would you recommend the speaker practice for the speech?

7. To what aspects of delivery should the speaker pay particular attention?

8. What questions might the speaker anticipate getting from the audience? How should the speaker prepare for those questions?

9. What changes would you suggest if the speaker is livestreaming the speech and also has an in-person audience?

10. What have you learned by completing this activity that you will apply to your own speeches?

CREDITS

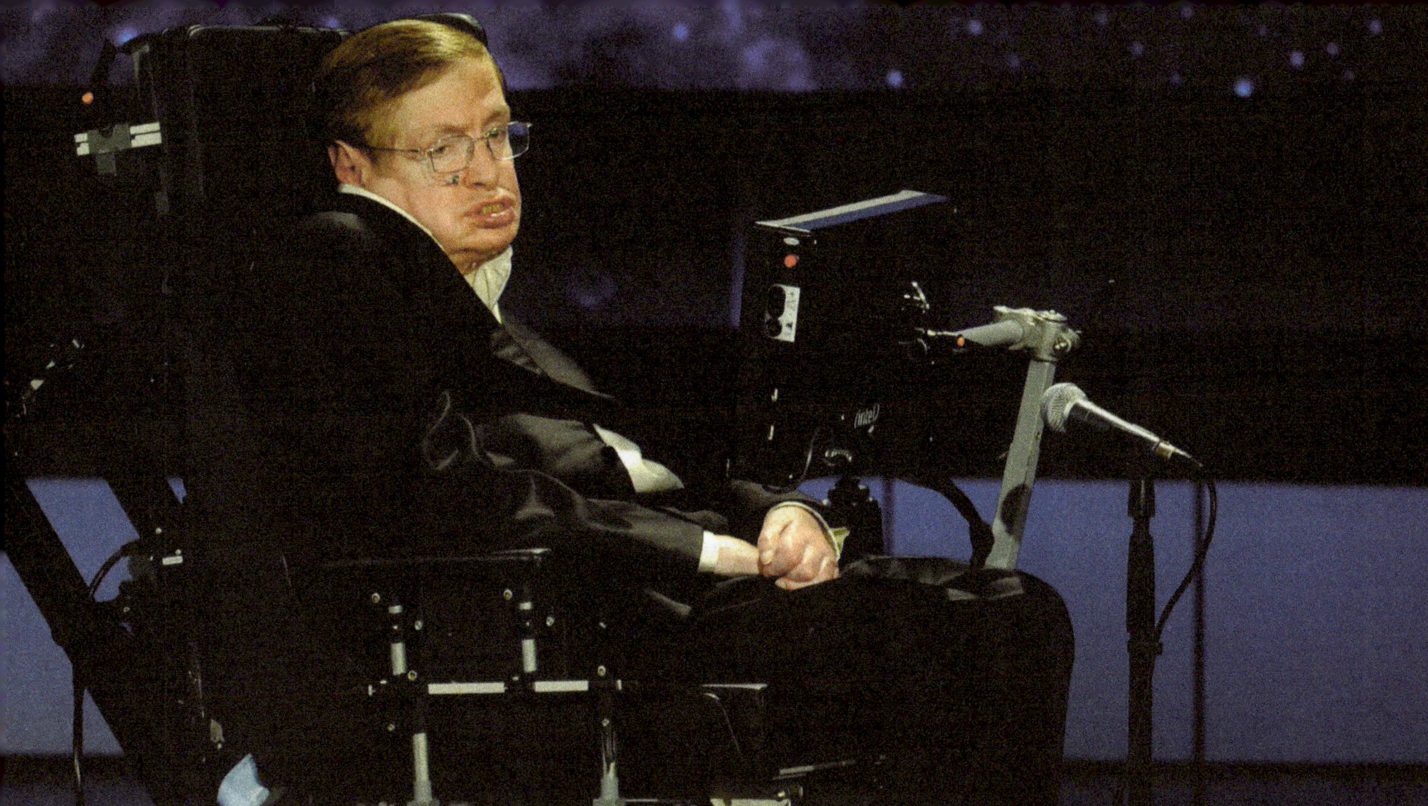

Speaking to Inform

I n 1948 the British Broadcasting Corporation (BBC) started a yearly radio lecture in which a well-known expert is invited to give a talk about a current topic of interest. When physicist Stephen Hawking presented his two-part talk on black holes, he discussed their nature in the universe. In the introduction he said:

> It is said that fact is sometimes stranger than fiction, and nowhere is that more true than in the case of black holes. Black holes are stranger than anything dreamed up by science fiction writers, but they are firmly matters of science fact. The scientific community was slow to realize that massive stars could collapse in on them-selves, under their own gravity, and how the object left behind would behave.[1]

Hawking went on to explain how scientists came to identify black holes, early controversies over the name itself, and doubts over their very existence. Then he encouraged listeners to imagine themselves encountering a black hole in their exploration of the universe:

A black hole has a boundary, called the event horizon. … Falling through the event horizon is a bit like going over Niagara Falls in a canoe. If you are above the falls, you can get away if you paddle fast enough, but once you are over the edge, you are lost. There's no way back. As you get nearer the falls, the current gets faster. This means it pulls harder on the front of the canoe than the back. There's a danger that the canoe will be pulled apart. It is the same with black holes. If you fall towards a black hole feet first, gravity will pull harder on your feet than your head, because they are nearer the black hole. The result is you will be stretched out longwise and squashed in sideways. If the black hole has a mass of a few times our sun you would be torn apart and made into spaghetti before you reached the horizon. However, if you fell into a much larger black hole, with a mass of a million times the sun, you would reach the horizon without difficulty. So, if you want to explore the inside of a black hole, make sure you choose a big one.[2]

informative speaking Speeches in which the speaker's goal is for audience members to learn more about a topic

With **informative speaking** the speaker wants the audience to learn more about a topic. Speakers may introduce audience members to an entirely new topic or provide in-depth information about a topic for which they have little knowledge. In Hawking's case, listeners may have heard of black holes, but likely did not know much about them. By comparing something the audience knew—going over a waterfall in a canoe—with something the audience did not know—how a black hole behaves—Hawking made a highly technical topic easier for his audience to understand. As he went on to discuss his own work in researching black holes, Hawking used clear definitions, explanations, and examples while keeping his listeners intrigued by the mystery of black holes.

This chapter discusses informative speaking in a pervasive communication environment, explains the characteristics of informative speaking, identifies and gives examples of the types of informative speaking, and presents guidelines to help you present an effective informative speech.

Speaking to Inform in a Pervasive Communication Environment

scenography How speakers perform their speeches as their presentations unfold in real time

Chapter 11 introduced you to the ways in which a pervasive communication environment has influenced ethics, information access and evaluation, and the diversity of the audience in public speaking. A pervasive communication environment has altered the context for informative speaking in three additional ways. First, a pervasive communication environment has altered how informative speakers create the speech's **scenography**, or how speakers perform their speeches as their presentations unfold in real time.[3] While all public speakers must manage the speech scene as they are speaking, the competing sources of information in a pervasive communication environment pose a special challenge for informative speakers. Why should audience members listen to the speaker? Why should they not just get out their phones and look up a topic? When speaking to inform, you go beyond setting the stage for listeners to learn and create a continuous scene in which they participate in the learning process. Speaking to inform in a pervasive communication environment requires you to do more than present information: You must invite your audience into this scene as key companions along this journey into your topic (Figure 15.1).

FIGURE 15.1 In a pervasive communication environment, speakers must invite the audience into the speech scene to capture and hold their attention.

In her lecture "How to Raise an Adult," given at the University of Washington, Julie Lythcott-Haims offered that invitation to her audience throughout her talk. At the start she informally polled her audience:

> I had no idea who would actually show up for a lecture by me at a place like this. So, raise your hand if you are on staff here. Yay. Okay. Raise your hand if you're a faculty member here. Yay. Okay. Raise your hand if you are a student. Here, awesome. Graduate student, undergraduate student. Are any of you alumni of this magnificent place? Right? Fabulous. What about community members? You're not in any other category. You're here. Excellent. … It is hard to figure out where to pitch a talk to everyone. Because you do range in age, I'm guessing from like 22 to 82. … So, I'm going to try to just land my thoughts here in a way that works some of the time for all of you. Okay? So, I stand before you with a set of accomplishments that [Yvette] read from my bio. But I think more important than knowing where I went to school where I worked is for you to know that I have been lost, and lonely and alone, and sad and frightened and terrified, and self-loathing and empty and unsure and insecure and arrogant, and mean, and scared and ashamed.[4]

In this example, observe how Lythcott-Haims encouraged her audience to participate, asking them to respond to her questions and recognizing their responses.

Then she invited them to learn more about her than what was included in the brief biography used to promote the lecture.

Second, unlike in pre-internet times, public speakers and audiences generally enjoy equal footing in information access even with the persistence of the digital divide. While not all speakers and audiences have equal access to digital resources, the ability to produce and consume information—the prosumer role discussed in Chapter 1—never has been greater. For informative speakers, this access has altered the speaking landscape from one of speakers as possessors of information to speakers as sharers of information.[5] That is, a speaker's authority on a topic comes from the effective presentation—sharing—of their research and knowledge, rather than serving as the sole owner of the information they are giving to the audience. Thus, in addition to collaborating with the audience to create the scene, Lythcott-Haims suggested ways in which they might identify with each other and participate together in this information-sharing event. Lythcott-Haims continued drawing in her audience with statements such as "I see some of you know what I'm talking about," "It's that inner voice in each of us, the one that speaks for our soul that aches to be heard," and "So now that I know you a little bit better, I am going to turn to some of my personal story is to give you a glimpse of how I learned the things that I'm trying to share with you this evening." At each point, Lythcott-Haims focused her audience's attention on the speaker-listener relationship and the ongoing sharing of information.

Finally, with the tremendous availability of information today, it might seem there is no need for informative speeches. Yet it is precisely the often-overwhelming amount of information that makes the informative speaker role especially important in a pervasive communication environment. As you learned in Chapter 12, not all information is created equal. The informative speaker must sort through pseudo or fake news, misinformation, disinformation, and other forms of faulty information to present accurate, credible, and reliable information to the audience. Lythcott-Haims referenced her long experience as a university administrator and dean to demonstrate the accuracy and reliability of her information while still connecting with her listeners. For instance, she told her audience:

> The last ten of my 14 years at Stanford were as the Dean of Freshmen. I had 1,700 of them every year, 17,000 over ten. … So, my job as Dean of Freshman was to care about my freshman, was to give a damn about them. Undergraduates are, I like to say, unfurling. They are becoming, they are unfolding into their adult selves, which is a process that can be a little ugly before it's beautiful. Remember, remember back for some of you, the journey back to the self at 18 is a longer journey than it is for others in the room. But we've all crossed that threshold as far as I can tell. Right? Remember yourself, doing that stuff of becoming your adult self. Not all of it is Instagram worthy.[6]

A pervasive communication environment challenges informative speakers to reach information-saturated and distracted audiences in ways that are meaningful and authentic. The characteristics of informative speaking suggest strategies to address those challenges.

Characteristics of Informative Speaking

All effective informative speeches educate the audience, elucidate the topic for the audience, and engage the audience (summarized in Table 15.1). Informative speakers, like Stephen Hawking in his lecture, strive to deepen audience members' comprehension of a topic by relating the topic to them in a clear and coherent fashion. Unlike persuasive speakers (Chapter 16), informative speakers avoid stating their opinions or trying to influence their audience's values, beliefs, and behaviors. For instance, if you wanted to give an informative speech on #SayTheirNames, you might discuss the history of this campaign to name and tell the stories of those injured or killed by law enforcement. In contrast, for a persuasive speech, you might encourage your audience to use the hashtag in their own social media posts, agree with the campaign's tactics, or carry a sign related to the campaign at a local racial injustice protest (Figure 15.2).

FIGURE 15.2 An informative speech on #SayTheirNames might educate the audience on the campaign's history, whereas a persuasive speech might encourage the audience to join an upcoming racial injustice protest.

TABLE 15.1 Characteristics of Informative Speaking

CHARACTERISTIC	AUDIENCE RESPONSE	OUTCOME
Educate the audience	"I want to learn something new."	The audience connects and applies the topic to their own experience.
Elucidate the topic for the audience	"I understand what the speaker is saying."	The audience recognizes the speaker as an expert on the topic.
Engage the audience	"I want to know more about this interesting topic."	The audience stays focused on the speaker and topic.

Informative Speaking Educates the Audience

When you educate your audience about a topic, you are offering them a learning opportunity. As shown in Figure 15.3, human **learning** involves becoming aware of sensory stimuli, selecting which stimuli to consider further, making connections with what the person already knows, and then applying it in some way.[6] Similar to the perception process discussed in Chapter 3, the learning process for your audience starts with sensing something in the environment through taste, touch, smell, sound, or sight, or some combination of the five senses. As a public speaker, you may not be able to appeal directly to all five of the audience's senses, but you can use words and images to stimulate their imaginations. When Professor Gina Bloom gave a talk about using videogame

learning A process that involves becoming aware of sensory stimuli, selecting which stimuli to consider further, making connections with what is known already, and then applying it in some way

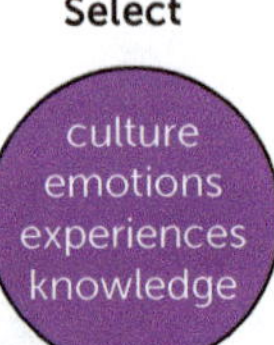

FIGURE 15.3 Overview of the learning process. Adapted from Greenleaf and Wells-Papanek (2005).

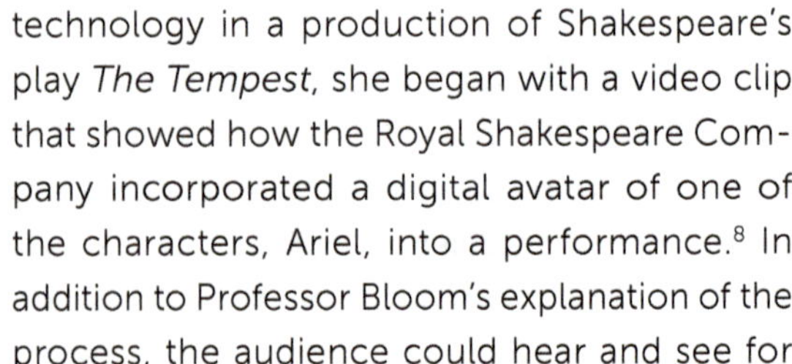

FIGURE 15.4 Nabiha Syed's vivid story of Ida B. Wells's pamphleteering in the 1800s enlightened her audience about the history of free speech and freedom of the press.

technology in a production of Shakespeare's play *The Tempest*, she began with a video clip that showed how the Royal Shakespeare Company incorporated a digital avatar of one of the characters, Ariel, into a performance.[8] In addition to Professor Bloom's explanation of the process, the audience could hear and see for themselves how it worked. Later, she gave a demonstration of the videogame *Play the Knave* so audience members could visualize more easily connections between a videogame and a Shakespearean play.

In the second step of the learning process, culture, emotions, past experiences, and knowledge serve as filters for sensory input. Because you want your audience to learn something new, you must consider what they already know about the topic, how they might feel about it, their cultural norms and expectations associated with the topic, and their experience related to the topic. Vice president and associate general counsel at BuzzFeed Nabiha Syed did just that when she gave a talk at Harvard Kennedy School's Shorenstein Center on Media, Politics, and Public Policy. She began with a brief story:

> I do think back to just ten years ago when I was eyeing media law as an area and a very esteemed, seasoned media lawyer looked at me and said, "Oh, Nabiha, it's such a shame you weren't alive in the '60s and '70s. That's when the real fight happened."[9]

Then she went on to tell a much lengthier story from the late 1880s about an early American pioneer in free speech and freedom of the press, Ida B. Wells (pictured in Figure 15.4), who

> self-published a pamphlet arguing that black people were lynched for competing economically with their white neighbors, for being joyful and loud in public, for not being as deferential as they should be. Many in Memphis were unsurprisingly unhappy with her pamphlet. A local newspaper carried this complaint. "The fact that a black scoundrel is able to live and utter such calumnies is a volume of evidence as to the wonderful patience of southern whites, but we've had enough of it." There was no commitment to free speech for Ida. None of these were empty words. Because when Ida went away for a business trip in Philadelphia, a mob burnt her office to the ground.[10]

In this way, Syed provided her audience with information they probably did not know beforehand and directed their attention to this important, yet overlooked, event in U.S. history. Syed's vivid description likely jarred her audience's emotions and brought to the forefront the reality of speaking out against the lynching of Black people in the late 19th century.

The third step in the learning process for the audience is connecting the new information to what they already know. This connection makes the topic more meaningful and memorable for listeners. For instance, in a speech on technology trends, Peter Warren Singer, an editor at *Popular Science* magazine, explained one trend this way:

> It's a very special kind of technology and it goes by lots of different buzzwords. So it used to be in the Rumsfeld-era Pentagon, "revolutionary technology." Now it's known as "disruptive technology." If you're outside the military, weirdly enough, you call it "a killer app." What are we talking about here?
>
> We're talking about technology that is not an evolutionary improvement—you know, the difference between iPhone 6 and iPhone 7—but a leap ahead. The difference between a regular phone and a smartphone. And what defines this is not just the idea that it gives you something that was a capability, that was science fiction a generation earlier, but how it provokes a whole new set of questions that were science fiction a generation earlier.[11]

Audience members likely were much more familiar with the term "killer app" than "revolutionary technology" or even "disruptive technology." Still, to frame the idea within his audience's experience, Singer defined the "leap ahead" that goes far beyond the latest iPhone to the science fiction of the past that is here today.

The final step in the learning process for the audience is demonstrating how this new information might be applied in some way. Effective informative speakers provide concrete examples that show the value of the information, suggest ways in which audience members might use the information in their own lives, or offer avenues for further exploration of the topic. All these strategies reinforce the speaker's points in the listeners' memories and complete the learning process. For example, later in his speech, Singer talked about 3D printing spare automotive parts as a disruptive technology, offering a real-world example:

> What happens as these spare parts are now available to be made by someone else, including the client? That bus there is Ollio. Ollio is a 3D-printed autonomous electric shuttle bus made by Local Motors, an Arizona-based automaker that crowd sources vehicle design. Every clause in that sentence is disruptive to the economy of cars: 3D printed, autonomous, electric, Arizona-based, crowd-source design. Again, this is not science fiction. Ollio is actually going to be operating in Washington, D.C., going back and forth at the Nationals baseball stadium.[12]

Singer's description of a bus with a whimsical name traveling around the nation's capital brought the idea of 3D printing to life. He showed the audience an application of this new technology, making it more memorable and less like science fiction.

Experience for yourself how informative speakers educate their audience and learn about where you live at the same time. In Box 15.1 on the following page, you will attend a community event that features an informative speaker and then consider a few questions afterward.

BOX 15.1: LIVE IT LOCAL

Getting Educated About Your Community

Nearly all communities have events featuring informative speeches, as with workshops, lectures, and talks, on a wide range of topics. Get educated about your community by attending a local event either in-person or online. Your city's or town's local government likely has an events page on its website. Also check your college's website for the campus's listing of events. In addition, community and campus libraries regularly host talks by authors and artists. Public meetings often include informative presentations on local issues. Learn about what's happening in your community.

After the presentation, consider the questions that follow on your own or use them for discussion with other attendees.

1. How did the speaker's message appeal to the audience's senses?

2. How did culture, emotions, past experiences, and your knowledge of the topic influence how you interpreted the speaker's message?

3. What did you think about as the speaker was talking? What connections did you make between what the speaker said and your own knowledge base?

4. What will you do with what you learned from the presentation? How will you apply this new information?

5. Overall, what was most memorable to you about the speech? Why does that stand out?

Informative Speaking Elucidates the Topic for the Audience

Elucidate, from the Latin *ēlūcidāre*, or "to enlighten," means to clarify or explain something.[13] When speakers elucidate topics for an audience, they provide clear and detailed information that paints a comprehensive picture, presenting both the right quantity and quality of information. Audience members respond to speakers who effectively elucidate a topic by thinking, "I understand what the speaker is saying."

Information refers to data and facts about your speech topic (Chapter 12). The pervasive communication environment in which you live also creates a pervasive *information* environment in which communicators are immersed in oceans of information. **Information overload** occurs when individuals have more information than they can interpret effectively. Experiencing information overload day after day has negative consequences, as with increased information stress and reduced feelings of happiness.[14] For example, at the start of the COVID-19 pandemic, the public was deluged with messages about the spread of the virus, how to avoid getting sick, ways to help others, instructions for sheltering in place, and a whole host of pandemic-related topics. When faced with too much information, and trying to sort out facts from fiction, individuals often turned to known sources that lacked credibility, such as social media, missing out on key information essential to their health.[15] Information overload can happen

information overload
Occurs when individual share more information than they can interpret effectively

to audience members when the speaker presents an overwhelming volume of information, irrelevant information, and highly complex information.[16] For example, when speakers show slide after slide filled with text, tell long and irrelevant stories, or use jargon and highly technical language, audiences experience mental exhaustion and then stop listening altogether.

In contrast, **information underload** occurs when individuals do not have enough information to consider and interpret the meaning of a message or topic.[17] While information overload overwhelms listeners, information underload underwhelms them, offering little in the way of learning. Audience members who experience information underload may become bored and tune out the speaker's message. In some respects, information underload is a more serious problem than information overload for informative speakers. If your speech lacks sufficient new information, you will seem unprepared and not an expert on your topic. You want to avoid both information overload and underload in your informative speeches (Figure 15.5).

In addition to having the right quantity of information, informative speakers must have high-quality information—reliable, valid, and current. **Reliable information** is verifiable and from a credible source. As you learned in Chapter 12, a credible source is one that is considered an authority on the topic. If you want to learn about local employment opportunities, for instance, then interviewing the director of your campus career information center would be a good place to start. **Valid information** is logical and reasonable. To assess the validity of your information, examine the underlying assumptions and the conclusions presented. **Current information** is up to date and provides the most-recent data on the topic. Even historical accounts are updated as new information is found. For instance, paleontologists once thought that birds were the first species on Earth to have feathers. However, a recent discovery of fossils in Russia led scientists to conclude that dinosaurs, which predated birds by millions of years, had feathers.[18] These dinosaurs did not fly, but the feathers likely allowed them to run faster.

Providing relevant, clear, and accurate information elucidates your topic for your audience. For example, in a lecture on translation and American Sign Language at Gallaudet University, Dr. Ruth Anna Spooner explained the differences between translating and interpreting, using this analogy:

> You go to a play, you have a stage, you have actors, you have lighting that's been carefully established for that play. You have different

information underload
Occurs when individuals lack sufficient information to consider and interpret a message's meaning

reliable information
Information that is verifiable and from a credible source

valid information Information that is logical and reasonable

current information
Information that is up to date and provides the most-recent data on the topic

FIGURE 15.5 Audiences remain attentive when you avoid information overload or information underload and provide them with the right amount of information.

props. The actors have been rehearsing for weeks on end, maybe several months before the production to make everything come together. Now you come in as an audience member and you see the perfected performance, right? That's translation. Interpreting is more like improv. You go to [an] improv show and you have an actor making it up as they go. Both are to be respected and both take quite a bit of skill, but they are in fact different and the main difference is time. With translation you have more time to delve into the text.[19]

FIGURE 15.6 In discussing American Sign Language, Dr. Ruth Anna Spooner explained that translating is like actors rehearsing in advance for a play and interpreting is on the fly like improv.

Dr. Spooner wanted to make clear from the start that the process of translating from or to American Sign Language is quite different from what interpreters do in real time (Figure 15.6). She went on to explain that translation is not about word-to-word equivalency. Rather, effectively translating something from one language to another involves considering the context of what is written or said and uncovering the meaning the author or speaker intends. Her initial analogy of translation as a well-rehearsed performance underscored the time, preparation, and research the process requires, creating the foundation for her additional explanation and enlightening her audience.

Informative Speaking Engages the Audience

When you engage your audience, you get and maintain their attention in meaningful ways that make your topic appealing and interesting. Although you are not trying to change listeners' behavior or thoughts on a topic as you would in persuasive speaking, you do want them thinking, "I want to know more." Your audience should feel involved and interested, waiting in anticipation for what you will say next. Moreover, reflecting on how you can engage your audience in your topic and adapting to their interests reduces your speaking anxiety and improves your public speaking abilities.[20]

Relevancy is key to engagement. That is, your strategies for gaining and maintaining your audience's attention must add to your topic, not distract or detract from it. For instance, starting out with a humorous joke might initially get your audience to focus on you, but if it has nothing to do with your topic, they will become confused and soon their minds will wander. Appropriateness also is essential to engagement. In a speech on violence in movies, one of our students showed a film clip of a particularly violent and gory scene. The images were so shocking that most of the audience paid little attention to the rest of the speech;

a few even left the room. Listeners became disengaged rather than engaged.

In his TED Talk "How to Speak So That People Want to Listen," sound consultant Julian Treasure proposes the HAIL model to engage an audience (Figure 15.7).[21] Effective speakers exhibit:

- *Honesty* by presenting information clearly and sincerely.

- *Authenticity* by offering a true representation of themselves.

- *Integrity* or trustworthiness by following through with what they say.

- *Love* by creaing a sense of goodwill and caring for others.

FIGURE 15.7 Listeners are engaged when speakers are honest, authentic, trustworthy, and demonstrate goodwill.

Treasure suggests that you and what you say are the most important tools for getting an audience to listen, rather than dazzling digital slides, spectacular video, or sensational feats of magic. Engaging your audience starts with you, speaking as your true self and presenting information that is accurate, clear, and in the audience's interest.

In addition to focusing on yourself as a tool for engaging your audience, presentation materials also serve a key function in keeping listeners' attention. As you learned in Chapter 14, presentation media such as digital slides and audio files can help explain a point or demonstrate an idea to an audience that words alone cannot. For example, in a lecture on algebra, Grant Franks, a tutor at St. John's College, began by following Treasure's approach to get the audience interested in his topic:

> This series of talks grew out of a remark that a former dean of the College made to me years ago. This person, whom I respect and admire greatly, said something I thought was seriously questionable. "Algebra is boring," he opined. "Our students love geometry," he said, "because its beauty strikes them immediately. But algebra is just a tool, a technique. Nobody wants to spend any class time studying it. It's just dull." I took those words as a challenge. … Algebraic structures have a real beauty, even if it takes a little work to notice and appreciate it.[22]

You likely can imagine many audience members nodding in agreement with the college dean, thinking that studying algebra is not all that exciting. But Franks's brief story left the audience wondering, "What would make algebra interesting?" Franks went on to ask that question and then hint at the answer:

> What, then, would make someone "in love with algebra?" If anything could do it, I think it would be Carl F. Gauss's construction of the heptadecagon. It's a beautiful result, and the pathway to it, while not perfectly smooth, requires only a few hours of preliminary work,

not years. Also, for this audience, it speaks directly to geometrical demonstrations that all St. John's students have encountered by the middle of the first semester of their freshman year, namely, the propositions of Book Four of Euclid's Elements where we see the construction of regular-sided polygons in circles. Euclid shows how to construct the equilateral triangle, the square, the pentagon—that's a hard one!—and the hexagon inside a given circle. Then, without explanation, he skips ahead to the fifteen-gon (Elements IV, 16). Then he stops. Why? Euclid, characteristically laconic, says nothing. Two thousand years later, a very young Carl Friedrich Gauss provided the answer.[23]

Franks used three strategies in this part of his lecture to maintain his audience's attention. He asked a rhetorical question his listeners probably were asking themselves: "What, then, would make someone 'in love with algebra'?" Then he connected his topic to his audience members' experience, referring to information they learned in one of their first classes at St. John's, equations for geometric figures. Finally, he suggested there is an answer to the rhetorical question, but he did not reveal it right away. At last, Franks provided the answer, integrating sounds and an image to illustrate his point:

> However, in order [to] reach that result, you need to come to terms with [horror suspense sound effect] *the square root of negative one.* [mad scientist laugh sound effect]
>
> Some people have difficulty accepting this number. They say, for instance, that they are put off by the fact that it doesn't exist. Which is ridiculous! You shouldn't let so trivial a problem prevent you from embracing this concept, for the square root of negative one, also denoted by the single letter *i*, is the gateway to an algebraic realm of amazing results.[24]

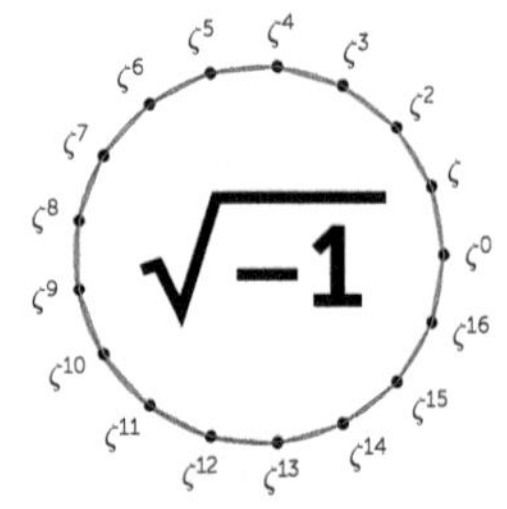

Excerpt from Grant Franks, "Meet Your New Best Friend: The Square Root of Negative One." Copyright © 2019 by St. John's College.

The sounds Franks used logically fit with his talk, as he built up suspense to what could make algebra intriguing. His simple image depicted the idea of negative one and reinforced his earlier reference to polygons within a circle. The sounds, images, and words worked together to engage the audience in a theoretical and complex topic.

As you probably have realized from these examples, there is no single strategy that will maintain your audience's focus on you and your message. Rather, you must use multiple tactics to sustain an engaging audience-centered speech. Answering the following questions will help you design your informative speech with your audience in mind:

- How can I show my audience my own interest in the topic?
- What aspects of my topic will my audience find intriguing?
- How can I encourage my audience to think about my topic in a new way?
- What might surprise my audience about my topic?
- What images will help visualize my topic for my audience?
- What sounds related to my topic will pique my audience's interest?

BOX 15.2: ETHICAL QUESTIONS

Speaker Interrupted

When Duke University President Vincent E. Price stood up to give a State of the University speech to alumni from the class of 1968, student protesters took to the stage. Using a bullhorn, students demanded action on a list of issues, including increased pay for the university's staff, improved services for immigrant students, and better support for students with disabilities. Some members of the audience cheered, while others heckled the hecklers and turned their backs. After about 15 minutes, the students left the stage and continued their protest outside. Later, a few alumni praised the protesters.[25] What do you think? Review the National Communication Association's ethical principles in Chapter 1. Should the students have taken over the event? Should they have used a bullhorn to silence the speaker? Consider the responses of the alumni. Should they have turned their backs? Should they have supported the students? What ethical principles apply to the situation? How should speakers in such a situation respond?

Keeping your audience's interests and knowledge base at the forefront of your planning as you develop your informative speech will increase the likelihood that your listeners will find your informative speech engaging. However, even the most engaging speaker can face listeners who refuse to actively listen. Consider the ethical dilemma presented in Box 15.2.

Types of Informative Speeches

Nearly all informative speeches are about objects, processes, events, or concepts, although some speeches may incorporate information from more than one category. For instance, in a speech about a new building on campus (object), you might talk about the ceremony commemorating its opening (event), background facts on how it came to be (process), or the structure's initial plans (concept). Still, these categories provide a useful way for you to think about informative speeches, especially when choosing your topic and phrasing your specific purpose.

Speeches About Objects

Speeches about objects present information on anything in the material world you can perceive with your senses, including physical places and spaces, people, animals, plants, and other living and nonliving things. Stop reading for a moment and consider the environment you are in right now. What is around you? What

speeches about objects Speeches that present information on anything in the material world you can perceive with your senses

do you notice with your senses? How could some of those things become an effective speech topic? Below are examples that show how informative speakers might frame speeches about objects.

EXAMPLE 15.1

Topic: Megabats

Title: "The Mighty Megabats"

General purpose: To inform

Specific purpose: To inform my audience about the social practices of megabats

Thesis: Megabats, also called fruit bats and flying foxes, are social animals that live in colonies, vocalize to communicate with each other, and form cooperative relationships.

EXAMPLE 15.2

Topic: Soft Robots

Title: "Soft and Cuddly Robots"

General purpose: To inform

Specific purpose: To inform my audience about robots that mimic living organisms

Thesis: Soft robots can have the flexibility of an octopus arm, the environmental adaptability of a flower, and the gentleness of a person's tender hug. (Figure 15.8)

EXAMPLE 15.3

Topic: Faith Ringgold

Title: "Faith Ringgold: Artist, Activist, Author"

General purpose: To inform

Specific purpose: To inform my audience about Faith Ringgold's work as an artist, activist, and author

Thesis: Faith Ringgold has worn many hats, including painter, quilter, sculptor, performance artist, feminist, and writer.

FIGURE 15.8 A speech to inform about an object could describe the characteristics of a soft robot that mimics living things.

Use your own interests and knowledge as the starting point for possible speech topics. Consider what your audience knows and what topics they might want to know more about. Think about the stories of famous and not-so-famous people you find compelling. Reflect on places you have visited and how they left a lasting memory. Identify your favorite food, pet, or gift from a friend. Apply the brainstorming and mind mapping techniques described in Chapter 11 to take these ideas and transform them into possible topics for your informative speech on an object.

Speeches About Processes

Speeches about processes explain to the audience how something works, how something is done, or how something was created. In speeches about processes, speakers typically present the steps or actions that lead to a final outcome or product. The emphasis is on the activity rather than the endpoint. Let's use the topic "soft robots" as an example. In a speech about objects, you might discuss the robots' qualities or characteristics, as in the previous section. In a speech about processes, you could talk about how soft robots are developed. The following are examples of additional speeches about processes.

speeches about processes Speeches that explain to the audience how something works, how something is done, or how something was created

EXAMPLE 15.4

Topic: Indoor Herb Garden

Title: "Five Steps to a Fragrant Indoor Herb Garden"

General purpose: To inform

Specific purpose: To inform my audience about the steps in creating an indoor herb garden

Thesis: Creating your own indoor herb garden requires choosing the right plants, potting them in the proper container, putting them in a sunny location, giving them the right amount of water, and harvesting a small amount at a time.

EXAMPLE 15.5

Topic: Hurricanes

Title: "How the Hurricane Naming System Was Created"

General purpose: To inform

Specific purpose: To inform my audience about how the current hurricane naming system came to be

Thesis: The current hurricane naming system developed by the World Meteorological Organization and the National Hurricane Center came about to increase community awareness of and preparedness for hurricanes.

EXAMPLE 15.6

Topic: Trees

Title: "The Rumors Are True—Trees Can Talk"

General purpose: To inform

Specific purpose: To inform my audience about how trees communicate

Thesis: Trees communicate with each other using underground fungal networks and above-ground chemical signals. (Figure 15.9)

Speeches about processes can prove challenging, especially if you are trying to demonstrate something complex. For instance, our students have attempted to explain how to play basketball, poker, Dungeons and Dragons, and any number of complex games, only to realize that a classroom speech does not offer sufficient time to cover all the rules, procedures, and mechanics associated with the games. In addition, some classmates had detailed knowledge of the games, while others knew little. The first group was bored, and the latter was confused. However, a speech on how basketball was started or the development of the early professional leagues likely is something your audience would not know and could be of interest to them. The opposite problem can occur as well, when speakers talk about processes that are well known or simple. For example, the COVID-19 pandemic brought with it a heightened attention to handwashing, with scores of public service announcements and brief news items on how to wash your hands properly. Student demonstration speeches on handwashing provided little new information, leaving classmates disinterested and restless. However, a speech explaining the process for developing new techniques for handwashing that do not involve soap and water or hand sanitizers, such as creating nano-aerosol mists, might pique the audience's interest. As with any speech, consider your speaking time limit and audience before making a final selection. Explore possible informative speech topics about processes in Box 15.3.

Speeches About Events

Speeches about events discuss an occurrence that has happened in the past, is happening in the present, or will happen in the future, as with a ceremony, festival, contest, or performance. The events may be public or private, corporate or community based, local or worldwide, a one-time or repeated occurrence. The examples that follow give you an idea of what you might cover in an informative speech about an event.

FIGURE 15.9 Although any audience likely has some knowledge of trees, giving a speech about how trees communicate with each other offers a fresh approach to the topic.

speeches about events
Speeches that discuss an occurrence that has happened in the past, is happening in the present, or will happen in the future

BOX 15.3: CONNECT & REFLECT

How Does That Work?

The internet abounds with sites that explain how things work, such as Explain That Stuff, How Everything Works, and How Stuff Works. In addition, TEDEd has more than 100 lessons on how things work. And YouTube has hundreds of thousands of videos that explain how things work. With all this information available, how can you choose a topic about a process that your audience easily can find out about online? Keep in mind that just because the information is out there does not mean your audience has accessed it. In addition, because you will research several sources, what you present will not be the same as a short article or video on the topic. Go to one of the sites listed above and choose a few topics that interest you. Then do additional research and identify the unique or novel information each source brings to your topic. What do you think your audience already knows about the topic? What new information could you present? How can you make it interesting? People naturally want to know how things work. Your informative speech on a process will tap into that human inclination.

EXAMPLE 15.7

Topic: The Chicxulub Impact Event

Title: "The Day the Oceans Died"

General purpose: To inform

Specific purpose: To inform my audience about what happened to marine life when the Chicxulub asteroid crashed into our planet

Thesis: The Chicxulub Impact Event occurred about 66 million years ago when an asteroid struck the Gulf of Mexico, turning the ocean water to acid and causing the extinction of marine life.

EXAMPLE 15.8

Topic: Family Reunions

Title: "Connecting and Reconnecting Those Familial Ties"

General purpose: To inform

Specific purpose: To inform my audience about the functions of family reunions

Thesis: Family reunions pay tribute to the meaning of family, give extended family members a place to meet and talk, and provide a cross-generational learning space.

FIGURE 15.10 An informative speech on earth day might describe the various events that take place on that day.

EXAMPLE 15.9

Topic: Earth Day

Title: "Celebrating Our Environment"

General purpose: To inform

Specific purpose: To inform my audience about Earth Day's activities

Thesis: Celebrated by more than a billion people worldwide every year on April 22, Earth Day activities have included drum chains, rallies, tree planting, teach-ins, and other actions in support of a clean environment. (Figure 15.10)

Events mark important personal, social, cultural, and natural occurrences. They bring meaning to what happens in life and help make sense of what can seem like a chaotic world. Because events often have rich historical narratives, they are fruitful topics for informative speeches. As you choose a topic for a speech about an event, consider what your audience likely already knows about it and then what more they would like to know.

Speeches About Concepts

speeches about concepts Speeches that address ideas, values, philosophies, principles, beliefs, theories, and similar abstract thinking

Speeches about concepts address ideas, values, philosophies, principles, beliefs, theories, and similar abstract thinking. Because speeches about concepts are not as concrete as the other types of informative speeches, they present a greater challenge to audience understanding. That is, speakers must be especially vigilant in offering comprehensive yet clear information. The examples that follow offer a few topic ideas for speeches about concepts.

EXAMPLE 15.10

Topic: Games as Art

Title: "The Beauty In Video Games"

General purpose: To inform

Specific purpose: To inform my audience about the beauty in video games

Thesis: Beauty in video games can be found in visual aesthetics, sound design, and narrative expressiveness.

EXAMPLE 15.11

Topic: The American Dream

Title: "Dreaming of the American Dream"

General purpose: To inform

Specific purpose: To inform my audience about the meaning of the American Dream

Thesis: The American Dream reflects the ideals of equality, liberty, democracy, and opportunity.

EXAMPLE 15.12

Topic: Friendship

Title: "Friends Are More Than a TV Show"

General purpose: To inform

Specific purpose: To inform my audience about the nature of friendship

Thesis: Friendship brings people together through respect, trust, affection, and loyalty.

Some topics evolve from concepts into something more concrete. For instance, the existence of black holes Stephen Hawking discussed in this chapter's opening example was a theory up until recently. However, with the Event Horizon Telescope's observations in 2017 and subsequent photograph of a black hole in 2019, the theory has become a tangible object (Figure 15.11). Similarly, 3D printing has gone from design sketches to an actual process that produces real things. And the idea for a graduation celebration becomes an event once the planning is complete.

Speeches about concepts allow for the greatest creativity of all informative speeches because they are so abstract. Their abstractness also makes this type of informative speech the most challenging to convey clearly to your audience. As you choose a topic for a speech about a concept, pay special attention to your time limit and your audience's knowledge of and viewpoints on your topic choices. Box 15.4 asks you to examine more closely the ambiguous nature of speeches about concepts.

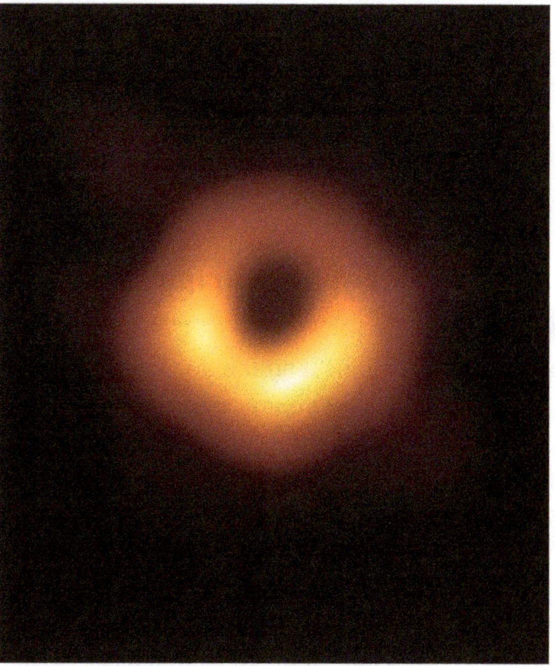

FIGURE 15.11 Concepts may evolve from ideas and theories to objects you can observe with your senses, as in the case of black holes. The Event Horizon Telescope, an international project, took the first photograph of a black hole and its shadow.

BOX 15.4: CONNECT & REFLECT

Engaging Contentious Topics

Some communication theorists argue that all communication is persuasive.[26] That is, by the topic you choose, the words you use, the sources you cite, and the presentation media you include in a speech, you are persuading your audience to view what it is you are talking about from your perspective. So is it possible to give an informative speech? We argue the answer is yes, but it does require that you keep your focus on educating your audience, elucidating your topic, and engaging your audience. For each of the topics listed below, consider how you can present an informative speech and avoid influencing your audience's beliefs, values, and attitudes about the subject.

- homelessness
- the Second Amendment
- welfare

- immigration
- health care/insurance
- violence in films
- fossil fuels
- capital punishment
- vaccinations
- democracy
- freedom
- paranormal activity

How successful were you? What strategies did you think of for keeping your speech informative rather than persuasive? Which topics were particularly challenging to present as an informative speech? Why? Are there some topics that are off limits for informative speaking? Why do you think that is the case?

Guidelines for Effective Informative Speaking

Educating the audience is central to all effective informative speaking. As the speaker, you want your audience to learn something new. The following guidelines will help you achieve that objective.

Become an Expert

When you are preparing to give an informative speech, think of yourself as a teacher: You are educating your audience about a topic. When you become an expert on your topic, you increase your confidence as a speaker (Chapter 14) and increase your ability to make thoughtful decisions about the best information for the audience. You will not present all you know about your topic in your speech. Recall Stephen Hawking's lecture on black holes. Because he is an expert on the topic, he can draw from his vast store of knowledge to present just the right amount and kind of information to his audience. In addition, when you know your topic well, you can explain points in greater depth if audience members seem confused. You will also be ready for any questions your listeners might ask. Review Chapter 12 for strategies on researching your speech.

Unlike in Stephen Hawking's case, your audience may not know you are an expert on your topic. You can indicate your expertise on your topic by discussing relevant personal experiences and referring to the credible sources of information for your speech. That is, use **credentialing**, the process of establishing your qualifications on a subject, to demonstrate your expertise. For example, in a speech on a campus food pantry, one of our students said, "The Spartan Food Pantry right here on campus is there for all students who face food insecurity. And I should know—the pantry was my lifeline when I was out of a job last semester." The student went on to say, "In my interview with the pantry's coordinator, I found out how the program works for students and how students can work for the program." Credentialing in two ways—personal experience and research—signaled to the audience the speaker was an expert on the topic. Learn more about researching your topic through interviews and developing your local expert network in Box 15.5.

credentialing The process in which speakers establish their qualifications on a topic

---

BOX 15.5: LIVE IT LOCAL

Develop Your Local Expert Network

As you learned in Chapter 12, expert interviews are an important information source in researching your speech topics. Experts personalize your speech and provide insights not possible from a newspaper article, book, or reference material. When you interview experts, you can ask for more detail or background on a topic, which you cannot do with other sources (Figure 15.12). How do you find local experts? You can start with your campus library, which may have a collection of publications by your institution's faculty. For example, the San José State University library has SJSU ScholarWorks, which houses scholarship produced by the university's students and faculty. Another campus resource is the media relations or similar office, which likely has a list of campus experts in a wide range of areas. Tap into the larger community's expertise as well. The reference librarian at your public library can suggest possible resources, as with a directory of local businesses. Neighborhood and professional groups often feature local experts in their fields. An online search with your topic and locale as key words should reveal people in your community who are knowledgeable about your topic. Once you have lined up the interview, review Chapter 12 for strategies on preparing questions. Use the interview to learn more about the people in your community and build you network by asking for the names of others who would be willing to talk with you about your topic. Meet the experts and become an expert.

FIGURE 15.12 Interviews with local experts offer one strategy for you to become an expert on your topic.

Organize for Clarity

Clarity in informative speaking applies both to the information you include and the order in which you present it. When you organize the research you have gathered for your informative speech, consider how your points fit together and how they will educate your audience about your topic, elucidate the topic for your audience, and engage your audience with your topic. Use your specific purpose and thesis to guide you in determining the best way to order your main points. As you learned in Chapter 13, the topical, chronological, narrative, and spatial patterns of organization work best with informative speeches. Table 15.2 gives examples of how you might apply those organizational patterns to the different types of speeches to inform.

Talk Up Your Topic

You can talk up your topic in two ways: demonstrating your own enthusiasm for it and personalizing the topic for your audience. That is, you show your audience that you care about your topic and you show them why they should care. The first question you want to ask yourself when selecting a topic is "How interested am I in the topic?" (Chapter 11). Your enthusiasm for your topic starts in this first stage of developing your speech. As you research your speech, identify resources and information that make your topic intriguing and engaging. When you present your speech, what you say and how you say it should reflect the passion you feel for the topic. For instance, in her talk "Black Tudors: Three Untold Stories," given at Gresham University in London, Dr. Miranda Kaufmann opened with this:

> I thought I knew the Tudors. I had "done" them at primary school, secondary school and university. But I was wrong. Because I didn't know about the Black Tudors. My journey began when I heard in a lecture that the Tudors began trading to Africa in the middle of the 16th century. This was news to me. I began to wonder what those Tudor sailors thought of the Africans they encountered, and vice versa. I went to the library and started reading. It was then that I found a reference to a document that left me dumbstruck. It said:
>
>> great numbers of … Blackamoores, which she is informed, Have crept into this Realme since the troubles between her highness and the King of Spaine.
>
> There were Africans living here in England! I had to find out more. These "great numbers" of Africans must have left some trace in the large body of records that survive from Tudor England. I began my search.[27]

Right from the start, Dr. Kaufmann's audience could sense her enthusiasm for her topic. She thought she knew the story of the Tudors, but it turned out she was wrong. She had much more to learn about this time in English history, and she wanted to share her findings with her audience.

Your excitement about a topic will take your audience only so far. They must feel that affinity, too. In choosing your topic, the second consideration is the audience (Chapter 11).

TABLE 15.2 Organizational Patterns for Types of Informative Speeches

	SPEECHES ABOUT OBJECTS	SPEECHES ABOUT PROCESSES	SPEECHES ABOUT EVENTS	SPEECHES ABOUT CONCEPTS
TOPIC	Exploring the Red Rocks of Sedona, Arizona	Roof-top Solar Panels	Memorable Moments of the Summer Olympic Games	My Idea for a Campus Commons
GENERAL PURPOSE	To inform	To inform	To inform	To inform
SPECIFIC PURPOSE	To inform my audience about the activities associated with Sedona, Arizona's red rock landscape	To inform my audience about how roof-top solar panels work	To inform my audience about memorable winning moments in the Summer Olympic Games	To inform my audience about my idea for a campus commons
THESIS	The red rocks of Sedona, Arizona, are home to artistic, spiritual, and recreational activities.	There is a four-step process for roof-top solar panels to convert sunshine into electricity you can use in your home.	The story of the Summer Olympic Games includes many memorable winning moments since they started in 1896.	Viewed clockwise, the campus commons would have four quadrants: north, east, south, and west.
ORGANIZATIONAL PATTERN	Topical	Chronological	Narrative	Spatial
MAIN POINTS	I. The red rocks of Sedona inspire the artist in everyone. II. The red rocks of Sedona are known for their spiritual and emotional powers. III. The red rocks of Sedona beckon visitors with many outdoor activities, such as hiking and mountain biking.	I. In the first step, the solar panels collect the sunlight and turns it into direct current (DC) electricity. II. In the second step, an inverter transforms the DC electricity into alternating current (AC) electricity that powers the lights and appliances in your home. III. In the third step, any extra electricity your roof-top solar panels produce goes into the power grid. IV. In the fourth step, the power grid supplies any additional electricity needed for your home, as when the sun is down.	I. The story of memorable Olympic winning moments begins in 1896 when Spyridon Louis won the marathon in the first of the modern-day Olympic Games, known then as the Games of the Olympiad, in Athens. II. The story continued in 1936 when Jesse Owens won four gold medals at the Olympic Games in Berlin. III. Women became part of the story in 1984 when Joan Benoit won the games' first women's marathon at the Olympic Games in Los Angeles. IV. Although not the end of the story, in 2016, Park In-bee of Korea became only the second female Olympic golf champion at the Olympic Games in Rio de Janeiro.	I. The north area of the campus commons would be for open discussions and debates. II. The east area of the campus commons would be for socializing and meeting friends. III. The south area of the campus commons would be for quiet reflection. IV. The west area of the campus commons would be for games, such as chess and checkers.

One question you ask yourself is "Can I spark my audience's interest in the topic?" If you made a good topic choice, you answered yes. When you present your speech, you turn that yes answer into action, showing your audience how they are connected to your topic. Dr. Kaufmann did just this when she went on to say:

> When I tell people about the Black Tudors, they often say: "Oh, you mean slaves?" Well, I don't. It is all too easy to assume that all Africans in Europe at this time were enslaved. We are bombarded with images of enslaved Africans, often dating to later periods. Most recently, in the film *12 Years a Slave* and the television series *Roots*. But we need to think of these images alongside the ones of Africans in early modern Europe. Then consider that people at the time made other assumptions. In 1572, Juan Gelofe, a forty-year-old African man, enslaved in a Mexican silver mine told an English sailor named William Collins "that England must be a good country as there were no slaves there." Collins replied that "it was true, that they were all freemen in England."[28]

Dr. Kaufmann challenged a common assumption, likely one that the audience held, and suggested why they would want to know more about the Black Tudors she studied. In pointing out a gap in the audience's knowledge, Dr. Kaufmann gave her audience a reason to listen. In addition, telling the stories of three of the Black Tudors she researched personalized the topic for her audience. That is, using words and images, she related the experiences of real people in rich detail, bringing them to life.

Align Presentation Resources With Your Purpose

Not all informative speeches require using presentation media, such as digital slides, video, and handouts, discussed in Chapter 14. Millions of speakers gave eloquent informative speeches before the invention of PowerPoint. Still, presentation media can facilitate audience members' retention of information presented in the speech.[29] In addition, if you are trying to explain a complex process or idea, a single image on a digital slide might take the place of several minutes of speaking.In his talk on the student social media team he cofounded, presented at the Twelfth International Conference on e-Learning and Pedagogies, Dr. Ted Coopman used the slide in Figure 15.13 to show his audience the team's various social media outlets. Rather than stating each one aloud, he gave a few examples and left the rest for the audience to read on the screen, as shown in Figure 15.14.

For some informative speeches, you must use presentation media to explain your points. For instance, in an informative TED Talk on sound in film, Tasos Frantzolas, the founder of Soundsnap.com, gave countless examples of how filmmakers use images and context to trick the brain into thinking sounds are something they are not.[30] He began his talk with three scenes of rain. Before showing the clips, he told the audience that the sound used for one of the scenes is frying bacon. But in reality, *all* the rainy scenes used the sound of frying bacon as a stand-in for rain. The video and audio Frantzolas included demonstrated his points in ways words alone could not.

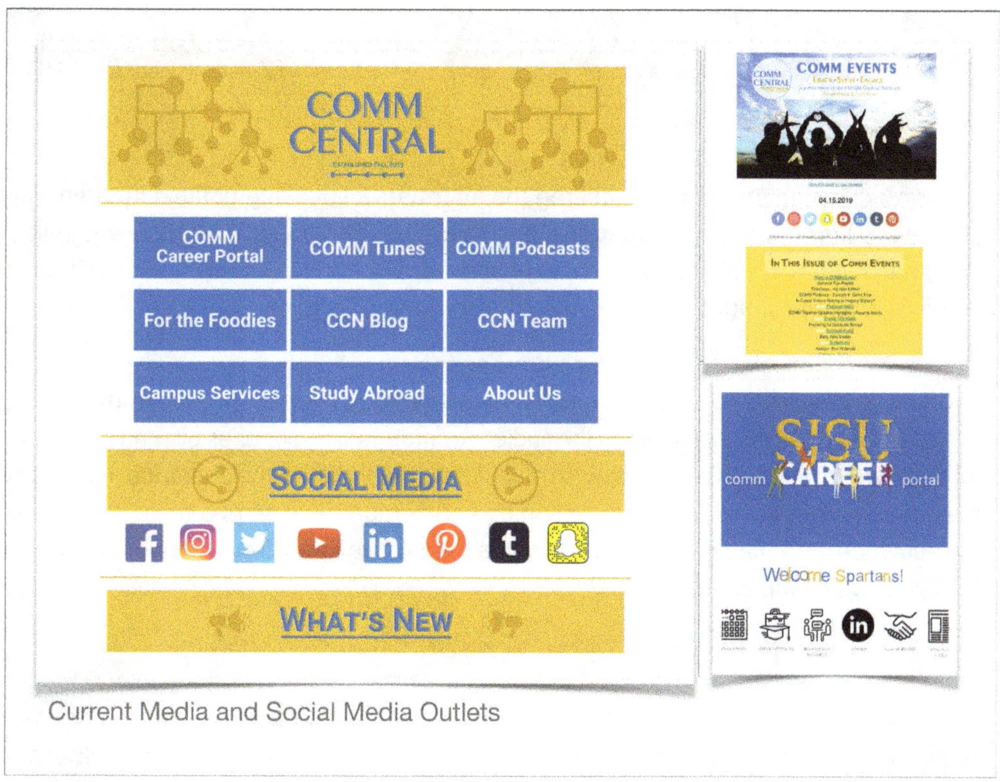

FIGURE 15.13 One well-designed digital slide can convey a wealth of information.

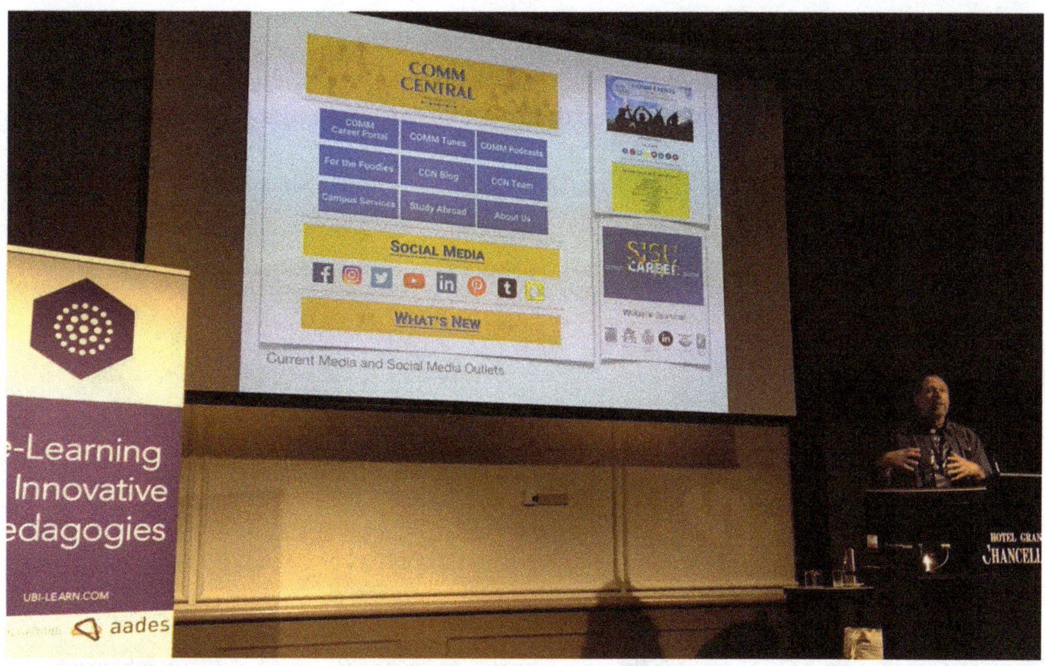

FIGURE 15.14 Dr. Ted Coopman presenting at the Twelfth International Conference on e-Learning and Pedagogies using the digital slide shown in Figure 15.13 to explain his topic.

CHAPTER REVIEW

- **Discuss the ways in which a pervasive communication environment has influenced the context for informative speaking.**

  A pervasive communication has altered the informative speaking context in three ways: (1) The speaker has a greater responsibility to identify and present credible and reliable information, (2) The speaker is a sharer of information, and (3) The speaker must attend to the speech's scenography.

- **Discuss the characteristics of informative speaking.**

  Informative speaking educates the audience, elucidates the topic for the audience, and engages the audience. In educating the audience, informative speeches offer an opportunity for listeners to learn something new. In elucidating the topic for the audience, informative speeches increase audience understanding and comprehension of a topic. In engaging the audience, informative speeches motivate listeners to keep their attention focused on the speaker and topic.

- **Identify and recognize examples of the types of informative speeches.**

  Informative speeches may be about objects, processes, events, or concepts. Speeches about objects center on things in the material world you can observe with your senses. Speeches about processes explain how things work, are done, or were created. Speeches about events are concerned with something that has happened, will happen, or is happening. Speeches about concepts address ideas, thoughts, principles, theories, and other abstract thinking. Discuss the guidelines for effective informative speaking.

- **Discuss the guidelines for effective informative speaking.**

  Effective informative speakers become experts on their topics, organize their speeches for clarity, talk up their topics, and align presentation resources with their speech purpose.

DISCUSS AND APPLY

Review this informative student speech on figs and answer the questions at the end.

SPEECH FOR ANALYSIS

Figs, Food, and Friendship

"If you have figs in your knapsack, everyone will want to be your friend." This Albanian proverb is as true today as it was thousands of years ago, for who doesn't like a sweet chewy fig? The other day my housemate asked me if I wanted a Fig Newton. Of course I said yes; I couldn't turn down that gesture of friendship.

For many Americans, a Fig Newton is the only time they've ever eaten a fig. But figs, which have been a part of the human diet for thousands of years, are so much more than the filling in a cake-like cookie. In fact, there are about 750 varieties of fig trees that produce hundreds of types of figs, some of them here on our campus. I won't tell you about all of them today, but I will tell you about the history of figs, common local figs, and how figs are used today.

The history of figs parallels the history of human agriculture. You might think wheat or barley was the first domesticated plant. However, scientific evidence points to the fig as the first plant of any kind that humans cultivated. An archeological dig of a prehistoric village in the Middle East found evidence of dried figs, suggesting that people were planting and tending fig trees more than 11,000 years ago. As one of the researchers, Dr. Bar-Yosef, said, these domesticated figs represent "the beginnings of agriculture."

Dr. Mike Shanahan, a rainforest ecology specialist, says that fig trees have a special place in human history. Dr. Shanahan argues, "No other plants have held such sway over human imagination. They feature in every major religion and have influenced kings and queens, scientists and soldiers." For example, *ficus religiosa*, or the sacred fig tree, is part of Buddhism and Hinduism. Egyptians buried dried figs with their dead to nourish them in the afterlife. And of course the fig tree is one of the trees planted in the Garden of Eden. I'm sure you remember how Eve and Adam made clothes out of fig leaves.

In his book *Sacred Seeds: New World Plants in Early Modern English Literature*, Professor Edward Test notes that 14th- and 15th-century European explorers referred to indigenous peoples of what are now Mexico, Central America, and the southern U.S. as "people of the figs." This fits with explorers' descriptions of the New World as a paradise and the fig's close association with the original paradise, the Garden of Eden. However, those indigenous people were not eating figs, as Europeans first introduced fig trees to the New World about 100 years later, starting in Florida and Virginia.

Fig trees quickly spread north and west as farmers and consumers grew to love the sweet and versatile fruit. An article in *Agricultural History* reports fig tree orchards flourished in Maryland, Mississippi, Alabama, Ohio, Illinois, North Carolina, and even a few New England states in the mid- to late-1800s. And by the early 1900s, fig trees were planted in Texas, Arizona, California, Oregon, Washington, and Utah. A circular from 1879 in Washington, DC, promoting the value of planting fig trees stated: "No other crop can be raised which will give so certain and so large returns in our Middle States and Northern States, as that delicious fruit, the fig." Although today nearly all figs grown for commercial use in the U.S. come from California, fig orchards still thrive in any locale with mild winters, like central Oregon.

Now that you know some of the fig's history, let's talk about specific types of figs. According to Shooting Star Nursery in the Rogue Valley area of Oregon, there are 11 varieties of edible figs that grow well here in the Pacific Northwest. The names of the figs often reflect what they look like or their origins. I'll tell you about three of them.

The first, the Desert King fig, is one of the most common. The fruit is large, juicy, and sweet; even the outside green skin tastes delicious. It's the king of fig trees in that it's the most productive. Planet Fig states that unlike some figs that trace their origins back hundreds or thousands of years, the first Desert King tree was identified in Madera, California, in 1930.

We had an enormous Desert King fig tree in the backyard of the house I grew up in. The tree would fruit in early summer and then we'd have figs for nearly every meal. Figs in oatmeal for breakfast. Fig and kale salad for lunch. Spice fig cobbler for dessert after dinner. We'd make and can fig sauce, fig chutney, fig syrup, and fig jam so that in the dead of winter we could have a taste of fig and a reminder of summer on even the coldest and dreariest days.

Another kind of fig, the Olympian, is even newer to the fig scene, discovered just 25 years ago near Olympia, Washington, and named by a retired environmental biologist. Interestingly, the tree itself was more than a century old, but hadn't been noticed among a grove of other fig trees. An article in the local paper *The News Tribune* describes this fig as "tangerine-sized and with violet skin and red flesh."

The last kind of fig I'll tell you about is the Oregon Prolific. From its name it would seem like the Oregon Prolific fig originated in Oregon, but it's actually from France. Also called the Marseille Fig and St. Anthony Fig, this was the kind of fig Thomas Jefferson brought to the U.S. The tree itself is small, more like a large shrub, but the fruit is exceptionally sweet. And if you want to try it out, you can, right here on our own campus's urban farm next to the club house.

What can you do with a fig besides eat it? Australian botanists David and Anna Wilson show how the fig truly is a friend to humans. For example, figs have been used in Peru and Brazil for gastrointestinal problems. Similarly, if you had an upset stomach in Bangladesh or Ethiopia, eating figs might be recommended. In Madagascar and Ghana, a fig tea with the leaves and stems is brewed to manage the symptoms associated with malaria. These researchers also found that fig trees were used to make barkcloth in Mexico, drums in India, and live fences in Cameroon.

Even more important than food, medicine, clothing, and fences, the fig tree is a friend to the planet and has an important role to play in slowing climate change. Rainforest ecologist Mike Shanahan observes that the hardy fig tree soaks up great quantities of carbon and has been used successfully in places such as Costa Rica, Thailand, and South Africa to repopulate over-logged forests. With their large leaves, fig trees provide natural shade umbrellas against the sun. Several species of fig trees are highly drought tolerant, producing sustainable food in increasingly arid places. And in Meghalaya, India, the rainiest place where people live, fig trees have been fashioned into living bridges that stand up to the worst monsoon rains—rains that no steel bridge could survive.

Today I've given you a brief history of the fig, which has been part of human agriculture for more than 11,000 years. Although figs originated in the Middle East, European explorers brought the first figs trees to America. Now figs are grown in most of the southeastern, southwestern, and western states, with California the top producer. As part of our campus urban farm, we have several mature fig trees. Three popular figs grown in the Pacific Northwest are the Desert King, Olympian, and Oregon Prolific. Finally, I've explained the use of figs and fig trees in human health and everyday life. Most important, fig trees contribute to managing and responding to climate change.

Figs and people go together, like friends for life. The next time someone offers you a fig or a Fig Newton, know that it's a gesture of friendship.

References

Condit, I. (1957). Fig history in the New World. *Agricultural History, 31*(2), 19–24.

Inspirational Stories. (2020). *Proverbs about figs.* https://inspirationalstories.com/proverbs/t/about-figs/

North Carolina State Cooperative Extension. (n.d.) *Ficus carica* "Marseille." https://plants.ces.ncsu.edu/plants/ficus-carica-marseille/

Planet Fig. (n.d.). *Desert king.* https://planetfig.com/cultivars/fcveng8588.html

Ponnekanti, R. (2014, October 22). Sweet on figs: Locally-discovered Olympian variety is for sale. *The News Tribune.* https://thenewstribune.com/living/article25888723.html

Shanahan, M. (2016, December 16). The nearly magical properties of fig trees. *Scientific American.* https://scientificamerican.com/guest-blog/the-nearly-magical-properties-of-fig-trees/

Shanahan, M. (2017, January 17). *The tree that shaped human history.* BBC Earth. https://bbc.com/earth/story/20170116-the-tree-that-shaped-human-history

Shooting Start Nursery. (2020). *Edible figs.* https://roguevalleynursery.com/plantlists/plant-of-the-week-archive/item/edible-figs

Test, E. M. (2019). *Sacred seeds: New world plants in early modern English literature.* University of Nebraska Press.

Wilford, J. N. (2006, June 1). Figs believed to be first cultivated fruit. *The New York Times.* https://nytimes.com/2006/06/01/science/01cnd-fig.html

Wilson, D., & Wilson, A. (2013). Figs as a global spiritual and material resource for humans. *Human Ecology, 41*(3), 459–464.

QUESTIONS FOR DISCUSSION

1. Review the organizational patterns for speeches to inform, discussed in Chapter 13. Which pattern did the speaker use? How do you know?

2. Review the discussion of introductions and conclusions in Chapter 13.

   a. How effective is the speaker's introduction? Why did you evaluate it that way?
   b. How effective is the speaker's conclusion? Provide support for your assessment.

3. How does the speaker meet the challenges of informative speaking in a pervasive communication environment? Give examples.

4. How does this speech fit the characteristics of informative speaking?

   a. How does the speaker educate the audience? Give examples.
   b. How does the speaker elucidate the topic for the audience? Give examples.
   c. How does the speaker engage the audience? Give examples.

5. What type of informative speech is this? How do you know?

6. How does the speaker follow the guidelines for informative speaking?

    a. How does the speaker demonstrate expertise? Give examples

    b. How is the speech organized for clarity? Explain your answer.

    c. How does the speaker talk up the topic? Give examples.

    d. How might the speaker use presentation resources? Give specific suggestions.

7. How informative did you find the speech? Why do you evaluate it that way?

8. If you were giving the speaker advice, what would you suggest for improvements?

9. What have you learned that you can apply in your own informative speaking?

CREDITS

CHAPTER 16

Speaking to Persuade

When actor Viola Davis spoke at Barnard University's commencement, she made this observation about two paths individuals take in their lives:

> The choice to think that your path is all about you and your success, how high you can climb in your career and your status. Or, the so-called "save the world" approach, where you have a vision for the world and, by God, you will change it because *you're different*. The first road requires you to mistake your presence for the event, to be in complete denial; and, the second requires you only to deny the really bad stuff. It requires you to forget racism, not see color, intersectionality, poverty. ... Forget anything in *me* that will get in the way. Forget *my* fear, *my* pain. ... Both result in well-intentioned, very bright, enthusiastic people doing NOTHING.[1]

Excerpt from Viola Davis, "2019 Barnard University Commencement Speech." Copyright © 2019 by Viola Davis.

Learning Outcomes

- Discuss the four characteristics of persuasion in a pervasive communication environment.

- Define and identify examples of the four means of persuasion.

- Define and identify examples of the different types of persuasive speeches.

- Describe the reasoning process in persuasive speaking.

- Explain how to use inductive and deductive reasoning to support your position in a persuasive speech.

- Identify examples of common fallacies in reasoning.

- Discuss the guidelines for effective persuasive speaking.

FIGURE 16.1 Viola Davis, shown here accepting a screen actors guild award for best female actor in a drama series, is well-known for her persuasive speaking skills.

persuasive speaking A type of speaking in which the speaker attempts to change or reinforce audience members' beliefs, values, attitudes, or behaviors

Then she argued for an alternative that leads to empathy and a passion for embracing change. Davis went on to talk about her own battles, failures, and triumphs and described her joy in serving others. She closed by telling her audience, "You can either leave something *for* people or you can leave something *in* people."

In **persuasive speaking** the speaker attempts to change or reinforce audience members' beliefs, values, attitudes, or behaviors. Viola Davis encouraged her audience not just to make a mark in history but to make a mark on their own history. She suggested a different way of thinking about history—acknowledging the positives and negatives of the past—and using that perspective to move forward into the world in the service of others (Figure 16.1).

This chapter addresses the fundamentals of persuasive speaking, including persuasion in a pervasive communication environment, how persuasive speakers appeal to their audiences, types of persuasive speeches, using reasoning to support your position, the kinds of audiences persuasive speakers face, and strategies that will help you develop effective persuasive speaking skills.

Speaking to Persuade in a Pervasive Communication Environment

As you learned in Chapter 1, a pervasive communication environment—the ability to communicate with anyone at anytime from anywhere—has fundamentally changed human interaction. Persuasive communication has evolved as well, from orators speaking in-person to small crowds in Aristotle's time to public speakers today reaching thousands and even millions of people via television, mainstream media websites, and social media. Consider Viola Davis's Barnard University commencement speech. Not only did she address the event's attendees but the online availability of her speech also extended her reach to anyone with a reliable internet connection or smartphone.

Four characteristics define persuasion today, summarized in Table 16.1.[2] First, the audience exercises agency in persuasive communication situations. That is, the audience takes an active role in the persuasive process, bringing with them their own knowledge, ideas, and viewpoints. Listeners no longer depend on persuaders alone for information. With personal digital devices, such as smartphones and tablets, audience members can fact check what a persuasive speaker says with ease or do a deep dive into the speaker's topic, before, after, and even during the speech. As a persuasive speaker, your task is to recognize the audience's resources and what they already know about the topic.

TABLE 16.1 Characteristics of Persuasion in a Pervasive Communication Environment

CHARACTERISTIC	IMPLICATIONS FOR PERSUASIVE SPEAKERS
Audiences exercise agency	Know what your audience knows; research audiences along with researching the topic
No single message will persuade an audience	Use multiple appeals and information sources
Much of persuasion is implied rather than stated	Actively engage the audience in interrogating information and drawing conclusions
Persuasion is everywhere, yet appears to be nowhere	Uncover the audience's underlying assumptions about everyday persuasive messages

Second, no single message will persuade an audience. Recall a recent large purchase you decided to make. More than likely, you did some research, talked with friends and family, and viewed several advertisements related to it. The combination of these messages—as well as your prior knowledge and own values and beliefs—pointed you to a particular conclusion. Because persuasion involves many factors, your job as a persuasive speaker is to tap into those sources and use them to your advantage in your speech. For instance, a speech on campus safety might include references to recent crime alerts sent out by the college's police and reports in the campus newspaper.

Third, much of persuasion is implied rather than stated (Figure 16.2). For example, many advertisers do not make explicit what they want their audiences to do. Instead, they associate images or words that appeal to the audience with their product. Automobile commercials rarely include facts about the vehicles, but instead show happy families on fun vacations or cute hamsters at the wheel, as with the Kia Soul marketing campaign.

FIGURE 16.2 What is the author's persuasive message with this painted rock? What information has the author left out? For example, to understand the message, the audience would need to know about the COVID-19 pandemic, the common use of "vax" for vaccination, and the campaign to vaccinate people against the coronavirus.

Because persuasive speakers cannot present all the information on a topic to their audience, they must decide what to include and what to leave out. Audiences expect this, as they are accustomed to filling in the information persuaders omit. For instance, late night talk show hosts' humor relies on the audience's knowledge of popular culture and current events to understand a joke.

Fourth, persuasion is everywhere, yet appears to be nowhere. Like fish in water, you are swimming in a sea of persuasive messages, but you do not notice them. Yet you are persuaded, often in ways you do not realize. Advertisements, in particular, permeate your communicative day—on social media, television, the radio, and websites. Your online activities are tracked, tallied, and analyzed to

logos The use of logical appeals to persuade an audience

ethos The use of appeals to the speaker's credibility or authority to persuade an audience

pathos The use of emotional appeals to persuade an audience

mythos The use of narrative appeals to persuade an audience

design specific messages targeted to your behaviors and habits. The next time you go to the grocery store, carefully observe your buying choices. Why do you choose one brand over another? You likely can trace back your decisions to the many advertisements you have encountered related to the product. The special challenge for persuasive speakers is to make their message stand out in a pervasive communication environment.

Means of Persuasion: Appealing to the Audience

FIGURE 16.3 Supreme Court Justice Sonia Sotomayor used pathos, logos, ethos, and mythos in connecting with and persuading her audience.

Aristotle argued that persuasive speakers have three means of persuading their audience: logos, ethos, and pathos. **Logos** refers to logical appeals or the evidence the speaker presents, **ethos** to the speaker's character and credibility, and **pathos** to the audience's emotions and sympathies. More recent scholars have added a fourth mode of proof, **mythos**, which relies on stories.[3] Ideally, persuasive speakers integrate all four kinds of appeals in persuading their audience. Justice Sonia Sotomayor did just that in her PEN World Voices Festival Arthur Miller Freedom to Write Lecture at Cooper Union in New York City (Figure 16.3). For instance, early in her speech, when she compared the challenges judges in other countries face to those in the United States, she used a logical appeal:

> Judges have been threatened with public censorship or removal, economically persecuted to ensure their decisions conform to the will or public opinion, and even jailed. ... In many countries judges are killed. We are fortunate that Article Three of the U.S. Constitution guarantees federal judges their positions for life absent impeachment for a high crime or misdemeanor, and that no diminution of our salaries can occur during our tenure.[4]

In this part of the speech, Justice Sotomayor cited the protections the U.S. Constitution offers judges, explaining the consequences of not having those protections.

Ethos rests on the speaker's reputation and authority on the speech topic as well as the audience's belief in that reputation and authority. In that sense, ethos exists in the relationship between speaker and audience.[5] An essential component of persuasive speaking, ethos can provide the motivating force to win over your audience.[6] For example, when two speakers present the same information, the speaker whom the audience views as more trustworthy also is viewed as more persuasive.[7] For Justice Sotomayor, audience members recognize her appointment to the U.S. Supreme Court as an indication of her positive reputation. In addition, her professional accomplishments demonstrate that she is a credible speaker to talk about the experience of serving as a judge. Moreover, in her speech she described her own path to that position, indicating her authority to talk about serving as a role model for others.

Justice Sotomayor began her speech with an emotional appeal to engage her audience right from the start:

> I feel so humbled to be in this great hall at a podium at which one of my favorite presidents spoke, President Lincoln, but among such distinguished writers. If— you must believe me because I say it—I'm in awe. Thank you for letting me be a part of tonight.[8]

Pathos, or emotional appeal, is particularly powerful in building an immediate and compelling connection with the audience. For example, an in-depth analysis of an especially popular TED Talk found that the speaker's use of pathos resonated both with the in-person audience and online viewers. In addition, emotional appeals made the scientific basis of the speaker's talk about the physical and neurological aspects of a stroke more approachable and personal.[9]

Still, much of Justice Sotomayor's speech relied on mythos, or stories, that exemplified her experiences as a judge and her path to the Supreme Court. In discussing the differences between privacy and secrecy, she told this story:

> But let me offer a simple example of the distinction between secrecy and privacy. I recently spoke at a very public event where Rita Moreno and I shared a stage and held a conversation in front of an audience of 800 people. To ensure Rita and I and the audience that we felt that we were in an intimate conversation, the press was not invited. It seems pointless today to exclude the press from any event when any member of the audience can tweet quotations. … But often the exclusion of the press can be solely an attempt to establish certain parameters of tone and context. So Rita and I planned to talk like girlfriends—we did—playing a certain role for the benefit of the audience present in the hall to listen in.[10]

The story Justice Sotomayor told of her on-stage conversation with Rita Moreno offered an experience with which the audience could relate and demonstrated that something could be private but not secret. Although you may use one appeal more than another in a persuasive speech, all four methods of proof—logos, ethos, pathos, and mythos—work together to convince your audience to support your position.[11]

Kentucky Governor Andy Beshear's daily updates on the novel coronavirus pandemic in spring 2020 provide another example of integrating the four methods of proof to persuade audiences. His five o'clock evening televised presentations became must-watch events for Kentucky residents. Media analysts compared Beshear's talks to President Franklin Delano Roosevelt's fireside chats, radio broadcasts the former president gave during World War II.[12] Governor Beshear offered facts about the rates of infection in the state (logos); emphasized the need for his administration to be transparent in its efforts to combat the virus (ethos); expressed both assurances as well as concern about Kentuckians' health (pathos); and related stories of first responders, health care providers, and other essential workers who were on the front lines of the epidemic as well as those stricken with the virus and their families (mythos). His persuasive appeals in those daily five o'clock broadcasts likely contributed to Kentucky residents following the governor's protocols to stay healthy at home, follow social-distancing guidelines, and wear face masks in public.

The examples of Justice Sotomayor and Governor Beshear reflect ethical uses of the four methods of proof. However, each can be used in unethical ways. For instance, speakers can inflate credentials (ethos), shade the truth about a source (logos), manipulate an audience's emotions (pathos), and embellish a story (mythos). In Box 16.1, you will consider how you might appeal to your audience—or not—in three situations.

BOX 16.1: ETHICAL QUESTIONS

Assessing Appeals

Review the National Communication Association's ethical principles introduced in Chapter 1 as you work through your responses to these ethical dilemmas in appealing to your audience.[13] First discuss the dilemma. Then assess your initial response after reviewing the circumstances of the situation.

Ethical Dilemma 1: You decide to give a speech on responsible pet ownership. To make your emotional appeal as strong as possible, you make up what you consider a harmless little story about some dogs that you say starved to death in your neighbor's backyard. You begin your speech by describing these dogs as they appeared when found nearly dead from neglect.

Discuss: Is it okay to make up the story to win the audience for a such a worthy cause? Explain your decision.

Circumstances for Ethical Dilemma 1: You know that thousands of pets in the United States die and are killed every year because of irresponsible pet ownership. You think this story is essential to convince your audience.

Discuss: How do the circumstances change your view of this ethical dilemma? What NCA ethical principle(s) did you apply to this dilemma? Explain your reasoning.

Ethical Dilemma 2: You're researching a speech on the effects of gambling using a standard search engine, such as Google or Bing. You find material that fits perfectly with your main argument.

Discuss: Do you use the material? Explain your decision.

Circumstances for Ethical Dilemma 2: The information is from a blog post

(continued)

on a personal website that includes the author's name and date of posting. The name is not familiar to you, and an online search reveals no information on the author.

Discuss: How do the circumstances change your view of this ethical dilemma? What ethical principle(s) did you apply to this dilemma? Explain your reasoning.

Ethical Dilemma 3: You're preparing a speech on mandatory vaccinations and feel strongly about your position on the issue. When researching your speech, you find both information that supports your position and information that does not. You want to use only the information with which you agree.

Discuss: Do you present the opposing information or not? Explain your decision.

Circumstances for Ethical Dilemma 3: This speech is worth 30% of your final grade in the class. You think presenting information from the opposing view will severely weaken your speech.

Discuss: How do the circumstances change your view of this ethical dilemma? What ethical principle(s) did you apply to this dilemma? Explain your reasoning.

Types of Persuasive Speeches

Persuasive speeches are categorized based on the nature of the issue or question they are addressing. Speakers may address a question of fact, value, or policy in their speeches. Table 16.2 provides a summary of questions types, which are discussed in greater detail below.

TABLE 16.2 Types of Questions

TYPE OF QUESTION	BRIEF DEFINITION	EXAMPLE
Question of fact	Whether something is true or false	Is ridership on the city's public transit system increasing?
Question of value	The quality of something	How practical is the city's public transit system?
Question of policy	What should/should not be done	Should the city expand its public transit system?

Speeches on Questions of Fact

When persuasive speakers address a **question of fact**, they seek to determine if something is true or false. For example, if a speaker wanted to talk about the use of surveillance cameras in public places, the speaker might ask, "Do surveillance cameras in public places reduce crime?" The speaker can verify if crime is reduced in areas where such cameras were introduced by examining crime rates from before and after the cameras were installed. The examples that follow show how persuasive speakers might address a question of fact.

question of fact A question that asks if something is true or false

EXAMPLE 16.1

Topic: Workplace Bullying

Question: Is workplace bullying increasing in U.S. organizations?

Title: "Workplace Bullying is on the Rise"

General purpose: To persuade

Specific purpose: To persuade my audience that workplace bullying is increasing

Thesis: Workplace bullying is increasing in every type of U.S. industry, including manufacturing, retail, government, and nonprofits.

EXAMPLE 16.2

Topic: Mars and Climate Change

Question: Does Mars provide a forecast of what will happen to Earth if our planet's greenhouse gases continue to increase?

Title: "The Future of Earth Is Mars"

General purpose: To persuade

Specific purpose: To convince my audience that what has happened to Mars will happen to Earth if our planet's carbon dioxide levels continue to rise

Thesis: Mars once was a habitable planet like Earth, but high levels of carbon dioxide, like the greenhouse gases on Earth, led to its current state as a frozen planet. (Figures 16.4 and 16.5)

FIGURES 16.4 AND 16.5 A persuasive speech on a question of fact might compare the current state of Mars to the future state of Earth.

EXAMPLE 16.3

Topic: Wonder Woman

Question: Was Wonder Woman the first comic book feminist?

Title: "Wonder Woman—The First Comic Book Feminist"

General purpose: To persuade

Specific purpose: To persuade my audience that Wonder Woman was the first feminist comic book superhero

Thesis: In contrast to previous depictions of women in comic books, Wonder Woman's qualities of strength, intelligence, independence, and courage made her the first feminist comic book superhero.

Speeches on Questions of Value

A **question of value** explores the subjective worth, quality, or condition of something, asking for evaluations of good or bad, beautiful or ugly, interesting or boring, and right or wrong, for instance. Ethical questions addressing the morality of behaviors and actions are questions of value. In the surveillance camera example, a speaker might ask, "Is it morally right to install surveillance cameras in public places?" Questions of value center on people's opinions and qualitative judgments, yet still require research because speakers want to get expert opinions and information. Questions of value often prove more complex and difficult to answer than questions of fact because the latter relies more on objective evidence and verifiable observations. Cultural beliefs and ideals play a prominent role in answering questions of value, so speakers must be particularly sensitive to their audience members' diverse backgrounds. Engaging in dialogue, which invites open deliberation and the free expression of ideas, as discussed in Chapter 1, facilitates a communication climate in which speakers and listeners can more openly discuss questions of value. The examples that follow show how a persuasive speaker might address a question of value.

question of value A question that asks for an evaluation of the subjective worth, quality, or condition of something

EXAMPLE 16.4

Topic: Unions For Student Athletes

Question: Are labor unions good for student athletes?

Title: "Student Athletes and Labor Unions"

General purpose: To persuade

Specific purpose: To persuade my audience that labor unions are good for student athletes

Thesis: Forming a labor union is good for student athletes because it helps protect their rights as college or university employees.

EXAMPLE 16.5 ━━━━━━━━━━━━━━━━━━━━━━━━━━━━

Topic: Income Inequality

Question: Is income inequality bad for the United States?

Title: "Income Inequality Hurts Everyone"

General purpose: To persuade

Specific purpose: To persuade my audience that income inequality is bad for the nation

Thesis: Income inequality—where the richer get richer and the poor get poorer—hurts our economy, innovation, educational progress, and prospects for the country's future.

EXAMPLE 16.6 ━━━━━━━━━━━━━━━━━━━━━━━━━━━━

Topic: Independent Bookstores

Question: Are independent bookstores the best place to buy a book?

Title: "Shop at Your Local Bookstore"

General purpose: To persuade

Specific purpose: To persuade my audience that independent, locally owned bookstores are the best places to buy books

Thesis: When you buy a book at your local independent bookstore, you get knowledgeable advice on choosing a book, you are more likely to find the book you want, and the money you spend supports your local community.

Speeches on Questions of Policy

question of policy A question that asks what course of action should be taken or how a problem should be solved

Speakers ask **questions of policy** when they are concerned about what course of action should be taken or how a problem should be solved. Questions of policy address what should be done, as with a speaker on surveillance cameras asking, "How should data from surveillance cameras be used?" (Figure 16.6.) The examples that follow show how a persuasive speaker might address a question of policy.

EXAMPLE 16.7 ━━━━━━━━━━━━━━━━━━━━━━━━━━━━

Topic: Family-Friendly Workplace Policies

Question: Should U.S. companies have policies that are more family friendly?

Title: "Making the U.S. Workplace More Family Friendly"

General purpose: To persuade

Specific purpose: To persuade my audience that U.S. companies should have policies that are more family friendly

Thesis: Familyfriendly workplace policies give parents more options to work, contribute to employee satisfaction, increase worker productivity, and raise the standard of living for average American families.

FIGURE 16.6 The type of persuasive speech you choose frames the question you ask about a topic, as with a speech on surveillance cameras: "Do surveillance cameras in public places reduce crime?" (fact) "Is it morally right to install surveillance cameras in public places?" (value) "How should data from surveillance cameras be used?" (policy).

EXAMPLE 16.8

Topic: Public Safety Funding

Question: Should some funding for police departments be reallocated to other community services?

Title: "Funding Public Safety in Our Community"

General purpose: To persuade

Specific purpose: To persuade my audience that some funding for local public safety should be reallocated to other community services, such as youth programs, mental health providers, food banks, and shelter for the unhoused

Thesis: Our community would benefit from reallocating some funding for the police department to community services in three ways: reduce reliance on the police for public health issues, increase education and social initiatives to prevent crime, and build stronger neighborhood coalitions.

EXAMPLE 16.9

Topic: Campus Food Nutrition

Question: Should the food served on our college campus be labeled with nutritional information?

Title: "Know What's in Your Campus Food"

General purpose: To persuade

Specific purpose: To persuade my audience that all food served on campus should have nutritional labels

Thesis: Our campus food should have nutritional labels so students know the calories, sodium, sugar, fat, and other basic information about what they are eating.

The type of question you ask influences your topic research (Chapter 12), idea organization (Chapter 13), and supporting materials you will provide for your position (Chapter 13). Examine how this process works in questions of policy in Box 16.2.

BOX 16.2: LIVE IT LOCAL
Weighing in on Community Policy

Go to your town, city, or county government's website. Find out when the governing body, such as the board of supervisors or city council, meets next. Review the agenda and identify the questions of policy the group faces. Choose one and research it so you have a solid working knowledge of the issues. Then, draft some comments you would like to make during the public discussion at the next meeting. You do not have to speak at the meeting, but you want to be prepared. Finally, attend the meeting, observing what participants talk about and noting their concerns. What strategies do speakers use to persuade listeners to their point of view? Did you decide to speak? What did you say? How persuasive were you? What have you learned about participating in community discussions of local policy issues?

The Reasoning Process in Persuasive Speaking

reasoning The process of using evidence or information to support claims

evidence The supporting materials—facts and statistics, examples, definitions, testimonies, and narratives—gathered in researching a speech

claim A declarative statement that may be either true or false

argument A claim or conclusion supported by reasons or the premise

conclusion The main assertion of an argument

premise The reason or support for an argument's conclusion

Persuasive speaking requires presenting good reasons for audience members to agree with your position. **Reasoning** involves the process of using evidence or information to support claims. **Evidence** refers to the supporting materials—facts and statistics, examples, definitions, testimonies, and narratives (Chapter 13)—you gathered in researching your speech (Chapter 12). A **claim** is a declarative statement that may be either true or false. For example, Justice Sotomayor made this claim: "Being able to imagine yourself in someone else's shoes [is] one of the skills that makes you a good judge."[14] Persuasive speakers make claims that they want their audiences to support. Good reasoning provides the foundation for effective persuasive speaking.

When persuasive speakers use reasoning to win over an audience, they give arguments that align with their position.[15] An **argument** presents a claim or conclusion supported by reasons or the premise. The **conclusion** is the argument's main assertion, whereas the **premise** is the reason or support for the conclusion.[16] Effective persuasive speakers include evidence based on their research (Chapter 12) to support their claims. Figure 16.7 shows how the elements of reasoning are related to persuasive speaking.

An argument may have multiple premises, but it can have only one conclusion. In Justice Sotomayor's speech, she first stated the conclusion and offered two primary premises or reasons:

Premise 1: You need to imagine yourself in the shoes of both parties that stand before you.

Premise 2: You need to be sensitive to how your words are perceived on all sides.

Conclusion: Being able to imagine yourself in someone else's shoes [is] one of the skills that makes you a good judge.

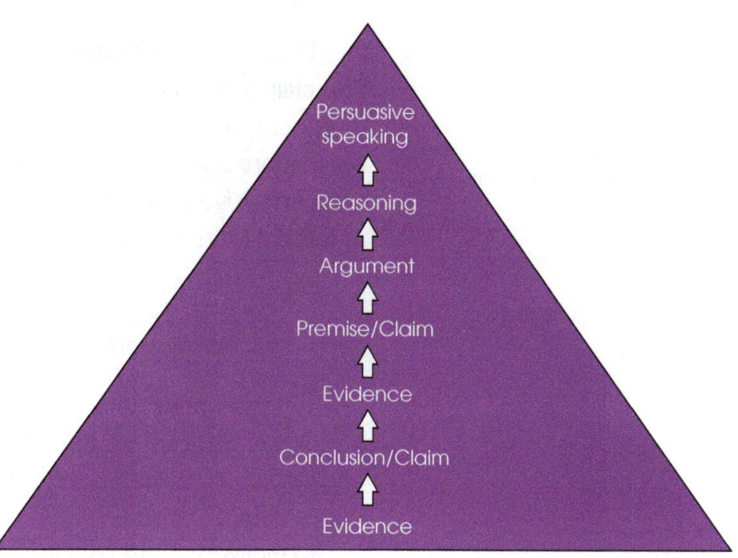

FIGURE 16.7 Persuasive speaking and the elements of reasoning.

indicator words Words that identify the parts of an argument

Both premises and conclusions are claims, but each type serves a different purpose. Premises provide reasons to support the conclusion; the conclusion tells you the overall assertion the speaker wants you to believe.

Indicator words can help you identify the parts of an argument. Sometimes the words are clearly stated; other times they are implied. Words that indicate a premise include *because, so, since, for, as, given that, as indicated by, as implied by, in that*, and *for the reason that*. Words indicating a conclusion include *therefore, thus, so, it follows that, as a result, consequently, hence*, and *accordingly*. In arguing against a plan to build a new health center at your school, you might say:

Premise 1: Because most students do not live on or near campus

Premise 2: In that students would have to drive or take public transportation to get to campus and use the new health center's services

Conclusion: It follows that a new health center should not be built on our campus.

However, speakers rarely present their arguments that explicitly. Often it is up to audience members to fill in the blanks and identify where those indicator words would go. For example, if Justice Sotomayor had used indicator words, she might have said:

Premise 1: *Because* you need to imagine yourself in the shoes of both parties that stand before you.

Premise 2: *Given that* you need to be sensitive to how your words are perceived on all sides.

Conclusion: *Therefore* being able to imagine yourself in someone else's shoes [is] one of the skills that makes you a good judge.

enthymeme An argument in which a premise or the conclusion is implied and not stated

Like indicator words, speakers may leave premises and conclusions unstated. The claims are still there, but left for the audience to fill in. When the premise or conclusion is implied, it is an argument that is called an **enthymeme**. When persuaders use an enthymeme, they are inviting the audience to participate in the persuasive process as members work through the missing pieces of the argument.[17] An example of an enthymeme with an implied premise is:

Premise: Fruits and vegetables are good for your health.

Conclusion: People need more fruits and vegetables in their diets.

The unstated premise is "People don't eat enough fruits and vegetables." The premise is obvious, so left unstated. Leaving a premise unstated is a way to engage audience members in an argument as they think about how the speaker reached the conclusion.[18]

Conclusions the audience already knows often are left unstated as well. Consider this example:

Premise 1: The city library is short staffed and lacks money to purchase books.

Premise 2: The new bond measure on the November ballot will provide the library with adequate funding.

The conclusion is obvious: Vote for the new bond measure. If a conclusion is easy for audience members to figure out, speakers often will leave the conclusion unstated. Encouraging audience members to reach the conclusion themselves is more persuasive than telling them what to do.[19]

Types of Reasoning

Persuasive speakers may use either deductive or inductive reasoning. The type of reasoning you use in your speech depends on your topic, purpose, and audience, as you will observe in the examples that follow.

Deductive Reasoning

deductive reasoning Reasoning that starts with a general principle and then applies it to a specific case

Deductive reasoning starts with a general principle and then moves to a specific case. For example:

Premise 1: All students who successfully complete a course in communication improve their communication skills.

Premise 2: Julie successfully completed a course in communication.

Conclusion: Julie improved her communication skills.

For deductive reasoning to work, both premises must be true. Review this example:

Premise 1: All people in this class are City College students.

Premise 2: Suzy is in this class.

Conclusion: Suzy is a City College student.

"Suzy is a City College student" seems like a reasonable conclusion *if* the major and minor premise are true. But there are some cases in which all the people in a class are *not* students. For instance, the teacher is not a student. In addition, some colleges allow members of the local community to audit a class without matriculating as a student. And other colleges have programs that allow students from another college to enroll in a class. As you can tell from this example, a faulty premise can lead to a faulty conclusion.

Inductive Reasoning

Inductive reasoning, also called reasoning from example, involves using specific examples or instances to support a general claim or principle. Think of inductive and deductive reasoning as opposite methods: The former starts with a specific case and moves to a general principle, and the latter starts with a general principle and applies it to a specific case, as shown in Figure 16.8.

Inductive reasoning takes two primary forms, enumerative and analogical.[20] With enumerative induction, the speaker uses observations about one or a few members of a group and generalizes to the entire group, as with this example:

Premise: Most people I know who live in the southern part of the United States are friendly.

Conclusion: All people from the southern part of the United States are friendly.

Enumerative inductive reasoning relies on numbers, either implied as with *most, some, all,* or stated, as with *95%, 4 out of 10, one third*. What is known about part of the group is used to draw a conclusion about all the members of the group. Pollsters use enumerative inductive reasoning to identify trends and patterns. For example, when the news media report that 40% of likely voters support Candidate X, 35% support Candidate Y, and the rest are undecided, those conclusions are based on the responses of only some voters. With enumerative reasoning, you are assuming part of the group is representative of the entire group.

inductive reasoning
Reasoning that involves using specific examples or instances to support a general claim or principle

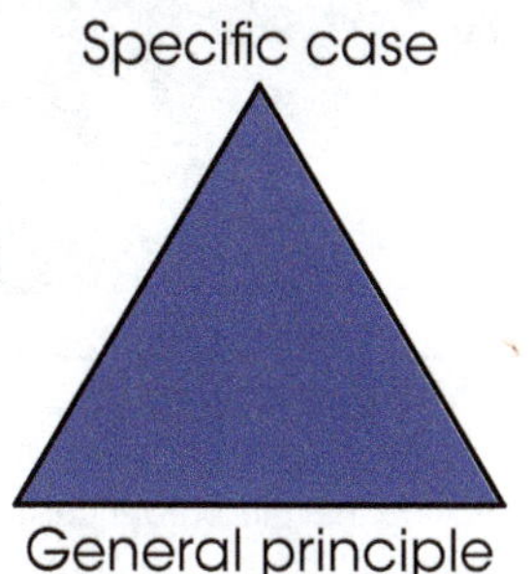

FIGURE 16.8 A comparison of deductive and inductive reasoning.

Analogical induction, also called argument by analogy, compares two things that share similar characteristics, as with this example:

Premise 1: City College is much like our college in that both are located in a metropolitan area, have a large number of commuter students, and have many of the same degree programs.

Premise 2: City College has instituted a successful online degree program in several areas, such as business and urban planning.

Conclusion: Therefore, an online degree program would be successful at our college.

For analogical reasoning to work, the two things compared must share sufficient qualities so the comparison makes sense. In the example, the colleges are similar in three ways. However, those may not be the most important areas of similarity for instituting an online program. For instance, students would need to have similar internet access and computer skills for this analogy to hold.

Deductive reasoning is more certain than inductive reasoning. With deductive reasoning, if the general principle and specific case are true, then the conclusion must be true.[21] If all border collies are smart and your dog is a border collie, for example, then she must be smart (Figure 16.9).

FIGURE 16.9 You could use deductive reasoning to conclude that this border collie is highly intelligent: All border collies are highly intelligent. My dog is a border collie. My dog is highly intelligent.

Inductive reasoning is based on generalizability—applying an example to an entire group. People naturally use this form of reasoning because they base conclusions on their own experiences. If you tried convincing an instructor to change your grade on a paper and the strategies you used were successful, you might think those strategies would work with all instructors. But inductive reasoning is only as good as the example or examples on which it is based. In addition, inductive reasoning always involves an element of probability—the conclusion likely is true based on the premises, but you cannot be absolutely certain.

Fallacies: Faulty Reasoning

A **fallacy** is a defective argument that reflects an error in reasoning.[22] Although a fallacy may be persuasive, its power to convince rests on false appeals rather than logic. Fallacies fall into two categories, those with irrelevant premises and those with unacceptable premises.[23]

1. Fallacies with irrelevant premises are statements unrelated to the conclusion. For instance, when speakers use the *ad hominem* fallacy, they attack the person who made a claim or presented evidence rather than the claim or evidence itself. Audiences can find *ad hominem* fallacies persuasive when they perceive a speaker has high status.[24] However, personal attacks fail to hold up to any kind of logical scrutiny.

2. Fallacies with unacceptable premises do not provide adequate support for the conclusion. For example, the false dilemma fallacy presents the audience with only two choices when additional ones are available. In a speech on addressing college graduates' rising debt, the speaker might argue that either tuition costs must be reduced or students will not be able to go to college. But there might be other solutions, such as changing graduation requirements so students take fewer units, expanding the work-study program, increasing scholarships, or basing the fee structure on a sliding scale that reflects students' ability to pay.

Knowing about common fallacies will help you identify them in your own and others' reasoning. Table 16.3 on the following page offers definitions and examples of the different types of fallacies you want to avoid as a persuasive speaker and quickly identify as an audience member.

Now that you have reviewed the various fallacies, test your skills in identifying them in Box 16.3 on page 405.

Guidelines for Effective Persuasive Speaking

Persuasive speaking requires your audience to do something, either by asking them to think another way or to take some sort of action. In addition to the strategies outlined so far in this chapter, the guidelines that follow will help you successfully meet your goal.

fallacy A defective argument that reflects an error in reasoning

TABLE 16.3 Common Fallacies in Reasoning

	FALLACY	BRIEF DEFINITION	EXAMPLE
Fallacies with irrelevant premises	**Composition**	Qualities of the parts are transferable to the whole	Each student on our team is an excellent student, so we will have a first-rate team for our presentation.
	Division	What is true of the group is true of the individual	Students in her major give the best speeches, so she will give a great speech.
	Red herring	Purposefully introducing an irrelevant issue	College students need to spend more time talking with each other, so we should ban all video games.
	Caricature (straw person)	Distorting a claim to more easily attack it	The people who oppose my plan support one that will turn our local parks into places only an elite few can use.
	Appeal to person (ad hominem)	Criticizing the person rather than the claim	He never does his research, so we should not support his proposal.
	Appeal to popularity (ad populum)	A claim must be true just because others support it	All universities are using this new learning management system, so we should use it too.
	Appeal to tradition	A claim must be true only because of long-standing practice	The best time for our college's commencement is Memorial Day weekend because we have always scheduled it then.
	Appeal to ignorance (ad ignorantiam)	A claim must be true because it has not been disproved	Your evidence does not prove that time travel is impossible, so it must be possible.
	Appeal to emotion	Using emotion as the basis of an argument	Without this new playground, neighborhood children will have no place to laugh, play, and have fun.
Fallacies with unacceptable premises	**Begging the question**	Using the conclusion as a premise	Our group is the best because we think we are the best.
	False dilemma	Providing only two alternatives when more exist	We either lay off teachers or increase class sizes.
	Slippery slope	Taking one step must lead to undesirable steps	If we do not allow people to park their cars in front of downtown stores, everyone will go to the mall where parking is plentiful.
	Hasty generalization	Inadequate sample size for a generalization	My friend's car tires were slashed in the parking garage, so it is not safe to park your car there.
	Faulty analogy	Things compared not sufficiently similar	Lecturing is the best way to teach because college students' brains are like sponges—they will soak up knowledge like water.

Know Your Audience

When you know your audience, you design your message to address your listeners' attitudes and interests. Briar Goldberg, director of speech coaching for TED Talk speakers, suggests you should ask yourself, "What gift am I giving my audience?"[25] Think about how you choose a gift for someone. You consider their interests, likes, dislikes, goals, and aspirations because you want to select a gift just for that person. That is what you want to do when you give a persuasive speech—design a message just for that audience (Figure 16.10).

As a persuasive speaker, you will encounter different types of audiences who will be more or less inclined to agree with you, more or less informed on your topic, and more or less interested in what you have to say. Research on audiences generally groups them into four main categories: supportive, hostile, uninformed, and apathetic. Follow these guidelines in developing effective arguments to support your position with each type of audience:

FIGURE 16.10 Asking "What gift can I give my audience?" will help you design an audience-centered persuasive speech.

- Although a supportive audience will view you and your position favorably, you do not want to take such an audience for granted. You still must present reasonable arguments that will encourage audience members to agree with you. When your listeners are supportive, you want to reinforce their commitment to your position. Using narratives and other engaging supporting materials will remind them of their identification with the topic.

- Establishing your ethos, or credibility, is especially important with a hostile audience that clearly disagrees with your position. Make clear that you understand all sides of the issue. Build on areas of common ground with the audience to encourage them to view you and your topic more favorably. Vivid examples will help listeners visualize the topic. Facts and statistics provide solid support.

Keep your tone conversational and avoid reacting nonverbally to any negative responses from the audience. By maintaining a friendly yet professional demeanor, your audience will view you as more approachable and less threatening.[26]

- With an uninformed audience that knows little about your topic, you first must educate them before you can persuade them. Use examples and narratives to personalize the topic. Avoid the temptation to overload your audience with facts and statistics. Explain how the topic and your position will benefit your audience.

- For an apathetic audience that knows about the topic but is disinterested, you must engage them from the first moment of your speech. Use supporting materials the audience members will find particularly compelling, such as statistics and examples directly related to them. In addition, have audience members visualize themselves as a character in a narrative related to your topic to keep their interest and persuade them to accept your position.[27] Give an especially dynamic and energetic presentation that integrates engaging presentation media.

These broad categories of audiences offer guidance in adapting your speech to their orientation on your topic. Still, nothing substitutes for a thorough analysis of your audience to get a sense of the mixture of values, beliefs, and viewpoints related to your topic that you will encounter. Complete the activity in Box 16.4 to learn more about the audiences in your community.

BOX 16.4: LIVE IT LOCAL
My Community, My Audiences

Who are the audiences that make up your local community? Identify a topic on which you would like to give a persuasive presentation at your local neighborhood association, church group, student organization, and similar groups. Choose three groups and draw up a plan for getting to know the audiences in those parts of your community. What might you find out from the groups' websites? From local newspapers? From each group's social media? From interviewing group members? Once you are satisfied with your audience research, consider giving a persuasive speech to one or all of the groups. How will you design your persuasive message for that audience?

Demonstrate Your Knowledge

Demonstrating your knowledge shows your audience you are an expert and you have done your research on the topic. The three strategies that follow are examples of ways to demonstrate your knowledge.

- *Use a variety of supporting materials.* As you research your speech, identify excellent evidence from a range of sources and types, as Viola Davis did in the examples included at the start of this chapter.

- *Cite authorities on the topic.* In his keynote address on privacy and secrecy at LA Hacks, Snapchat CEO Evan Spiegel referred to New York University professor Helen Niessenbaum's work on privacy and technology and quoted novelist Milan Kundera on privacy and the disappearance of the individual. These references indicated that Spiegel had researched his topic.

- *Refer to your own experiences.* Your experience with a topic is part of your knowledge base. If you are trying to convince your audience to exercise more and you regularly work out, let your audience know. Even when your audience is familiar with your background, reminding them of your experience with the topic reinforces your authority on it. For example, Justice Sotomayor referred to her experience as a Supreme Court justice several times in her PEN Lecture discussed earlier this chapter.

Engage Your Ethos

Although demonstrating your knowledge on a topic does enhance your credibility, you need to do more than that to show your audience you are trustworthy. The strategies that follow offer suggestions for using ethos to appeal to your audience.

- *Treat all sides of the issue fairly.* Avoid appearing biased and one-sided. Show that while you favor a particular position on the topic, you appreciate others' opinions.

- *Give due consideration to opposing viewpoints.* Make clear that you understand the arguments in opposition to your own. Be prepared to support the weaknesses in your position.

- *Present sound arguments.* Speakers who present well-constructed arguments are viewed as more credible than those who use faulty reasoning.[28] Review your arguments for unsubstantiated claims and fallacies and correct those errors in reasoning before you give your speech.

- *Use language your audience understands.* Avoid excessive jargon. Define any concepts or terms your audience may find confusing.

- *Have a conversation.* Talk *with* your audience rather than at them. Make them feel included in your speech. When you show your listeners you respect them, they are more likely to respect you (Figure 16.11).[29]

FIGURE 16.11 Your speeches will be more persuasive when you make your audience feel included and respected.

Give a Polished Presentation

Whenever you give a speech, you want to give a polished presentation. This is especially true for persuasive speeches. The guidelines that follow won't guarantee you will persuade your audience, but they will increase the likelihood of achieving your goal.

- *Practice, practice, practice.* Rehearse your speech thoroughly so you know it well and can easily adjust if you need to make accommodations based on audience feedback.

- *Articulate your words and avoid vocalized pauses.* Clearly speaking, pronouncing words correctly, and avoiding disfluencies and vocalized pauses demonstrate your command of the topic. When speakers are difficult to understand, make errors in pronunciation, and use "ahs" and "ums," the audience interprets those behaviors as lack of knowledge and preparation.[30]

- *Turn up the energy.* Because you are trying to motivate your audience to agree with you, they must feel energized to take that step. A lively and dynamic presentation will help set a tone that can make your audience feel more inclined to agree with you.

- *Use simple yet elegant presentation media.* Integrate intriguing and appealing presentation media that support your arguments.

CHAPTER REVIEW

- **Discuss the four characteristics of persuasion in a pervasive communication environment.**

  Audiences exercise agency in persuasive communication situations, no single message will persuade an audience, much of persuasion is implied rather than stated, and persuasion is everywhere yet appears to be nowhere.

- **Define and identify examples of the four means of persuasion.**

  Aristotle identified three means of persuasion or modes of proof: logos, or logical appeals; ethos, or appeals to the speaker's character; and pathos, or emotional appeals. In addition, more recent scholars have suggested a fourth means of persuasion, mythos, or narrative appeals.

- **Define and identify examples of the different types of persuasive speeches.**

  Persuasive speeches address questions of fact, value, or policy. A question of fact examines whether something is true or false. A question of value considers the subjective worth of something, such as good/bad, right/wrong, important/unimportant. A question of policy is concerned with an action to be taken about something.

- **Describe the reasoning process in persuasive speaking.**

  Reasoning is the process of using evidence or information to support claims. Claims are declarative statements that may be true or false. Speakers use arguments, which include at least one premise and a conclusion, to support their position on a topic.

- **Explain how to use inductive and deductive reasoning to support your position in a persuasive speech.**

  Speakers may use deductive reasoning in which a general principle is used to support a specific case or inductive reasoning in which a specific case is used to represent a group. Enumerative induction uses observations about one or a few members of a group and generalizes to the entire group. Analogical induction compares two things that share similar characteristics.

- **Identify examples of common fallacies in reasoning.**

  The two common groups of fallacies are fallacies with irrelevant premises and fallacies with unacceptable premises.

- **Discuss the guidelines for effective persuasive speaking.**

  Knowing your audience, demonstrating your knowledge, engaging your ethos, and giving a polished presentation are the key guidelines for effective persuasive speaking.

DISCUSS AND APPLY

Review this student speech given for a persuasive speaking assignment and then answer the questions at the end.

GET READY TO VOLUNTEER

I come from a family of volunteers. Every week, my mom delivers meals for Meals on Wheels. On Saturdays she tutors people for the GED, the exam you can take if you didn't finish high school. My dad chairs our local friends of the library group, organizing book drives and giving library tours to middle school children. We've fostered puppies and kittens for our town's animal shelter since I can remember. My older sister served for 2 years in the Peace Corps as a health volunteer, working with families and community groups to promote healthy nutrition practices. My younger brother, who's a senior in high school, helps out at our neighborhood's community garden project and last summer taught art classes for a kids' summer camp. Now he's tutoring grade schoolers in math online because the schools are closed.

And then there's me. I always was too busy with sports, school, my friends, and just doing whatever I wanted to volunteer. Plus, I figured my family was doing enough. That all changed with the COVID-19 pandemic and the governor's stay-at-home orders. Suddenly, all sports activities were canceled, my classes were online, and I couldn't hang out in-person with my friends. There's only so much time you can chat on FaceTime. Then I read about the shortage of face masks for first responders, health care providers, and others in critical fields. I realized there was something I could do. I have basic sewing skills, my family has a sewing machine, and I certainly had time. I became a volunteer, sewing special cloth face masks with a pocket for a filter that have been donated to essential workers.

Volunteers are needed for every aspect of life and every task you can think of. There are so many ways college students can help. Today I'll tell you about the need for volunteers, how that need might be met, the benefits of volunteering, and specific volunteer opportunities available right here in the Bay Area.

A recent study by the Corporation for National and Community Service, the federal agency that oversees AmeriCorps and Senior Corps, found that about 30% of adults in the U.S. do some sort of organized volunteer work for a total of nearly 7 billion volunteer hours, worth almost $700 billion dollars. Just over 20% of volunteers are under 24 years old. And those numbers don't include the millions of people contributing billions of hours to informal volunteering as they help out family, friends, and neighbors.

What city is ranked first in the U.S. for millennial volunteers? If you guessed San José, you're right. Volunteerhub reports our city is a hotbed of young people volunteering. In addition, out of 51 large metropolitan areas, San José-Sunnyvale-Santa Clara is fifth in the nation for number of volunteers and the top one in California, ahead of San Francisco, San Diego, Sacramento, and LA.

However, in spite of those impressive statistics, CityLab reported that the number of people who volunteer in the U.S. is on the decline. That is, today, fewer people are volunteering more hours. Commute times, financial stress, and the rise of the gig economy all contribute to lower rates of volunteerism. And yet, there is so much need for volunteers, especially during this pandemic. VolunteerMatch estimates more than 500 volunteers are needed in San José alone, and the organization's website lists more than 250,000 online volunteer opportunities available in the U.S. right now. And those are just the volunteers needed to respond to the COVID-19 pandemic.

As college students, we can help. If every person in this class volunteers, and each of you recruits two people to volunteer, and each of them recruits two people to volunteer, and they do the same—I'm sure you've figured out where I'm going with this. With a little bit of networking and recruiting, we'd soon fill those 500 plus volunteer positions needed in San José. Of course, that is just a start, but it is a start and one way each person in this classroom can make a difference to those in need.

Imagine feeling healthier, happier, and less stressed. You will when you volunteer. Volunteering will benefit you in many ways, from connecting with your community and helping land a job to feeling better physically and emotionally. Volunteering does have its downsides, but I'll tell you about the upsides first.

For those of you who have volunteered, you've likely experienced many of the benefits that come with helping others. Research has found that when you volunteer in your community, you're more likely to know and interact with your neighbors and feel a part of the place you live. Volunteering can decrease feelings of depression and loneliness, as you reach out and help others. A study of 1,500 volunteers in Texas found that volunteering increased feelings of well-being, overall satisfaction with life, and mental and physical health.

There's no doubt serving as a volunteer contributes positively to your emotional health. University students in Canada who volunteered at a food bank reported that they felt a sense of freedom to do their work, a sense of belonging and community, and a sense of fulfillment. But mostly, students felt the satisfaction of knowing they were helping others. As one student in the study said, "Yes, it is fun to enjoy the time there … But for me it was important to offer my help at 100% and to leave my personal joy aside." Also, a study of college students who volunteered in health care organizations found that the students developed greater resilience, flexibility, and confidence, as well as improving specific skills. For instance, one student said: "I can now talk in large groups of people; I am more confident with that. Public speaking was definitely one issue I had before."

From a practical standpoint, volunteering gives you an added boost in the job search process. A study of student volunteers at the University of Minnesota in Morris found that their volunteer experiences led directly to important career opportunities. As these students discovered, volunteering can contribute to the development of key professional skills and present avenues for networking. An article on the job search site Monster noted that recruiters view job candidates with volunteer experience as having passion and ambition. In a recent survey of U.S. companies by the multinational accounting firm Deloitte, the vast majority of employers—more than 80%—reported that they were more likely to hire job candidates with volunteer experience and that experience would make up for other areas in which candidates might be lacking. Your volunteer work will make you stand out from other applicants.

However, there can be drawbacks to volunteering. You have to be careful of over-volunteering or choosing volunteer opportunities that aren't a good fit for you. Psychologist Joachim Krueger at Brown University suggests that if you place too much value on volunteering, you'll become overcommitted and burned out. In addition, you want to find volunteer positions that match your interests and skills. Also, you already may be feeling overburdened with work, school, and family. You don't want volunteering to add to your stress level. Finally, sometimes your expectations might not meet with your experience volunteering, and that can be disappointing.

Still, the benefits of volunteering are clear. When you volunteer, you'll have improved physical and emotional health, greater confidence, an increased skill set, and a resume boost. Those are some of the benefits you'll get from volunteering.

Are you ready to volunteer? I'll tell you about three ways to get started. First, consider enrolling in a service-learning course. For example, one of my friends took COMM 157SL, Community Action/Communication Service, last semester. She worked with elementary school children at the Sacred Heart Community Center in downtown San José. Not only do you get a meaningful volunteer experience, you get upper-division GE credit.

Second, check with local nonprofit organizations you're interested in supporting and find out what kind of volunteer help they need. Even with the current stay-at-home orders, there still are ways you can help out. For example, I contacted the Santa Cruz SPCA because I love dogs. They needed help with their social media, so I am posting about the pets they have for adoption on their Facebook page.

Third, go online to a volunteer listing website, such as Oakland-based VolunteerMatch or Great-Nonprofits. If you want to volunteer in your community, type in your location and you'll get a list of volunteer openings. Scroll through and identify those you think will fit with your interests and skills. There's also HandsOn Bay Area, which lists volunteer opportunities, corresponding organizations, and specific dates and times when help is needed. Or you can apply for online volunteer work—VolunteerMatch lists more than 650,000 opportunities in the U.S.

During the time of this pandemic, it's easy just to focus on yourself. But by volunteering, you can help others and improve your health, skills, and job prospects. I've explained three ways to find volunteer positions—sign up for a service-learning course, contact a local nonprofit related to your interests, or go online to a volunteer opportunities website and search for a position that's a good fit for you.

Last week, I found out the masks I sewed were sent to 50 workers at our local grocery store chain. A few of the employees posted selfies of them wearing my masks on the store's Twitter feed. Of all the things I've done in the past month, I'm the proudest of this. You could feel that way, too. Step up for your community and volunteer.

References

Corporation for National and Community Service. (n.d.). *City rankings by volunteer rate*. https://nationalservice.gov/vcla/city-rankings-volunteer-rate

Corporation for National and Community Service. (2018, November 13). *Volunteering in U.S. hits record high; worth $167 billion*. https://nationalservice.gov/newsroom/press-releases/2018/volunteering-us-hits-record-high-worth-167-billion

Deloitte Development LLC. (2017). *Deloitte volunteer impact research: Measuring important aspects of corporate community engagement*. https://deloitte.com/us/en/pages/about-deloitte/articles/citizenship-deloitte-volunteer-impact-research.html

HandsOn Bay Area. (2020). *Volunteer opportunity calendar*. https://handsonbayarea.org/calendar.

Haski-Leventhal, D., Paull, M., Young, S., Maccallum, J., Holmes, K., Omari, M., Scott, R., & Alony, I. (2020). The multidimensional benefits of university student volunteering: psychological contract, expectations, and outcomes. *Nonprofit and Voluntary Sector Quarterly*, *49*(1), 113–133.

Krueger, J. (2019). The vexing volunteer's dilemma. *Current Directions in Psychological Science*, *28*(1), 53–58.

Neumann, A. (n.d.). *How volunteering can help you get a job*. Monster. https://monster.com/career-advice/article/five-ways-volunteering-can-help-you-get-a-job-hot-jobs

Poon, L. (2019, September 11). *Why Americans stopped volunteering*. Bloomberg CityLab. https://citylab.com/life/2019/09/volunteer-opportunities-charitable-giving-national-service/597856/

Skulan, N. (2018). Staffing with students. *Digital Library Perspectives*, *34*(1), 32–44.

VolunteerHub. (2020). *25 volunteer statistics that will blow your mind*. https://volunteerhub.com/blog/25-volunteer-statistics

VolunteerMatch. (2020). *During these uncertain times, how can we help?* https://volunteermatch.org/covid19

Williamson, I., Wildbur, D., Bell, K., Tanner, J., & Matthews, H. (2018). Benefits to university students through volunteering in a health context: A new model. *British Journal of Educational Studies*, *66*(3), 383–402.

Yeung, J. W. K., Zhang, Z., & Kim, T. Y. (2017). Volunteering and health benefits in general adults: Cumulative effects and forms. *BMC Public Health*, *18*(1), 1–8.

Yuriev, A. (2019). Exploring dimensions of satisfaction experienced by student volunteers. *Leisure/Loisir*, *43*(1), 55–78.

QUESTIONS FOR DISCUSSION

1. Review the organizational patterns for speeches to persuade discussed in Chapter 13. Which pattern did the speaker use? How do you know?

2. In what ways does the speech reflect the four characteristics of persuasion? Give examples.

3. What means of persuasion or appeals to the audience does the speaker use? Give examples.

4. Is the speaker addressing a question of fact, value, or policy? How do you know?

5. What is the main claim the speaker is making? On what premises is the speaker basing the claim?

6. What type(s) of reasoning does the speaker use? Give examples.

7. How persuasive did you find the speech? Why do you evaluate it that way?

8. If you were giving the speaker advice, what would you suggest for improvements?

9. What have you learned that you can apply in your own persuasive speaking?

CREDITS

GLOSSARY

A

accenting   Using nonverbal communication to stress or emphasize parts of verbal communication; to accentuate

accommodating style   Acquiesces to let the other person win; shows little concern for your own interests but high concern for your partner's interests

adaptors   Gestures related to comfort and less centered on communication

aesthetic needs   Desires for balance, order, and beauty

affect displays   Gestures that share emotions

affective trust   The emotional bond and reciprocity of care and concern group members feel toward each other

appearance filter   Your interpretations of how a person "appears" to you; largely physical appearance

appreciative listening   A type of listening that involves listening for enjoyment

argument   A claim or conclusion supported by reasons or the premise

arrangement   One of the five canons of rhetoric, arrangement is the order in which the speaker organizes the ideas in a speech

artifacts   Objects associated with a culture, such as buildings, artwork, and clothing; Use of physical objects to communicate, such as items that we wear or have in our surroundings

associative friendships   Form due to an association, or physical connection, with another person; might be closer to friendly relations than friendships

attitudes/values filter   Concerned with the psychology underlying the person's behavior, their worldview, and shared psychological similarities the two of you have

autocratic rule   A type of decision making where the leader alone makes the decision for the group

autonomy versus connection dialectic   Competing desires to be both independent and interdependent

avoiding couple   Works to avoid conflicts; they sometimes believe that by giving issues time the conflict will decrease or disappear; thus, it is better to not confront it

avoiding style   Seeks to not have the conflict; demonstrates little concern for individual interests (i.e., one's own or the other person's)

B

belongingness needs   What you need to feel connected with others, accepted, and loved

body language   Nonverbal communication

C

cause-effect pattern of organization   A pattern of organization in which the main points in the body of a speech are arranged in a way that attempts to show that one event or series of events led to a particular outcome or set of outcomes

chronemics   Use of time to communicate

chronological pattern of organization   A pattern of organization in which the main points in the body of a speech are arranged based on when they occurred in time

claim   A declarative statement that may be either true or false

coculture   A set of people with a distinct set of behavior and beliefs that differentiate them from a larger culture of which they are a part

code switching   Move back and forth between ways of speaking and behaving to adapt to different cocultures

cognitive needs   Desires associated with gathering information, exploring the world, and better understanding something

cognitive offloading   The use of external devices, such as smartphones, to reduce an individual's mental efforts and extend mental capabilities

cognitive trust   A logical reason to believe in other group members

collaborating style   Tries to maximize outcomes; shows high concern for both your interests and your partner's interests; wants both to win

collective communication competence   Occurs when individuals share an ability to interact effectively and appropriately, adapting to each other and various contexts

collectivist culture   Prize loyalty to the group and cooperation

common knowledge effect   Occurs when you overemphasize common knowledge held by a group of people and underemphasize knowledge held by certain individuals

communal identity   The identity groups or networks people form over time

communication   The exchange of socially shared symbols that represent interactants' thoughts, ideas, and feelings

communication climate   The tone of interaction among group members

communication competence   Occurs when an individual demonstrates effectiveness in achieving goals and appropriateness in adapting to norms and contexts

communication filter   Using others' communication to try to assess what they think and feel; their internal thinking

communication imperative   The inherent social nature of human beings that drives them to communicate with others using the tools available

community of practice   Occurs when group members share ways of working and learning together that help them strive toward common goals

competing style   Defines conflict as a competition and seeks to gain—goal is to win; shows high concern for your interests and low regard for your partner's interests

complementing   Aligning nonverbal and verbal communication; to support

complete-sentence outline   Written in full sentences, this outline includes all the aspects of a speech, such as the thesis, main points, transitions, and bibliography

compromising style   Each participant gets part of what wanted—the goal is to ensure each benefits some; shows some degree of concern for both your interests and your partner's interests

conclusion   The main assertion of an argument

conflict   Occurs when individuals are interdependent and perceive that they have incompatible goals

conflict management styles   Competing, accommodating, avoiding, compromising, and collaborating; discussed in terms of how much concern is shown for two types of interests— your own and your relational partner's

connotative meaning   Personal or informal meanings that you associate with a word or other symbol; related to the reference in the meaning triangle

consensual families   Interest centers in both conversation and conformity; children are encouraged to share their views and participate in making decisions, though parents make the final decisions, and may use their parents' worldview to evaluate information

consensus   A decision-making process where the solution is what is acceptable to all members

consultative rule   The group's leader makes the decision after discussing it with the group's members

content listening   A type of listening in which you want to understand the essence of the person's message

contradicting   Misaligning nonverbal and verbal communication; to deviate

convergence   The intersection of two different entities that previously were separate

coordination   The principle of outlining that refers to how the points of your speech are related to each other, with each point having the same level or degree of importance

copyright   A legal framework designed to protect the rights of creators over their works

correlation   A statistical term that indicates when two variables are related to each other

creativity   Communicating with others to develop alternative, novel ideas and solutions to problems

credentialing   The process in which speakers establish their qualifications on a topic

critical listening   A type of listening in which the goal is to evaluate or assess the other person's message

critical thinking   Purposefully interpreting, analyzing, and evaluating information to solve problems and make decisions

cultural appreciation   Demonstrated by wanting to learn about another culture, build intercultural relationships, and increase understanding; broad view focused on understanding

cultural appropriation   Taking something from within a culture and using it for your benefit, often with little understanding of its cultural significance; narrow view (e.g., interest in one item or aspect, rather than understanding culture more fully)

cultural indicators   Help define key aspects of culture; several important indicators are vocabulary, practices, rituals, artifacts, and stories

culture   A set of values or beliefs informing behavior [and a] collective construction that transcends individual preferences

current information   Information that is up to date and provides the most-recent data on the topic

D

deductive reasoning   Reasoning that starts with a general principle and then applies it to a specific case

deepfakes   Videos and other digital content that have been manipulated using artificial intelligence (AI) or similar technology to create false images that appear authentic

defensive communication   The use of personal attacks and similar unproductive communication strategies that discourages individuals from freely expressing themselves

definition   The meaning of a word or phrase

deliberative practice   A process in which you identify goals, get feedback on your performance, and always explore ways to improve

delivery   One of the five canons of rhetoric, delivery involves using the voice and body to present a speech to an audience

democratic leadership   Group members vote on who will serve as leader

denotative meaning   The explicit, formal, or "dictionary" definition of a word or other symbol; related to the referent in the meaning triangle

designated leader   An elected or appointed group leader

dialogic listening   A type of listening that involves four components: listening with heart and mind, listening for and to culture, listening for power inequities, and listening for dialectics

dialogic virtuosity   Occurs when communicators are passionate about listening to and understanding the other person's perspective as well as clearly articulating their own

dialogue   Communication that invites all participants to express their viewpoints while at the same time retaining an openness to other perspectives

digital divide   The gap between those who have access to new communication technologies and those who do not due to social, economic, educational, and geographic inequities

digital literacy   The ability to use information and communication technologies to find, evaluate, create, and communicate information, requiring both cognitive and technical skills

direct speech   Language that is specific, detailed, and forthright

discovery listening   A process in which communicators identify specific listening problems, determine what causes those problems, and develop ways to address those problems

distributed leadership   Occurs when group members share leadership responsibilities, completing tasks and maintaining relationships when the need arises

diversity   Individuals with different attributes contributing their unique experiences to a group

dual-culture approach   Suggests that gender differences in language occur because women and men are members of different cocultures and thus have different goals in interaction

dyadic phase of dissolution   Make your partner aware of your relational dissatisfaction and assessment of the relationship

E

emblems   Gestures with culturally agreed upon meanings that represent words or phrases

emotional intelligence   The ability to perceive emotions, link emotion to thought, recognize the complexities of emotion, and manage one's own and others' emotions

emotional labor   Occurs when workers must monitor and control their own emotions as part of their job

empathic listening   A type of listening that involves withholding judgment, advice, suggestions, and evaluation

empathy   Understanding a message from the sender's perspective

empowerment   A cooperative form of power that seeks to enhance the capabilities and the influence of group members

enacted identity   The aspects of the self an individual express to others when they interact with them

enthymeme   An argument in which a premise or the conclusion is implied and not stated

esteem needs   Desires associated with having others recognize your achievements and competency

ethic of care   An approach to ethical behavior that involves providing support to others and not causing harm

ethic of responsibility   An approach to ethical behavior based on justice and equity

ethics   Making judgments based on moral principles to differentiate between right and wrong

ethos   The use of appeals to the speaker's credibility or authority to persuade an audience

event schemata   Cognitive structures that provide scripts for common occasions and situations

evidence   The supporting materials—facts and statistics, examples, definitions, testimonies, and narratives—gathered in researching a speech

example   Something that is used to represent an entire group or class of things

expert rule   Occurs when groups rely on a professional in a specified topic area to make a decision for the group

extemporaneous speaking   Giving a researched, organized, and practiced speech in a conversational style

external distractions   Anything in the environment that interferes with listening

extrinsic motivators   Incentives external to individuals that cause them to behave in a certain way

F

face and eyes   Types of nonverbal communication that includes eye contact and facial expressions; frequently used to communicate interest and emotion

fact   Something that is known to be true based on experience and observation

fair use doctrine   A legal principle that allows the limited use of an author's work if attribution is given to the source

fallacy   A defective argument that reflects an error in reasoning

first place group   The family, which is a social, involuntary, planned group

followership   The activities of those who are subordinate to a group's leader

forming phase   The first phase of group development when group members first join together

G

general purpose   The speaker's overall goal for the speech

grave-dressing phase of dissolution   Develop explanations for the relationship's end

groupthink   A process in which group members agree on a course of action or solution to a problem without fully discussing the advantages and disadvantages of the range of alternatives

H

haptics   Use of touch to communicate; includes playfulness, liking/positive emotion, control, ritual, hybrid, task related, and accidental

hate speech   Language that attacks people based on perceptions that they belong to a specific group, such as a particular race, sexual orientation, or religion

hearing   The physiological reception of aural stimuli or sounds

hesitation markers   Function to lessen the impact of language

hidden agenda   An individual's concealed motivations or plans that underlie the person's actions

hostile couple   Has frequent arguments that use negative communication behavior; disagreements are cutting, harmful, sarcastic, critical, and insulting—showing disdain for their partner; defensiveness is high, and listening and an interest in the other's views are low

I

identity   An individual's view of the self and the image of the self presented to others

identity gap   A disjuncture between or among the layers of identity

identity shift   Something people experience when feedback from their online self-presentation leads to changes in their identity

illustrators   Gestures that illustrate what is said; typically do not have clear meanings distinct from the verbal messages that they accompany

immediate memory   Memory you retain for just enough time to determine if you should continue processing the information

impromptu speaking   Creating a speech on the spot with little or no planning

indicator words   Words that identify the parts of an argument

indirect speech   Speech that relies on subtlety and the listener understanding social conventions to infer meaning

individualist culture   Prize self-attainment, competition, and personal growth

inductive reasoning   Reasoning that involves using specific examples or instances to support a general claim or principle

information   Facts and data compiled about something

information literacy   Recognizing when information is needed and having the ability to locate, evaluate, and effectively use the needed information

information overload   Occurs when individuals share more information than they can interpret effectively

information underload   Occurs when individuals lack sufficient information to consider and interpret a message's meaning

informative speaking   Speeches in which the speaker's goal is for audience members to learn more about a topic

integrity   The implementation of ethical practices

intensifiers   Function to amplify the conversational thrust of a statement

internal distractions   Physical sensations, emotions, and thoughts that interfere with listening

interpersonal communication   Message exchanges (message production and interpretation) between two people or a small group of people who are interdependent and mutually aware of each other; can be online or face-to-face and involve verbal and nonverbal communication

interpersonal relationships   Exist across contexts (e.g., family, friendship, etc.); can vary in duration as well as level of self-disclosure and intimacy; are characterized by a common connection, interpersonal communication, reciprocal benefits, and adherence to or renegotiation of cultural expectations

intimate interaction zone   Used for interactions with individuals we are very close to (romantic partners); typically under 1.5 feet and may involve touching

intrapsychic phase of dissolution   Individual phase where you consider the costs and benefits of the relationship and try to determine whether you want to remain in the relationship or leave it

intrinsic motivators   Incentives that encourage people to take action based on forces internal to the individual

invention   One of the five canons of rhetoric, invention is developing the ideas and arguments for a speech

involuntary relationships   Relationships in which you do not select your relational partner or partners; for example, you do not select your family members

K

key words   Terms and phrases that represent the essence of an idea

keyword   A word used to find information when using search software, such as an internet search engine or library database

kinesics   Use of movement to communicate

L

laissez-faire families   Little interest is placed on conversation or conformity; there tends to be less communication and conversational engagement as well as less emphasis on influencing others' thoughts or making decisions for them

language   The system we use to convey our meaning through symbols (words)

learning   A process that involves becoming aware of sensory stimuli, selecting which stimuli to consider further, making connections with what is known already, and then applying it in some way

listening   The process of receiving, interpreting, understanding, evaluating, and responding to messages

listening anxiety   Occurs when communicators feel inadequate in receiving and processing messages

listening fatigue   Occurs when communicators become physically and mentally weary due to a listening situation

listening fidelity   The degree of similarity between the sender's intended message and the receiver's interpretation of that message

listening reciprocity   Occurs when individuals listen to each other openly, completely, and thoughtfully, withholding judgment until the speaker is finished

logos   The use of logical appeals to persuade an audience

long-term memory   The main storage area for incoming messages

M

maintenance roles   Behaviors that facilitate constructive relationships among group members

majority rule   Making a decision whereby the alternative with the most votes wins

manuscript speaking   Giving a speech from a document written in advance

meaning triangle   Explains the relationship between objects or ideas (referent), words (symbols), and understandings (reference/ thought)

memorized speaking   Writing out a speech word-for-word in advance, committing it memory, and reciting it to the audience

memory   One of the five canons of rhetoric, memory is the speaker's depth of knowledge about a topic and ability to remember what to say when giving a speech

message interpretation   Processing what your conversational partner communicated

message production   Choosing the best words to get your ideas across to others

metasearch engine   A search engine that that uses several search engines at the same time

microexpressions   Facial expressions that occur rapidly, may go unnoticed, and unconsciously reveal information

mind mapping   A way to diagram ideas to visually represent how they are related

mindful communication   Involves the active and conscious processing of current information during interactions with others

mindless communication   The outcome of habitual and routinized interaction involving minimal processing of new information

monotone   Little variety in a speaker's rate, pitch, and volume

Monroe's motivated sequence pattern of organization   A pattern of organizing the main points of a speech into five steps: attention, need, satisfaction, visualization, and action

mutual awareness   All interactants are aware of the other and can adapt their communication accordingly

mythos   The use of narrative appeals to persuade an audience

N

narrative   A story or anecdote

narrative pattern of organization   A pattern of organizing the main points in the body of a speech to tell the story of a topic as a series of dramatic events

noise   Distractions that interfere with the listening process

nonverbal communication   Ways we communicate beyond words; sometimes called body language

nonverbal symbols   Representations of thoughts, ideas, and feelings without the use of language, as with gestures, eye contact, body movement, and touch

norming phase   The third phase of group development when groups create norms of interaction that guide how the group functions

norms   Rules that are socially enforced

O

online disinhibition effect   The tendency to self-disclose at higher levels when interacting online than when interacting in-person

openness versus closedness dialectic   Tensions surrounding how much information to share in a relationship

oral citations   Acknowledgments of the sources for supporting materials

orientation to conformity   A dimension used to conceptualize family communication; conformity focuses on the degree of overlap in family members' worldviews (e.g., opinions, values, etc.)

orientation to conversation   A dimension used to conceptualize family communication; conversation focuses on whether families encourage an open exchange of ideas

P

paralanguage   Use of features associated with the voice, but not words, to communicate; also called vocalics

parallelism   The principle of outline that refers to using a consistent numbering system and consistent language for your main and supporting points

participative leadership   The leader shares power with group members, empowering them to take initiative in their work and in making decisions

pathos   The use of emotional appeals to persuade an audience

perception   Sensing, selecting, organizing, interpeting and recalling verbal and nonverbal messages

performing phase   The fourth phase of group development when the group reaches goal attainment

person schemata   Cognitive structures concerned with specific people you know and particular types of people

personal identity   An individual's self-concept

personal interaction zone   Used for interacting with family and friends; usually 1.5–3 feet

person-based relationships   Relationships in which you recognize the others as distinct human beings, as with friends and family members

persuasive speaking   A type of speaking in which the speaker attempts to change or reinforce audience members' beliefs, values, attitudes, or behaviors

pervasive communication environment   A context in which individuals have the ability to use multiple access points to tap into an integrated communication structure with text, audio, video, and voice capabilities

phubbing   Gazing at your phone rather than attending to the people you are with

physiological needs   What you must have to keep your body functioning, such as food, water, and sleep

plagiarism   Occurs when you represent someone else's work as your own

pluralist families   Priority is placed on conversation and not on conformity; there is considerable discussion of ideas and opinions, with family members making their own conclusions; independent thinking is valued and supported

power   A force that gives one communicator the ability to influence others to take an action that would not be taken without that motivating force

power distance   How those with less power view power inequalities

powerful speech   Language that conveys social power; often attributed to men

powerless speech   Language associated with a subordinate social position; often attributed to women

practices   Ways tasks are accomplished; may or may not coincide with the rules or guidelines for getting things done or how things work

pragmatics   How to construct messages that are socially and culturally appropriate

predictability versus novelty dialectic   Competing desires for both stability and excitement

premise   The reason or support for an argument's conclusion

presentation media   Audio and visual materials that support the points or arguments of a speech

presentation outline   A brief outline with key words and phrases used when giving a speech

problem-solution pattern of organization   A pattern of organizing the main points in the body of a speech to show that a problem exists and then suggest how to solve it

prosumer   An individual who is both a communication producer and consumer

protective families   High interest in conformity but low interest in conversation; parents shoulder the responsibility, protect children from decisions, and expect agreement, and children may learn to accept the authority of others and to avoid engaging in conflict

proxemics   Use of space or distance to communicate

pseudo news   Also called fake news, these messages are disinformation that purposefully mimics the form of news stories, but the content lacks accuracy and credibility

pseudolistening   Occurs when communicators appear to be listening, but are not fully processing the message

public good   Outcomes that benefit everyone

public interaction zone   Used in situations such as public speeches or presentations; 12 or more feet between communicators

Q

question of fact   A question that asks if something is true or false

question of policy   A question that asks what course of action should be taken or how a problem should be solved

question of value   A question that asks for an evaluation of the subjective worth, quality, or condition of something

R

rapport talk   Seeks connection and agreement with others; often associated with women

reasoning   The process of using evidence or information to support claims

receptive friendships   Form despite differences in power; tangible benefits may seem to flow more in one direction

reciprocal friendships   Form between equals; benefits from the relationship are shared equally

reference   Your thoughts about a referent, such as how your regard your iPad; one of three parts of the meaning triangle

referent   A concept or entity, such as an iPad; one of three parts of the meaning triangle

regulating   Using nonverbal communication to start, indicate speaking order and length, and end verbal conversations; to synchronize

regulators   Gestures that help to regulate conversation

relational dialectics   Competing wants or needs that exist simultaneously in relationships, pull us in multiple directions, and play out through discursive struggles

relational dissolution model   Contains five phases: intrapsychic, dyadic, social, grave-dressing, and resurrection

relational filtering model   Developing relationships involves assessments and selections; four filters through which potential partners must pass

relational identity   The aspect of identity that evolves in the relationships individuals form through communication with others

relational schemata   Cognitive structures that provide a frame for romantic, kinship, friendship, work, and other relationships

reliable information   Information that is verifiable and from a credible source

repeating   Using nonverbal communication to repeat your verbal statements; to reinforce

report talk   Emphasizes individuality and talents; often associated with men

resurrection phase of dissolution   Takes place once healing has occurred; reflect on what they have learned

rituals   Cultural events that happen regularly; some are grand occasions that mark major transitions, such as a college graduation ceremony, and others are daily activities, such as how you greet people when you get to class and how the instructor ends the class each day

role ambiguity   Occurs when a group member is unsure of the responsibilities and behaviors associated with a role, how performance of the role is evaluated, and what will happen if the responsibilities are not met

role schemata   Cognitive structures that provide information about how you and others should behave based on social categories

role-based relationships   Relationships in which people communicate with each other in terms of the job or function they fulfill, such as instructor, student, salesperson, customer, server, diner

roles filter   Attention to the behaviors the person uses in performing the formal and informal roles assigned to them or that they choose to fulfill; expand your impressions

S

safety needs   What you need to feel safe, secure, and free from harm

scenography   How speakers perform their speeches as their presentations unfold in real time

schemata   Cognitive structures representing basic categories of information grouped together by differences and similarities

search engine   Software program that searches for information on the web and then compiles it based on specific rules or parameters

self-actualization needs   Desires associated with wanting to achieving your full potential

self-centered roles   Behaviors that interfere with the group reaching its goal and foster poor relationships among group members

self-disclosure   The information individuals reveal about themselves to others

self schemata   Cognitive structures associated with your view of self, including your beliefs, attitudes, goals, and emotions

semantics   The relationship between words and their meanings

semiotics   The study of how you make meaning from, or interpret, communication

social awareness   The recognition of connections with others in shared online experiences

social cohesion   Occurs when group members are attracted to the group and want to remain in it

social exchange theory   Examines relationship rewards and costs to determine likelihood of continuing relationship and satisfaction

social facilitation   Occurs when individuals put forth more effort in a group than on their own

social interaction zone   Used in contexts such as workplaces; usually 4–11 feet

social loafing   Occurs when individuals put forth less effort in a group than on their own

social phase of dissolution   People outside the relationship learn of the concerns and that the relationship may end; concerns go from private to public

source credibility   Refers to the trustworthiness, soundness, and believability of an information source

spatial pattern of organization   An organizational pattern in which the main points in the body of a speech are arranged based on where they are in physical proximity to each other

specialized search engine   A search engine designed to access specific types of information

specific purpose   The particular objective for a speech topic that the speaker wants to achieve with the audience

speech anxiety   The fear of speaking before an audience

speech code   Socially constructed system of rules, terms, and meanings that guide communication within a particular group

speech to entertain   A speech in which the speaker encourages audience members to enjoy themselves as they consider the lighter side of a topic

speech to inform   A speech in which the speaker wants the audience to learn more about a topic

speech to persuade   A speech in which the speaker wants to influence audience members' beliefs, values, attitudes, or behaviors

speeches about concepts   Speeches that address ideas, values, philosophies, principles, beliefs, theories, and similar abstract thinking

speeches about events   Speeches that discuss an occurrence that has happened in the past, is happening in the present, or will happen in the future

speeches about objects   Speeches that present information on anything in the material world that you can perceive with your senses

speeches about processes   Speeches that explain to the audience how something works, how something is done, or how something was created

statistic   A numerical representation of observations, often involving a comparison of variables

stereotype   Categorizing people based on the social group to which an individual thinks people belong

stories and myths   Transmit cultural values and beliefs in implicit ways; may recount how the group was started, reveal anecdotes about group members, or relate tales of overcoming adversity that let others learn about the culture; also used to reinforce connectedness

storming phase   The second phase of group development when groups encounter conflict and tensions

style   One of the five canons of rhetoric, style refers to the language the speaker uses in a speech

subordination   The principle of outlining that refers to the hierarchical ordering of points, from large to small or general to specific

substituting   Using nonverbal communication instead of verbal communication; to replace

supportive communication   Messages that encourage others to freely express their ideas

symbol   Representations, such as words and signs, used to convey meanings; arbitrary; one of three parts of the meaning triangle

symptoms   Behaviors that convey information about a physical state

syntactics   The rules within a language that govern word order

T

tag questions   Function to encourage conversation and agreement

task cohesion   Involves good working relationships among group members and agreement on desired outcomes and ways to achieve those outcomes

task roles   Behaviors associated with moving the group toward its goal and accomplishing specific objectives or assignments that must be completed

technological imperative   The assumption that technology controls human behavior

testimony   Information from an expert or authority on a topic

thesis   A single declarative sentence that expresses the essence of a speech

thinking to listen   When communicators think to listen, they expand the topic, adding in their own interpretations and thoughts on what the other person is saying

thinking to speak   When communicators think to speak, they reduce the topic, identifying the key points they want to convey with their message

third place groups   Social groups which may be voluntary or involuntary

thought-speech differential   The difference between the rate senders can produce messages and receivers can process those messages

topical pattern of organization   An organizational pattern in which the main points in the body of a speech are arranged based on major themes or elements associated with a topic

transcendence needs   Desires associated with setting aside personal concerns and focusing on something greater than one's own self

transformational leadership   This approach to leadership is one in which the leader promotes change in the individual and group

U

unanimity   A decision in which there is complete agreement among all group members

universal design   Creating materials and environments that people can access regardless of ability

V

valid information   Information that is logical and reasonable

validating couple   Uses positive communication behavior, such as compromising or collaborating; they show concern for each other and search for areas of agreement and ways to establish common ground

verbal communication   Using language to communicate

verbal symbols   Representations of thoughts, ideas, and feelings with oral and written language

vocabulary   One way cultural members define themselves, build identity, and distinguish who is and is not in the group

vocal characteristics   One of two types of paralanguage; includes speech volume, rate, pitch, quality, articulation, resonance, and accent

vocal clarity   How clearly speakers articulate and how correctly they pronounce their words

vocal fluency   Speaking with natural pauses and avoiding long hesitations and verbal fillers

vocal inferences   One of two types of paralanguage; used when searching for what to say or attempting to delay communicating; also called verbal fillers

vocal quality   The rate, pitch, and volume variety of a speaker's voice

vocal volume   The loudness or softness of a speaker's voice

volatile couple   Has heated arguments, and conversations may escalate quickly to conflicts; when in conflict, they may powerfully assert their position but simultaneously show affection for their partner

voluntary relationships   You choose your relational partner or partners

W

web directory   An index of websites that professionals and experts review and categorize by topics and subtopics

words   Key type of symbol on which language relies

working memory   Also known as short-term memory; involves the temporary storage and processing of information

working outline   The outline that reflects the initial planning for a speech and includes the main points and possible subpoints using key phrases

ENDNOTES

CHAPTER 1

1. McLuhan (1994).
2. Kahan (2019, April 8).
3. McClain et al. (2021, September 11).
4. Perrin (2019, May 31).
5. Perrin and Turner (2019, August 20).
6. Vogels (2020, September 20).
7. Auxier and Anderson (2020, March 16).
8. Vogels (2020, September 20).
9. Finley (2020, April 9).
10. Alliance for Affordable Internet (2020, December).
11. Bell (2010, February 15); Danielle (2019); Dickson (2016, June 21); LaFrance (2014, July 28); Romm (2015, October 15), Plato (360 BCE).
12. Storm, Stone, and Benjamin (2016).
13. Redcay and Schilbach (2019); Schilbach et al. (2013); Syal and Anderson (2013).
14. Balbi and Kittler (2016).
15. Koeze and Popper (2020, April 7).
16. Bowman and Targowski (1987); Shannon (148); Shannon and Weaver (1949).
17. Reddy (1979).
18. Narula (2006).
19. Schramm (1954).
20. Putnam, Phillips, and Chapman (1996).
21. Narula (2006).
22. Berlo (1960).
23. Chen (2019); Diehl and Prins (2008); Hasler and Friedman (2012).
24. O'Dea (2021, August 4); BankMyCell (2021); O'Dea (2019, August 6); Pew Research Center (2021, April 7).
25. Coopman (2009).
26. CNN Wire (2019, August 9); Wall Street Journal (2019, August 19); Vogels et al. (2020, April 30).
27. Melke and Young (2012); Winocur (2009).
28. Kunst (2020, November 19); O'Dea (2021, March 19).
29. Al Balushi (2019); Rijt and Hoffman (2014); Zargaran et al. (2018).
30. Nelson, Simek, and Maschke (2017); Askew (2012).
31. Watzlawick, Beavin, and Jackson (1967).
32. Motley (1990).
33. Cronkhite (1986).
34. Kellermann (1992).
35. National Communication Association (1999).
36. Johannesen, Valde, and Whedbee (2008).
37. Chawla and Sharma (2019); Foley, Hang-yue, and Wong (2005).
38. National Communication Association (1999).
39. Snellman (2015).
40. Diamond et al. (2016); Mangan (2015); Rojas (2016, Febuary 20); Silveira and Hudson (2015).
41. Allan and Madden (2011).
42. Arnett (2001).
43. Atouba, Carlson, and Lammers (2019); Levine (1994).
44. Hammond, Anderson, and Cissna (2003); Paquette, Sommerfeldt, and Kent (2015).
45. National Communication Association (1999).
46. Perrin and Atske (2021, April 2).
47. Schumacher and Kent (2020, April 2).
48. Anderson, Rainie, and Vogels (2021, February 18).
49. Allen (2014); Stroud and Pye (2013).
50. Lipari (2014).
51. Teven et al. (2010).
52. Babin, Palazzolo, and Rivera (2012).
53. Arroyo and Segrin (2011).
54. McManus and Donovan (2012).
55. Canary and MacGregor (2008).
56. Dunleavy and Martin (2010).
57. Hays Research (2019).
58. Madlock (2008); Steele and Plenty (2015).
59. Thompson (2009).
60. Harper (2019).

CHAPTER 2

1. Contreras (2013, May 12).
2. Ly (2019, February 26).
3. Stephey (2009, December 9).
4. Herman (2016, December 2); Ka'ili (2016, December 6); Gumbel (2012, September 3).
5. Bradsher (2004, October 13).
6. Choi (2012).
7. Castells (2001); Morley (2013, May).
8. Croucher, Sommier, and Rahmani (2015); Dooly and Rubinstein (2018); Gopalkrishnan (2019).
9. Castells (2001), 36–37.
10. McElroy (1999).
11. Frey and Sunwolf (2005).
12. Coopman and Meidlinger (2000).
13. Southern Poverty Law Center (2021, February 1).
14. McNamee, Peterson, and Pena (2010); Perry and Olsson (2009).
15. Berlet (2004); Duffy (2003); Lee and Leets (2002).
16. Southern Poverty Law Center (2021, February 1).
17. Orbe (1998).
18. Herakova (2012); Loehwing and Motter (2012).
19. Phillipsen (1997).
20. Bassett (2012).
21. Edgerly (2011).
22. Castells (2001).
23. Samovar, et al. (2013).

24. Hornsey et al. (2005); Tjosvold, Law, and Sun (2003); Vaidyanathan, Aggarwal, and Kozłowski (2013).
25. Markus and Kitayama (1991).
26. Samovar et al. (2013).
27. Hofstede, Hofstede, and Minkov (2010).
28. Waters and Lo (2012).
29. Poepsel (2018); Bagdikian (2004).
30. Hsiao-Cheng (2019); Lenard and Balint (2020); Young (2010).
31. Mueller, Dirks, and Picca (2007).
32. Grinberg (2011, October 26); Liu (2019, October 30).
33. Phillips and Baker (2017, October 30).
34. Phillips and Baker (2017, October 30); Vincenty (2020, October 13).
35. Blumenstyk, (2013, March 15); Cahalan et al. (2016); Zinshteyn (2016, April 25).

36. Sargent and Sue-Chan (2001).
37. Rock and Grant (2016, November 4); Rock, Grant, and Grey (2016, September 22).
38. Hobman, Bordia, and Gallois (2004).
39. Neuliep (2012).
40. Mohammed and Angell (2004).
41. Akiba and Miller (2004).
42. Gray (1992).
43. Wood (2002).
44. Burleson et al. (2011); Tannen (1990).
45. Kimbrough et al. (2013).
46. Kapidzic and Herring (2011).
47. Hofstede, Hofstede, and Minkov, M. (2010).
48. Balboni and Caon (2014).
49. Bright and Gambrell (2017); Ross (in press).

CHAPTER 3

1. Kaplan (2013, November 20); NBC Bay Area Staff (2016, February 22).
2. Zimmermann and Emspak (2017, June 27).
3. Wagner and Boczkowski (2019).
4. Lazer et al. (2018).
5. Gilbert (2019, November 6); Silverman and Pham (2018, December 28).
6. Van Heekeren (2020).
7. Tandoc, Lim, and Ling (2018).
8. Sanderson et al. (2020, April 7); Suciu (2020, April 8).
9. Pennycook, Cannon, and Rand (2018).
10. Correia, Jerónimo, and Gradim (2019); Torres, Gerhart, and Negahban (2018).
11. Buchanan and Benson (2019).
12. Shao (2020, January 17).
13. Foley (2020, March 23).
14. Hameleers, et al. (2020).
15. Liv and Greenbaum (2020).
16. Basol, Roozenbeek, and Van Der Linden (2020).
17. Chui et al. (2012, July).
18. Clement (2019, August 9).
19. Burke (2019, January 24).
20. Holmes (2019, February 19).
21. SlickText (2020).
22. Newberry (2019, October 22).
23. Ominicore Blog (2019, September 6).
24. Koeze and Popper (2020, April 7).
25. Liu, Li, and Diao (2020); Xu (2017).
26. Aagaard (2019).
27. Wiradhany, van Vugt, and Nieuwenstein (2020); Madore et al. (2020).
28. Uncapher, and Wagner (2018); Wang and Tchernev (2012).
29. Coopman (1997).
30. Baldwin (1992); Samosh (2019).
31. Coover and Murphy (2000); Proctor (2016).
32. Pitts, Jr. (2014, February 14).
33. Onishi and Méheut (2020, April 9).
34. Graham-Harrison, Oltermann, and Jones (2020, April 26).
35. Kellas(2010); Koerner and Fitzpatrick (2002).
36. Winter and Reed (2020).
37. Schiff (2020, August 13).
38. History.com staff (2020, Septebmer 20).
39. Brownell (2017).
40. Lee (2019); Pérez-Fuentes, Jurado, and Linares (2018).
41. Bostrom and Waldhart (1988).

42. Melis, et al. (2016).
43. Brownell (2018); Cotton and Ricker (2021); Moriya (2019); Plancher, Mazeres, and Vallet (2019); Sandry, Zuppichini, and Ricker (2020).
44. Comello (2009); Huston (2009).
45. Alfasi (2019).
46. Modica (2019).
47. Carr and Foreman (2016); Prieto-Arranz, Juan-Garau, and Jacob (2013); Turkle (1996); Walther et al. (2011).
48. de Vries and van Elferen (2010).
49. Chen and Li (2017); Walther (1996); Walther and Whitty (2021).
50. Walther et al (2021).; Walther and Whitty (2021).
51. Palupi (2019).
52. Blackhart et al. (2021); Hance, Blackhart, and Dew (2018).
53. Yau and Reich (2019).
54. Günsoy et al. (2020).
55. Wang et al. (2010).
56. Walther (2007).
57. Ledbetter and Finn (2018).
58. Ito, Yang, and Li (2021).
59. Boyle and O'Sullivan (2016).
60. Yang and Lee (2020).
61. Hecht (2009); Hecht et al. (2004); Weaver et al. (2021).
62. Fish et al (2021); Pearson et al. (2010).
63. Lewis et al. (2018).
64. Mejia-Smith and Gushue (2017).
65. Hecht and Choi (2012).
66. Kisner (2018, January 23); Patel (2013, December).
67. Palomares and Eun-Ju (2010).
68. Messinger et al. (2019).
69. Sevelius et al. (2020).
70. Jelinek et al. (2020).
71. Adichie (2009, July).
72. Ballantine, Lin, and Veer (2015); Dupuis, Khadeer, and Huang (2017); Page, Harper, and Frobenius (2015); Zentz (2021).
73. Bergquist et al. (2019).
74. Lammers, Atouba, and Carlson (2013).
75. Bergquist, et al. (2019).
76. Amado, Snyder, and Gutchess (2020); Jung (2011); Jung (2013).
77. Ramsey, Knight, and Knight (2019).
78. Compton (2019).
79. Widiyanti (2020).

80. Nazione, Perrault, and Keating (2019).
81. Derlega et al. (2008); Rabin (2019).
82. Li, Cho, and Goh (2019).
83. Pechmann et al. (2020).
84. Forga (2011).
85. Zhang and Fu (2020).
86. Rabin (2019).
87. Rosenfeld (2000).
88. Chen and Nakazawa (2012); Li et al. (2021).
89. Mathews, Derlega, and Morrow (2006).
90. Dotterer and Day (2019).
91. Leaper (2019).
92. Tian (2013).
93. Hinson and Sword (2019).
94. Gentina and Chen (2019).
95. Clark-Gordon et al. (2019).
96. Rains (2014).
97. Levontin and Yom-Tov (2017).
98. Gibbs, Ellison, and Lai (2011).
99. Hollenbaugh and Everett (2013).
100. Jebbour and Mouaid (2019); Song, Kim, and Luo (2016).
101. Song, Kim, and Park (2019).
102. Stokoe (2009).
103. Jang and Stefanone (2011).
104. Malloch and Taylor (2019).
105. Weisel and King (2007).
106. Kroll and Stieglitz (2021).
107. Dieckmann (2000).
108. Jeske, Lippke, and Shultz (2019).

CHAPTER 4

1. Congressional Black Caucus (2017, April 3).
2. Jan et al. (2020, June 13); Enwemeka (2020, July 13).
3. Lipetz, Kluger, and Bodie (2020).
4. Jonsdottir and Fridriksdottir (2020); Rane (2011).
5. Bregenzer et al. (2020); Castro et al. (2018); Kristinsson and Snorrason (2019).
6. Flynn, Valikoski, and Grau (2008).
7. Vickery (2018).
8. Ala-Kortesmaa and Isotalus (2015); Bodie et al. (2012); Cooper and Buchanan (2010); Hawkins and Fillion (1999); Meldrum (2011).
9. Baird and Parayitam (2019); Coffelt, Grauman, and Smith (2019); Cyphert et al. (2019); Kirtley Johnston, and Reed (2017).
10. Brink and Costigan (2015); Schmidt (2018).
11. Janusik and Wolvin (2009).
12. Brownell (2017); Purdy (1991); Wolvin and Coakley (2000).
13. Bentley (2000).
14. Fernbank (2003); Campbell (2003).
15. Pitardi and Marriott (2021).
16. Lee, Lee, and Sheehan (2020).
17. Langer, König, and Hemsing (2020).
18. Pradhan, Findlater, and Lazar (2019).
19. Pence and Vickery (2012); Bommelje, Houston, and Smither (2003).
20. Bostrom (2011); Bostrom and Waldhart (1988).
21. Holman et al. (2021); Hicks amd Tharpe (2002).
22. Sessions et al. (2020).
23. Lee (2020, April 27).
24. Ellis (2004); Hamouda (2012); King and Behnke (2004); Ledbetter and Schrodt (2008).
25. Powers and Bodie (2004); Powers and Sawyer (2011); Powers and Witt (2008).
26. Sawyer et al. (2014).
27. Burleson (2011).
28. Wise et al. (2012); Wise et al. (2013).
29. Young, Oram, and Napier (2019).
30. McIntosh (2000); Rose and Smith (2000).
31. Haynes (2014).
32. Brownell (2017).
33. Dipper, Black, and Bryan (2005); Wessel-Tolvig and Paggio (2016).
34. Gearhart, Denham, and Bodie (2014); Welch and Mickelson (2020).
35. Brownell (2017); Cornwell and Orbe (1999); Purdy (1991).
36. Schäfer and Eerola (2020).
37. Vickery, Keaton, and Bodie (2015).
38. Floyd (2014).
39. Lipari (2010).
40. Hanson (2019).
41. Jonsdottir and Kristinsson (2020).
42. Itzchakov and Kluger (2017).
43. Stewart and Koenig Kellas (2020).
44. Arnett (2019); Purdy and Manning (2015).
45. Alerby (2019).
46. Johnson et al. (2003).
47. Verouden and Van Der Sanden (2016).
48. Janusik and Imhof (2017).
49. Kim (2003); Zohoori (2013).
50. Díaz-Vera and Caballero (2013); Hess, Blaison, and Kafetsios (2016); Liu (2012).
51. Cornwell and Orbe (1999).
52. Pearce and Pearce (2000).
53. Thompson et al. (2004).
54. Thompson et al. (2004).
55. Thompson et al. (2004).
56. Petrie and Carrel (1976); Vandergrift (2005).
57. Froemming and Penington (2011); Rane (2011).
58. Brown and Strand (2019); Sarampalis et al. (2009).
59. Burleson (2011).
60. Brownell (2017).
61. Reinhard et al. (2011).
62. National Communication Association (1999).
63. Bodie et al. (2015); Jones, Bodie, and Hughes (2019); Weger, Bell, and Robinson (2014).
64. Brownell (2017); Forward, Czech, and Lee (2011); Gibb (1961).
65. High and Solomon (2016).
66. Gist-Mackey, Wiley, and Erba (2018).
67. Bodie and Jones (2012); Bodie, Vickery, and Gearhart (2013).
68. Davis (2019).
69. High and Young (2018).
70. Koenig Kellas et al. (2021).
71. Huston, Garland, and Farb (2011); Manusov et al., (2020).
72. Jones et al. (2019).
73. Bauer et al. (2017).
74. Burleson (2011); Huston et al.; King and Sawyer (1998).
75. Anderson (2012); Taraban et al. (2016).
76. Brownell (2013); Cornwell and Orbe (1999); Lipari (2004).
77. Browne et al. (2014).
78. Bodie (2013).
79. Rouse and Al-Maqbali (2014).

80. Bregman and Haythornthwaite (2003); Bulu (2012); Thulin and Vilhelmson (2017).
81. Marzban and Isazadeh (2012); Tanewong (2019).
82. Bommelje, Houston, and Smither (2003).
83. Blankenstein et al. (2011); Caspersz and Stasinska (2015).
84. McClain et al. (2021, September 1).
85. Lee (2020); Nadler (2020).
86. Vanden Abeele and Postma-Nilsenova (2018).
87. Treasure (2011).

CHAPTER 5

1. Sports e-cyclopedia (n.d.).
2. Chaney, Burke, and Burkley (2011); Change the Mascot (n.d.).; Fryberg et al. (2008); Halbritter (2013, December 27); Kim-Prieto et.al (2010); Proud to Be (2013); Rothman (2014, March 26); American Psychological Association (2005); Toensing (2014, April 11).
3. Goddard (2005); Gugliotta (2005, October 3); *The Washington Post* (2013, October 9).
4. Brady (2013, October 25; The Leadership Conference on Civil and Human Rights (2013, December 12); Maske, Kane, and Vargas (2014, February 9); O'Connor (2013, October 30); Editorial Board (2019, August 17); Anwar (2020, February 4); Fryberg et al. (2020).
5. Bonesteel (2021, January 22); Waldron (2020, December 16).
6. Ogden and Richards (1989).
7. Greenslade (2013, September 9); King (2013, September 6).
8. King (2013, September 6).
9. O'Conner (2013, October 30); Plotz (2013, August 8).
10. Howar (2021, March 10); Moye (2021, February 22).
11. Hayawaka and Hayawaka (1991).
12. Juster (1989).
13. Chomsky, Belletti, and Rizzi (2002).
14. Piantadosa, Tily, and Gibson (2012).
15. Crystal (2005); McWhorter (2003); McWhorter (2011).
16. Chaika (2007); De Smit (2012).
17. Aitchison (2001); Graddol (2004).
18. Cingel and Sundar (2012); Crystal (2009); Tagliamonte and Denis (2008); Varnhagen et al. (2010).
19. Plester, Wood, and Bell (2008); Plester, Wood, and Joshi (2009).
20. McWhorter (2013, February).
21. Chandler (2007); Eco (1976).
22. Kearns (2011); Kreidler (2014).
23. Culicover (2013); den Dikken (2013); van Vilin, Jr. (2005).
24. Ariel (2010); Cummings (2005); Yule (1996).
25. Sedaris (1999).
26. Du and King (2013); Hofstede (1980); Pascual et al. (2012).
27. Economidou-Kogetsidis (2004); Economidou-Kogetsidis (2010); Tannen (1981).
28. O'Connor (2013, October 30).
29. Waltman (2003); Waltman and Haas (2007); Waltman and Haas (2010).
30. Lakoff (2004).
31. Leaper and Robnett (2011).
32. Freeman and McElhinny (1996).
33. Tannen (1996).
34. Algeo (2006); Holmes (1983); Holmes (1995); Tottie and Hoffmann (2006).
35. Eckert and McConnell-Ginet (1992); Holmes and Meyerhoff (2003); Montgomery (1995).

CHAPTER 6

1. Bowman (2020); Mandal (2014); Mehrabian (2009); Mehrabian and Wiener (1967).
2. Ekman (1965); Knapp, Hall, and Horgan (2013).
3. Argyle, Alkema, and Gilmour (1971); Waxer (1977).
4. Jaffe and Feldstein (1970); Wiemann and Knapp (2006).
5. Siegman and Feldstein (1978).
6. Marchak (2013).
7. Jiang et al. (2017).
8. D'agostino and Bylund (2014).
9. Elfenbein and Ambady (2002); Ekman (1970); Ekman (1973); Ekman (2003); Ekman (2009); Ekman and Friesen (1971); Ekman and Friesen (1974); Ekman, Friesen, and Ellsworth (1972); Ekman, Friesen, and Tomkins (1971); Ekman and Oster (1979); Ekman, Sorenson, and Friesen (1969).
10. Todorov, Baron, and Oosterhof (2008).
11. Roter et al. (2006).
12. de Groot et al. (2012).
13. Ekman (2009); Hurley et al. (2014).
14. Bailenson (2021).
15. Andersen (1999).
16. Hall (1959).
17. Hall (1966); Hall (1968).
18. Jones (1999); Jones and Yarbrough (1985).
19. Burgoon, Guerrero, and Floyd (2010).
20. Andersen (1999).
21. Floyd and Guerrero (2006); Trager (1958); Trager (1960); Trager (1961).
22. Van Zant and Berger (2020).
23. Ballard and Seibold (2000).
24. Lakoff and Johnson (1980).
25. Egland et al. (1997).
26. Andersen (1999).
27. Martin, Gnoth, and Strong (2009).
28. Bruneau (1974); Bruneau (1990); Bruneau and Ishii (1988).
29. Matsumoto and Hwang (2013).
30. Koerner (2003, March 28).
31. Matsumoto and Hwang (2013).
32. LaFrance and Vial (2016); Hall and Gunnery (2013); Shields (1987).
33. LaFrance and Vial (2016).
34. West and Zimmerman (1987).
35. Vingerhoets and Scheirs (2000).
36. Deaux and Major (1987); LaFrance and Vial (2016).

CHAPTER 7

1. Candland (1995); Hargie (2011); Leary (2001); Williams and Zadro (2001).
2. Lee, Cadigan, and Rhew (2020); Loades et al. (2020); Pietrabissa and Simpson (2020); Smith and Lim (2020).
3. Berger (2008); Berger (2005); Bylund, Peterson, and Cameron (2012); Manning (2014).
4. "Annual Career-Builder Survey" (2016); "Association of American Colleges and Universities (AAC&U) Survey" (2021); Brun (2010); National Association of Colleges and Employers (2016); Singh (2014).
5. Baumeister and Leary (1994); Dorrance Hall and Shebib (2020); Ellis and Ledbetter (2015); Fonseca et al. (2020); Okdie and Ewoldsen (2018); Ryan and Deci (2017); Twiselton et al. (2020).
6. Knap (1978); Knapp and Vangelisti (2009).
7. Cook and Rice (2001); Cook and Rice (2006); Emerson (1976); Harvey and Wenzel (2006); Homans, (1961); Noller (2006).
8. Duck (2007); Duck (1998); Duck (1989); Duck (2011).
9. Guerrero, Andersen, and Afifi (2007).
10. Canary and Stafford (1992); Stafford and Canary (1991).
11. Canary, Stafford, and Semic (2002); Ogolsky and Bowers (2013).
12. Canary and Stafford (1992); Canary and Stafford (2001).
13. Duck (2011).
14. Davis (1973)
15. Duck (1982); Rollie and Duck (2006).
16. Gottman (1995); Hooper et al. (2017).
17. Baxter (1988); Baxter and Braithwaite (2008); Baxter and Montgomery (1996); Baxter, Scharp, and Thomas (2021); Brown, Werner, and Altman (1998); Halliwell (2015); Johnson et al. (2018); Ngai and Singh (2018); Montgomery and Baxter (1998); Rawlins (1998); Sahlstein Parcel and Baker (2018).
18. Baxter and Simon (1993).
19. Baxter (2011).
20. Baxter et al. (2004).
21. Thomas (1976); Thomas and Kilmann (1974).
22. Gamble (2000).
23. Cole (2001).
24. McCornack and Levine (1990).
25. Miller, Mongeau, and Sleight (1986).

CHAPTER 8

1. Vangelisti (1993); Vangelisti (2013).
2. Burke, Woszidlo, and Segrin (2013); Burleson and Kunkel (2002); Hart, Newell, and Olsen (2003); Kamaruddin et al. (2012); Ledbetter (2010); Satir (1983); Satir (1988); Satir and Baldwin (1983), Satir et al. (1991); Satir, Stachowiak, Taschman (1975); Schrodt et al (2008); Wang, Roaché, and Pusateri (2019).
3. Gerhardt (2016).
4. Fitzpatrick and Ritchie (1994); Koerner, and Fitzpatrick (2006); Koerner and Fitzpatrick (2002); Samek and Rueter (2011).
5. Koerner and Fitzpatrick (2002).
6. Rueter and Koerner (2008); Steinberg (2001).
7. Diamond, Allen, and Butterworth (2012); Droser (2017); Floyd and Morman (2014); Harris and González (2015); Soliz and Phillips (2018); Suter (2016).
8. Gazso and McDaniel (2015); Jones-Wild (2012).
9. Rawlins (1992).
10. Rawlins (1992).
11. Reisman (1979); VanLear, Koerner, and Allen (2006).
12. Kurth (1970).
13. Johnson et. al (2003).
14. Helm (2017); Rawlins (1992); Wiseman (1986).
15. Amati et al. (2018); Blieszner, Ogletree, and Adams (2019); Collins and Madsen (2006); Demir, Orthel, and Andelin (2013); Denworth (2020); Maunder and Monks (2019); McCabe (2016); León-Jiménez (2020); Sakyi et al. (2015); Stadtfeld et al. (2019).
16. Hargie (2011).
17. McBride and Bergen (2015); McBride and Kirby (2009); Whitman and Mandeville (2019).
18. Barsade and O'Neill (2014); Dumas, Phillips, and Rothbard (2013); Dumas and Sanchez-Burks (2015); Hood, Cruz, and Bachrach (2017); Ingram and Zou (2008); Methot et al. (2016); Pillemer and Rothbard (2018); Sias and Gallagher (2009); Sias et.al. (2004).
19. Braithwaite and Holt-Lunstad (2017); Gómez-López, Viejo, and Ortega-Ruiz (2019); Umberson and Montez (2010); Waldinger and Schulz (2010).
20. Canary and Dainton (2003); Stafford (2003).
21. Canary and Stafford (1994); Dainton and Stafford (1993); Stafford and Canary (1991); Stafford, Dainton, and Haas (2000).
22. Ayres (1983); Dainton (2008); Dainton and Gross (2008); Dainton and Stafford (1993); Dindia and Baxter (1987); Goodboy and Bolkan (2011).
23. Fitzpatrick and Winke (1979); Guerrero (1994); Koerner and Fitzpatrick (2002); Sereno, Welch, and Braaten (1987); Sillars et al. (1982).
24. Umberson et al. (2006).
25. Burgess and Burgess (1996); Siegert and Stamp (1994).
26. Alberts (1989); Caughlin and Vangelisti (2006); Gottman and Levenson (2000); Heavey (1995); Lloyd (1990); Pike and Sillars (1985); Sabourin, Infante, and Rudd (1993); Ting-Toomey (1983).
27. Delatorre and Wagner (2019); Gottman (1993); Gottman (1998); Gottman, and Levenson (1992); Hample and Richards (2018).
28. Holt-Lunstad, Smith, and Layton (2010); Weir (2018, March).
29. Holt-Lunstad, Smith, and Layton (2010).
30. Perissinotto, Stijacic, and Covinsky (2012).
31. Holt-Lunstad, Robles, and Sbarra (2017).
32. Khaleque and Rohner (2018); Lang and Fingerman (2004).
33. Waldinger and Schulz (2016).
34. Chen et al. (2017); Russek and Schwartz, (1997).
35. Blieszner and Roberto (2004); Hartup, and Stevens (1999); Nicolaisen and Thorsen (2017); Rawlins (1992); Rawlins (2009).
36. Rawlins (1994).

37. Chopik (2017).
38. Chopik (2017); Giles et al. (2005).
39. Pearce, Machin, and Dunbar (2021).
40. Huo et al. (2020); Rook and Charles (2017).
41. Luong, Charles, and Fingerman (2011).
42. Magnuson et al. (2012, November 14–16); Walker et al. (2011, March 23–27).
43. Montgomery and Sorell (1997).
44. Ben-Shlomo et al. (1993); Hu and Goldman (1990); Kaplan and Kronick (2006); Robards et al. (2012).
45. Hu and Goldman (1990); Kaplan and Kronick (2006).
46. Rowe, Stannard, and The Spice Girls (1996).

CHAPTER 9

1. Kentucky Student Environment Coalition (n.d.).
2. Maslow (1954); Maslow (2000); Maslow, Stephens, and Heil (1971); Maslow (1999).
3. Rowan (1988).
4. Hanley and Abell (2002).
5. Washington Military Department (2018).
6. Arakaki (2020, March 29).
7. Drouin et al. (2018); Hosein, Ramanau, and Jones (2010).
8. Boyle (2013); Parsons (2019); Sharma and Khadka (2019); Zhu and Stephens (2019).
9. Gross et al. (2015); Solanki, McPartlan, and Xu (2019).
10. Schiavio et al. (2019).
11. Lichtenstein (2019).
12. Kamini (2012).
13. Çelebi and Polat (2018).
14. Gazzale (2019).
15. Sherman and Burns (2015).
16. Basuki et al. (2015).
17. Ho et al. (2015).
18. Holdsworth and Quinn (2012); MacNeela and Gannon (2014).
19. Bostrom (1970).
20. Bostrom (1970).
21. Moreland (2010).
22. Melero, Hernandez-Leo, and Manatunga (2015).
23. Gully, Devine, and Whitney (2012).
24. Garcia (2020, January 27).
25. Treen et al. (2016).
26. Soboroff, Kelley, and Lovaglia (2020).
27. Cruz, Boster, and Rodriguez (1997).
28. Panek et al. (2018).
29. Kivlighan, Jr., London, and Miles (2012); Purvanova (2013); Sidorenkov, Borokhovski, and Kovalenko (2018).
30. Haas, Sypher, and Sypher (1992).
31. Armfield (2011); National Urban League (2020).
32. Black Lives Matter (n.d.).
33. Cha, Cichy, and Kim (2011); Mcbey, Karakowsky, Ng (2017); Vecina et al. (2013).
34. Malone (2020, March 27).
35. California Coastal Commission (2019).
36. De Wever et al. (2008).
37. Benne and Sheats (1948).
38. Lehmann-Willenbrock, Beck, and Kauffeld (2016).
39. Ruch et al. (2018).
40. Gander et al. (2018).
41. Northouse (2018).
42. Choi (2018).
43. Tang (2019).
44. İnce (2018).
45. Tang (2019).
46. Jiang and Chen (2018); Mascareño Rietzschel, and Wisse (2020).
47. Chen et al. (2018).
48. Li, Liu, and Luo (2018); Wazir, and Khan (2018).
49. Boak et al. (2015); Feng et al. (2017); Karriker, Madden, and Katell (2017).
50. Fransen et al. (2020).
51. Savage et al. (2018).
52. Gist (2020).
53. Tapia and Polonskaia (2020).
54. Violanti and Ray (2021).
55. DigglesCreative (2021).
56. Grubaugh and Flynn (2018); Musinguzi et al. (2018); Zijl et al. (2020).
57. Endacott, Hartwig, and Yu (2017); Kauffeld and Lehmann-Willenbrock (2012); Lehmann-Willenbrock et al. (2015).
58. Nowels (2021, February 27).
59. Kindarto, Zhu, and Gardner (2020).
60. Carrington, Combe, and Mumford (2019).
61. Pujiastuti et al. (2020).
62. Kelley (1992).
63. Khan et al. (2020).
64. Larsson and Nielsen (2021).
65. Varpio and Teunissen (2021).
66. Schindler (2015); Violanti and Ray (2021).
67. Espey (2018); Graham and Jones (2019); Lvina, Johns, and Vandenberghe (2018); Driskell et al. (2017); Wilderom et al. (2015); Yue, Fong, and Li (2019).
68. Evans and Dion (2012); Leo et al. (2015).
69. Croy and Eva (2018); Woerkom and Sanders (2010).
70. Rodríguez-Sánchez et al. (2017).
71. Hansen (2016).
72. Graham and Jones (2019).
73. Hardy, Eys, and Carron (2005); Michalisin, Karau, and Tangpong (2004).
74. McLaren and Spink (2018); Soldan (2010); Spink et al. (2005).
75. Eys and Carron (2001); Kang and Seibold (2018).
76. Ebbers and Wijnberg (2017).
77. Zhang, Jiang, and Jin (2020).
78. Dunlop and Lee (2004).
79. Liu, Xiao, and Wang (2020).
80. Gatling et al. (2017).
81. Junikka et al. (2017).
82. Ladbury and Hinsz (2018).
83. Kosfeld and Rustagi (2015).
84. Abrams, de Moura, and Travaglino (2013); Kamau (2013); Pavitt et al. (2006).
85. Weng and Carlsson (2015).
86. Pfattheicher, Böhm, and Kesberg (2018).
87. Meng, Fulk, and Yuan (2015).
88. Hjertø and Paulsen (2016).
89. Coetzer (2015); Mayer, Salovey, and Caruso (2004).
90. Wilkinson (2015).
91. Quisenberry (2018).
92. Edelman and Van Knippenberg (2018); Kunalic et al. (2016); Lone and Lone (2018).
93. Dartey-Baah and Mekpor (2017); Yung-Shui and Tung-Chun (2009).

94. Chang, Sy, and Choi (2012); Fallon et al. (2014).
95. Herpertz, Schütz, and Nezlek (2016).
96. Boyraz (2019).
97. Mundt et al. (2016).
98. Glikson and Erez (2019).
99. Boyraz (2019).
100. Da'as (2020); Dowell, Morrison, and Heffernan (2015); Erdem and Ozen (2003).
101. Anderson-Hanley et al. (2011); Edwards et al. (2018).
102. Seitchik and Harkins (2014); Xiangyu et al. (2014); Zhu, Singh, and Wang (2019).
103. Teng and Luo (2015).
104. Bououd et al. (2016).
105. Ding and Ding (2008); Hall and Buzwell (2013).
106. Lam (2015).
107. Culiberg (2019).

CHAPTER 10

1. Sakellaropoulou (2020, July 23); Siliezar (2020, May 13).
2. Bellis (2008).
3. National Association of Colleges and Employers (2019, November).
4. LinkedIn Learning (2020).
5. Catmull (2014); Parolin and Pellegrinelli (2020).
6. Ness and Glăveanu (2019).
7. Rosso (2014); Shin (2014); Yong, Sauer, and Mannix (2014); Zhang, Li, and Reynolds (2020).
8. Syrett and Lammiman (2002).
9. Zhang, Li, and Yan (2020, August 2).
10. Naidoo and Vernillo (2014); Weller (2020).
11. Marzocchi and Ramlogan (2019); Wang et al. (2013).
12. Hod, Basil-Shachar, and Sagy (2018).
13. Van Damme et al. (2019).
14. You (2020).
15. Pillay et al. (2020).
16. Motro, Spoelma, and Ellis (2020, June 18).
17. Sözbilir (2018); Yang, Lee, and Cheng (2017).
18. Taylor (2013).
19. Gerard (2005).
20. Hunter et al. (2007).
21. Thonney and Montgomery (2019).
22. Courtright (1978).
23. Badie (2010); Matusitz and Breen (2012); Valine (2018).
24. Forsyth (2020).
25. Riccobono, Bruccoleri, and Größler (2016); Szanto (2017); Tsikerdekis (2013).
26. Bailin and Battersby (2016); Mok and Morris (2010).
27. Drummond (2012); Sunstein and Hastie (2015); Waddell, Roberto, and Yoon (2013).
28. Cole and Ji (2014); Espey (2018).
29. Chow (2018); Li et al. (2017); Smith and Barr (2008).
30. Dipasquale and Hunter (2018).
31. Robert III (2020).
32. Mendelberg and Karpowitz (2016); Mendelberg, Karpowitz, and Goedert (2014).
33. Kimura and Katayama (2013).
34. Lei et al. (2010).
35. Rast, Hogg, and Giessner (2013).
36. Aravena (2019); Raus, Haita, and Lazăr (2012).
37. Ismail, Zainuddin, and Ibrahim (2010).
38. Bahrami et al. (2012).
39. Mojtahedi, Ioannou, and Hammond (2018).
40. Farmer, Bissière, and Benkirane (2018).
41. Pütz (2019).
42. Zhang et al. (2019).
43. Pérez et al. (2018).
44. Mumby (1988).
45. Kim, Sutton, and Gong (2013); Liu, Wang, and Yao (2019); Müllern and Nordin (2012); Tuuli and Rowlinson (2009).
46. Cavazotte, Hartman, and Bahiense (2014); Hofstede (2001).
47. Luo, Wang, and Tong (2020); Sheng-Min and Jian-Qiao (2013).
48. West and Turner (2020).
49. Rahim (2017).
50. Kilmann and Thomas (1977).

CHAPTER 11

1. Obama (2017, January 6).
2. TED Conferences (n.d.).
3. Stob (2015).
4. National Association of Colleges and Employers (2016).
5. Majid et al. (2019).
6. LinkedIn Learning (2020).
7. Hansen (2003).
8. Turnitin, (2017, June 7).
9. Strauss (2014, July 26).
10. Carroll (2014, July 14).
11. Borchers (2013).
12. Tapia (2001).
13. Davies (2011).
14. Vijayavalsalan (2016).
15. Henning (2012).
16. Mazer and Titsworth (2012).
17. Vaughn and Parry (2013).
18. Owusu and Adade-Yeboah (2014).
19. Miller and Pessoa (2016).
20. Groody and Măcelaru (2018).
21. Gargouri and Naatus (2017); Kernan, Basch, and Cadorett (2018); Paxman (2011).
22. Buzan (2014).

CHAPTER 12

1. Allen (2020, June 26).
2. LaGarde and Hudgins (2018).
3. Allen et al. (2020).
4. Tsfati et al. (2020).
5. Mitchell et al. (2019, June 5).
6. American Library Association (2013).
7. Becker (2018).
8. American Library Association (2020b).
9. Eshet-Alkalai (2012).
10. American Library Association (2020a).
11. Becker (2018, p. 4).

12. Hamburg, O'Brien, and Vladut (2019); Oggero, Rossi, and Ughetto (2020).
13. Jose (2016).
14. Bejaković and Mrnjavac (2020).
15. Kahne and Bowyer (2019); Martens and Hobbs (2015).
16. Rosenbaum, Johnson, and Deane (2018).
17. Foresman (2017).
18. Klump (2020, January 19).
19. National Aeronautics and Space Administration (2020, August 21).
20. Buis (2019, July 12).
21. Clement (2020, June 18).
22. Cash and Stewart (2018).
23. Society of Professional Journalists (2014, Spetember 6).
24. Yeung (2020, August 12).
25. Letić (2021, March 19).
26. Feldman (2019, June 26).
27. "Read the entire U.S. Constitution on the National Archives website at The Constitution of the United States: A transcription." (n.d.)
28. Goh (2013).
29. Turnitin (2012).
30. Neville (2016).

CHAPTER 13

1. Landon (2020, March 20).
2. Sonam et al. (2019).
3. Mitchell and Oiphant (2020, March 18).
4. Hersman (2014).
5. Whitme (2020, April 20).
6. National Communication Association (1999).
7. Thomsen (2020, July 16).
8. Mboya (2014).
9. Le Guin (2014, November 20).
10. Spiegel (2014, April 11).
11. Dahlstrom (2014).
12. Hoeken and Sinkeldam (2014).
13. Bing and Fink (2012).
14. Yousafzai (2014, December 10).
15. Weber and Wirth (2014).
16. Williams (2013, May 11).
17. MacIntyre, MacIntyre, and Carre (2010).
18. Geiger, Grossman, and Schrader (2019).
19. Gilligan-Lee (2020).
20. Vigen (2015).
21. Landon (2020, March 20).
22. German et al. (2013).
23. de Smet et al. (2014).

CHAPTER 14

1. Berklee College of Music (2019, May 14).
2. McClish (2016); Scenters-Zapica and Cos (2003).
3. Dwyer and Davidson (2012).
4. Bodie (2010).
5. LeFebvre, LeFebvre, and Allen (2018).
6. Hunter, Westwick, and Haleta (2014).
7. McDermott (2014).
8. MacIntyre and Thivierge (1995).
9. Hertz, Kerkhof, and van Woerkum (2016).
10. Mayer (2009).
11. Gallo (2010).
12. Fay, Page, and Serfaty (2010); Goberman, Hughes, and Haydock (2011).
13. Chen, Mak, and Fujita (2015); Goberman, Hughes, and Haydock (2011); Hertz, Kerkhof, and van Woerkum (2016).
14. Ayres and Sonandré (2003); Choi, Honeycutt, M., and Bodie (2015); Honeycutt, Choi, and DeBerry (2009).
15. Shi, Brinthaupt, and Mccree (2015).
16. Jackson et al. (2017).
17. Witt et al. (2006).
18. McDermott (2014).
19. Nordin and Broeckelman-Post (2019).
20. Chadwick et al. (2016).
21. Wolfe (2019, May).
22. United States Access Board (1995, December 7).
23. LaGrande (2019, March 29).
24. Gallo (2010).
25. Mortaji (2018).
26. Centers for Disease Control and Prevention (2019, September).
27. Hearing Loss Association of America (2018, May).
28. American Federation for the Blind (2019).
29. Lipkin and Haddad (2018); Coopman (2003); Jenks (2019); Malleson (2018).
30. Rodero, Mas, and Blanco (2014).
31. McCreary (2019, September 5).
32. Nelson (2017).
33. Carpenter (2012); Christenfeld, (1995); Galili, Amir, and Gilboa-Schechtman (2013).
34. Charoenruk and Olson (2018); Goberman, Hughes, and Haydock (2011).
35. Daly and Redlick (2016).
36. Gallo (2010).

CHAPTER 15

1. Hawking (2016, January 26).
2. Hawking (2016, January 26).
3. Rossette-Crake (2019).
4. Lythcott-Haims (2019, October 1).
5. Rossette-Crake (2020).
6. Lythcott-Haims (2019, October 1).
7. Greenleaf and Wells-Papanek (2005).
8. Bloom (2019, April 22).
9. Syed (2018, November 14).
10. Syed (2018, November 14).
11. Singer (2017, June 21).
12. Singer (2017, June 21).
13. Collins Dictionary (n.d.).
14. Mariamdaran and Veloo (2017).
15. Bawden and Robinson (2020).
16. Haase et al. (2016); Peter (2018).
17. O'Reilly (1980); Zinko et al. (2019).
18. University of Bristol (2019, June 3).
19. Spooner (2017, November 10).
20. Stewart et al. (2017).

21. Treasure (2011).
22. Franks (2019, September 9).
23. Franks (2019, September 9).
24. Franks (2019, September 9).
25. Stancill (2018, April 14).

26. Hogan (2012).
27. Kaufmann (2019, October 17).
28. Kaufmann (2019, October 17).
29. Kedrowicz and Taylor (2016).
30. Frantzolas (2016, February).

CHAPTER 16

1. Davis (2019, May 20).
2. Borchers (2013).
3. Asen (2015); Jasinski (2001).
4. Sotomayor (2013, May 5).
5. Meyer (2019).
6. Baumlin and Meyer (2018).
7. Appel and Mara (2013).
8. Sotomayor (2013, May 5).
9. di Carlo (2015).
10. Sotomayor (2013, May 5).
11. Tindale (2011).
12. Hodge (2020, March 20).
13. National Communication Association (1999).
14. Sotomayor (2013, May 5).
15. Saint-Dizier de Almeida et al. (2016); Nettel and Roque (2012).

16. Campbell, Huxman, and Burkholder (2015).
17. Heracleous, Paroutis, and Lockett (2019).
18. Shank (2018); Williams (2003).
19. Smith (2007); Waisanen (2015).
20. Vaughn (2018).
21. Rapp and Wagner (2013).
22. Jasinski (2001).
23. Vaughn (2018).
24. Budzynska and Witek (2014).
25. Goldberg (2019, October 29).
26. Fodor and Wick (2009).
27. de Graaf et al. (2012).
28. Hosman and Siltanen (2011).
29. Gallo (2014).
30. Carpenter (2012).

REFERENCES

Aagaard, J. (2019). Multitasking as distraction: A conceptual analysis of media multitasking research. *Theory & Psychology, 29*(1), 87–99. https://doi.org/10.1177/0959354318815766

Abrams, D., de Moura, G., & Travaglino, G. A. (2013). A double standard when group members behave badly: Transgression credit to ingroup leaders. *Journal of Personality & Social Psychology, 105*(5), 799–815.

Adichie, C. N. (2009, July). *The danger of a single story* [Video]. TED Conferences. https://www.ted.com/talks/chimamanda_ngozi_adichie_the_danger_of_a_single_story?language=en

Aitchison, J. (2001). *Language change: Progress or decay?* (3rd ed.). Cambridge University Press.

Akiba, D., & Miller, F. (2004). The expression of cultural sensitivity in the presence of African Americans: An analysis of motives. *Small Group Research, 35*, 623–642.

Ala-Kortesmaa, S., & Isotalus, P. (2015). Professional listening competence promoting well-being at work in the legal context. *International Journal of Listening, 29*(1), 30–49.

Al Balushi, A. (2019). The ethics and legality of using personal smartphones to take medical photographs. *Sultan Qaboos University Medical Journal, 19*(2), E99–E102.

Alberts, J. K. (1988). An analysis of couples' conversational complaints. *Communication Monographs, 55*, 184–197.

Alerby, E. (2019). Places for silence and stillness in schools of today: A matter for educational policy. *Policy Futures in Education, 17*(4), 530–540.

Alfasi, Y. (2019). The grass is always greener on my friends' profiles: The effect of Facebook social comparison on state self-esteem and depression. *Personality and Individual Differences, 147*, 111–117.

Algeo, J. (2006). *British or American English?* Cambridge University Press.

Allen, A. M. (2020, June 26). *Fighting racism: How to restructure society so it's open to all* [Webinar]. Berkeley News. https://news.berkeley.edu/2020/06/26/berkeley-talks-transcript-amani-allen-and-camara-jones/

Allen, B. J. (2014). Communication, diversity, and ethics in higher education. In R. C. Arnett & P. Arneson (Eds.), *Philosophy of communication ethics: Alterity and the other* (pp. 108–113). Rowman & Littlefield.

Allen, J., Howland, B., Mobius, M., Rothschild, D., & Watts, D. J. (2020). Evaluating the fake news problem at the scale of the information ecosystem. *Science Advances, 6*(14). https://doi.org/10.1126/sciadv.aay3539

Alliance for Affordable Internet. (2020, December). *The affordability report 2020.* https://a4ai.org/affordability-report/report/2020/

Amado, S., Snyder, H. R., & Gutchess, A. (2020). Mind the gap: The relation between identity gaps and depression symptoms in cultural adaptation. *Frontiers in Psychology, 11*, 1156–1156. https://doi.org/10.3389/fpsyg.2020.01156

Amati, V., Meggiolaro, S., Rivellini, G., & Zaccarin, S. (2018). Social relations and life satisfaction: The role of friends. *Genus, 74*, 7.

American Federation for the Blind. (2019). *Facts and figures on adults with vision loss.* American Foundation for the Blind. https://afb.org/research-and-initiatives/statistics/adults

American Library Association. (2013). *Digital literacy, libraries, and public policy: Report of the office for information technology policy's digital literacy task force.* https://ala.org

American Library Association. (2020). *Digital literacy.* https://literacy.ala.org/digital-literacy

American Library Association. (2020). *Information literacy.* https://literacy.ala.org/information-literacy

American Psychological Association. (2005). *Summary of the recommendation recommending retirement of American Indian mascots.* http://www.apa.org/pi/oema/resources/indian-mascots.aspx

Andersen, P. A. (1999). *Nonverbal communication: Forms and functions.* Mayfield.

Anderson, J., Rainie, L., & Vogels, E. A. (2021, February 18). *Experts say the "new normal" in 2025 will be far more tech-driven, presenting more big challenges.* Pew Research Center. https://pewresearch.org/internet/2021/02/18/experts-say-the-new-normal-in-2025-will-be-far-more-tech-driven-presenting-more-big-challenges/

Anderson, W. (2012). Mindful listening instruction: does it make a difference. *Contributions to Music Education, 39*, 13–30

Anderson-Hanley, C., Snyder, A. L., Nimon, J. P., & Arciero, P. J. (2011). Social facilitation in virtual reality-enhanced exercise: Competitiveness moderates exercise effort of older adults. *Clinical Interventions in Aging, 6*, 275–280

Annual Career-Builder Survey (2016). CareerBuilder. https://press.careerbuilder.com/

Anwar, Y. (2020, February 4). Washington Redskins' name, Native mascots offend more than previously reported. *Berkeley News.* https://news.berkeley.edu/2020/02/04/native- mascots-survey/

Appel, M., & Mara, M. (2013). The persuasive influence of a fictional character's trustworthiness. *Journal of Communication, 63*(5), 912–932.

Arakaki, M. (2020, March 29). *UH Mānoa student-led group creates mask sewing kits aimed at saving lives*. University of Hawai'i News. https://hawaii.edu/news/2020/03/29/student-mask-sewing-kits/

Aravena, F. (2019). Destructive leadership behavior: An exploratory study in Chile. *Leadership and Policy in Schools, 18*(1), 83–96.

Argyle, M. Alkema, F., & Gilmour, R. (1971). The communication of friendly and hostile attitudes by verbal and non-verbal signals. *European Journal of Social Psychology, 1*(3), 385–401.

Ariel, M. (2010). *Defining pragmatics*. Cambridge University Press.

Armfield, F. (2011). *Eugene Kinckle Jones: The National Urban League and Black social work, 1910–1940*. University of Illinois Press.

Arnett, R. C. (2001). Dialogic civility as pragmatic ethical praxis: An interpersonal metaphor for the public domain. *Communication Theory, 11*, 315–338.

Arnett, R. C. (2019). Response: Dialogic listening as attentiveness to place and space. *International Journal of Listening, 33*(3), 181–187.

Arroyo, A., & Segrin, C. (2011). The relationship between self- and other-perceptions of communication competence and friendship quality. *Communication Studies, 62*(5), 547–562.

Asen, R. (2015). Critical engagement through public sphere scholarship. *Quarterly Journal of Speech: Centennial Issue, 101*(1), 132–144.

Askew, A. L. (2012). Iethics: How cloud computing has impacted the rules of professional conduct. *North Dakota Law Review, 88*(2), 453–475.

Association of American Colleges and Universities (AAC&U) Survey (2021). AACU. https://www.aacu.org/research/how-college-contributes-to-workforce-success

Atouba, Y. C., Carlson, E. J., & Lammers, J. C. (2019). Directives and dialogue: Examining the relationship between participative organizational communication practices and organizational identification among IT workers. *International Journal of Business Communication, 56*(4), 530–559.

Auxier, B., & Anderson, M. (2020, March 16). *As schools close due to the coronavirus, some U.S. students face a digital "homework gap."* Pew Research Center. https://pewresearch.org/fact-tank/2020/03/16/as-schools-close-due-to-the-coronavirus-some-u-s-students-face-a-digital-homework-gap/

Ayres, J. (1983). Strategies to maintain relationships: Their identification and perceived usage. *Communication Quarterly, 31*, 62–67.

Ayres, J., & Sonandré, D. M. A. (2003). Performance visualization: Does the nature of the speech model matter? *Communication Research Reports, 20*, 260–268.

Babin, E. A., Palazzolo, K. E., & Rivera, K. (2012). Communication skills, social support, and burnout among advocates in a domestic violence agency. *Journal of Applied Communication Research, 40*(2), 147–166.

Badie, D. (2010). Groupthink, Iraq, and the war on terror: Explaining US policy shift toward Iraq. *Foreign Policy Analysis, 6*(4), 277–296

Bagdikian, B. H. (2004). *The new media monopoly* (Rev. ed.). Beacon Press.

Bahrami, S. S., Rajaeepour, S. S., Keyvanara, M. M., Raesei, A. R., & Kazemi, I. I. (2012). A study of the relationship between managers' decision making styles and organizational health in Isfahan University of Medical Sciences. (English). *Iran Occupational Health, 9*(3), 96–104.

Bailenson, J. N. (2021). Nonverbal overload: A theoretical argument for the causes of Zoom fatigue. *Technology, Mind, and Behavior, 2*(1).

Bailin, S., & Battersby, M. (2016). Fostering the virtues of inquiry. *Topoi, 35*(2), 367–374.

Baird, A., & Parayitam, S. (2019). Employers' ratings of importance of skills and competencies college graduates need to get hired. *Education Training, 61*(5), 622–634.

Balbi, G., & Kittler, J. (2016). One-to-one and one-to-many dichotomy: Grand theories, periodization, and historical narratives in communication studies. *International Journal of Communication, 10*, 1971–1990.

Balboni, P. E., & Caon, F. (2014). A performance-oriented model of intercultural communicative competence. *Journal of Intercultural Communication, 35*(1). https://immi.se/intercultural/nr35/balboni.html

Baldwin, M. W. (1992). Relational schemas and the processing of social information. *Psychological Bulletin, 112*, 461–484.

Ballantine, P., Lin, Y., & Veer, E. (2015). The influence of user comments on perceptions of Facebook relationship status updates. *Computers in Human Behavior, 49*, 50–55.

Ballard, D., & Seibold, D. (2000). Time orientation and temporal variation across work groups: Implications for group and organizational communication. *Western Journal of Communication, 64*, 218–242.

Barsade, S. G., & O'Neill, O. A. (2014). What's love got to do with it? A longitudinal study of the culture of companionate love and employee and client outcomes in a long-term care setting. *Administrative Science Quarterly, 59*, 551–598.

Basol, M., Roozenbeek, J., & Van Der Linden, S. (2020). Good news about bad news: Gamified inoculation boosts confidence and cognitive immunity against fake news. *Journal of Cognition, 3*(1), 2.

Bassett, D. R. (2012). Notions of identity, society, and rhetoric in a speech code of science among scientists and engineers working in nanotechnology. *Science Communication, 34*(1), 115–159.

Basuki, Y., Akbar, R., Pradono, & Miharja, M. (2015). ICT and social relationship engagement: Women's online communities in Indonesia. *Procedia—Social and Behavioral Sciences, 184*(C), 245–251.

Bauer, A. A., Loy, L. S., Masur, P. K., & Schneider, F. M. (2017). Mindful instant messaging: Mindfulness and autonomous motivation as predictors of well-being in smartphone communication. *Journal of Media Psychology: Theories, Methods, and Applications, 29*(3), 159–165.

Baumeister, R. F., & Leary, M. R. (1994). The need to belong: Desire for interpersonal attachments as a fundamental human motivation. *Psychological Bulletin, 117*, 497–529.

Baumlin, J., & Meyer, C. (2018). Positioning ethos in/for the twenty-first century: An introduction to histories of ethos. *Humanities, 7*(3), 78.

Bawden, D., & Robinson, L. (2020). Information overload: An overview. In *Oxford encyclopedia of political decision making*. Oxford University Press. https://openaccess.city.ac.uk/id/eprint/23544/

Baxter, L., Braithwaite, D., Bryant, L., & Wagner, A. (2004). Stepchildren's perceptions of the contradictions in communication with stepparents. *Journal of Social and Personal Relationships, 21*, 447–467.

Baxter, L. A. (1988). A dialectical perspective of communication strategies in relationship development. In S. Duck (Ed.), *Handbook of personal relationships* (pp. 257–273). Wiley.

Baxter, L. A. (2011). *Voicing relationships: A dialogic perspective*. SAGE.

Baxter, L. A., & Braithwaite, D. O. (2008). Relational dialectics theory. In L. A. Baxter & D. O. Braithwaite (Eds.), *Engaging theories in interpersonal communication: Multiple perspectives* (pp. 349–361). SAGE.

Baxter, L. A., & Montgomery, B. M. (1996). *Relating: Dialogues and dialectics*. Guilford.

Baxter, L. A., Scharp, K. M., & Thomas, L. J. (2021). Relational dialectics theory. *Journal of Family Theory & Review, 13*, 7–20.

Baxter, L. A., & Simon, E. P. (1993). Relationship maintenance strategies and dialectical contradictions in personal relationships. *Journal of Social and Personal Relationships, 10*, 225–242.

Becker, B. W. (2018). Information literacy in the digital age: Myths and principles of digital literacy. *School of Information Student Research Journal, 7*(2). https://scholarworks.sjsu.edu/slissrj/vol7/iss2/2

Bejaković, P., & Mrnjavac, Z. (2020). The importance of digital literacy on the labour market. *Employee Relations: The International Journal, 42*(4), 921–932.

Bell, V. (2010, February 15). Don't touch that dial! A history of media technology scares, from the printing press to Facebook. *Slate*. https://slate.com/technology/2010/02/a-history-of-media-technology-scares-from-the-printing-press-to-facebook.html

Bellis, M. (2008). *The first disposable cell phone: Randice-Lisa Altschul creates the world's first disposable cell phone*. About. https://inventors.about.com

Benne, K. D., & Sheats, P. (1948). Functional roles of group members. *Journal of Social Issues, 3*(2), 41–49.

Ben-Shlomo, Y., Smith, G. D., Shipley, M., & Marmot, M. G. (1993). Magnitude and causes of mortality differences between married and unmarried men. *Journal of Epidemiology and Community Health, 47*, 200–205.

Bentley, S. C. (2000). Listening in the 21st century. *International Journal of Listening*, 129–142.

Berger, C. R. (2005). Interpersonal communication: Theoretical perspectives, future prospects. *Journal of Communication, 55*, 415–447.

Berger, C. R. (2008). Interpersonal communication. In W. Donsbach (Ed.), *The international encyclopedia of communication* (pp. 3671–3682). Wiley-Blackwell.

Bergquist, G., Soliz, J., Everhart, K., Braithwaite, D., & Kreimer, L. (2019). Investigating layers of identity and identity gaps in refugee resettlement experiences in the Midwestern United States. *Western Journal of Communication, 83*(3), 383–402. https://doi.org/10.1080/10570314.2018.1552009

Berklee College of Music. (2019, May 14). Missy Elliott—Berklee College of Music commencement address 2019 [Video]. YouTube. https://youtube.com/watch?v=9uiZJgKxlY8

Berlet, C. (2004). Hate, oppression, repression, and the apocalyptic style: Facing complex questions and challenges. *Journal of Hate Studies, 3*, 145–166.

Berlo, D. K. (1960). *The process of communication*. Holt, Rinehart and Winston.

Bing, H., & Fink, E. L. (2012). How do statistical and narrative evidence affect persuasion?: The role of evidentiary features. *Argumentation & Advocacy, 49*(1), 39–58.

Blackhart, G. C., Hernandez, D. K., Wilson, E., & Hance, M. A. (2021). The impact of rejection sensitivity on self-disclosure within the context of online dating. *Cyberpsychology, Behavior and Social Networking*. Advance online publication. https://doi.org/10.1089/cyber.2020.0257

Black Lives Matter. (n.d.). *About*. https://blacklivesmatter.com/about/

Blankenstein, F., Dolmans, D. M., Vleuten, C., & Schmidt, H. G. (2011). Which cognitive processes support learning during small-group discussion? The role of providing explanations and listening to others. *Instructional Science, 39*(2), 189–204.

Blieszner, R., Ogletree, A. M., & Adams, R. G. (2019). Friendship in later life: A research agenda. *Innovation in Aging, 3*, igz005.

Blieszner, R., & Roberto, K. A. (2004). Friendship across the life span: Reciprocity in individual and relationship development. In F. R. Lang & K. L. Fingerman (Eds.), *Advances in personal relationships. Growing together: Personal relationships across the lifespan* (pp. 159–182). Cambridge University Press.

Bloom, G. (2019, April 22). *Rough magic: Performing Shakespeare through gaming technology*. Folgerpedia. https://folgerpedia.folger.edu

Blumenstyk, G. (2013, March 15). "Yes We Must"-coalition steps up to advocate for needy students—and their colleges. Chronicle of Higher Education. https://www.chronicle.com/article/yes-we-must-coalition-steps-up-to-advocate-for-needy-students-and-their-colleges/

Boak, G., Dickens, V., Newson, A., & Brown, L. (2015). Distributed leadership, team working and service improvement in healthcare. *Leadership in Health Services, 28*(4), 332–344.

Bodie, G. D. (2010). A racing heart, rattling knees, and ruminative thoughts: Defining, explaining, and treating public speaking anxiety. *Communication Education, 59*(1), 70–105.

Bodie, G. D. (2013). The role of thinking in the comforting process: An empirical test of a dual-process framework. *Communication Research, 40*(4), 533–558.

Bodie, G. D., & Jones, S. M. (2012). The nature of supportive listening II: The role of verbal person centeredness and nonverbal immediacy. *Western Journal of Communication, 76*(3), 250–269.

Bodie, G. D., St. Cyr, K., Pence, M., Rold, M., & Honeycutt, J. (2012). Listening competence in initial interactions I: Distinguishing between what listening is and what listeners do. *International Journal of Listening, 26*(1), 1–28.

Bodie, G. D., Vickery, A. J., Cannava, K., & Jones, S. M. (2015). The role of "active listening" in informal helping conversations: Impact on perceptions of listener helpfulness, sensitivity, and supportiveness and discloser emotional improvement. *Western Journal of Communication, 79*(2), 151–173

Bodie, G. D., Vickery, A. J., & Gearhart, C. C. (2013). The nature of supportive listening I: Exploring the relation between supportive listeners and supportive people. *International Journal of Listening, 27*(1), 39–49.

Bommelje, R., Houston, J. M., & Smither, R. (2003). Personality characteristics of effective listeners: A five factor perspective. *International Journal of Listening, 17,* 32–46.

Bommelje, R., Houston, J. M., & Smither, R. (2003). Personality characteristics of effective listeners: A five factor perspective. *International Journal of Listening, 17,* 32–46.

Bonesteel, M. (2021, January 22). The Braves have resisted a name change, but Hank Aaron's death renews calls for "the Hammers." *The Washington Post.* https://www.washingtonpost.com/sports/2021/01/22/braves-name-change-hammers/

Borchers, T. A. (2013). *Persuasion in the media age* (3rd ed.). Waveland.

Bostrom, R. N. (1970). Patterns of communicative interaction in small groups. *Speech Monographs, 37,* 257–263.

Bostrom, R. N. (2011). Rethinking conceptual approaches to the study of 'listening'. *International Journal of Listening, 25*(1/2), 10–26.

Bostrom, R. N., & Waldhart, E. S. (1988). Memory models and the measurement of listening. *Communication Education, 37,* 1–13.

Bostrom, R. N., & Waldhart, E. S. (1988). Memory models and the measurement of listening. *Communication Education, 37,* 1–18.

Bououd, I., Skandrani, S., Boughzala, I., & Makhlouf, M. (2016). Impact of object manipulation, customization and social loafing on competencies management in 3D virtual worlds. *Information Systems Frontiers, 18*(6), 1191–1203.

Bowman, J. M. (2020). *Nonverbal communication: An applied approach.* SAGE.

Bowman, J. P., & Targowski, A. S. (1987). Modeling the communication process: The map is not the territory. *Journal of Business Communication, 24*(4), 21–34.

Boyle, A., & O'Sullivan, L. (2016). Staying connected: Computer-mediated and face-to-face communication in college students' dating relationships. *Cyberpsychology Behavior and Social Networking, 19*(5), 299–307.

Boyle, M. P. (2013). Psychological characteristics and perceptions of stuttering of adults who stutter with and without support group experience. *Journal of Fluency Disorders, 38*(4), 368–381.

Boyraz, M. (2019). Faultlines as the "Earth's crust": The role of team identification, communication climate, and subjective perceptions of subgroups for global team satisfaction and innovation. *Management Communication Quarterly, 33*(4), 581–615. https://doi.org/10.1177/0893318919860799

Bradsher, K. (2004, October 13). Disney tailors Hong Kong park for cultural differences. *New York Times.* https://www.nytimes.com/2004/10/13/business/worldbusiness/disney-tailors-hong-kong-park-for-cultural.html

Brady, E. (2013, October 25). DC clergy join push to change Redskins name. *Religion News Service.* http://www.religionnews.com/2013/10/25/clergy-members-join-push-to-change-football-teams-name/

Braithwaite, S., & Julianne Holt-Lunstad, J. (2017). Romantic relationships and mental health. *Current Opinion in Psychology, 13,* 120–125.

Bregenzer, A., Milfelner, B., Sarotar Zizek, S., & Jimenez, P. (2020). Health-promoting leadership and leaders' listening skills have an impact on the employees' job satisfaction and turnover intention. *International Journal of Business Communication.* Advance online publication. https://doi.org/10.1177/2329488420963700

Bregman, A., & Haythornthwaite, C. (2003). Radicals of presentation: Visibility, relation, and co-presence in persistent conversation. *New Media & Society, 5,* 117–140.

Bright, A., & Gambrell, J. (2017). Calling in, not calling out: A critical race framework for nurturing cross-cultural alliances in teacher candidates. In J. Keengwe (Ed.), *Handbook of research on promoting cross-cultural competence and social justice in teacher education* (pp. 217–235). IGI Global.

Brink, K. E., & Costigan, R. D. (2015). Oral communication skills: Are the priorities of the workplace and AACSB- accredited business programs aligned? *Academy of Management Learning & Education, 14*(2), 205–221.

Brown, B. B., Werner, C. M., & Altman, I. (1998). Choice points for dialecticians: A dialectical-transactional perspective on close relationships. In B. Montgomery & L. Baxter (Eds.), *Dialectical approaches to studying personal relationships* (pp. 137–154). Lawrence Erlbaum.

Brown, V., & Strand, J. (2019). Noise increases listening effort in normal-hearing young adults, regardless of working memory capacity. *Language, Cognition and Neuroscience, 34*(5), 628–640.

Browne, C., Mokuau, N., Ka'opua, L., Kim, B., Higuchi, P., & Braun, K. (2014). Listening to the voices of Native Hawaiian elders and 'Ohana caregivers: Discussions on aging, health, and care preferences. *Journal of Cross-Cultural Gerontology, 29*(2), 131–151.

Brownell, J. (2013). Robert Bostrom's contribution to listening in organizational contexts. *International Journal of Listening, 27*(2), 101–103.

Brownell, J. (2017). *Listening: Attitudes, principles, and skills* (6th ed.). Taylor & Francis.

Brownell, J. (2018). *Listening: Attitudes, principles, and skills* (6th ed.). Routledge.

Brun, J. P. (2010). *Missing pieces: 7 ways to improve employee well-being and organizational effectiveness.* Palgrave Macmillan.

Bruneau, T. (1974). Time and nonverbal communication. *Journal of Popular Culture, 8,* 658–666.

Bruneau, T. (1990). Chronemics: The study of time in human interaction. In J. DeVito & M. Hecht (Eds.), *The nonverbal reader* (pp. 301–311). Waveland Press.

Bruneau, T., & Ishii, S. (1988). Communicative silence: East and west. *World Communication, 17,* 1–33.

Buchanan, T., & Benson, V. (2019). Spreading disinformation on Facebook: Do trust in message source, risk propensity, or personality affect the organic reach of "fake news"? *Social Media + Society, 5*(4). https://doi.org/10.1177/2056305119888654

Budzynska, K., & Witek, M. (2014). Non-inferential aspects of ad hominem and ad baculum. *Argumentation, 28*(3), 301–315.

Buis, A. (2019, July 12). *Nope, the Earth isn't cooling.* NASA: Global Climate Change: Vital Signs of the Planet. https://climate.nasa.gov/blog/2893/nope-earth-isnt-cooling

Bulu, S. (2012). Place presence, social presence, co-presence, and satisfaction in virtual worlds. *Computers & Education, 58*(1), 154–161.

Burgess, H., & Burgess, G. (1996). Constructive confrontation: A transformative approach to intractable conflict. *Conflict Resolution Quarterly, 13*, 305–322.

Burgoon, J. K., Guerrero, L. K., & Floyd, K. (2010). *Nonverbal communication.* Pearson.

Burke, K. (2019, January 24). 107 texting statistics that answer all your questions. *Text Request.* https://textrequest.com/blog/texting-statistics-answer-questions/

Burke, T. J., Woszidlo, A., & Segrin, C. (2013). The intergenerational transmission of social skills and psychosocial problems among parents and their young adult children. *Journal of Family Communication, 13*, 77–91.

Burleson, B. R. (2011). A constructivist approach to listening. *International Journal of Listening, 25*(1/2), 27–46.

Burleson, B. R., Hanasono, L. K., Bodie, G. D., Holmstrom, A. J., McCullough, J. D., Rack, J. J., & Rosier, J. (2011). Are gender differences in responses to supportive communication a matter of ability, motivation, or both? Reading patterns of situation effects through the lens of a dual-process theory. *Communication Quarterly, 59*(1), 37–60.

Burleson, B. R., & Kunkel, A. (2002). Parental and peer contributions to the emotional support skills of the child: From whom do children learn to express support? *Journal of Family Communication, 2*, 79–97.

Buzan, T. (2014). *Mind maps for business: Using the ultimate thinking tool to revolutionise how you work* (2nd ed.). Pearson Education.

Bylund, C. L., Peterson, E. B., & Cameron, K. A. (2012). A practitioner's guide to interpersonal communication theory: An overview and exploration of selected theories. *Patient Education and Counseling, 87*, 261–267.

Cahalan, M., Perna, L., Yamashita, M., Ruiz, R., & Franklin, K. (2016). *Indicators of higher education equity in the United States: 2016 historical trend report.* Pell Institute for the Study of Opportunity in Higher Education, Council for Opportunity in Education (COE), Alliance for Higher Education and Democracy of the University of Pennsylvania (PennAHEAD).

California Coastal Commission. (2019). *Meetings: Rules and procedures.* https://coastal.ca.gov/meetings/rules-procedures

Campbell, K. K., Huxman, S. S., & Burkholder, T. R. (2015). *The rhetorical act: Thinking, speaking, and writing critically* (5th ed.). Cengage Learning.

Campbell, S. W. (2003). Listening to the voices in an online class. *Qualitative Research Reports in Communication, 4*, 9–15.

Canary, D. J., & Dainton, M. (2003). *Maintaining relationships through communication: Relational, contextual, and cultural variations.* Lawrence Erlbaum.

Canary, D. J., & MacGregor, I. (2008). Differences that make a difference in assessing student communication competence. *Communication Education, 57*(1), 41–63.

Canary, D. J., & Stafford, L. (1992). *Relational maintenance strategies* and equity in marriage. *Communication Monographs, 59*, 243–267.

Canary, D. J., & Stafford, L. (1992). Relational maintenance strategies and equity in marriage. *Communication Monographs, 59*, 243–267.

Canary, D. J., & Stafford, L. (1994). Maintaining relationships through strategic and routine interactions. In D. J. Canary & L. Stafford (Eds.), *Communication and relational maintenance* (pp. 1–22). Emerald Publishing.

Canary, D. J., & Stafford, L. (2001). Equity in the preservation of personal relationships. In J. H. Harvey & A. Wenzel (Eds.), *Close romantic relationships: Maintenance and enhancement* (pp. 133–153). Lawrence Erlbaum.

Canary, D. J., Stafford, L., & Semic, B. A. (2002). A panel study of the associations between maintenance strategies and relational characteristics. *Journal of Marriage and the Family, 64*, 395–406

Candland, D. K. (1995). *Feral children and clever animals: Reflections on human nature.* Oxford University Press.

Carpenter, C. (2012). A meta-analysis and an experiment investigating the effects of speaker disfluency on persuasion. *Western Journal of Communication, 76*(5), 552–569.

Carpenter, C. J. (2012). A meta-analysis and an experiment investigating the effects of speaker disfluency on persuasion. *Western Journal of Communication, 76*(5), 552–569.

Carr, C. T., & Foreman, A. C. (2016). Identity shift III: Effects of publicness of feedback and relational closeness in computer-mediated communication. *Media Psychology, 19*(2), 334–358.

Carrington, D. J., Combe, I. A., & Mumford, M. D. (2019). Cognitive shifts within leader and follower teams: Where consensus develops in mental models during an organizational crisis. *The Leadership Quarterly, 30*(3), 335–350. https://doi.org/10.1016/j.leaqua.2018.12.002

Carroll, E. C. (2014, July 14). Mansfield school chief quits amid plagiarism charges. *Boston Globe.* https://www.bostonglobe.com/metro/2014/07/16/mansfield-school-superintendent-announces-she-will-step-down-amid-plagiarism-controversy/MQwQxRSzcsSTbRATQuS4yO/story.html

Cash, W. B., & Stewart, C. W. (2018). *Interviewing: Principles and practices* (15th ed.). McGraw-Hill.

Caspersz, D., & Stasinska, A. (2015). Can we teach effective listening? An exploratory study. *Journal of University Teaching and Learning Practice, 12*(4). https://ro.uow.edu.au/jutlp/vol12/iss4/2

Castells, M. (2001). *The internet galaxy.* Oxford University Press.

Castro, D. R., Anseel, F., Kluger, A. N., Lloyd, K. J., & Turjeman-Levi, Y. (2018). Mere listening effect on creativity and the mediating role of psychological safety. *Psychology of Aesthetics, Creativity, and the Arts, 12*(4), 489–502. https://doi.org/10.1037/aca0000177

Catmull, E. E. (2014). *Creativity, Inc.: Overcoming the unseen forces that stand in the way of true inspiration*. Random House.

Caughlin, J. P., & Vangelisti, A. L. (2006). Conflict in dating and romantic relationships. In J. Oetzel & S. Ting-Toomey (Eds.), *The SAGE handbook of conflict communication* (pp. 129–157). SAGE.

Cavazotte, F., Hartman, N. S., & Bahiense, E. (2014). Charismatic leadership, citizenship behaviors, and power distance orientation: Comparing Brazilian and U.S. workers. *Cross-Cultural Research, 48*(1), 3–31.

Çelebi, Ö., & Polat. A. (2018). "The Women of Future": An authentic self-help group study with Syrian women. *Kocaeli Üniversitesi Sağlık Bilimleri Dergisi, 4*(3), 81–87.

Centers for Disease Control and Prevention. (2019, September). *Disability impacts us all*. https://cdc.gov/ncbddd/disabilityandhealth/infographic-disability-impacts-all.html

Cha, J., Cichy, R., & Kim, S. (2011). Commitment and volunteer-related outcomes among private club board and committee member volunteer leaders. *Journal of Hospitality & Tourism Research, 35*(3), 308–333.

Chadwick, A. E., Zoccola, P. M., Figueroa, W. S., & Rabideau, E. M. (2016). Communication and stress: Effects of hope evocation and rumination messages on heart rate, anxiety, and emotions after a stressor. *Health Communication, 31*(12), 1447–1459.

Chaika, E. (2007). *Language: The social mirror* (4th ed.). Heinle.

Chandler, D. (2007). *Semiotics: The basics* (2nd ed.). Routledge.

Chaney, J., Burke, A., & Burkley, E. (2011). Do American Indian mascots = American Indian people? Examining implicit bias towards American Indian people and American Indian mascots. *American Indian and Alaska Native Mental Health Research, 18*(1), 42–62.

Chang, J., Sy, T., & Choi, J. (2012). Team emotional intelligence and performance: Interactive dynamics between leaders and members. *Small Group Research, 43*(1), 75–104.

Change the Mascot. (n.d.). *About*. http://www.changethemascot.org

Charoenruk, N., & Olson, K. (2018). Do listeners perceive interviewers' attributes from their voices and do perceptions differ by question type? *Field Methods, 30*(4), 312–328.

Chawla, S., & Sharma, R. (2019). Enhancing women's well-being: the role of psychological capital and perceived gender equity, with social support as a moderator and commitment as a mediator. *Frontiers in Psychology, 10*. https://doi.org/10.3389/fpsyg.2019.01377

Chen, E., Brody, G. H., & Miller, G. E. (2017). Childhood close family relationships and health. *American Psychologist, 72*, 555–566.

Chen, H. (2019). An explorative study on the virtual world: Investigating the avatar gender and avatar age differences in their social interactions for help-seeking. *Information Systems Frontiers, 22*, 911–925. https://doi.org/10.1007/s10796-019-09904-2

Chen, H., & Li, X. (2017). The contribution of mobile social media to social capital and psychological well-being: Examining the role of communicative use, friending and self-disclosure. *Computers in Human Behavior, 75*, 958–965.

Chen, J., Mak, R., & Fujita, S. (2015). The effect of combination of video feedback and audience feedback on social anxiety. *Behavior Modification, 39*(5), 721–739.

Chen, Y., & Nakazawa, M. (2012). Measuring patterns of self-disclosure in intercultural friendship: Adjusting differential item functioning using multiple-indicators, multiple-causes models. *Journal of Intercultural Communication Research, 41*(2), 131–151.

Chen, Y., Ning, R., Yang, T., Feng, S., & Yang, C. (2018). Is transformational leadership always good for employee task performance? Examining curvilinear and moderated relationships. *Frontiers of Business Research in China, 12*(1), 1–28.

Choi, C. W., Honeycutt, J. M., & Bodie, G. D. (2015). Effects of imagined interactions and rehearsal on speaking performance. *Communication Education, 64*(1), 25–44.

Choi, K. (2012). Disneyfication and localisation: The cultural globalisation process of Hong Kong Disneyland. *Urban Studies, 49*, 383–397.

Choi, W. (2018). From Occupy Wall Street to the Umbrella Movement: Leaderless uprising explained. *Perspectives on Global Development and Technology, 17*(1–2), 220–235.

Chomsky, N., Belletti, A., & Rizzi, L. (2002). *On nature and language*. Cambridge University Press.

Chopik, W. J. (2017). Associations among relational values, support, health, and well-being across the adult lifespan. *Personal Relationships, 24*, 408–422.

Chopik, W. J. (2017). Associations among relational values, support, health, and well-being across the adult lifespan. *Personal Relationships, 24*, 408–422.

Chow, I. H. (2018). Cognitive diversity and creativity in teams: The mediating roles of team learning and inclusion. *Chinese Management Studies, 12*(2), 369–383.

Christenfeld, N. (1995). Does it hurt to say um? *Journal of Nonverbal Behavior, 19*(3), 171–186.

Chu, J. (2002, March 18). Happily ever after? *Time*. http://content.time.com/time/subscriber/article/0,33009,218398,00.html

Chui, M., Manyika, J., Bughin, J., Dobbs, R., Roxburgh, C., Sarrazin, H., Sands, G., & Westergren, M. (2012, July). *The social economy: Unlocking value and productivity through social technologies*. McKinsey Global Institute.

Cingel, D. P, & Sundar, S. S. (2012). Texting, techspeak, and tweens: The relationship between text messaging and English grammar skills. *New Media & Society, 14*(8), 1304–1320.

Clark-Gordon, C., Bowman, N., Goodboy, A., & Wright, A. (2019). Anonymity and online self-disclosure: A meta-analysis. *Communication Reports, 32*(2), 98–111.

Clement, J. (2019, August 9). *Number of sent and received e-mails per day worldwide from 2017 to 2023*.

Statista. https://statista.com/statistics/456500/daily-number-of-e-mails-worldwide/

Clement, J. (2020, June 18). *Global market share of search engines 2010–2020*. Statista. https://statista.com/statistics/216573/worldwide-market-share-of-search-engines

Cline, B. J. (2013). The science and sanity of listening. *ETC: A Review of General Semantics, 70*(3), 247–259.

Clokie, T., & Fourie, E. (2016). Graduate employability and communication competence: Are undergraduates taught relevant skills? *Business and Professional Communication Quarterly, 79*(4), 442–463.

CNN Wire. (2019, August 29). Blocking social media would be 'the end of the open internet of Hong Kong.' It also wouldn't work.

Coetzer, G. (2015). Emotional versus cognitive intelligence: Which is the better predictor of efficacy for working in teams? *Journal of Behavioral and Applied Management, 16*(2), 116–133.

Coffelt, T., Grauman, D., & Smith, F. (2019). Employers' perspectives on workplace communication skills: The meaning of communication skills. *Business and Professional Communication Quarterly, 82*(4), 418–439.

Cole, D., & Ji, Z. (2014). Diversity and collegiate experiences affecting self-perceived gains in critical thinking. *JGE: The Journal of General Education, 63*(1), 15–34.

Cole, T. (2001). Lying to the one you love: The use of deception in romantic relationships. *Journal of Social and Personal Relationships, 18*(1), 107–129.

Collins, W. A., & Madsen, S. D., (2006). Personal relationships in adolescence and early adulthood. In A. L. Vangelisti & D. Perlman (Eds.), *The Cambridge handbook of personal relationships* (pp. 191–210). Cambridge University Press.

Collins Dictionary. (n.d.). Elucidate. In *CollinsDictionary.com dictionary*. https://collinsdictionary.com/us/dictionary/english/elucidate

Comello, M. G. (2009). William James on "possible selves": Implications for studying identity in communication contexts. *Communication Theory, 19*(3), 337–350.

Compton, C. (2019). Communicatively navigating identities, silence, and social protest. *Communication Studies, 70*(4), 412–432.

Congressional Black Caucus. (2017, April 3). *CBC launches HBCU Tour #CBCOnTheYard*. https://cbc.house.gov/news/documentsingle.aspx?DocumentID=434

Contreras, R. (2013, May 12). No 'Day of the Dead' trademark for Disney. *San Jose Mercury News*, E1, E5.

Cook, K. S., & Rice, E. (2006). Social exchange theory. In J. DeLamater (Ed.), *Handbook of social psychology* (pp. 53–76). Springer.

Cook, K. S., & Rice, E. R. W. (2001). Exchange and power: Issues of structure and agency. In J. H. Turner (Ed.), *Handbook of sociological theory* (pp. 699–719). Springer.

Cooper, L. O., & Buchanan, T. (2010). Listening competency on campus: A psychometric analysis of student listening. *International Journal of Listening, 24*(3), 141–163.

Coopman, S. (2003). Communicating disability: Metaphors of oppression, metaphors of empowerment. In P. Kalbfleisch (Ed.), *Communication yearbook 27* (pp. 337–394). Lawrence Erlbaum.

Coopman, S. J. (1997). Personal constructs and communication in interpersonal and organizational contexts. In G. Neimeyer & R. Neimeyer (Eds.) *Advances in personal construct psychology* (Vol. 4, pp. 101–147). JAI Press.

Coopman, S. J., & Meidlinger, K. (2000). Power, hierarchy and change: The stories of a Catholic parish staff. *Management Communication Quarterly, 13*, 567–625.

Coopman, T. M. (2009). Toward a pervasive communication environment perspective. *First Monday, 14*(1). https://doi.org/10.5210/fm.v14i1.2277

Coover, G. E., & Murphy, S. T. (2000). The communicated self: Exploring the interaction between self and social context. *Human Communication Research, 26*, 125–147.

Cornwell, N. C., & Orbe, M. P. (1999). Critical perspectives on hate speech: The centrality of 'dialogic listening.' *International Journal of Listening, 13*, 75–96

Correia, J. C., Jerónimo, P., & Gradim, A. (2019). Fake news: Emotion, belief and reason in selective sharing in contexts of proximity. *Brazilian Journalism Research, 15*(3), 590–613.

Cotton, K., & Ricker, T. J. (2021). Working memory consolidation improves long-term memory recognition. *Journal of Experimental Psychology. Learning, Memory, and Cognition, 47*(2), 208–219. https://doi.org/10.1037/xlm0000954

Courtright, J. A. (1978). A laboratory investigation of groupthink. *Communication Monographs, 45*, 229–246.

Cronkhite, G. (1986). On the focus, scope, and coherence in the study of human symbolic activity. *Quarterly Journal of Speech, 72*, 231–246.

Croucher, S. M., Sommier, M., & Rahmani, D. (2015). Intercultural communication: Where we've been, where we're going, issues we face. *Communication Research and Practice, 1*, 71–87.

Croy, G., & Eva, N. (2018). Student success in teams: Intervention, cohesion and performance. *Education Training, 60*(9), 1041–1056.

Cruz, M. G., Boster, F. J., & Rodriguez, J. I. (1997). The impact of group size and proportion of shared information on the exchange and integration of information in groups. *Communication Research, 24*, 291–313.

Crystal, D. (2005). *How language works: How babies babble, words change meaning, and languages live or die*. Avery.

Crystal, D. (2009). *Txtng: The gr8 db8*. Oxford University Press.

Culiberg, B. (2019). Reaping the fruits of another's labor: The role of moral meaningfulness, mindfulness, and motivation in social loafing. *Journal of Business Ethics, 160*(3), 713–727.

Culicover, P. W. (2013). *Explaining syntax: Representations, structures, and computation*. Oxford University Press.

Cummings, L. (2005). *Pragmatics: A multidisciplinary perspective*. Edinburgh University Press.

Curşeu, P., Jansen, R. G., & Chappin, M. H. (2013). Decision rules and group rationality: Cognitive gain or standstill?. *Plos ONE, 8*(2), 1–7.

Cyphert, D., Holke-Farnam, C., Dodge, E., Lee, W., & Rosol, S. (2019). Communication activities in the 21st century business environment. *Business and Professional Communication Quarterly, 82*(2), 169–201.

Da'as, R. (2020). Participation in decision making and affective trust among the teaching staff: A 2-year cross-lagged structural equation modeling during implementation reform. *International Journal of Educational Reform, 29*(1), 77–97.

D'agostino, T. A., & Bylund, C. L. (2014). Nonverbal accommodation in health care communication. *Health Communication, 29*(6), 563–573.

Dahlstrom, M. (2014). Using narratives and storytelling to communicate science with nonexpert audiences. *Proceedings of the National Academy of Sciences of the United States of America, 111*, 13614–13620.

Dainton, M. (2008). The use of maintenance behaviors as a mechanism to explain the decline in marital satisfaction among parents. *Communication Reports, 21*, 33–45.

Dainton, M., & Gross, J. (2008). The use of negative behaviors to maintain relationships. *Communication Research Reports, 25*, 179–191.

Dainton, M., & Stafford, L. (1993). Routine maintenance behaviors: A comparison of relationship type, partner similarity and sex differences. *Journal of Social and Personal Relationships, 10*, 255–271.

Dainton, M., & Stafford, L. *(1993).* Routine maintenance behaviors: A comparison of relationship type, partner similarity and sex differences. *Journal of Social and Personal Relationships, 10*, 255–271.

Daly, J. A., & Redlick, M. H. (2016). Handling questions and objections affects audience judgments of speakers. *Communication Education, 65*(2), 164–181.

Danielle, T. (2019). *9 times in history when everyone freaked out about new technology.* Ranker. https://ranker.com/list/ historical-freak-outs-about-new-technology/ taeyura

Dartey-Baah, K., & Mekpor, B. (2017). Emotional intelligence: Does leadership style matter? Employees' perception in Ghana's banking sector. *International Journal of Business, 22*(1), 41–54.

Davies, M. (2011). Concept mapping, mind mapping and argument mapping: What are the differences and do they matter? *Higher Education, 62*(3), 279–301.

Davis, M. S. (1973). *Intimate relations.* Free Press.

Davis, S. (2019). When sistahs support sistahs: A process of supportive communication about racial microaggressions among Black women. *Communication Monographs, 86*(2), 133–157.

Davis, V. (2019, May 20). *Viola Davis, remarks as delivered.* Barnard. https://barnard.edu/commencement/ archives/2019/davis-remarks

Deaux, K., & Major, B. (1987). Putting gender into context: An interactive model of gender-related behavior. *Psychological Review, 94*, 369–389.

de Graaf, A., Hoeken, H., Sanders, J., & Beentjes, J. J. (2012). Identification as a mechanism of narrative persuasion. *Communication Research, 39*(6), 802–823.

de Groot, J. H. B., Smeets, M. A. M., Kaldewaij, A., Duijndam, M. J. A., & Semin, G. R. (2012). Chemosignals communicate human emotions. *Psychological Science, 23*(11), 1417–1424.

Delatorre, M., & Wagner, A. (2019). How do couples disagree? An analysis of conflict resolution profiles and the quality of romantic relationships. *Revista Colombiana de Psicología, 28*, 91–108.

Demir, M., Orthel, H., & Andelin, A. K. (2013). Friendship and happiness. In S. A. David, I. Boniwell, & A. C. Ayers, *Oxford handbook of happiness* (pp. 860–870). Oxford University Press.

den Dikken, M. (2013). *The Cambridge handbook of generative syntax.* Cambridge University Press.

Denworth, L. (2020). *Friendship: The evolution, biology, and extraordinary power of life's fundamental bond.* W. W. Norton.

Derlega, V. J., Winstead, B. A., Mathews, A., & Braitman, A. L. (2008). Why does someone reveal highly personal information? Attributions for and against self-disclosure in close relationships. *Communication Research Reports, 25*(2), 115–130.

de Smet, M. R., Brand-Gruwel, S., Leijten, M., & Kirschner, P. A. (2014). Electronic outlining as a writing strategy: Effects on students' writing products, mental effort and writing process. *Computers & Education, 78*, 352–366.

De Smit, H. (2012). The course of actualization. *Language, 88*(3), 601–633.

de Vries, I., & van Elferen, I. (2010). The musical *Madeleine*: Communication, performance, and identity in musical ringtones. *Popular Music & Society, 33*(1), 61–74.

De Wever, B., Schellens, T., van Keer, H., & Valcke, M. (2008). Structuring asynchronous discussion groups by introducing roles: Do students act in line with assigned roles? *Small Group Research, 39*, 770–794.

Diamond, A. B., Callahan, S. T., Chain, K. F., & Solomon, G. S. (2016). Qualitative review of hazing in collegiate and school sports: Consequences from a lack of culture, knowledge and responsiveness. *British Journal of Sports Medicine, 50*(3), 149–153.

Diamond, L. M., Allen, K., & Butterworth, M. R. (2012). The family relationships of sexual minorities. In A. Vangelisti (Ed.), *The Routledge handbook of family communication* (2nd ed., pp. 176–189). Routledge.

Díaz-Vera, J. E., & Caballero, R. (2013). Exploring the feeling-emotions continuum across cultures: Jealousy in English and Spanish. *Intercultural Pragmatics, 10*(2), 265–294.

di Carlo, G. S. (2015). Pathos as a communicative strategy for online knowledge dissemination: The case of TED Talks. *3L, Language, Linguistics, Literature, 21*(1), 23–34.

Dickson, M. (2016, June 21). *The Victorians had the same concerns about technology as we do.* The Conversation. https://theconversation.com/ the-victorians-had-the-same-concerns-about-technology-as-we-do-60476

Dieckmann, L. E. (2000). Private secrets and public disclosures: The case of battered women. In S. Petronio (Ed.), *Balancing the secrets of private disclosures* (pp. 275–286). Lawrence Erlbaum.

Diehl, W. C., & Prins, E. (2008). Unintended outcomes in Second Life: Intercultural literacy and cultural identity in a virtual world. *Language & Intercultural Communication, 8*, 101–118.

DigglesCreative. (2021). *Brand vision worksheet.* https:// digglescreative.com/blog/create-your-brand-vision.html

Dindia, K., & Baxter, L. A. (1987). Strategies for maintaining and repairing marital relationships. *Journal of Social and Personal Relationships, 4*, 143–158.

Ding, H., & Ding, X. (2008). Project management, critical praxis, and process-oriented approach to teamwork. *Business Communication Quarterly, 71*, 456–471.

Dipasquale, J., & Hunter, W. J. (2018). Critical thinking in asynchronous online discussions: A systematic review. *Canadian Journal of Learning and Technology, 44*(2). https://cjlt.ca/index.php/cjlt/article/view/27782

Dipper, L., Black, M., & Bryan, K. (2005). Thinking for speaking and thinking for listening: The interaction of thought and language in typical and non-fluent comprehension and production. *Language and Cognitive Processes, 20*, 417–441.

Dooly, M., & Rubinstein, C. V. (2018). Bridging across languages and cultures in everyday lives: An expanding role for critical intercultural communication. *Language and Intercultural Communication, 18*, 1–8.

Dorrance Hall, E., & Shebib, S. J. (2020). Interdependent siblings: Associations between closest and least close sibling social support and sibling relationship satisfaction. *Communication Studies, 71*, 612–632.

Dotterer, A., & Day, E. (2019). Parental knowledge discrepancies: Examining the roles of warmth and self-disclosure. *Journal of Youth and Adolescence, 48*(3), 459–468.

Dowell, D., Morrison, M., & Heffernan, T. (2015). The changing importance of affective trust and cognitive trust across the relationship lifecycle: A study of business-to-business relationships. *Industrial Marketing Management, 44*, 119–130.

Driskell, T., Driskell, J. E., Burke, C. S., & Salas, E. (2017). Team roles: A review and integration. *Small Group Research, 48*(4), 482–511.

Droser, V. (2017). (Re)Conceptualizing family communication: Applying Deetz's conceptual frameworks within the field of family communication. *Journal of Family Communication, 17*, 89–104.

Drouin, M., Reining, L., Flanagan, M., Carpenter, M., & Toscos, T. (2018). College students in distress: Can social media be a source of social support? *College Student Journal, 52*(4), 494–504.

Drummond, C. K. (2012). Team-based learning to enhance critical thinking skills in entrepreneurship education. *Journal of Entrepreneurship Education, 15*, 57–63.

Du, H., & King, R. B. (2013). Placing hope in self and others: Exploring the relationships among self-construals, locus of hope, and adjustment. *Personality & Individual Differences, 54*(3), 332–337.

Duck, S. (1982). A topography of relationship disengagement and dissolution. In S. W. Duck (Ed.), *Personal relationships 4: Dissolving personal relationships* (pp. 1–29). Academic Press.

Duck, S. (1989). *Relating to others.* Open University Press.

Duck, S. (1998). *Human relationships* (3rd ed.). SAGE.

Duck, S. (2007). *Human relationships* (4th ed.). SAGE.

Duck, S. (2011). *Rethinking relationships.* SAGE.

Duffy, M. E. (2003). Web of hate: A fantasy theme analysis of the rhetorical vision of hate groups online. *Journal of Communication Inquiry, 27*, 291–312.

Dumas, T. L., Phillips, K. W., & Rothbard, N. P. (2013). Getting closer at the company party: Integration experiences, racial dissimilarity, and workplace relationships. *Organization Science, 24*, 1377–1401.

Dumas, T. L., & Sanchez-Burks, J. (2015). The professional, the personal, and the ideal worker: Pressures and objectives shaping the boundary between life domains. *Academy of Management Annals, 9*, 803–843.

Dunleavy, K., & Martin, M. M. (2010). Instructors' and students' perspectives of student nagging: Frequency, appropriateness, and effectiveness. *Communication Research Reports, 27*(4), 310–319.

Dunlop, P. D., & Lee, K. (2004). Workplace deviance, organizational citizenship behavior, and business unit performance: The bad apples do spoil the whole barrel. *Journal of Organizational Behavior, 25*, 67–80.

Dupuis, M., Khadeer, S., & Huang, J. (2017). "I got the job!": An exploratory study examining the psychological factors related to status updates on Facebook. *Computers in Human Behavior, 73*, 132–140.

Dwyer, K., & Davidson, M. (2012). Is public speaking really more feared than death? *Communication Research Reports, 29*(2), 99–107.

Ebbers, J., & Wijnberg, N. (2017). Betwixt and between: Role conflict, role ambiguity and role definition in project-based dual-leadership structures. *Human Relations, 70*(11), 1342–1365.

Eckert, P., & McConnell-Ginet, S. (1992). Think practically and look locally: Language and gender as community-based practice. *Annual Review of Anthropology, 21*(1), 461–490.

Eco, U. (1976). *A theory of semiotics.* Indiana University Press.

Economidou-Kogetsidis, M. (2004). Requestive directness in intercultural communication: Greek and English. In J. Leigh & E. Loo (Eds.), *Outer limits: A reader in communication and behaviour across cultures* (pp. 25–48). Language Australia.

Economidou-Kogetsidis, M. (2010). Cross-cultural and situational variation in requesting behaviour: Perceptions of social situations and strategic usage of request patterns. *Journal of Pragmatics, 42*(8), 2262–2281.

Edelman, P., & Van Knippenberg, D. (2018). Emotional intelligence, management of subordinate's emotions, and leadership effectiveness. *Leadership & Organization Development Journal, 39*(5), 592–607.

Edgerly, L. (2011). Difference and political legitimacy: Speakers' construction of 'citizen' and 'refugee' personae in talk about Hurricane Katrina. *Western Journal of Communication, 75*(3), 304–322.

Editorial Board. (2019, August 17). Schools in two states just retired Native American mascots. The D.C. NFL team should take note. *The Washington Post.* https://www.washingtonpost.com/opinions/schools-in-2-states-just-retired-native-american-mascots-the-dc-nfl-team-should-take-note/2019/08/17/9368827a-8c53-11e9-adf3-f70f78c156e8_story.html?noredirect=on

Edwards, A., Dutton-Challis, L., Cottrell, D., Guy, J., & Hettinga, F. (2018). Impact of active and passive social facilitation on self-paced endurance and sprint exercise: Encouragement augments performance and motivation to exercise. *BMJ Open*

Sport—Exercise Medicine, 4(1), E000368. https://doi.org/10.1136/bmjsem-2018-000368

Edwards, R. (2011). Listening and message interpretation. *International Journal of Listening, 25*(1/2), 47–65.

Egland, K. I., Stelzner, M. A., Andersen, P. A., & Spitzberg, B. S. (1997). Perceived understanding, nonverbal communication, and relational satisfaction. In J. E. Aitken & L. J. Shedletsky (Eds.), *Intrapersonal communication processes* (pp. 386–396). Speech Communication Association.

Ekman, P. (1965). Communication through nonverbal behavior: A source of information about an interpersonal relationship. In S. S. Tomkins & C. E. Izard (Eds.), *Affect, cognition, and personality* (pp. 390–442). Springer.

Ekman, P. (1970). Universal facial expressions of emotions. *California Mental Health Research Digest, 8(4)*, 151–158.

Ekman, P. (2003). *Emotions revealed* (2nd ed.). Times Books.

Ekman, P. (2009). Lie catching and micro expressions. In C. Martin (Ed.), *The philosophy of deception.* Oxford University Press.

Ekman, P. (Ed.). (1973). *Darwin and facial expression: A century of research in review.* Academic Press.

Ekman, P., & Friesen, W. V. (1971). Constants across culture in the face and emotion. *Journal of Personality and Social Psychology, 17,* 124–129.

Ekman, P., & Friesen, W. V. (1974). Nonverbal behavior and psychopathology. In R. J. Friedman & M. Katz (Eds.), *The psychology of depression: Contemporary theory and research* (pp. 3–31). Winston and Sons.

Ekman, P., Friesen, W. V., & Ellsworth, P. (1972). *Emotion in the human face: Guidelines for research and an integration of findings.* Pergamon Press.

Ekman, P., Friesen, W. V., & Tomkins, S. S. (1971). Facial affect scoring technique: A first validity study. *Semiotica, 3,* 37–58.

Ekman, P., & Oster, H. (1979). Facial expressions of emotion. *Annual Review of Psychology, 30,* 527–554.

Ekman, P., Sorenson, E. R., & Friesen, W. V. (1969). Pan-cultural elements in facial displays of emotion. *Science, 164*(3875), 86–88.

Ekman P. (2009). Darwin's contributions to our understanding of emotional expressions. *Philosophical Transactions of the Royal Society of London B, Biological Sciences, 364*(1535), 3449–3451

Elfenbein, H. A., & Ambady, N. (2002). On the universality and cultural specificity of emotion recognition: A meta-analysis. *Psychological Bulletin, 128*(2), 205–235.

Ellis, K. (2004). The impact of perceived teacher confirmation on receiver apprehension, motivation, and learning. *Communication Education, 53,* 1–20.

Ellis, N. K., & Ledbetter, A. M. (2015). Why might distance make the heart grow fonder?: A relational turbulence model investigation of the maintenance of long distance and geographically close romantic relationships. *Communication Quarterly, 63,* 568–585.

Emerson, R. M. (1976). Social exchange theory. *Annual Review of Sociology, 2,* 335–362.

Endacott, C., Hartwig, R., & Yu, C. (2017). An exploratory study of communication practices affecting church leadership team performance. *Southern Communication Journal, 82*(3), 129–139.

Enwemeka, Z. (2020, July 13). 'Keep the momentum': How *Mass. companies say they'll start—and continue—to fight racism.* WBUR. https://wbur.org/bostonomix/2020/07/13/massachusetts-businesses-racism-protests

Erdem, F., & Ozen, J. (2003). Cognitive and affective dimensions of trust in developing team performance. *Team Performance Management, 9,* 131–135.

Eshet-Alkalai, Y. (2012). Thinking in the digital era: A revised model for digital literacy. *Issues in Informing Science & Information Technology Education, 9,* 267–276.

Espey, M. (2018). Diversity, effort, and cooperation in team-based learning. *The Journal of Economic Education, 49*(1), 8–21.

Espey, M. (2018). Enhancing critical thinking using team-based learning. *Higher Education Research & Development, 37*(1), 15–29.

Evans, C. R., & Dion, K. L. (2012). Group cohesion and performance: A meta-analysis. *Small Group Research, 43*(6), 690–701.

Eys, M. A., & Carron, A. V. (2001). Role ambiguity, task cohesion, and task self-efficacy. *Small Group Research, 32,* 356–373.

Fallon, C. K., Panganiban, A., Wohleber, R., Matthews, G., Kustubayeva, A. M., & Roberts, R. (2014). Emotional intelligence, cognitive ability and information search in tactical decision-making. *Personality & Individual Differences, 6524*–6529.

Farmer, Y., Bissière, M., & Benkirane, A. (2018). Impacts of authority and unanimity on social conformity in online chats about climate change. *Canadian Journal of Communication, 43*(2), 265–279.

Fay, N., Page, A. C., & Serfaty, C. (2010). Listeners influence speakers' perceived communication effectiveness. *Journal of Experimental Social Psychology, 46*(4), 689–692.

Feldman, B. (2019, June 26). Piracy is back. *Intelligencer.* https://nymag.com/intelligencer/2019/06/piracy-is-back.html

Feng, Y., Hao, B., Iles, P., & Bown, N. (2017). Rethinking distributed leadership: Dimensions, antecedents and team effectiveness. *Leadership & Organization Development Journal, 38*(2), 284–302.

Fernbank, F. (2003). Legends on the net: An examination of computer-mediated communication as a locus of oral culture. *New Media & Society, 5,* 29–45.

Finley, K. (2020, April 9). When school is online, the digital divide grows greater. *Wired.* https://wired.com/story/school-online-digital-divide-grows-greater/

Fish, J., Aguilera, R., Ogbeide, I. E., Ruzzicone, D. J., & Syed, M. (2021). When the personal is political: Ethnic identity, ally identity, and political engagement among indigenous people and people of color. *Cultural Diversity & Ethnic Minority Psychology, 27*(1), 18–36. https://doi.org/10.1037/cdp0000341

Fitzpatrick, M. A., & Ritchie, L. D. (1994). Communication schemata within the family: Multiple perspectives on family interaction. *Human Communication Research, 20,* 275–301.

Fitzpatrick, M. A., & Winke, J. (1979). You always hurt the one you love: Strategies and tactics in interpersonal conflict. *Communication Quarterly, 27,* 3–11.

Floyd, K. (2014). Empathic listening as an expression of interpersonal affection. *International Journal of Listening, 28*(1), 1–12.

Floyd, K., & Morman, M. T. (Eds.). (2014). *Widening the family circle: New research on family communication.* SAGE.

Floyd K., & Guerrero, L. K. (2006). *Nonverbal communication in close relationships.* Lawrence Erlbaum.

Flynn, J., Valikoski, T., & Grau, J. (2008). Listening in the business context: Reviewing the state of research. *International Journal of Listening, 22*(2), 141–151.

Fodor, E. M., & Wick, D. P. (2009). Need for power and affective response to negative audience reaction to an extemporaneous speech. *Journal of Research in Personality, 43*(5), 721–726.

Foley, J. (2020, March 23). *10 deepfake examples that terrified and amused the internet.* Creative Bloq. https://creativebloq.com/features/deepfake-examples

Foley, S., Hang-yue, N., & Wong, A. (2005). Perceptions of discrimination and justice: Are there gender differences in outcomes? *Group & Organization Management, 30,* 421–450.

Fonseca, A. L., Ye, T., Koyama, J., Curran, M., & Butler, E. A. (2020). A theoretical model for understanding relationship functioning in intercultural romantic couples. *Personal Relationships, 27,* 760–784.

Foresman, G. (2017). *The critical thinking toolkit.* Wiley-Blackwell.

Forgas, J. P. (2011). Affective influences on self-disclosure: Mood effects on the intimacy and reciprocity of disclosing personal information. *Journal of Personality & Social Psychology, 100*(3), 449–461.

Forsyth, D. R. (2020). Group-level resistance to health mandates during the COVID-19 pandemic: A groupthink approach. *Group Dynamics, 24*(3), 139–152. https://doi.org/10.1037/gdn0000132

Forward, G. L., Czech, K., & Lee, C. (2011). Assessing Gibb's supportive and defensive communication climate: An examination of measurement and construct validity. *Communication Research Reports, 28*(1), 1–15.

Franks, G. H. (2019, September 9). *Meet your new best friend: The square root of negative one.* St. John's College Digital Archives. https://digitalarchives.sjc.edu/items/show/6721

Fransella, F. (2016). What is a construct? In D. A. Winter & N. Reed (Eds.), *The Wiley handbook of personal construct psychology* (pp. 32–38). Wiley Blackwell.

Fransen, K., Haslam, S. A., Steffens, N. K, & Boen, F. (2020). Standing out from the crowd: Identifying the traits and behaviors that characterize high-quality athlete leaders. *Scandinavian Journal of Medicine & Science in Sports, 30*(4), 766–786. https://doi.org/10.1111/sms.13620

Frantzolas, T. (2016, February). *Everything you hear on film is a lie* [Video]. TED Conferences. https://ted.com/talks/tasos_frantzolas_everything_you_hear_on_film_is_a_lie

Freeman, R., & McElhinny, B. (1996). Language and gender. In S. L. McKay & N. H. Hornberger (Eds.), *Sociolinguistics and language teaching.* Cambridge University Press.

Frey, L., & Sunwolf. (2005). The symbolic-interpretive perspective of group life. In M. S. Poole & A. B. Hollingshead (Eds.), *Theories of small groups: Interdisciplinary perspectives* (pp. 185–239). SAGE.

Froemming, K. J., & Penington, B. A. (2011). Emotional triggers: Listening barriers to effective interactions in senior populations. *International Journal of Listening, 25*(3), 113–131.

Fryberg, S. A., Eason, A. E., Brady, L., Jessop, N., & Lopez, J. (2020). *Unpacking the mascot debate: Native American identification predicts opposition to Native mascots.* https://doi.org/10.31234/osf.io/d5gte

Fryberg, S. A., Markus, H. R., Oyserman, D., & Stone, J. M. (2008). Of warrior chiefs and Indian princesses: The psychological consequences of American Indian mascots. *Basic and Applied Social Psychology, 30*(3), 208–218.

Galili, L., Amir, O., & Gilboa-Schechtman, E. (2013). Acoustic properties of dominance and request utterances in social anxiety. *Journal of Social & Clinical Psychology, 32*(6), 651–673.

Gallo, C. (2010). *The presentation secrets of Steve Jobs: How to be insanely great in front of any audience.* McGraw-Hill.

Gallo, C. (2014). *Talk like TED: The 9 public-speaking secrets of the world's top minds.* St Martin's Press.

Gamble, W. (2000). *Children and lying: Study focuses on why.* The University of Arizona College of Agriuculture and Life Sciences. https://cals.arizona.edu/pubs/general/resrpt2000/childrenlying.pdf

Gander, F., Ruch, W., Platt, T., Hofmann, J., & Elmer, T. (2018). Current and ideal team roles: Relationships to job satisfaction and calling. *Translational Issues in Psychological Science, 4*(3), 277–289.

Garcia, M. (2020, January 27). Science soon resumes on cosmic ray detector, crew packs cargo ship for departure. *NASA.* https://blogs.nasa.gov/spacestation/tag/spacewalk/

Gargouri, C., & Naatus, M. K. (2017). An experiment in mind-mapping and argument-mapping: Tools for assessing outcomes in the business curriculum. *Journal of Business Education and Scholarship of Teaching, 11*(2), 39–78.

Gatling, A., Shum, C., Book, L., & Bai, B. (2017). The influence of hospitality leaders' relational transparency on followers' trust and deviance behaviors: Mediating role of behavioral integrity. *International Journal of Hospitality Management, 62,* 11–20.

Gazso, A., & McDaniel, S. A. (2015). Families by choice and the management of low income through social supports. *Journal of Family Issues, 36,* 371–395.

Gazzale, L. (2019). Motivational implications leading to the continued commitment of volunteer firefighters. *International Journal of Emergency Services, 8*(2), 205–220.

Gearhart, C. C., Denham, J. P., & Bodie, G. D. (2014). Listening as a goal-directed activity. *Western Journal of Communication, 78*(5), 668–684.

Geiger, S., Grossman, P., & Schrader, U. (2019). Mindfulness and sustainability: Correlation or causation? *Current Opinion in Psychology, 28,* 23–27.

Gentina, E., & Chen, R. (2019). Digital natives' coping with loneliness: Facebook or face-to-face? *Information & Management, 56*(6), 103–138. https://doi.org/10.1016/j.im.2018.12.006

Gerard, G. (2005). Creating new connections: Dialogue and improv. In B. Banathy & P.M. Jenlink (Eds.),

Dialogue as a means of collective action (pp. 335–356). Kluwer Academic/Plenum.

Gerhardt, C. (2016). Family of procreation. In C. L. Shehan (Ed.), *The Wiley Blackwell encyclopedia of family studies* (Vol. 3). Wiley Blackwell. https://doi.org/10.1002/9781119085621.wbefs223

German, K. M., Gronbeck, B. E., Ehninger, D., & Monroe, A. H. (2013). *Principles of public speaking* (18th ed.). Pearson.

Gibb, J. R. (1961). Defensive communication. *Journal of Communication, 11*, 141–148.

Gibbs, J. L., Ellison, N. B., & Lai, C. (2011). First comes love, then comes Google: An investigation of uncertainty reduction strategies and self-disclosure in online dating. *Communication Research, 38*(1), 70–100.

Gilbert, B. (2019, November 6). The 10 most-viewed fake-news stories on Facebook in 2019 were just revealed in a new report. *Business Insider.* https://businessinsider.com/most-viewed-fake-news-stories-shared-on-facebook-2019-2019-11

Giles, L. C., Glonek, G. F. V., Luszcz, M. A., & Andrew, G. R. (2005). Effect of social networks on 10-year survival in very old Australians: The Australian longitudinal study of aging. *Journal of Epidemiology & Community Health, 59*(7), 574–579.

Gilligan-Lee, C. (2020). Causing trouble. *New Scientist, 246*(3279), 32–35.

Gist, M. (2020). *The extraordinary power of leader humility.* Berrett-Koehler Publishers.

Gist-Mackey, A., Wiley, M., & Erba, J. (2018). "You're doing great. Keep doing what you're doing": Socially supportive communication during first-generation college students' socialization. *Communication Education, 67*(1), 52–72.

Glikson, E., & Erez, M. (2019). The emergence of a communication climate in global virtual teams. *Journal of World Business, 55*(6). https://doi.org/10.1016/j.jwb.2019.101001

Goberman, A., Hughes, S., & Haydock, T. (2011). Acoustic characteristics of public speaking: Anxiety and practice effects. *Speech Communication, 53*(6), 867–876.

Goddard, I. (2005). "I am a red-skin": The adoption of a Native American expression (1769–1826). *European Review of Native American Studies, 19*(2), 1–20.

Goh, E. (2013). Plagiarism behavior among undergraduate students in hospitality and tourism education. *Journal of Teaching in Travel & Tourism, 13*(4), 307–322.

Goldberg, B. (2019, October 29). *Before your next presentation or speech, here's the first thing you must think about.* TED. https://ideas.ted.com/before-your-next-presentation-or-speech-heres-the-first-thing-you-must-think-about

Gómez-López, M., Viejo, C., & Ortega-Ruiz, R. (2019). Well-being and romantic relationships: A systematic review in adolescence and emerging adulthood. *International Journal of Environmental Research and Public Health, 16*(13), 2415.

Goodboy, A., & Bolkan, S. (2011). Attachment and the use of negative relational maintenance behaviors in romantic relationships. *Communication Research Reports, 28*, 327–336.

Gopalkrishnan, N. (2019). Cultural competence and beyond: Working across cultures in culturally

dynamic partnerships. *The International Journal of Community and Social Development, 1*, 28–41.

Gottman, J. M. (1993). The roles of conflict engagement, escalation, and avoidance in marital interaction: A longitudinal view of five types of couples. *Journal of Consulting and Clinical Psychology, 61*, 6–15.

Gottman, J. M. (1995). *Why marriages succeed or fail: And how you can make yours last.* Simon & Schuster.

Gottman, J. M. (1998). Psychology and the study of marital processes. *Annual Review of Psychology, 49*, 169–197.

Gottman, J. M., & Levenson, R. W. (1992). Marital processes predictive of later dissolution: Behavior, physiology, and health. *Journal of Personal and Social Psychology, 63*, 221–233.

Gottman, J. M., & Levenson, R. W. (2000). The timing of divorce: Predicting when a couple will divorce over a 14-year period. *Journal of Marriage and the Family, 62*, 737–745.

Graddol, D. (2004). The future of language. *Science, 303*(5662), 1329–1331.

Graham, C. M., & Jones, N. (2019). Impact of a social network messaging app on team cohesiveness and quality of completed team projects in an undergraduate team project. *Journal of Educational Technology Systems, 47*(4), 539–553.

Graham-Harrison, E., Oltermann, P., & Jones, S. (2020, April 26). Coronavirus face masks: Why covering up is becoming the new normal. *The Guardian.* https://theguardian.com/world/2020/apr/26/coronavirus-face-masks-why-covering-up-is-becoming-the-new-normal

Gray, J. (1992). *Men are from Mars, women are from Venus: A practical guide to improving communication and getting what you want in your relationships.* HarperCollins.

Greenleaf, R. K., & Wells-Papanek, D. (2005). *Memory, recall, the brain & learning: Improve student learning outcomes by engaging learners in visual & nonlinguistic strategies, activities, & organizers.* Greenleaf & Papanek Publications.

Greenslade, R. (2013, September 9). *US sports reporters refuse to use name of Washington Redskins.* Greenslade Blog. http://www.theguardian.com/media/greenslade/2013/sep/09/washington-redskins-us-press-publishing

Grinberg, E. (2011, October 26). *'We're a culture, not a costume' this Halloween.* CNN. https://www.cnn.com/2011/10/26/living/halloween-ethnic-costumes/index.html

Groody, D., & Măcelaru, M. (2018). Mind-mapping migration: Understanding the deeper contours of a contentious debate. *Transformation: An International Journal of Holistic Mission Studies, 35*(2), 77–90.

Gross, D., Iverson, E., Willett, G., & Manduca, C. (2015). Broadening access to science with support for the whole student in a residential liberal arts college environment. *Journal of College Science Teaching, 44*(4), 99–107.

Grubaugh, M. L., & Flynn, L. (2018). Relationships among nurse manager leadership skills, conflict management, and unit teamwork. *JONA: The Journal of Nursing Administration, 48*(7/8), 383–388.

Guerrero, L. K. (1994). "I'm so mad I could scream": The effects of anger expression on relational satisfaction and communication competence. *Southern Communication Journal, 59*(2), 125–141.

Guerrero, L. K., Andersen, P. A., & Afifi, W. A. (2007). *Close encounters: Communication in relationships* (2nd ed.). Sage.

Gugliotta, G. (2005, October 3). A linguist's alternative history of 'redskin.' *The Washington Post.* http://www.washingtonpost.com/wp-dyn/content/article/2005/10/02/AR2005100201139.html

Gully, S. M., Devine, D. J., & Whitney, D. J. (2012). A meta-analysis of cohesion and performance: Effects of level of analysis and task interdependence. *Small Group Research, 43*(6), 702–725.

Gumbel, P. (2012, September 3). The not so magic kingdom: Why Disneyland Paris has yet to pay off for Disney. *Time.* http://content.time.com/time/subscriber/article/0,33009,2122686-2,00.html

Günsoy, C., Okten, I. O., Cross, S. E, & Saribay, S. A. (2020). Cultural differences in self-expression on Facebook: A comparison of facebook status updates in Turkey and the U.S.A. *International Journal of Human-Computer Interaction, 36*(18), 1775–1782. https://doi.org/10.1080/10447318.2020.1794623

Haas, J., Sypher, B. D., & Sypher, H. (1992). Do shared goals really make a difference? *Management Communication Quarterly, 6,* 166–179.

Haase, R. F., Ferreira, J. A., Fernandes, R. I., Santos, E. R., & Jome, L. M. (2016). Development and validation of a revised measure of individual capacities for tolerating information overload in occupational settings. *Journal of Career Assessment, 24*(1), 130–144.

Halbritter, R. (2013, December 27). Year of reckoning for the Washington 'R-word.' *The Washington Post.* http://www.washingtonpost.com/opinions/year-of-reckoning-for-the-washington-r-word/2013/12/27/80723674-6e78-11e3-aecc-85cb037b7236_story.html

Hall, D., & Buzwell, S. (2013). The problem of free-riding in group projects: Looking beyond social loafing as reason for non-contribution. *Active Learning in Higher Education, 14*(1), 37–49.

Hall, E. T. (1959). *The silent language.* Doubleday.

Hall, E. T. (1966). *The hidden dimension.* Anchor Books.

Hall, E. T. (1968). Proxemics. *Current Anthropology, 9*(2), 83–95.

Hall, J. A., & Gunnery, S. D. (2013). *Gender differences in nonverbal communication.* In J. A. Hall & M. L. Knapp (Eds.), *Handbooks of communication science. Nonverbal communication* (pp. 639–669). De Gruyter Mouton.

Halliwell, D. (2015). Extending relational dialectics theory: Exploring new avenues of research. *Annals of the International Communication Association, 39,* 67–95.

Hamburg, I., O'Brien, E., & Vladut, G. (2019). Entrepreneurial learning and AI literacy to support digital entrepreneurship. *Balkan Region Conference on Engineering and Business Education, 1*(1), 132–144.

Hameleers, M., Powell, T., Van Der Meer, T., & Bos, L. (2020). A picture paints a thousand lies? The effects and mechanisms of multimodal disinformation and rebuttals disseminated via social media. *Political Communication, 37*(2), 281–301.

Hammond, S. C., Anderson, R., & Cissna, K. N. (2003). The problematics of dialogue and power. In P. J. Kalbfleisch (Ed.), *Communication Yearbook 27* (pp. 125–157). Lawrence Erlbaum.

Hamouda, A. (2012). Listening comprehension problems—Voices from the classroom. *Language in India, 12*(8), 1–49.

Hample, D., & Richards, A. (2018). Personalizing conflict in different interpersonal relationship types. *Western Journal of Communication, 83,* 1–20.

Hance, M., Blackhart, G., & Dew, M. (2018). Free to be me: The relationship between the true self, rejection sensitivity, and use of online dating sites. *The Journal of Social Psychology, 158*(4), 421–429.

Hanley, S. J., & Abell, S. C. (2002). Maslow and relatedness: Creating an interpersonal model of self-actualization. *Journal of Humanistic Psychology, 42*(4), 37–57.

Hansen, B. (2003). Combating plagiarism. *Congressional Quarterly, 13,* 775–782.

Hansen, D. (2016). Cohesion in online student teams versus traditional teams. *Journal of Marketing Education, 38*(1), 37–46.

Hanson, K. (2019). Beauty "therapy": The emotional labor of commercialized listening in the salon industry. *International Journal of Listening, 33*(3), 148–153.

Hardy, J. Eys, M. A., & Carron, A. V. (2005). Exploring the potential disadvantages of high cohesion in sports teams. *Small Group Research, 36,* 166–187.

Hargie, O. (2011). *Skilled interpersonal interaction: Research, theory, and practice.* Routledge.

Harper, D. (2019). *Online etymology dictionary.* https://etymonline.com.

Harris, T. M., & González, A. (2015). Constructing a cultural identity through family communication. In L. H. Turner & R. West (Eds.), *The SAGE handbook of family communication* (pp. 26–40). Sage.

Hart, C. H., Newell, L. D., & Olsen, S. F. (2003). Parenting skills and social-communicative competence in childhood. In J. O. Greene & B. R. Burleson (Eds.), *Handbook of communication and social interaction skills* (pp. 753–797). Lawrence Erlbaum.

Hartup, W., & Stevens, N. (1999). Friendships and adaptation across the life span. *Current Directions in Psychological Science, 8*(3), 76–79.

Harvey, J. H., & Wenzel, A. (2006). Theoretical perspectives in the study of close relationships. In A. L. Vangelisti & D. Perlman (Eds.), *The Cambridge handbook of personal relationships* (pp. 35–49). Cambridge University Press.

Hasler, B., & Friedman, D. (2012). Sociocultural conventions in avatar-mediated nonverbal communication: A cross-cultural analysis of virtual proxemics. *Journal of Intercultural Communication Research, 41*(3), 238–259.

Hawking, S. (2016, January 26). *Do black holes have hair?* BBC News World Service. https://bbc.co.uk/programmes/p03fx1hd

Hawkins, K. W., & Fillion, B. P. (1999). Perceived communication skills needs for work group. *Communication Research Reports, 16,* 167–174.

Hayawaka, S. I., & Hayawaka, A. R. (1991). *Language in thought and action* (5th ed.). Houghton Mifflin Harcourt.

Haynes, S. (2014). Effectiveness of communication strategies for deaf or hard of hearing workers in group settings. *Work, 48*(2), 193–202.

Hays Research. (2019). *Top 10 global skills shortage*. https://hays.cn/en/press-releases/HAYS_014336

Hearing Loss Association of America. (2018, May). *Hearing loss facts and statistics*. https://www.hearingloss.org/wp-content/uploads/HLAA_HearingLoss_Facts_Statistics.pdf

Heavey, C. L., Christensen, A., & Malamuth, N. M. (1995). The longitudinal impact of demand and withdrawal during marital conflict. *Journal of Consulting and Clinical Psychology, 65*, 797–801.

Hecht, M., & Choi, H. (2012). The communication theory of identity as a framework for health message design. In H. Cho (Ed.), *Health communication message design: Theory and practice* (pp. 137–152). SAGE.

Hecht, M. L. (2009). Communication theory of identity. In S. Littlejohn & K. Foss (Eds.), *Encyclopedia of communication theory* (pp. 140–142). SAGE.

Hecht, M. L., Warren, J., Jung, J., & Krieger, J. (2004). Communication theory of identity. In W. B. Gudykunst (Ed.), *Theorizing about intercultural communication* (pp. 257–278). SAGE.

Helm, B. (2017). Friendship. In E. N. Zalta (Ed.), *The Stanford encyclopedia of philosophy*. https://plato.stanford.edu/archives/fall2017/entries/friendship/

Henning, Z. (2012). Knowing your audience: Demographic and situational topic analysis. *Journal of the Communication, Speech & Theatre Association of North Dakota, 25*, 61–63.

Heracleous, L., Paroutis, S., & Lockett, A. (2019). Rhetorical enthymeme: The forgotten trope and its methodological import. *European Management Review, 17*(1), 311–326.

Herakova, L. L. (2012). Nursing masculinity: Male nurses' experiences through a co-cultural lens. *Howard Journal Of Communications, 23*(4), 332–350.

Herman, D. (2016, December 2). How the story of "Moana" and Maui holds up against cultural truths. *Smithsonian Magazine*. https://www.smithsonianmag.com/smithsonian-institution/how-story-moana-and-maui-holds-against-cultural-truths-180961258/

Herpertz, S., Schütz, A., & Nezlek, J. (2016). Enhancing emotion perception, a fundamental component of emotional intelligence: Using multiple-group SEM to evaluate a training program. *Personality and Individual Differences, 95*, 11–19.

Hersman, D. P. (2014). I saw a woman today. *Vital Speeches of the Day, 80*, 149–151.

Hertz, B., Kerkhof, P., & van Woerkum, C. (2016). PowerPoint slides as speaking notes. *Business & Professional Communication Quarterly, 79*(3), 348–359.

Hess, U., Blaison, C., & Kafetsios, K. (2016). Judging facial emotion expressions in context: The influence of culture and self-construal orientation. *Journal of Nonverbal Behavior, 40*(1), 55–64.

Hicks, C. B., & Tharpe, A. M. (2002). Listening effort and fatigue in school-age children with and without hearing loss. *Journal of Speech, Language, and Hearing Research, 45*, 573–584.

High, A., & Solomon, D. (2016). Explaining the durable effects of verbal person-centered supportive communication: Indirect effects or invisible support?: Durable effects of VPC support. *Human Communication Research, 42*(2), 200–220.

High, A. C., & Young, R. (2018). Supportive communication from bystanders of cyberbullying: Indirect effects and interactions between source and message characteristics. *Journal of Applied Communication Research, 46*(1), 28–51.

Hinson, K., & Sword, B. (2019). Illness narratives and Facebook: Living illness well. *Humanities, 8*(2), 106. https://doi.org/10.3390/h8020106

History.com. (2020, September 25). *Boston Tea Party*. history.com/topics/american-revolution/boston-tea-party

Hjertø, K. B., & Paulsen, J. M. (2016). Beyond collective beliefs: Predicting team academic performance from collective emotional intelligence. *Small Group Research, 47*(5), 510–541.

Ho, C., Liao, T., Huang, S., & Chen, H. (2015). Beyond environmental concerns: Using means-end chains to explore the personal psychological values and motivations of leisure/recreational cyclists. *Journal of Sustainable Tourism, 23*(2), 234–254.

Hobman, E. V., Bordia, P., & Gallois, C. (2004). Perceived dissimilarity and work group involvement: The moderating effects of group openness to diversity. *Group & Organization Management, 29*, 560–587.

Hod, Y., Basil-Shachar, J., & Sagy, O. (2018). The role of productive social failure in fostering creative collaboration: A grounded study exploring a classroom learning community. *Thinking Skills and Creativity, 30*, 145–159.

Hodge, R. (2020, March 20). *How Kentucky became a surprising leader in flattening the curve on COVID-19*. CNET. https://cnet.com/news/how-kentucky-became-a-surprising-leader-in-flattening-the-curve-on-covid-19/

Hoeken, H., & Sinkeldam, J. (2014). The role of identification and perception of just outcome in evoking emotions in narrative persuasion. *Journal of Communication, 64*(5), 935–955.

Hofstede, G. (1980). *Culture's consequences*. SAGE.

Hofstede, G. (2001). *Culture's consequences: Comparing values, behaviors, institutions, and organizations across cultures*. SAGE.

Hofstede, G., Hofstede, G. J., & Minkov, M. (2010). *Cultures and organizations: Software of the mind: Intercultural cooperation and its importance for survival*. McGraw Hill.

Hogan, J. (2012). Persuasion in the rhetorical tradition. In J. P. Dillard & L. Shen (Eds.), *The SAGE handbook of persuasion: Developments in theory and practice* (pp. 2–19). SAGE. https://doi.org/10.4135/9781452218410.n1

Holdsworth, C., & Quinn, J. (2012). The epistemological challenge of higher education student volunteering: 'reproductive' or 'deconstructive' volunteering? *Antipode, 44*(2), 386–405.

Hollenbaugh, E. E., & Everett, M. K. (2013). The effects of anonymity on self-disclosure in blogs: An application of the online disinhibition effect. *Journal of Computer-Mediated Communication, 18*(3), 283–302.

Holman, J. A., Hornsby, B. W. Y., Bess, F. H., & Naylor, G. (2021). Can listening-related fatigue influence well-being? Examining associations between hearing loss, fatigue, activity levels and well-being. *International Journal of Audiology, 60*(2), 47–59. https://doi.org/10.1080/14992027.2020.1853261

Holmes, J. (1983). The functions of tag questions. *English Language Research Journal, 3*, 40–65.

Holmes, J. (1995). *Women, men and politeness.* Longman.

Holmes, J., & Meyerhoff, M. (Eds.). (2003). *The handbook of language and gender.* Blackwell.

Holmes, R. (2019, February 19). *We now see 5,000 ads a day … and it's getting worse. LinkedIn Pulse.* https://linkedin.com/pulse/have-we-reached-peak-ad-social-media-ryan-holmes/

Holt-Lunstad, J., Robles, T. F., & Sbarra, D. A. (2017). Advancing social connection as a public health priority in the United States. *The American Psychologist, 72*, 517–530.

Holt-Lunstad, J., Smith, T. B., & Layton, J. B. (2010). Social relationships and mortality risk: A meta-analytic review. *PLoS Medicine, 7*(7), e1000316.

Homans, G. (1961). *Social behavior: Its elementary forms.* Harcourt Brace Jovanovich.

Homsey, D. M., & Sandel, T. (2012). The code of food and tradition: Exploring a Lebanese (American) speech code in practice in Flatland. *Journal of Intercultural Communication Research, 41*(1), 59–80.

Honeycutt, J. M., Choi, C. W., & DeBerry, J. R. (2009). Communication apprehension and imagined interactions. *Communication Research Reports, 26*(3), 228–236.

Hood, A. C., Cruz, K. S., & Bachrach, D. G. (2017). Conflicts with friends: A multiplex view of friendship and conflict and its association with performance in teams. *Journal of Business and Psychology, 32*, 73–86.

Hooper, A., Spann, C., McCray, T., & Kimberly, C. (2017). Revisiting the basics: Understanding potential demographic differences with John Gottman's four horsemen and emotional flooding. *The Family Journal, 25*, 224–229.

Hornsey, M. J., Jetten J., McAuliffe B. J., & Hogg, M. A. (2005). The impact of individualist and collectivist group norms on evaluations of dissenting group members. *Journal of Experimental Social Psychology, 42*, 57–68.

Hosein, A., Ramanau, R., & Jones, C. (2010). Learning and living technologies: A longitudinal study of first-year students' frequency and competence in the use of ICT. *Learning, Media & Technology, 35*(4), 403–418.

Hosman, L. A., & Siltanen, S. A. (2011). Hedges, tag questions, message processing, and persuasion. *Journal of Language & Social Psychology, 30*(3), 341–349.

Howard, S. C. (2021, March 10). Why Jeep's response to the Cherokee Nation is harmful and disrespectful. *Fortune.* https://fortune.com/2021/03/10/cherokee-nation-jeep-name-change/

How many phones are in the world? (2021). BankMyCell. https://bankmycell.com/blog/how-many-phones-are-in-the-world

Hsiao-Cheng, H. (2019). Moving from cultural appropriation to cultural appreciation. *Art Education, 72*, 8–13.

Hu, Y. R., & Goldman, N. (1990). Mortality differentials by marital-status—an international comparison. *Demography, 27*, 233–250.

Hunter, K., Westwick, J., & Haleta, L. (2014). Assessing success: The impacts of a fundamentals of speech course on decreasing public speaking anxiety. *Communication Education, 63*(2), 124–135.

Hunter, S. T., Friedrich, T. L., Bedell-Avers, K. E., & Mumford, M. D. (2007). Creative cognition in the workplace: An applied perspective. In T. Davila, M. J. Epstein, & R. Shelton (Eds.), *The creative enterprise: Managing innovative organizations and people* (Vol. 2, pp. 171–192). Praeger.

Huo, M., Fuentecilla, J. L., Birditt, K. S., & Fingerman, K. L. (2020). Empathy and close social ties in late life. *The Journals of Gerontology. Series B, Psychological Sciences and Social Sciences, 75*, 1648–1657.

Hurley, C. M., Anker, A. E., Frank, M. G., Matsumoto, D., & Hwang, H. C. (2014). Background factors predicting accuracy and improvement in micro expression recognition. *Motivation and Emotion, 38*(5).

Huston, D. C., Garland, E. L., & Farb, N. S. (2011). Mechanisms of mindfulness in communication training. *Journal of Applied Communication Research, 39*(4), 406–421.

Huston, D. C., Garland, E. L., & Farb, N. S. (2011). Mechanisms of mindfulness in communication training. *Journal of Applied Communication Research, 39*(4), 406–421.

Huston, Z. (2009). The best imitation of myself: Communication and the arbitrary construction of the self. *American Communication Journal, 11*(3), 1–15.

Imhof, M. (2001). How to listen more efficiently: Self-monitoring strategies in listening. *International Journal of Listening, 15*, 2–19.

İnce, F. (2018). The effect of democratic leadership on organizational cynicism: A study on public employees. *İşletme Araştırmaları Dergisi, 10*(2), 245–253.

Ingram, P., & Zou, X. (2008). Business friendships. *Research in Organizational Behavior, 28*, 167–184.

Ismail, A., Zainuddin, N., & Ibrahim, Z. (2010). Linking participative and consultative leadership styles to organizational commitment as an antecedent of job satisfaction. *UNITAR E-Journal, 6*(1), 11–26.

Ito, K., Yang, S., & Li, L. M. W. (2021). Changing Facebook profile pictures to dyadic photos: Positive association with romantic partners' relationship satisfaction via perceived partner commitment. *Computers in Human Behavior, 120*, 106748. https://doi.org/10.1016/j.chb.2021.106748

Itzchakov, G., & Kluger, A. N. (2017). Can holding a stick improve listening at work? The effect of listening circles on employees' emotions and cognitions. *European Journal of Work and Organizational Psychology, 26*(5), 663–676. https://doi.org/10.1080/1359432X.2017.1351429

Jackson, B., Compton, J., Thornton, A., & Dimmock, J. (2017). Re-thinking anxiety: Using inoculation messages to reduce and reinterpret public speaking fears. *PLoS One, 12*(1), 1–18.

Jaffe, J., & Feldstein, S. (1970). *Rhythms of dialogue.* Academic Press.

Jan, T., McGregor, J., Merle, R., & Tiku, N. (2020, June 13). As big corporations say 'Black lives matter,' their track records raise skepticism. *The Washington Post.* https://washingtonpost.com/business/2020/06/13/after-years-marginalizing-black-employees-customers-corporate-america-says-black-lives-matter

Jang, C., & Stefanone, M. A. (2011). Non-directed self-disclosure in the blogosphere. *Information, Communication & Society, 14*(7), 1039–1059.

Janis, I. L. (1982). *Groupthink* (2nd ed.). Houghton Mifflin.

Janusik, L., & Imhof, M. (2017). Intercultural listening: Measuring listening concepts with the LCI-R. *International Journal of Listening, 31*(2), 80–97.

Janusik, L. A., & Wolvin, A. D. (2009). 24 hours in a day: A listening update to the time studies. *International Journal of Listening, 23*(2), 104–120.

Jasinski, J. (2001). *Sourcebook on rhetoric: Key concepts in contemporary rhetorical studies.* SAGE.

Jebbour, M., & Mouaid, F. (2019). The impact of teacher self-disclosure on student participation in the university English language classroom. *International Journal on Teaching and Learning in Higher Education, 31*(3), 424–436.

Jelinek, S., Toor, F., Becker, K., Smith, K., Schindel, B., Puoplo, N., Hebert, C., Winik, B., Mann, N., Ambler, L., Birbiglia, C., Stengel, A., Hodo, L. N., Tenore, C., & Katz, C. (2020). Building inclusive healthcare for LGBTQ+ youth: Improving the collection and utilization of patients' sexual orientation and gender identity (SOGI) information, preferred names and gender pronouns in a pediatric clinic. *Journal of Scientific Innovation in Medicine, 3*(3). https://doi.org/10.29024/jsim.76

Jenks, A. (2019). Crip theory and the disabled identity: Why disability politics needs impairment. *Disability & Society, 3*, 449–469.

Jeske, D., Lippke, S., & Shultz, K. (2019). Predicting self-disclosure in recruitment in the context of social media screening. *Employee Responsibilities and Rights Journal, 31*(2), 99–112.

Jiang, J., Borowiak, K., Tudge, L., Otto, C., & von Kriegstein, K. (2017). Neural mechanisms of eye contract when listening to another person talking. *Social Cognitive and Affective Neuroscience, 12*(2), 319–328.

Jiang, Y., & Chen, C. C. (2018). Integrating knowledge activities for team innovation: Effects of transformational leadership. *Journal of Management, 44*(5), 1819–1847.

Johannesen, R. L., Valde, K. S., & Whedbee, K. E. (2008). *Ethics in human communication* (6th ed.). Waveland.

Johnson, A. J., Wittenberg, E., Villagran, M. M., Mazur, M., & Villagran, P. (2003). Relational progression as a dialectic: Examining turning points in communication among friends. *Communication Monographs, 70*, 230–249.

Johnson, I. W., Pearce, C. G., Tuten, T. L., & Sinclair, L. (2003). Self-imposed silence and perceived listening effectiveness. *Business Communication Quarterly, 66*(1), 23–45.

Johnson, W. B., Jensen, K. C., Sera, H., & Cimbora, D. M. (2018). Ethics and relational dialectics in mentoring relationships. *Training and Education in Professional Psychology, 12*, 14–21.

Jones, S., Bodie, G., & Hughes, S. (2019). The impact of mindfulness on empathy, active listening, and perceived provisions of emotional support. *Communication Research, 46*(6), 838–865.

Jones, S. E., & Yarbrough, A. E. (1985). A naturalistic study of the meanings of touch. *Communication Monographs, 52*(1), 19–56.

Jones-Wild, R. (2012). Reimagining families of choice. In S. Hines & Y. Taylor (Eds.), *Sexualities: Past reflections, future directions* (pp. 149–167). Palgrave Macmillan.

Jonsdottir, I. J., & Fridriksdottir, K. (2020). Active listening: Is it the forgotten dimension in managerial communication? *International Journal of Listening, 34*(3), 178–188. https://doi.org/10.1080/10904018.2019.1613156

Jonsdottir, I. J., & Kristinsson, K. (2020). Supervisors' active-empathetic listening as an important antecedent of work engagement. *International Journal of Environmental Research and Public Health, 17*(21), 7976. https://doi.org/10.3390/ijerph17217976

Jose, K. (2016). Digital literacy matters. Increasing workforce productivity through blended English language programmes. *Higher Learning Research Communications, 6*(4). https://doi.org/10.18870/hlrc.v6i4.352

Jung, E. (2011). Identity gap: Mediator between communication input and outcome variables. *Communication Quarterly, 59*(3), 315–338.

Jung, E. (2013). Delineation of a threefold relationship among communication input variables, identity gaps, and depressive symptoms. *Southern Communication Journal, 78*(2), 163–184.

Junikka, J., Molleman, L., Van Den Berg, P., Weissing, F., & Puurtinen, M. (2017). Assortment, but not knowledge of assortment, affects cooperation and individual success in human groups. *PLoS ONE, 12*(10), E0185859. https://doi.org.libaccess.sjlibrary.org/10.1371/journal.pone.0185859

Juster, N. (1989). *The phantom tollbooth.* Bullseye Books.

Kahan, J. (2019, April 8). It's time for a new approach for mapping broadband data to better serve Americans. *Microsoft.* https://blogs.microsoft.com/on-the-issues/2019/04/08/its-time-for-a-new-approach-for-mapping-broadband-data-to-better-serve-americans/

Kahne, J., & Bowyer, B. (2019). Can media literacy education increase digital engagement in politics? *Learning, Media and Technology, 44*(2), 211–224

Ka'ili, T. O. (2016, December 6). *Goddess Hina: The missing heroine from Disney's Moana.* Huffington Post. https://www.huffpost.com/entry/goddess-hina-the-missing-heroine-from-disney%-CA%BCs-moana_b_5839f343e4b0a79f7433b6e5.

Kamaruddin, N., Omar, S. Z., Hassan, S., Hassan, M. A., Mee, A. A. C., & D'Silva, J. L. (2012). Impact of time spent in parents-children communication on children misconduct. *American Journal of Applied Sciences, 9*, 1818–1823.

Kamau, C. (2013). What does being initiated severely into a group do? The role of rewards. *International Journal of Psychology, 48*(3), 399–406.

Kamini, S. S. (2012). Empowerment of the urban women through SHGs in Coimbatore District—A study. *Language in India, 12*(7), 367–384.

Kang, P., & Seibold, D. (2018). The emergence and management of role ambiguity in a collegiate basketball team: An organizational tensions perspective. *Communication & Sport, 6*(4), 499–517.

Kapidzic, S., & Herring, S. C. (2011). Gender, communication, and self-presentation in teen chatrooms revisited: Have patterns changed? *Journal of Computer-Mediated Communication, 17*(1), 39–59.

Kaplan, R. M., & Kronick, R. G. (2006). Marital status and longevity in the United States population. *Journal of Epidemiology and Community Health, 60*, 760–765.

Kaplan, T. (2013, November 20). San Jose State students accused of tormenting black roommate are charged with hate crimes. *San Jose Mercury News.* https://www.mercurynews.com/2013/11/20/san-jose-state-students-accused-of- tormenting-black-roommate-are-charged-with-hate-crimes/

Karriker, J., Madden, L., & Katell, L. (2017). Team composition, distributed leadership, and performance: It's good to share. *Journal of Leadership & Organizational Studies, 24*(4), 507–518.

Kauffeld, S., & Lehmann-Willenbrock, N. (2012). Meetings matter: Effects of team meetings on team and organizational success. *Small Group Research, 43*(2), 130–158.

Kaufmann, M. (2019, October 17). *Black Tudors: Three untold stories.* Gresham College. https://gresham.ac.uk/lectures-and-events/black-tudors

Kearns, K. (2011). *Semantics* (2nd ed.). Palgrave Macmillan.

Kedrowicz, A. A., & Taylor, J. L. (2016). Shifting rhetorical norms and electronic eloquence. *Journal of Business and Technical Communication, 30*(3), 352–377. https://doi.org/10.1177/1050651916636373

Kellas, J. (2010). Transmitting relational worldviews: The relationship between mother-daughter memorable messages and adult daughters' romantic relational schemata. *Communication Quarterly, 58*(4), 458–479.

Kellermann, K. (1992). Communication: Inherently strategic and primarily automatic. *Communication Monographs, 59*, 288–300.

Kelley, R. E. (1992). *The power of followership.* Doubleday.

Kennedy, P. T. (2010). *Local lives and global transformations: Towards world society.* Palgrave Macmillan.

Kentucky Student Environment Coalition. (n.d.). *About.* https://kystudentenvironmentalcoalition.org

Kernan, W., Basch, C., & Cadorett, V. (2018). Using mind mapping to identify research topics: A lesson for teaching research methods. *Pedagogy in Health Promotion, 4*(2), 101–107.

Khaleque, A., & Rohner, R. P. (2018). *Intimate relationships across the lifespan: Formation, development, enrichment, and maintenance.* Praeger.

Khan, S. N., Abdullah, A. M., Busari, A. H., Mubushar, M., & Khan, I. U. (2020). Reversing the lens: The role of followership dimensions in shaping transformational leadership behaviour; mediating role of trust in leadership. *Leadership & Organization Development Journal, 41*(1), 1–18. https://doi.org/10.1108/LODJ-03-2019-0100

Khan, S. N., Abdullah, S. M., Busari, A. H., Mubushar, M., & Khan, I. U. (2020). Reversing the lens: The role of followership dimensions in shaping transformational leadership behaviour. *Leadership & Organization Development Journal, 41*(1), 1–18. https://www.doi.org/10.1108/LODJ-03-2019-0100

Kilmann, R. H., & Thomas, K. W. (1977). Developing a forced-choice measure of conflict-handling behavior: The "mode" instrument. *Educational and Psychological Measurement, 37*(2), 309–325.

Kim, H., Sutton, K., & Gong, Y. (2013). Group-based pay-for-performance plans and firm performance: The moderating role of empowerment practices. *Asia Pacific Journal of Management, 30*(1), 31–52.

Kim, M. (2003). Cultural and school-grade differences in Korean and White American children's narrative skills. *International Review of Education, 49*, 177–190.

Kimbrough, A. M., Guadagno, R. E., Muscanell, N. L., & Dill, J. (2013). Gender differences in mediated communication: Women connect more than do men. *Computers In Human Behavior, 29*(3), 896–900.

Kim-Prieto, C., Goldstein, L. A., Okazaki, S., & Kirschner, B. (2010). Effect of exposure to an American Indian mascot on the tendency to stereotype a different minority group. *Journal of Applied Social Psychology, 40*(3), 534–553.

Kimura, K., & Katayama, J. (2013). Outcome evaluations in group decision making using the majority rule: An electrophysiological study. *Psychophysiology, 50*(9), 848–857.

Kindarto, A., Zhu, Y., & Gardner, D. G. (2020). Full range leadership styles and government IT team performance: The critical roles of follower and team competence. *Public Performance & Management Review, 43*(4), 889–917. https://doi.org/10.1080/15309576.2020.1730198

King, P. (2013, September 6). A note from me about the use of the nickname "Redskins." *Sports Illustrated.* http://mmqb.si.com/2013/09/06/eli-manning-new-york-giants-dallas-cowboys/2/

King, P. E., & Behnke, R. R. (2004). Patterns of state anxiety in listening performance. *Southern Communication Journal, 70*, 72–80

King, P. E., & Sawyer, C. R. (1998). Mindfulness, mindlessness, and communication instruction. *Communication Education, 47*, 326–336.

Kirtley Johnston, M., & Reed, K. (2017). Listening environment and the bottom line: How a positive environment can improve financial outcomes. *International Journal of Listening, 31*(2), 71–79.

Kisner, J. (2018, January 23). How a new technology is changing the lives of people who can't speak. *The Guardian.* https://theguardian.com/news/2018/jan/23/voice-replacement-technology-adaptive-alternative-communication-vocalid

Kivlighan, D. M., Jr., London, K., & Miles, J. R. (2012). Are two heads better than one? The relationship between number of group leaders and group members, and group climate and group member benefit from therapy. *Group Dynamics, 16*(1), 1–13.

Klump, M. (2020, January 19). Global warming is a hoax. *The Daily Telegram.* https://lenconnect.com/opinion/20200119/margaret-klump-global-warming-is-hoax

Knapp, M. L. (1978). *Social intercourse: From greeting to goodbye.* Allyn & Bacon.

Knapp, M. L., Hall, J. A., & Horgan, T. G. (2013). *Nonverbal communication in human interaction* (8th ed.). Cengage.

Knapp, M. L., & Vangelisti, A. L. (2009). *Interpersonal communication and human relationships.* Pearson.

Koenig Kellas, J., Baker, J., Cardwell, M., Minniear, M., & Horstman, H. K. (2021). Communicated perspective-taking (CPT) and storylistening: Testing the impact of CPT in the context of friends telling stories of difficulty. *Journal of Social and*

Personal Relationships, 38(1), 19–41. https://doi.org/10.1177/0265407520955239

Koerner, A., & Fitzpatrick, M. (2002). Understanding family communication patterns and family functioning: The roles of conversation orientation and conformity orientation. *Annals of the International Communication Association, 26,* 36–65.

Koerner, A., & Fitzpatrick, M. (2006). Family communication patterns theory: A social cognitive approach. In D. O. Braithwaite and L. A. Baxter (Eds.), *Engaging theories in family communication: Multiple perspectives* (pp. 50–65). SAGE.

Koerner, A. F., & Fitzpatrick, M. A. (2002). Toward a theory of family communication. *Communication Theory, 12,* 70–91.

Koerner, A. F., & Fitzpatrick, M. A. (2002). You never leave your family in a fight: The impact of family of origin on conflict-behavior in romantic relationships. *Communication Studies, 53,* 234–251.

Koerner, B. (2003, March 28). What does a "thumbs up" mean in Iraq? *Slate.* https://slate.com/news-and-politics/2003/03/what-does-a-thumbs-up-mean-in-iraq.html

Koeze, E., & Popper, N. (2020, April 7). The virus changed the way we internet. *The New York Times.* https://nyti.ms/2XePBWp

Kosfeld, M., & Rustagi, D. (2015). Leader punishment and cooperation in groups: Experimental field evidence from commons management in Ethiopia. *American Economic Review, 105*(2), 747–783.

Kreidler, C. W. (2014). *Introducing English semantics* (2nd ed.). Routledge.

Kristinsson, K., Jonsdottir, I. J., & Snorrason, S. K. (2019). Employees' perceptions of supervisors' listening skills and their work-related quality of life. *Communication Reports, 32*(3), 137–147.

Kroll, T., & Stieglitz, S. (2021). Digital nudging and privacy: Improving decisions about self-disclosure in social networks. *Behaviour & Information Technology, 40*(1), 1–19. https://doi.org/10.1080/0144929X.2019.1584644

Kunalic, B., Jusic, J., Rovcanin, D., Mujkic, A., & Pajevic, M. (2016). Impact of emotional intelligence on leader effectiveness. *Ekonomski Vjesnik, 29*(2), 297–310.

Kunst, A. (2020, November 19). *Internet access by type in the U.S. 2020.* Statista. https://statista.com/forecasts/997163/

Kurth, S. B. (1970). Friendship and friendly relations. In G. J. McCall, M. M. McCall, N. K. Denzin, G. D. Suttles, & S. B. Kurth (Eds.), *Social relationships.* Aldine.

Ladbury, J. L., & Hinsz, V. B. (2018). How the distribution of member expectations influences cooperation and competition in groups: A social relations model analysis of social dilemmas. *Personality and Social Psychology Bulletin, 44*(10), 1502–1518.

LaFrance, A. (2014, July 28). In 1858, people said the telegraph was 'Too fast for the truth.' *The Atlantic.* https://theatlantic.com/technology/archive/2014/07/in-1858-people-said-the-telegraph-was-too-fast-for-the-truth/375171

LaFrance, M., & Vial, A. C. (2016). Gender and nonverbal behavior. In D. Matsumoto, H. C. Hwang, & M. G. Frank (Eds.), *APA handbook of nonverbal communication* (pp. 139–161). American Psychological Association.

LaGarde, J., & Hudgins, D. (2018). *Fact vs. fiction: Teaching critical thinking skills in the age of fake news.* International Society for Technology in Education.

LaGrande, S. (2019, March 29). *My student plagiarized a Ted Talk: And here's what I learned as a result.* Medium. https://medium.com/the-philosophers-stone/my-student-plagiarized-a-ted-talk-8a82a9606ed6

Lakoff, G., & Johnson, M. (1980). *Metaphors we live by.* University of Chicago Press.

Lakoff, R. T. (2004). *Language and woman's place: Text and commentaries* (Rev. ed.). Oxford University Press.

Lam, C. (2015). The role of communication and cohesion in reducing social loafing in group projects. *Business and Professional Communication Quarterly, 78*(4), 454–475.

Lammers, J. C., Atouba, Y. L., & Carlson, E. J. (2013). Which identities matter? A mixed-method study of group, organizational, and professional identities and their relationship to burnout. *Management Communication Quarterly, 27*(4), 503–536.

Landon, E. (2020, March 20). *Dr. Emily Landon speaks about COVID-19 at Illinois governor's press conference.* University of Chicago Medicine. https://uchicagomedicine.org/forefront/coronavirus-disease-covid-19/emily-landon-speaks-about-covid-19-at-illinois-governors-press-conference

Lang, F. R., & Fingerman, K. L. (Eds.). (2004). *Growing together: Personal relationships across the lifespan (Advances in personal relationships).* Cambridge University Press.

Langer, M., König, C. J., & Hemsing, V. (2020). Is anybody listening? The impact of automatically evaluated job interviews on impression management and applicant reactions. *Journal of Managerial Psychology, 35*(4), 271–284. https://doi.org/10.1108/JMP-03-2019-0156

Larsson, M., & Nielsen, M. F. (2021). The risky path to a followership identity: From abstract concept to situated reality. *International Journal of Business Communication, 58*(1), 3–30. https://doi.org/10.1177/2329488417735648

Lazer, D., Baum, M., Benkler, Y., Berinsky, A., Greenhill, K., Menczer, F., … Zittrain, J. (2018). The science of fake news. *Science, 359*(6380), 1094–1096.

Leaper, C. (2019). Young adults' conversational strategies during negotiation and self-disclosure in same-gender and mixed-gender friendships. *Sex Roles, 81*(9), 561–575.

Leaper, C., & Robnett, R. D. (2011). Women are more likely than men to use tentative language, aren't they? A meta-analysis testing for gender differences and moderators. *Psychology of Women Quarterly, 35*(1), 129–142.

Leary, M. R. (2001). Toward a conceptualization of interpersonal rejection. In M. R. Leary (Ed.), *Interpersonal rejection* (pp. 3–20). Oxford University Press.

Ledbetter, A., & Finn, A. (2018). Perceived teacher credibility and students' affect as a function of instructors' use of PowerPoint and email. *Communication Education, 67*(1), 31–51.

Ledbetter, A. M. (2010). Family communication patterns and communication competence as predictors of online communication attitude: Evaluating a dual pathway model. *Journal of Family Communication, 10*, 99–115.

Ledbetter, A. M., & Schrodt, P. (2008). Family communication patterns and cognitive processing: Conversation and conformity orientations as predictors of informational reception apprehension. *Communication Studies, 59*(4), 388–401.

Lee, C. (2019). The effect of team emotional intelligence on team process and effectiveness. *Journal of Management and Organization, 25*(6), 844–859.

Lee, C. M., Cadigan, J. M., & Rhew, I. C. (2020). Increases in loneliness among young adults during the COVID-19 pandemic and association with increases in mental health problems. *The Journal of Adolescent Health, 67*, 714–717.

Lee, E., & Leets, L. (2002). Persuasive storytelling by hate groups online: Examining its effects on adolescents. *American Behavioral Scientist, 45*, 927–957.

Lee, K., Lee, K. Y., & Sheehan, L. (2020). Hey Alexa! A magic spell of social glue?: Sharing a smart voice assistant speaker and its impact on users' perception of group harmony. *Information Systems Frontiers, 22*(3), 563–583. https://doi.org/10.1007/s10796-019-09975-1

Lee, N. (2020, April 27). *Why is video conferencing so exhausting?* Engadget. https://engadget.com/online-conferencing-video-chat-fatigue-172357939.html

LeFebvre, L., LeFebvre, L. E., & Allen, M. (2018). Training the butterflies to fly in formation: Cataloguing student fears about public speaking. *Communication Education, 67*(3), 348–362.

Le Guin, U. K. (2014, November 20). Speech at the National Book Awards. *The Guardian*. https://theguardian.com/books/2014/nov/20/ursula-k-le-guin-national-book-awards-speech

Lehmann-Willenbrock, N., Beck, S. J., & Kauffeld, S. (2016). Emergent team roles in organizational meetings: Identifying communication patterns via cluster analysis. *Communication Studies, 67*(1), 37–57.

Lehmann-Willenbrock, N., Meinecke, A., Rowold, J., & Kauffeld, S. (2015). How transformational leadership works during team interactions: A behavioral process analysis. *The Leadership Quarterly, 26*(6), 1017–1033.

Lei, C., Wu, T., Wang, L., & Jia, J. (2010). Fast convergence in language games induced by majority rule. *Physica A, 389*(19), 4046–4051.

Lenard, P. T., & Balint, P. (2020). What is (the wrong of) cultural appropriation? *Ethnicities, 20*, 331–352.

Leo, F., Gonzalez-Ponce, I., Sanchez-Miguel, P., Ivarsson, A., & Garcia-Calvo, T. (2015). Role ambiguity, role conflict, team conflict, cohesion and collective efficacy in sport teams: A multilevel analysis. *Psychology of Sport & Exercise, 20*, 60–66.

León-Jiménez, S. (2020). "This brings you to life" The impact of friendship on health and well-being in old age: The case of La Verneda Learning Community. *Research on Ageing and Social Policy, 8*, 192–215.

Letić, J. (2021, March 19). *Piracy is back: Piracy statistics for 2021.* DataProt. https://dataprot.net/statistics/piracy-statistics/

Levine, L. (1994). Listening with spirit and the art of team dialogue. *Journal of Organizational Change Management, 7*(1), 61–73. Levontin, L., & Yom-Tov, E.

(2017). Negative self-disclosure on the web: the role of guilt relief. *Frontiers in Psychology, 8*, https://doi.org/10.3389/fpsyg.2017.01068

Lewis, J., Raque-Bogdan, T., Lee, S., & Rao, M. (2018). Examining the role of ethnic identity and meaning in life on career decision-making self-efficacy. *Journal of Career Development, 45*(1), 68–82.

Li, C., Lin, C., Tien, Y., & Chen, C. (2017). A multilevel model of team cultural diversity and creativity: The role of climate for inclusion. *The Journal of Creative Behavior, 51*(2), 163–179.

Li, G., Liu, H., & Luo, Y. (2018). Directive versus participative leadership: Dispositional antecedents and team consequences. *Journal of Occupational & Organizational Psychology, 91*(3), 645–664

Li, L. M. W., Chen, Q., Gao, H., Li, W., & Ito, K. (2021). Online/offline self-disclosure to offline friends and relational outcomes in a diary study: The moderating role of self-esteem and relational closeness. *International Journal of Psychology, 56*(1), 129–137. https://doi.org/10.1002/ijop.12684

Li, P., Cho, H., & Goh, Z. (2019). Unpacking the process of privacy management and self-disclosure from the perspectives of regulatory focus and privacy calculus. *Telematics and Informatics, 41*, 114–125.

Lichtenstein, M. (2019). "Younger people want to do it themselves"—Self-actualization, commitment, and the reinvention of community. *Qualitative Sociology, 42*(2), 181–203.

LinkedIn Learning. (2020). *4th annual 2020 workplace learning report: L&D in a new decade: Taking the strategic long view.* https://learning.linkedin.com/resources/workplace-learning-report

Lipari, L. (2004). Listening for the other: Ethical implications of the Buber-Levinas encounter. *Communication Theory, 14*, 122–141.

Lipari, L. (2010). Listening, thinking, being. *Communication Theory, 20*(3), 348–362.

Lipari, L. (2014). Ethics, kairos, and akroasis: An essay on time and relation. In R. C. Arnett & P. Arneson (Eds.), *Philosophy of communication ethics: Alterity and the other* (pp. 87–106). Rowman & Littlefield.

Lipetz, L., Kluger, A., & Bodie, G. (2020). Listening is listening is listening: Employees' perception of listening as a holistic phenomenon. *International Journal of Listening, 34*(2), 71–96.

Lipkin, J., & Haddad, R. (2018). "Be fierce, believe in yourself, and find your people": A roundtable discussion about disability and performance. *Journal of Literary & Cultural Disability Studies, 12*(2), 221–238.

Liu, M. (2012). Same path, different experience: Culture's Influence on attribution, emotion, and interaction goals in negotiation. *Journal of Asian Pacific Communication, 22*(1), 97–119.

Liu, M. (2019, October 30). A culture, not a costume. *The Washington Post*. https://www.washingtonpost.com/nation/2019/10/30/culture-not-costume/

Liu, Y., Wang, S., & Yao, X. (2019). Individual goal orientations, team empowerment, and employee creative performance: A case of cross-level interactions. *The Journal of Creative Behavior, 53*(4), 443–456.

Liu, Y., Xiao, H., & Wang, D. (2020). Just because I like you: Effect of leader–member liking on workplace deviance. *Social Behavior and Personality, 48*(3), 1–10.

Liu, Z., Li, T., & Diao, H. (2020). The influence of multi-assessment model on the motivation of adults' English listening and speaking learning. *Theory and Practice in Language Studies, 10*(9), 1086–1093. https://doi.org/10.17507/tpls.1009.11

Liv, N., & Greenbaum, D. (2020). Deep fakes and memory malleability: False memories in the service of fake news. *AJOB Neuroscience, 11*(2), 96–104.

Lloyd, S. A. (1990). Conflict in premarital relationships: Differential perceptions of males and females. *Family Relations, 38*, 290–294.

Loades, M. E., Chatburn, E., Higson-Sweeney, N., Reynolds, S., Shafran, R., Brigden, A., Linney, C., McManus, M. N., Borwick, C., & Crawley, E. (2020). Rapid systematic review: The impact of social isolation and loneliness on the mental health of children and adolescents in the context of COVID-19. *Journal of the American Academy of Child and Adolescent Psychiatry, 59*, 1218–1239.

Loehwing, M., & Motter, J. (2012). Cultures of circulation: Utilizing co-cultures and counterpublics in intercultural new media research. *China Media Research, 8*(4), 29–38.

Lone, M. A., & Lone, A. H. (2018). Does emotional intelligence predict leadership effectiveness? An exploration in non-western context. *South Asian Journal of Human Resource Management, 5*(1), 28–39.

Luo, S., Wang, J., & Tong, D. Y. K. (2020). Does power distance necessarily hinder individual innovation? A moderated-mediation model. *Sustainability, 12*(6). https://doi.org/10.3390/su12062526

Luong, G., Charles, S. T., & Fingerman, K. L. (2011). Better with age: Social relationships across adulthood. *Journal of Social and Personal Relationships, 28*, 9–23.

Lvina, E., Johns, G., & Vandenberghe, C. (2018). Team political skill composition as a determinant of team cohesiveness and performance. *Journal of Management, 44*(3), 1001–1028.

Ly, I. (2019, February 26). Disney under fire for cultural appropriation over "Hakuna Matata" trademark. *Daily Blog*. http://www.ipbrief.net/2019/02/26/disney-under-fire-for-cultural-appropriation-over-hakuna-matata-trademark/

Lythcott-Haims, J. (2019, October 1). *How to raise an adult* [Presentation]. The Graduate School Public Lectures Series, University of Washington, Seattle, WA, United States. https://grad.washington.edu/public-lecture-series/lectures-library M. Bucholtz, (Ed.). Oxford University Press.

MacIntyre, P. D., & Thivierge, K. A. (1995). The effects of audience pleasantness, audience familiarity, and speaking contexts on public speaking anxiety and willingness to speak. *Communication Quarterly, 43*, 456–466.

MacIntyre, V. A., MacIntyre, P. D., & Carre, G. (2010). Heart rate variability as a predictor of speaking anxiety. *Communication Research Reports, 27*(4), 286–297.

MacNeela, P., & Gannon, N. (2014). Process and positive development: An interpretative phenomenological analysis of university student volunteering. *Journal of Adolescent Research, 29*(3), 407–436.

Madlock, P. E. (2008). Employee satisfaction: An examination of supervisors' communication competence. *Human Communication, 11*(1), 87–100.

Madore, K. P., Khazenzon, A. M., Backes, C. W., Jiang, J., Uncapher, M. R., Norcia, A. M., & Wagner, A. D. (2020). Memory failure predicted by attention lapsing and media multitasking. *Nature, 587*(7832), 87–91. https://doi.org/10.1038/s41586-020-2870-z

Magnuson, J., Jones, R., Hendricks, L., Hart, J. L., & Walker, K. L. (2012, November 14–16). *Comics, coffee, and communication: Examining later-life men's breakfast groups* [Paper presentation]. Annual Meeting of the National Communication Association, Orlando, FL, United States.

Majid, S., Eapen, C. M., Ei Mon Aung, & Oo, K. T. (2019). The Importance of soft skills for employability and career development: Students and employers' perspectives. *IUP Journal of Soft Skills, 13*(4), 7–39.

Malleson, T. (2018). Interdependency: The fourth existential insult to humanity. *Contemporary Political Theory, 17*(2), 160–186.

Malloch, Y. Z., & Taylor, L. D. (2019). Emotional self-disclosure in online breast cancer support groups: Examining theme, reciprocity, and linguistic style matching. *Health Communication, 34*(7), 764–773.

Malone, S. (2020, March 27). How I've been using free virtual Alcoholics Anonymous meetings to connect and stay sober while in COVID-19 isolation. *Business Insider*. https://businessinsider.com/alcoholics-anonymous-aa-meetings-zoom-covid-19-coronavirus-2020-3

Mandal, F. B. (2014). Nonverbal communication in humans. *Journal of Human Behavior in the Social Environment, 24*(4), 417–421.

Mangan, K. (2015). One hazing death, dozens of charges, and questions about making a frat come clean. *Chronicle of Higher Education, 62*(5), 17.

Manning, J. (2014). A constitutive approach to interpersonal communication studies. *Communication Studies, 65*, 432–440.

Manusov, V., Stofleth, D., Harvey, J., & Crowley, J. (2020). Conditions and consequences of listening well for interpersonal relationships: Modeling active-empathic listening, social-emotional skills, trait mindfulness, and relational quality. *International Journal of Listening, 34*(2), 110–126.

Marchak, F. M. (2013). Detecting false intent using eye blink measures. *Frontiers in Psychology, 4*, 736.

Mariamdaran, S., & Veloo, A. (2017). Relationship between information overload syndrome (IOS) and stress management of post graduate students. *Paradigms, 11*(2), 253–258.

Markus, H. R., & Kitayama, S. (1991). Culture and the self: Implications for cognition, emotion, and motivation. *Psychological Review, 98*, 224–253. https://doi.org/10.1037/0033-295X.98.2.224

Martens, H., & Hobbs, R. (2015). How media literacy supports civic engagement in a digital age. *Atlantic Journal of Communication, 23*(2), 120–137.

Martin B. A. S., Gnoth, J., & Strong, C. (2009). Temporal construal in advertising: The moderating role of temporal orientation and attribute importance in consumer evaluations. *Journal of Advertising, 38*(3), 5–20.

Marzban, A., & Isazadeh, F. (2012). Discovery listening and explicit strategy-based-instruction models' effect on the Iranian intermediate EFL listening comprehension. *Procedia-Social and Behavioral Sciences, 46*(C), 5435–5439.

Marzocchi, C., & Ramlogan, R. (2019). Forsaking innovation: Addressing failure and innovation behaviour variety. *Technology Analysis & Strategic Management, 31*(12), 1462–1473.

Mascareño, J., Rietzschel, E., & Wisse, B. (2020). Envisioning innovation: Does visionary leadership engender team innovative performance through goal alignment? *Creativity and Innovation Management, 29*(1), 33–48.

Maske, M., Kane, P., & Vargas, T. (2014, February 9). NFL faces pressure from Congress to change Redskins' name. *The Washington Post.* http://www.washingtonpost.com/blogs/football-insider/wp/2014/02/09/nfl-faces-pressure-from-congress-to-change-redskins-name/

Maslow, A. H. (1954). *Motivation and personality.* Harper.

Maslow, A. H. (1999). *Toward a psychology of being* (3rd ed.). Wiley & Sons.

Maslow, A. H. (2000). *The Maslow business reader.* John Wiley.

Maslow, A. H., Stephens, D. C., & Heil, G. (1971). *Maslow on management.* John Wiley.

Mathews, A., Derlega, V. J., & Morrow, J. (2006). What is highly personal information and how is it related to self- disclosure decision-making? The perspective of college students. *Communication Research Reports, 23*(2), 85–92.

Matsumoto, D., & Hwang, H. C. (2013). Cultural similarities and differences in emblematic gestures. *Journal of Nonverbal Behavior, 37*, 1–27.

Matusitz, J., & Breen, G. (2012). An examination of pack journalism as a form of groupthink: A theoretical and qualitative analysis. *Journal of Human Behavior in the Social Environment, 22*(7), 896–915.

Maunder, R., & Monks, C. P. (2019). Friendships in middle childhood: Links to peer and school identification, and general self-worth. *The British Journal of Developmental Psychology, 37*, 211–229.

Mayer, J. D., Salovey, P., & Caruso, D. R. (2004). Emotional intelligence: Theory, findings, and implications. *Psychological Inquiry, 15*, 197–215.

Mayer, R. E. (2009). *Multimedia learning.* Cambridge University Press.

Mazer, J. P., & Titsworth, S. (2012). Passion and preparation in the basic course: The influence of students' ego-involvement with speech topics and preparation time on public-speaking grades. *Communication Teacher, 26*(4), 236–251.

Mboya, S. (2014). When empowerment is realized. *Vital Speeches of the Day, 80*, 272–274.

Mcbey, K., Karakowsky, L., & Ng, P. (2017). Can I make a difference here? The impact of perceived organizational support on volunteer commitment. *Journal of Management Development, 36*(8), 991–1007.

McBride, M. C., & Bergen, K. M. (2015). Work spouses: Defining and understanding a "new" relationship. *Communication Studies, 66*, 487–508.

McBride, M. C., & Kirby, E. L. (2009). "You're totally her work husband": Managing misperceptions of the "work-spouse" relationship. In E. L. Kirby & M. C. McBride (Eds.), *Gender actualized: Cases in communicatively constructing realities* (pp. 133–140). Kendall Hunt.

McCabe, J. (2016). Friends with academic benefits. *Contexts, 15*(3), 22–29.

McClain, C., Vogels, E. A., Perrin, A., Sechopoulos, S., & Rainie, L. (2021, September 1). *The internet and the pandemic.* https://www.pewresearch.org/internet/2021/09/01/the-internet-and-the-pandemic/

McClish, G. (2016). "A kind of eloquence of the body": Quintilian's advice on delivery for the twenty-first-century rhetor. *Advances in the History of Rhetoric, 19*(2), 172–187.

McCornack, S. A., & Levine, T. R. (1990). When lies are uncovered: Emotional and relational outcomes of discovered deception. *Communication Monographs, 57*, 119–138.

McCreary, G. (2019, September 5). *How to project your normal speaking voice.* wikiHow. https://wikihow.com/Project-Your-Normal-Speaking-Voice

McDermott, M. (2014). *Speak with courage: 50+ insider strategies for presenting with confidence.* Bedford/St. Martin's.

McElroy, J. H. (1999). *American beliefs: What keeps a big country and a diverse people united.* Ivan R. Dee.

McIntosh, A. (2000). When the deaf and the hearing interact: Communication features, relationships, and disability issues. In D. O. Braithwaite & T. L. Thompson, *Handbook of communication and people with disabilities* (pp. 353–368). Lawrence Erlbaum.

McLaren, C. D., & Spink, K. S. (2018). Examining communication as information exchange as a predictor of task cohesion in sport teams. *International Journal of Sport Communication, 11*(2), 149–162.

McLuhan, M. (1994). *Understanding media: The extensions of man.* MIT Press.

McManus, T. G., & Donovan, S. (2012). Communication competence and feeling caught: Explaining perceived ambiguity in divorce-related communication. *Communication Quarterly, 60*(2), 255–277.

McNamee, L. G., Peterson, B. L., & Pena, J. (2010). A call to educate, participate, invoke and indict: Understanding the communication of online hate groups. *Communication Monographs, 77*(2), 257–280.

McWhorter, J. (2003). *The power of babel: A natural history of language.* Harper Perennial

McWhorter, J. (2011). *What language is (and what it isn't and what it could be).* Gotham.

McWhorter, J. (2013, February). *Txting is killing language. JK!!!* [Video]. TED Conferences. http://www.ted.com/talks/john_mcwhorter_txtng_is_killing_language_jk

Mehrabian, A. (2009). Nonverbal communication (3rd ed.). Aldine Transaction.

Mehrabian, A., & Wiener, M. (1967). Decoding of inconsistent communications. *Journal of Personality and Social Psychology, 6*(1), 109–114.

Mejia-Smith, B., & Gushue, G. (2017). Latina/o college students' perceptions of career barriers: Influence of ethnic identity, acculturation, and self-efficacy. *Journal of Counseling & Development, 95*(2), 145–155.

Meldrum, H. (2011). The listening practices of exemplary physicians. *International Journal of Listening, 25*, 145–160.

Melero, J., Hernandez-Leo, D., & Manatunga, K. (2015). Group-based mobile learning: Do group size and sharing mobile devices matter? *Computers in Human Behavior, 44*, 377–385.

Melis, A., Grocke, P., Kalbitz, J., & Tomasello, M. (2016). One for you, one for me: Humans' unique turn-taking skills. *Psychological Science, 27*(7), 987–996.

Melke, G., & Young, S. (2012). *Media convergence: Networked digital media and everyday life.* Palgrave Macmillian.

Mendelberg, T., & Karpowitz, C. F. (2016). Power, gender, and group discussion. *Political Psychology, 37*(Suppl. 1), 23–60.

Mendelberg, T., Karpowitz, C. F., & Goedert, N. (2014). Does descriptive representation facilitate women's distinctive voice? How gender composition and decision rules affect deliberation. *American Journal of Political Science, 58*(2), 291–306.

Meng, J., Fulk, J., & Yuan, Y. C. (2015). The roles and interplay of intragroup conflict and team emotion management on information seeking behaviors in team contexts. *Communication Research, 42*(5), 675–700.

Messinger, P., Ge, X., Smirnov, K., Stroulia, E., & Lyons, K. (2019). Reflections of the extended self: Visual self-representation in avatar-mediated environments. *Journal of Business Research, 100,* 531–546.

Methot, J. R., LePine, J. A., Podsakoff, N. P., & Christian, J. S. (2016). Are workplace friendships a mixed blessing? Exploring tradeoffs of multiplex relationships and their associations with job performance. *Personnel Psychology, 69,* 311–355.

Meyer, C. (2019). From Wounded Knee to sacred circles: Oglala Lakota ethos as "haunt" and "wound." *Humanities, 8*(1). https://doi.org/10.3390/h8010036

Michalisin, M. D., Karau, S. J., & Tangpong, C. (2004). Top management team cohesion and superior industry returns. *Group & Organization Management, 29,* 125–140.

Mikkelson, A. C., York, J. A., & Arritola, J. (2015). Communication competence, leadership behaviors, and employee outcomes in supervisor-employee relationships. *Business & Professional Communication Quarterly, 78*(3), 336–354.

Miller, G. R., Mongeau, P. A., & Sleight, C. (1986). Fudging with friends and lying to lovers: Deceptive communication in personal relationships. *Journal of Social and Personal Relationships, 3,* 495–512.

Miller, R. T., & Pessoa, S. (2016). Where's your thesis statement and what happened to your topic sentences? Identifying organizational challenges in undergraduate student argumentative writing. *TESOL Journal, 7*(4), 847–873.

Mitchell, A., Gottfried, J., Stocking, G., Walker, M., & Fedeli, S. (2019, June 5). *Many Americans say made-up news is a critical problem that needs to be fixed.* Pew Research Center. https://journalism.org/2019/06/05/many-americans-say-made-up-news-is-a-critical-problem-that-needs-to-be-fixed

Mitchell, A., & Oiphant, J. B. (2020, March 18). *Americans immersed in COVID-19 news.* Pew Research Center.

Mitchell, A., & Oliphant, J. B. (2020). *Americans Immersed in COVID-19 News; Most think media are doing fairly well covering it.* Pew Research Center. https://journalism.org/2020/03/18/americans-immersed-in-covid-19-news-most-think-media-are-doing-fairly-well-covering-it/

Modica, C. (2019). Facebook, body esteem, and body surveillance in adult women: The moderating role of self-compassion and appearance-contingent self-worth. *Body Image, 29,* 17–30. https://doi.org/10.1016/j.bodyim.2019.02.002

Mohammed, S., & Angell, L. C. (2004). Surface- and deep-level diversity in workgroups: Examining the moderating effects of team orientation and team process on relationship conflict. *Journal of Organizational Behavior, 25,* 1015–1039. Mojtahedi, D., Ioannou, M., & Hammond, L. (2018). Group size, misinformation and unanimity influences on co-witness judgements. *The Journal of Forensic Psychiatry & Psychology, 29*(5), 844–865.

Mok, A., & Morris, M. W. (2010). An upside to bicultural identity conflict: Resisting groupthink in cultural in-groups. *Journal of Experimental Social Psychology, 46*(6), 1114–1117.

Montgomery, B. M., & Baxter, L. A. (Eds.). (1998). *Dialectical approaches to studying personal relationships.* Lawrence Erlbaum.

Montgomery, M. (1995). *An introduction to language and society.* Routledge.

Montgomery, M., & Sorell, G. T. (1997). Differences in love attitudes across family life stages. *Family Relations, 46,* 55–61.

Moreland, R. L. (2010). Are dyads really groups? *Small Group Research, 41*(2), 251–267.

Moriya, J. (2019). Visual-working-memory training improves both quantity and quality. *Journal of Cognitive Enhancement, 3*(2), 221–232. https://doi.org/10.1007/s41465-018-00120-5

Morley, D. (2013, May). On living in a techno-globalised world: Questions of history and geography. *Telematics & Informatics, 30*(2), 61–65.

Mortaji, L. (2018). University students' perceptions of videotaping as a teaching tool in a public speaking course. *European Scientific Journal, 14*(8), 102.

Motley, M. T. (1990). On whether one can(not) not communicate: an examination via traditional communication postulates. *Western Journal of Speech Communication, 54,* 1–20.

Motro, D., Spoelma, T. M., & Ellis, A. P. J. (2020, June 18). Incivility and creativity in teams: Examining the role of perpetrator gender. *Journal of Applied Psychology.* Advance online publication. https://dx.doi.org/10.1037/apl0000757.

Moye, D. (2021, February 22). Cherokee Nation wants Jeep to stop using tribe's name on SUVs. *Huffington Post.* https://www.huffpost.com/entry/cherokee-nation-jeep-cherokee_n_6034187f-c5b673b19b6a8c7d

Mueller, J. C., Dirks, D., & Picca, L. H. (2007). Unmasking racism: Halloween costuming and engagement of the racial other. *Qualitative Sociology, 30,* 315–335.

Müllern, T., & Nordin, A. (2012). Revisiting empowerment: A study of improvement work in health care teams. *Quality Management in Health Care, 21*(2), 81–92.

Mumby, D. K. (1988). *Communication and power in organizations: Discourse, ideology and domination.* Ablex.

Mundt, M., Agneessens, F., Tuan, W., Zakletskaia, L., Kamnetz, S., & Gilchrist, V. (2016). Primary care team communication networks, team climate, quality of care, and medical costs for patients with diabetes: A cross-sectional study. *International Journal of Nursing Studies, 58*, 1–11.

Musinguzi, C., Namale, L., Rutebemberwa, E., Dahal, A., Nahirya-Ntege, P., & Ketitiinwa, A. (2018). The relationship between leadership style and health worker motivation, job satisfaction and teamwork in Uganda. *Journal of Healthcare Leadership, 10*, 21–32.

Nadler, R. (2020). Understanding "Zoom fatigue": Theorizing spatial dynamics as third skins in computer-mediated communication. *Computers and Composition, 58*, 102613. https://doi.org/10.1016/j.compcom.2020.102613

Naidoo, S., & Vernillo, A. T. (2014). Adapting a community of practice model to design an innovative ethics curriculum in healthcare. *Medical Principles & Practice, 23*, 60–68.

Narula, U. (2006). *Models of communication*. Atlantic.

National Aeronautics and Space Administration. (2020, August 21). *Global climate change: Vital signs of the planet*. https://climate.nasa.gov/causes

National Association of Colleges and Employers. (2016). *Job outlook 2016: The attributes employers want to see on new college graduates' resumes*.

National Association of Colleges and Employers. (2019, November). *Job outlook 2020*. https://nace.org

National Association of Colleges and Employers. (2020, January 16). *The top attributes employers want to see on resumes*. https://naceweb.org/about-us/press/2020/the-top-attributes-employers-want-to-see-on-resumes/

National civil and human rights coalition calls for Washington football team to drop offensive name [Press release]. (2013, December 12). The Leadership Conference on Civil and Human Rights. http://www.civilrights.org/press/2013/washington-football-team-name-change-resolution.html

National Communication Association. (1999). *NCA credo for ethical communication*. https://natcom.org/advocacy-public-engagement/public-policy/public-statements

National Communication Association. (1999). *NCA credo for ethical communication*. https://www.natcom.org/sites/default/files/pages/1999_Public_Statements_NCA_Credo_for_Ethical_Communication_November.pdf

National Urban League. (2020). *Mission and history*. https:// nul.org/index.php/mission-and-history

Nazione, S., Perrault, E., & Keating, D. (2019). Finding common ground: Can provider-patient race concordance and self-disclosure bolster patient trust, perceptions, and intentions? *Journal of Racial and Ethnic Health Disparities, 6*(5), 962–972.

NBC Bay Area Staff. (2016, February 22). 3 San Jose State students guilty of battery on black freshman, but not hate crime. *NBC Bay Area*. https://nbcbayarea.com/news/local/Jury-Verdict-in-San-Jose-State-Hate-Crime-Against-Black-Roommate-369701611.html

Nelson, J. (2017). *The voice exercise book: A guide to healthy and effective voice use*. National Theatre Publishing.

Nelson, S., Simek, J., & Maschke, M. (2017). Technology and cybersecurity policies help lawyers manage technology. *The Computer & Internet Lawyer, 34*(7), 14–36.

Ness I. J., & Glăveanu V. (2019) Polyphonic orchestration: The dialogical nature of creativity. In R. Beghetto & G. Corazza (Eds.), *Dynamic perspectives on creativity* (pp. 189–206). Springer.

Nettel, A., & Roque, G. (2012). Persuasive argumentation versus manipulation. *Argumentation, 26*(1), 55–69.

Neuliep, J. W. (2012). The Relationship among intercultural communication apprehension, ethnocentrism, uncertainty reduction, and communication satisfaction during initial intercultural interaction: An extension of Anxiety and Uncertainty Management (AUM) Theory. *Journal of Intercultural Communication Research, 41*(1), 1–16.

Neville, C. (2016). *Complete guide to referencing and avoiding plagiarism* (3rd ed.). Open University Press.

Newberry, C. (2019, October 22). 37 Instagram stats that matter to marketers in 2020. *Hootsuit Blog*. https://blog.hootsuite.com/instagram-statistics/

Ngai, C. S.-B., & Singh, R. G. (2018). Using dialectics to build leader-stakeholder relationships: An exploratory study on relational dialectics in Chinese corporate leaders' web-based messages. *International Journal of Business Communication, 55*, 3–29.

Nicolaisen, M., & Thorsen, K. (2017). What are friends for? Friendships and loneliness over the lifespan—from 18 to 79 years. *The International Journal of Aging and Human Development, 84*(2), 126–158.

Noller, P. (2006). Bringing it all together: A theoretical approach. In A. L. Vangelisti & D. Perlman (Eds.), *The Cambridge handbook of personal relationships* (pp. 769–789). Cambridge University Press.

Nordin, K., & Broeckelman-Post, M. A. (2019). Can I get better? Exploring mindset theory in the introductory communication course. *Communication Education, 68*(1), 44–60.

Northouse, P. G. (2018). *Leadership: Theory and practice* (8th ed.). SAGE.

Nöth, W. (1995). *Handbook of semiotics*. Indiana University Press.

Nowels, M. (2021, February 27). Former San Jose State trainer found responsible for sexual abuse claims made by athletes. *San Jose Mercury News*. https://mercurynews.com/2021/02/27/former-san-jose-state-trainer-reportedly-found-responsible-for-sexual-abuse-claims-made-by-swimmers/amp/

Obama, M. (2017, January 6). *The first lady honors the 2017 school counselor of the year* [Video]. YouTube. https://youtube.com/watch?v=SXrPIWYw8fE

O'Conner, L. (2013, October 30). San Francisco Chronicle to ban Washington Redskins team name. Huffington Post. http://www.huffingtonpost.com/2013/10/30/san-francisco-chronicle-redskins_n_4179212.html

O'Dea, S. (2021, August 4). *Smartphones in the U.S.—Statistics & facts*. Statista. https://www.statista.com/topics/2711/us-smartphone-market/

O'Dea, S. (2021, August 6). *Smartphone users worldwide 2016–2021*. Statista. https://statista.com/statistics/330695/number-of-smartphone-users-worldwide

O'Dea, S. (2021, March 19). *Smartphone users in the United States 2018–2025*. Statista. https://statista.com/statistics/201182/forecast-of-smartphone-users-in-the-u/

Ogden, C. K., & Richards, I. A. (1989). *The meaning of meaning: A study of the influence of language upon thought and of the science of symbolism*. Harcourt Brace Jovanovich.

Oggero, N., Rossi, M. C., & Ughetto, E. (2020). Entrepreneurial spirits in women and men. The role of financial literacy and digital skills. *Small Business Economics, 55*(2), 313–327.

Ogolsky, B. G., & Bowers, J. R. (2013). A meta-analytic review of relationship maintenance and its correlates. *Journal of Social and Personal Relationships, 30*, 343–367.

Okdie, B. M., & Ewoldsen, D. R. (2018). To boldly go where no relationship has gone before: Commentary on interpersonal relationships in the digital age. *Journal of Social Psychology, 158*, 508–513.

Ominicore Blog. (2019, September 6). Snapchat by the numbers: Stats, demographics & fun facts. https://omnicoreagency.com/snapchat-statistics/

Onishi, N., & Méheut, C. (2020, April 9). Mask-wearing is a very new fashion in Paris (and a lot of other places). *The New York Times.* https://nytimes.com/2020/04/09/world/europe/virus-mask-wearing.html

Orbe, M. (1998). *Constructing co-cultural theory: An explication of culture, power, and communication.* SAGE.

O'Reilly, C. (1980). Individuals and information overload in organizations: Is more necessarily better? *Academy of Management Journal, 23*(4), 684–696

Owusu, E., & Adade-Yeboah, A. (2014). Thesis statement: A vital element in expository essays. *Journal of Language Teaching and Research, 5*(1), 56–62.

Page, R., Harper, R., & Frobenius, M. (2013). From small stories to networked narrative: The evolution of personal narratives in Facebook status updates. *Narrative Inquiry, 23*(1), 192–213.

Palomares, N. A., & Eun-Ju, L. (2010). Virtual gender identity: The linguistic assimilation to gendered avatars in computer-mediated communication. *Journal of Language & Social Psychology, 29*(1), 5–23.

Palupi, P. (2019). Selective self-presentation on video-mediated communication: A study of hyperpersonal communication. *Mediator: Jurnal Komunikasi, 12*(1), 102–112. https://doi.org/10.29313/mediator.v12i1.4509

Panek, E., Hollenbach, C., Yang, J., & Rhodes, T. (2018). The effects of group size and time on the formation of online communities: Evidence from Reddit. *Social Media + Society, 4*(4). https://doi.org/10.1177/2056305118815908

Paquette, M., Sommerfeldt, E. J., & Kent, M. L. (2015). Do the ends justify the means? Dialogue, development communication, and deontological ethics. *Public Relations Review, 41*(1), 30–39.

Parolin, L. L, & Pellegrinelli, C. (2020). Unpacking distributed creativity: Analysing sociomaterial practices in theatre artwork. *Culture & Psychology, 26*(3), 434–453.

Parsons, C. (2019). Hystersisters: Messages of social support in an online health support group for women. *Journal of Communication in Healthcare, 12*(2), 121–133.

Pascual, A., Oteme, C., Samson, L., Wang, Q., Halimi-Falkowicz, S., Suochet, L., Girandola, F., Guéguen, N., & Joule, R.-V. (2012). Cross-cultural investigation of compliance without pressure: The "you are free to …" technique in France, Ivory Coast, Romania, Russia, and China. *Cross-Cultural Research, 46*(4), 394–416.

Patel, R. (2013, December). *Synthetic voice, as unique as fingerprints* [Video]. TED Conferences. https://www.ted.com/talks/rupal_patel_synthetic_voices_as_unique_as_fingerprints?language=en

Pavitt, C., Zingerman, V., Towey, E., & McFeeters, C. (2006). Group communication during resource dilemmas: 2. Effects of harvest limit and reward asymmetry. *Communication Research, 33*(1), 64–91.

Paxman, C. G. (2011). Map your way to speech success! Employing mind mapping as a speech preparation technique. *Communication Teacher, 25*(1), 7–11.

Pearce, E., Machin, A. & Dunbar, R. I. M. (2021). Sex differences in intimacy levels in best friendships and romantic partnerships. *Adaptive Human Behavior and Physiology, 7,* 1–16.

Pearce, W. B., & Pearce, K. A. (2000). Combining passions and abilities: Toward dialogic virtuosity. *Southern Communication Journal, 2 & 3,* 161–175.

Pearson, J. C., Semlak, J. L., Western, K. J., & Herakova, L. L. (2010). Answering a call for service: An exploration of family communication schemata and ethnic identity's effect on civic engagement behaviors. *Journal of Intercultural Communication Research, 39*(1), 49–68.

Pechmann, C., Yoon, K. E., Trapido, D., & Prochaska, J. J. (2020). Perceived costs versus actual benefits of demographic self-disclosure in online support groups. *Journal of Consumer Psychology, 31*(3), 450–477. https://doi.org/10.1002/jcpy.1200

Pence, M. E., & Vickery, A. J. (2012). The roles of personality and trait emotional intelligence in the active-empathic listening process: Evidence from correlational and regression analyses. *International Journal of Listening, 26*(3), 159–174.

Pennycook, G., Cannon, T., & Rand, D. (2018). Prior exposure increases perceived accuracy of fake news. *Journal of Experimental Psychology: General, 147*(12), 1865–1880.

Pérez, I. J., Cabrerizo, F. J., Alonso, S., Dong, Y. C., Chiclana, F., & Herrera-Viedma, E. (2018). On dynamic consensus processes in group decision making problems. *Information Sciences, 459,* 20–35.

Pérez-Fuentes, M., Jurado, M., & Linares, José J. G. (2018). *Emotional intelligence: Perceptions, interpretations and attitudes.* Nova Science.

Perissinotto, C. M., Stijacic, C. I., & Covinsky, K. E. (2012). Loneliness in older persons: A predictor of functional decline and death. *Archives of Internal Medicine, 172,* 1078–1084.

Perrin, A. (2019, May 31). *Digital gap between rural and nonrural America persists. Pew Research Center.* https://pewresearch. org/fact-tank/2019/05/31/digital-gap-between-rural-and-nonrural-america-persists/

Perrin, A., & Atske, S. (2021, April 2). *7% of Americans don't use the internet. Who are they?* Pew Research Center. https://pewresearch.org/fact-tank/2021/04/02/7-of-americans-dont-use-the-internet-who-are-they

Perrin, A., & Turner, E. (2019, August 20). *Smartphones help blacks, Hispanics bridge some—but not all—digital gaps with whites. Pew Research Center.* https://pewresearch.org/fact-tank/2019/08/20/smartphones-help-blacks-hispanics-bridge-some-but-not-all-digital-gaps-with-whites/

Perry, B., & Olsson, P. (2009). Cyberhate: The global-ization of hate. *Information & Communications Technology Law, 18*(2), 185–199.

Peter, G. R. (2018). Information overload in the informa-tion age: A review of the literature from business administration, business psychology, and related disciplines with a bibliometric approach and framework development. *Business Research, 11*(4), 1–44.

Petrie, C. R., Jr., & Carrel, S. D. (1976). The relationship of motivation, listening capability, initial informa-tion, and verbal organizational ability to lecture comprehension and retention. *Communication Monographs, 43,* 187–194.

Pew Research Center. (2021, April 7). *Mobile fact sheet.* https://www.pewresearch.org/internet/fact-sheet/mobile

Pfattheicher, S., Böhm, R., & Kesberg, R. (2018). The advantage of democratic peer punishment in sustaining cooperation within groups. *Journal of Behavioral Decision Making, 31*(4), 562–571.

Phillips, A., & Baker, A. (2017, October 30). *Cultural appropriation or appreciation?* The University of Utah. https://attheu.utah.edu/facultystaff/cultural-appropriation-or-appreciation/

Phillipsen, G. (1997). A theory of speech codes. In G. Phillipsen & T. Albrecht (Eds.), *Developing commu-nication theory* (pp. 119–156). SUNY Press.

Piantadosa, S., Tily, H., & Gibson, E. (2012). The com-municative function of ambiguity in language. *Cognition, 122*(3), 280–291.

Pietrabissa, G., & Simpson, S. G. (2020). Psychological consequences of social isolation during COVID-19 outbreak. *Frontiers in Psychology, 11.*

Pike, G. R., & Sillars, A. L. (1985). Reciprocity of marital communication. *Journal of Social and Personal Relationships, 2,* 303–332.

Pillay, N., Park, G., Kim, Y. K., & Lee, S. (2020). Thanks for your ideas: Gratitude and team creativity. *Organi-zational Behavior and Human Decision Processes, 156,* 69–81.

Pillemer, J., & Rothbard, N. P. (2018). Friends without benefits: Understanding the dark sides of work-place friendship. *Academy of Management Review, 43,* 635–660.

Pitardi, V., & Marriott, H. R. (2021). Alexa, she's not human but … Unveiling the drivers of consumers' trust in voice-based artificial intelligence. *Psy-chology & Marketing, 38*(4), 626–642. https://doi.org/10.1002/mar.21457

Pitts, L., Jr. (2014, February 14). Just what does the word 'thug' mean to you? *San Jose Mercury News,* A17.

Plagiarism: Facts & stats. (2017, June 7). Turnitin. https://plagiarism.org/article/plagiarism-facts-and-stats

Plancher, G., Mazeres, F., & Vallet, G. T. (2019). When motion improves working memory. *Memory, 27*(3), 410–416. https://doi.org/10.1080/09658211.2018.1510012

Plato. (360 BCE). *Phaedrus* (B. Jowett, Trans.). https://classics.mit.edu/Plato/phaedrus.html

Plester, B., Wood, C., & Bell, V. (2008). Txt message n school literacy: Does texting and knowledge of text abbreviations adversely affect children's literacy attainment? *Literacy, 42*(3), 137–144.

Plester, B., Wood, C., & Joshi, P. (2009). Exploring the relationship between children's knowledge of text message abbreviations and school literacy

outcomes. *British Journal of Developmental Psy-chology, 27*(1), 145–161.

Plotz, D. (2013, August 8). *Why Slate will no longer refer to Washington's NFL team as the Redskins.* Slate. http://www.slate.com/articles/sports/sports_nut/2013/08/washington_redskins_nickname_why_slate_will_stop_referring_to_the_nfl_team.html

Poepsel, M. (2018). *Media, society, culture, and you.* Rebus Community.

Powers, W., & Sawyer, C. R. (2011). Towards a theoretical framework for listening fidelity research. *Human Communication, 14*(3), 313–326.

Powers, W. G., & Bodie, G. D. (2003). Listening fidelity: Seeking congruence between cognitions of the listener and the sender. *International Journal of Listening, 17,* 19–31.

Powers, W. G., & Witt, P. L. (2008). Expanding the the-oretical framework of communication fidelity. *Communication Quarterly, 56*(3), 247–267.

Pradhan, A., Findlater, L., & Lazar, A. (2019). "Phantom friend" or "just a box with information": Personi-fication and ontological categorization of smart speaker-based voice assistants by older adults. *Proceedings of the ACM on Human-Com-puter Interaction, 3*(CSCW), 1–21. https://doi.org/10.1145/3359316

Prieto-Arranz, J., Juan-Garau, M., & Jacob, K. (2013). Re-imagining cultural identity: transcultural and translingual communication in virtual third-space environments. *Language, Culture & Curriculum, 26*(1), 19–35.

Proctor, H. (2016). Personal construct psychology, soci-ety, and culture: A Review. In D. A. Winter & N. Reed (Eds.), *The Wiley handbook of personal construct psychology* (pp. 152–165). Wiley Blackwell.

Proud to Be. (2013). *Change the mascot* [Video]. YouTube. http://www.youtube.com/watch?v=mR-tbOxlhvE

Pujiastuti, E. T., Budiarti, H. N., Sulaeman, K. A., Fatu-rachman, D., & Setiawati, R. (2020). Influence of Leadership, loyalty, and followership on team performance of P.T. Gaharu Galangan Interna-tional. *International Journal of Marine Engineering Innovation and Research, 5*(2), 81–91. https://doi.org/10.12962/j25481479.v5i2.7027

Purdy, M. (1991). What is listening? In D. Borisoff & M. Purdy (Eds.), *Listening in everyday life: A personal and professional approach* (pp. 3–19). University Press of America.

Purdy, M. W., & Manning, L. M. (2015). Listening in the multicultural workplace: A dialogue of theory and practice. *International Journal of Listening, 29*(1), 1–11.

Purvanova, R. K. (2013). The role of feeling known for team member outcomes in project teams. *Small Group Research, 44*(3), 298–331.

Putnam, L. L., Phillips, N., & Chapman, P. (1996). Meta-phors of communication and organization. In. S. R. Clegg, C. Hardy, & W. R. Nord (Eds.), *Handbook of organization studies* (pp. 375–408). SAGE.

Pütz, O. (2019). Common understandings of and con-sensus about collective action: The transformation of specifically vague proposals as a collective achievement. *Human Studies, 42*(3), 483–512.

Quisenberry, W. (2018). Exploring how emotional intel-ligence contributes to virtual teams: Interpretive analysis of a phenomenological study. *European Scientific Journal, 14*(5), 19–39.

Rabin, C. (2019). Self-disclosure to peers by young adult cancer survivors. *Psycho-oncology, 28*(1), 181–186.

Rabin, C. (2019). Self-disclosure to peers by young adult cancer survivors. *Psycho-oncology, 28*(1), 181–186.

Rahim, M. A. (2017). *Managing conflict in organizations* (4th ed.). Routledge.

Rains, S. A. (2014). The implications of stigma and anonymity for self-disclosure in health blogs. *Health Communication, 29*(1), 23–31.

Ramsey, M. C., Knight, R. A., & Knight, M. L. (2019). Student identification and communication instruction: An examination of identity gaps as predictors of communication satisfaction and teacher apprehension. *Communication Studies, 70*(5), 620–632.

Rane, D. B. (2011). Good listening skills make efficient business sense. *IUP Journal of Soft Skills, 5*(4), 43–51.

Rapp, C., & Wagner, T. (2013). On some Aristotelian sources of modern argumentation theory. *Argumentation, 27*(1), 7–30.

Rast, D. E., Hogg, M. A., & Giessner, S. R. (2013). Self-uncertainty and support for autocratic leadership. *Self & Identity, 12*(6), 635–649.

Raus, A., Haita, M., & Lazâr, L. (2012). Hierarchy of needs, perception and preference for leadership styles within a police educational institution. *Transylvanian Review of Administrative Sciences, 8*(35), 238–255.

Rawlins, W. K. (1989). A dialectical analysis of the tensions, functions, and strategic challenges of communication in young adult friendships. *Annals of the International Communication Association, 12*, 157–189.

Rawlins, W. K. (1992). *Friendship matters: Communication, dialectics, and the life course*. Aldine De Gruyter.

Rawlins, W. K. (1994). Being there and growing apart: Sustaining friendships during adulthood. In D. J. Canary & L. Stafford (Eds.), *Communication and relational maintenance* (pp. 275–294). Academic Press.

Rawlins, W. K. (2009). *The compass of friendship: Narratives, identities, and dialogues*. SAGE.

Read the entire U.S. Constitution on the National Archives website at The Constitution of the United States: A transcription. https://archives.gov/founding-docs/constitution-transcript

Redcay, E., & Schilbach, L. (2019). Using second-person neuroscience to elucidate the mechanisms of social interaction. *Nature Reviews Neuroscience, 20*(8), 495–505.

Reddy, M. J. (1979). The conduit metaphor: A case of frame conflict in our language about language. In A. Ortony (Ed.), *Metaphor and thought* (2nd ed., pp. 164–201). Cambridge University Press.

Reinhard, M., Sporer, S. L., Scharmach, M., & Marksteiner, T. (2011). Listening, not watching: Situational familiarity and the ability to detect deception. *Journal of Personality and Social Psychology, 101*(3), 467–484.

Reisman, J. (1981). Features of adult friendships. In S. Duck & R. Gilmour (Eds.), *Personal relationships 2: Developing personal relationships* (pp. 205–230). Academic Press.

Reisman, J. M. (1979). *Anatomy of friendship*. Lewis.

Riccobono, F., Bruccoleri, M., & Größler, A. (2016). Groupthink and project performance: The influence of personal traits and interpersonal ties. *Production and Operations Management, 25*(4), 609–629.

Rijt, R., & Hoffman, S. (2014). Ethical considerations of clinical photography in an area of emerging technology and smartphones. *Journal of Medical Ethics, 40*(3), 211–212.

Robards, J., Evandrou, M., Falkingham, J., & Vlachantoni, A. (2012). Marital status, health and mortality. *Maturitas, 73*, 295–299.

Robert, H. M. III. (2020). *Robert's rules of order* (12th ed.). Hachette.

Rock, D., & Grant, H. (2016, November 4). Why diverse teams are smarter. *Harvard Business Review.* https://hbr.org/2016/11/why-diverse-teams-are-smarter

Rock, D., Grant, H., & Grey, J. (2016, September 22). Diverse teams feel less comfortable—and that's why they perform better. *Harvard Business Review.* https://hbr.org/2016/09/diverse-teams-feel-less-comfortable-and-thats-why-they-perform-better

Rodero, E., Mas, L., & Blanco, M. (2014). The influence of prosody on politicians' credibility. *Journal of Applied Linguistics & Professional Practice, 11*(1), 89–111.

Rodríguez-Sánchez, A. M., Devloo, T., Rico, R., Salanova, M., & Anseel, F. (2017). What makes creative teams tick? Cohesion, engagement, and performance across creativity tasks: A three-wave study. *Group & Organization Management, 42*(4), 521–547.

Rojas, R. (2016, February 20). College student's death prompts an inquiry into hazing. *New York Times.*

Rollie, S. S., & Duck, S. W. (2006). Stage theories of marital breakdown. In J. H. Harvey & M. A. Fine (Eds.), *Handbook of divorce and dissolution of romantic relationships* (pp. 176–193). Lawrence Erlbaum.

Romm, C. (2015, October 15). Americans are more afraid of robots than death. *The Atlantic.* https://theatlantic.com/technology/archive/2015/10/americans-are-more-afraid-of-robots-than-death/410929/.

Rook, K. S., & Charles, S. T. (2017). Close social ties and health in later life: Strengths and vulnerabilities. *The American Psychologist, 72*, 567–577.

Rose, H. M., & Smith, A. R. (2000). Signing sound/sounding sight: The "violence" of deaf-hearing communication. In D. O. Braithwaite & T. L. Thompson, *Handbook of communication and people with disabilities* (pp. 369–388). Lawrence Erlbaum.

Rosenbaum, J. E., Johnson, B. K., & Deane, A. E. (2018). Health literacy and digital media use: Assessing the Health Literacy Skills Instrument—Short Form and its correlates among African American college students. *Digital Health, 4*. https://doi.org/10.1177/2055207618770765

Rosenfeld, L. R. (2000). Overview of the ways privacy, secrecy, and disclosure are balanced in today's society. In S. Petronio (Ed.), *Balancing the secrets of private disclosures* (pp. 3–17). Lawrence Erlbaum.

Ross, L. J. (2020, November 19). What if instead of calling people out, we called in them in? *New York Times.* https://www.nytimes.com/2020/11/19/

style/loretta-ross-smith-college-cancel-culture. html

Ross, L. J. (in press). *Calling in the calling out culture: Detoxing our movement*. Simon & Schuster.

Rossette-Crake, F. (2019). *Public speaking and the new oratory: A guide for non-native speakers*. Palgrave Macmillan.

Rossette-Crake, F. (2020). 'The new oratory': Public speaking practice in the digital, neoliberal age. *Discourse Studies, 22*(5), 571–589. https://doi.org/10.1177/1461445620916363

Rosso, B. D. (2014). Creativity and constraints: Exploring the role of constraints in the creative processes of research and development teams. *Organization Studies, 35*(4), 551–585.

Roter, D. L., Frankel, R. M., Hall, J. A., & Sluyter, D. (2006). The expression of emotion through nonverbal behavior in medical visits. *Journal of General Internal Medicine, 21*(Suppl. 1), S28–S34.

Rothman, N. (2014, March 26). *Olbermann: If Redskins isn't 'racist,' why doesn't Dan Snyder call Native Americans that? Mediaite*. http://www.mediaite.com/tv/olbermann-if-redskins-isnt-racist-why-doesnt-dan-snyder-call-native-americans-that/

Rouse, R. A., & Al-Maqbali, M. (2014). Identifying nurse managers' essential communication skills: An analysis of nurses' perceptions in Oman. *Journal of Nursing Management, 22*(2), 192–200.

Rowan, J. (1998). Maslow amended. *Journal of Humanistic Psychology, 38*(1), 81–92.

Rowe, M., Stannard, R., & The Spice Girls. (1996). Wannabe [Recorded by The Spice Girls]. On *Spice* [CD]. Strongroom.

Ruch, W., Gander, F., Platt, T., & Hofmann, J. (2018). Team roles: Their relationships to character strengths and job satisfaction. *The Journal of Positive Psychology, 13*, 190–199.

Rueter, M. A., & Koerner, A. F. (2008). The effect of family communication patterns on adopted adolescent adjustment. *Journal of Marriage and the Family, 70*, 715–727.

Russek, L. G., & Schwartz, G. E. (1997). Feelings of parental caring predict health status in midlife: A 35-year follow-up of the Harvard Mastery of Stress Study. *Journal of Behavioral Medicine, 20*, 1–13.

Ryan, R. M., & Deci, E. L. (2017). *Self-determination theory: Basic psychological needs in motivation, development, and wellness*. Guilford.

Sabourin, T. C., Infante, D. A., & Rudd, J. (1993). Verbal aggression in marriages: A comparison of violent, distressed but nonviolent, and nondistressed couples. *Human Communication Research, 20*, 245–267.

Sahlstein Parcell, E., & Baker, B. M. A. (2018). Relational dialectics theory: A new approach for military and veteran-connected family research. *Journal of Family Theory & Review, 10*, 672–685.

Saint-Dizier de Almeida, V., Colletta, J., Auriac-Slusarczyk, E., Specogna, A., Simon, J., Fiema, G., & Luxembourger, C. (2016). Studying activities that take place in speech interactions: A theoretical and methodological framework. *International Journal of Qualitative Studies in Education, 29*(5), 686–713.

Sakellaropoulou, R. (2020, July 23). *Research in the time of a pandemic: SENTINEL—A proactive, early warning system to pre-empt future pandemics*. Springer Nature. https://springernature.com/gp/researchers/the-source/blog/blogposts-communicating-research/research-in-the-time-of-a-pandemic-sentinel-pandemics-preemption/18201894

Sakyi, K. S., Surkan, P. J., Fombonne, E., Chollet, A., & Melchior, M. (2015). Childhood friendships and psychological difficulties in young adulthood: An 18-year follow-up study. *European Child & Adolescent Psychiatry, 24*, 815–826.

Samek, D. R., & Rueter, M. A. (2011). Associations between family communication patterns, sibling closeness, and adoptive status. *Journal of Marriage and the Family, 73*, 1015–1031.

Samosh, J. (2019). What is workplace incivility? An investigation of employee relational schemas. *Organization Management Journal, 16*(2), 81–97. https://doi.org/10.1080/15416518.2019.1604197

Samovar, L. A., Porter, R. E., McDaniel, E. R., & Roy, C. S. (2013). *Communication between cultures* (8th ed.). Wadsworth.

Sanderson, Z., Aslett, K., Godel, W., Persily, N., Nagler, J., Bonneau, R., & Tucker, J. (2020, April 7). It's not easy for ordinary citizens to identify fake news: And fake coronavirus news is no exception. *The Washington Post*. https://washingtonpost.com/politics/2020/04/07/its-not-easy-ordinary-citizens-identify-fake-news

Sandry, J., Zuppichini, M. D., & Ricker, T. J. (2020). Attentional flexibility and prioritization improves long-term memory. *Acta Psychologica, 208*, 1–14. https://doi.org/10.1016/j.actpsy.2020.103104

Sarampalis, A., Kalluri, S., Edwards, B., & Hafter, E. (2009). Objective measures of listening effort: Effects of background noise and noise reduction. *Journal of Speech, Language & Hearing Research, 52*(5), 1230–1240.

Sargent, L. D., & Sue-Chan, C. (2001). Does diversity affect group efficacy? The intervening role of cohesion and task interdependence. *Small Group Research, 32*, 426–450.

Sarwari, A. Q., Wahab, M. N., Mat Said, M. H., & Abdul Aziz, N. A. (2018). Assessment of the characteristics of interpersonal communication competence among postgraduate students from different cultures. *Journal of Intercultural Communication,* (47). https://immi.se/intercultural

Satir, V. (1983). *Conjoint family therapy* (3rd ed.). Science and Behavior Books.

Satir, V. (1988). *The new peoplemaking*. Science and Behavior Books.

Satir, V., & Baldwin, M. (1983). *Satir step by step: A guide to creating change in families*. Science and Behavior Books.

Satir, V., Banmen, J., Gerber, J., & Gomori, M. (1991). *The Satir model*. Science and Behavior Books.

Satir, V., Stachowiak, L., & Taschman, H. A. (1975). *Helping families to change*. Tiffany.

Savage, M., Storkholm, M. H., Mazzocato, P., & Savage, C. (2018). Effective physician leaders: an appreciative inquiry into their qualities, capabilities and learning approaches. *BMJ Leader, 2*(3), 95–102. https://doi.org/10.1136/leader-2017-000050

Sawyer, C. R., Gayle, K., Topa, A., & Powers, W. G. (2014). Listening fidelity among native and nonnative English-speaking undergraduates as a function of listening apprehension and gender. *Communication Research Reports, 31*(1), 62–71.

Scenters-Zapica, J. T., & Cos, G. C. (2003). A new canon for a new rhetoric education. In J. Petraglia & D. Bahri (Eds.). *The realms of rhetoric: The prospects for rhetoric education* (pp. 61–71). State University of New York Press.

Schäfer, K., & Eerola, T. (2020). How listening to music and engagement with other media provide a sense of belonging: An exploratory study of social surrogacy. *Psychology of Music, 48*(2), 232–251. https://doi.org/10.1177/0305735618795036

Schiavio, A., van der Schyff, D., Gande, A., & Kruse-Weber, S. (2019). Negotiating individuality and collectivity in community music. A qualitative case study. *Psychology of Music, 47*(5), 706–721.

Schiff, S. (2020, August 13). The Boston Tea Party was more than that. It was a riot. *The New York Times.* https://nytimes.com/2020/08/13/opinion/protests-monuments-history.html

Schilbach, L., Timmermans, B., Reddy, V., Costall, A., Bente, G., Schlicht, T., & Vogeley, K. (2013). Toward a second-person neuroscience. *Behavioral & Brain Sciences, 36*(4), 393–414.

Schindler, J. H. (2015). *Followership: What it takes to lead.* Business Expert Press.

Schmidt, G. (2018). Listening Is essential: An experiential exercise on listening behaviors. *Management Teaching Review, 3*(4), 287–294.

Schramm, W. (1954). *The process and effects of mass communication.* University of Illinois.

Schrodt, P., Witt, P. L., & Messersmith, A. S. (2008). A meta-analytical review of family communication patterns and their associations with information processing, behavioral, and psychosocial outcomes. *Communication Monographs, 75,* 248–269.

Schumacher, S., & Kent, N. (2020, April 2). *8 charts on internet use around the world as countries grapple with COVID-19.* Pew Research Center. https://pewresearch.org/fact-tank/2020/04/02/8-charts-on-internet-use-around-the-world-as-countries-grapple-with-covid-19/

Sedaris, D. (1999). *Me talk pretty one day.* Little, Brown.

Seitchik, A. E., & Harkins, S. G. (2014). The effects of nonconscious and conscious goals on performance. *Basic & Applied Social Psychology, 36*(2), 99–110.

Sereno, K. K., Welch, M., & Braaten, D. (1987). Interpersonal conflict: Effects of variations in manner of expressing anger and justification for anger upon perceptions of appropriateness, competence, and satisfaction. *Journal of Applied Communication Research, 15,* 128–143.

Sessions, H., Nahrgang, J. D., Newton, D. W., & Chamberlin, M. (2020). I'm tired of listening: The effects of supervisor appraisals of group voice on supervisor emotional exhaustion and performance. *Journal of Applied Psychology, 105*(6), 619–636. https://doi.org/10.1037/apl0000455

Sevelius, J. M., Chakravarty, D., Dilworth, S. E., Rebchook, G., & Neilands, T. B. (2020). Gender affirmation through correct pronoun usage: Development and validation of the transgender women's importance of pronouns (TW-IP) Scale. *International Journal of Environmental Research and Public Health, 17*(24), 9525. https://doi.org/10.3390/ijerph17249525

Shalev, S. (2011). Solitary confinement and supermax prisons: A human rights and ethical analysis. *Journal of Forensic Psychology Practice, 11,* 151.

Shank, B. (2018). Maximizing the heuristic potential of the enthymeme. *English Journal, 107*(3), 75–80.

Shannon, C. E. (1948). A mathematical theory of communication. *The Bell System Technical Journal, 27,* 379–423, 623–656.

Shannon, C. E., & Weaver, W. (1949). *The mathematical theory of communication.* University of Illinois.

Shao, G. (2020, January 17). *What 'deepfakes' are and how they may be dangerous.* NBC. https://nbc.com/2019/10/14/what-is-deepfake-and-how-it-might-be-dangerous.html

Sharma, S., & Khadka, A. (2019). Role of empowerment and sense of community on online social health support group. *Information Technology & People, 32*(6), 1564–1590.

Sheng-Min, L., & Jian-Qiao, L. (2013). Transformational leadership and speaking up: Power distance and structural distance as moderators. *Social Behavior & Personality: An International Journal, 41*(10), 1747–1756.

Sherman, J., & Burns, H. (2015). 'Radically different learning': Implementing sustainability pedagogy in a university peer mentor program. *Teaching in Higher Education, 20*(3), 231–243.

Shi, X., Brinthaupt, T., & Mccree, M. (2015). The relationship of self-talk frequency to communication apprehension and public speaking anxiety. *Personality and Individual Differences, 75,* 125–129.

Shields, S. (1987). Women, men, and the dilemma of emotion. In P. Shaver & C. Hendrick (Eds.), *Review of personality and social psychology: Vol. 7. Sex and gender* (pp. 229–251). SAGE.

Shin, Y. (2014). Positive group affect and team creativity: Mediation of team reflexivity and promotion focus. *Small Group Research, 45*(3), 337–364

Sias, P. M., & Gallagher, E. (2009). Developing, maintaining and disengaging from workplace friendships. In R. Morrison & S. Wright (Eds.,), *Friends and enemies in organizations* (pp. 78–100). Palgrave Macmillan.

Sias, P. M., Heath, R. G., Perry, T., Silva, D., & Fix, B. (2004). Narratives of workplace friendship deterioration. *Journal of Social and Personal Relationships, 21,* 321–340.

Sidorenkov, A., Borokhovski, E., & Kovalenko, V. (2018). Group size and composition of work groups as precursors of intragroup conflicts. *Psychology Research and Behavior Management, 11,* 511–523.

Siegert, J. R., & Stamp, G. H. (1994). "Our first big fight" as a milestone in the development of close relationships. *Communication Monographs, 61,* 345–360.

Siegman, A. W., & Feldstein, S. (1978). *Nonverbal behavior and communication.* Lawrence Erlbaum.

Siliezar, J. (2020, May 13). *Responding to this pandemic, preparing for the next.* The Harvard Gasette. https://news.harvard.edu/gazette/story/2020/05/pardis-sabetis-work-on-infectious-disease-coronavirus/

Sillars, A. L., Coletti, S. F., Parry, D., & Rogers, M. A. (1982). Coding verbal conflict tactics: Nonverbal and perceptual correlates of the "avoidance-distributive-integrative" distinction. *Human Communication Research, 9,* 83–95.

Silveira, J. M., & Hudson, M. W. (2015). Hazing in the college marching band. *Journal of Research in Music Education, 63*(1), 5–27.

Silverman, C., & Pham, S. (2018, December 28). *These are 50 of the biggest fake news hits on Facebook In 2018.* BuzzFeedNews. https://buzzfeednews.com/article/craigsilverman/facebook-fake-news-hits-2018

Singer, P. W. (2017, June 21). *NextTech: The future of technology, security, and threats.* Todd Lecture Series, Norwich University. https://tls.norwich.edu/p-w-singer-lecture-transcript

Singh, A. K. (2014). Role of interpersonal communication in organizational effectiveness. *International Journal of Research in Management and Business Studies, 1*(4), 36–39.

Smith, B. J., & Lim, M. H. (2020). How the COVID-19 pandemic is focusing attention on loneliness and social isolation. *Public Health Research & Practice, 30*(2), 3022008.

Smith, R., & Barr, S. (2008). Towards education inclusion in a contested society: From critical analysis to creative action. *International Journal of Inclusive Education, 12*, 401–422.

Smith, V. J. (2007). Aristotle's classical enthymeme and the visual argumentation of the twenty-first century. *Argumentation & Advocacy, 43*(3/4), 114–123.

Snellman, C. L. (2015). Ethics management: How to achieve ethical organizations and management? *Business, Management & Education, 13*(2), 336–357.

Soboroff, S., Kelley, C., & Lovaglia, M. (2020). Group size, commitment, trust, and mutual awareness in task groups. *The Sociological Quarterly, 61*(2), 334–346.

Society of Professional Journalists. (2014, September 6). *SPJ code of ethics.* https://spj.org/ethicscode.asp

Solanki, S., McPartlan, P., & Xu, D. (2019). Success with EASE: Who benefits from a STEM learning community? *PLoS One, 14*(3), 1–20.

Soldan, Z. (2010). Group cohesiveness and performance: The moderating effect of diversity. *International Journal of Interdisciplinary Social Sciences, 5*(4), 155–168.

Soliz, J. & Phillips, K. E. (2018). Toward a more expansive understanding of family communication: Considerations for inclusion of ethnic-racial and global diversity. *Journal of Family Communication, 18*, 5–12

Sonam, C., Prasad, Y., Anwar, S., & Kumar, C. (2019). Mathematical modelling and analysis of plastic waste pollution and its impact on the ocean surface. *Journal of Ocean Engineering and Science, 5*(2), 136–163. https://doi.org/10.1016/j.joes.2019.09.005

Song, H., Kim, J., & Luo, W. (2016). Teacher–student relationship in online classes: A role of teacher self-disclosure. *Computers in Human Behavior, 54*, 436–443. https://doi.org/10.1016/j.chb.2015.07.037

Song, H., Kim, J., & Park, N. (2019). I know my professor: Teacher self-disclosure in online education and a mediating role of social presence. *International Journal of Human-Computer Interaction, 35*(6), 448–455. https://doi.org/10.1080/10447318.2018.1455126

Sotomayor, S. (2013, May 5). *2013 PEN World Voices Festival Arthur Miller Freedom to Write lecture* [Video]. C-SPAN. https://c-span.org/video/?312669-1/sonia-sotomayor-freedom-write-lecture

Southern Poverty Law Center. (2021, February 1). *The year in hate and extremism 2020: Hate groups became more difficult to track amid COVID amid COVID and migration to online networks.* https://www.splcenter.org/news/2021/02/01/year-hate-and-extremism-2020-hate-groups-became-more-difficult-track-amid-covid-and

Sözbilir, F. (2018). The interaction between social capital, creativity and efficiency in organizations. *Thinking Skills and Creativity, 27*, 92–100.

Spiegel, E. (2014, April 11). *2014 LA Hacks keynote* [Keynote address]. LA Hacks, Los Angeles, CA, United States. https://blog.snapchat.com/post/82635264882/2014-la-hacks-keynote

Spink, K. S., Nickel, D., Wilson, K., & Odnokon, P. (2005). Using a multilevel approach to examine the relationship between task cohesion and team task satisfaction in elite ice hockey players. *Small Group Research, 36*, 539–554.

Spooner, R. A. (2017, November 10). *Literary translation in signed language.* Department of Interpretation and Translation Colloquium Lecture Series. https://gallaudet.edu/Documents/Department-of-Interpretation/Colloquium%20Lecture%20Series%20-%20Transcripts/2_RUTH%20ANNA%20SPOONER_FINAL.pdf

Sports e-cyclopedia. (n.d.). Washington Redskins. In *Sportse-cyclopedia.com dictionary.* http://www.sportsecyclopedia.com/nfl/washington/redskins.html

Stadtfeld, C., Voros, A., Elmer, T., Boda, Z., & Raabe, I. J. (2019). Integration in emerging social networks explains academic failure and success. *Proceedings of the National Academy of Sciences of the United States of America (PNAS), 116*, 792–797. https://www.pnas.org/content/116/3/792

Stafford, L. (2003). Maintaining romantic relationships: Summary and analysis of one research program. In D. J. Canary & M. Dainton (Eds.), *Maintaining relationships through communication: Relational, contextual, and cultural variations* (pp. 51–78). Lawrence Erlbaum.

Stafford, L., & Canary, D. (1991). Maintenance strategies and romantic relationship type, gender and relational characteristics. *Journal of Social and Personal Relationships, 8*, 217–242.

Stafford, L., & Canary, D. J. (1991). Maintenance strategies and romantic relationship type, gender and relational characteristics. *Journal of Social and Personal Relationships, 8*, 217–242.

Stafford, L., Dainton, M., & Haas, S. (2000). Measuring routine and strategic relational maintenance: Scale revision, sex versus gender roles, and the prediction of relational characteristics. *Communication Monographs, 67*, 306–323.

Stancill, J. (2018, April 14). Duke protesters disrupt president's speech to alumni. *The News & Observer.* https://newsobserver.com/news/local/education/article208912519.html

Steele, G. A., & Plenty, D. (2015). Supervisor–subordinate communication competence and job and communication satisfaction. *International Journal of Business Communication, 52*(3), 294–318.

Steinberg, L. (2001). We know some things: Parent-adolescent relationships in retrospect and prospect. *Journal of Research on Adolescence, 11*, 1–19.

Stephey, M. J. (2009, December 9). Top 10 Disney controversies. *Time*. https://entertainment.time.com/2009/12/09/top-10-disney-controversies/slide/all/

Stewart, C., McConnell, J., Stallings, L., & Roscoe, R. (2017). An initial exploration of students' mindsets, attitudes, and beliefs about public speaking. *Communication Research Reports, 34*(2), 180–185.

Stewart, J., & Koenig Kellas, J. (2020). Co-constructing uniqueness: An interpersonal process promoting dialogue. *Atlantic Journal of Communication, 28*(1), 5–21. https://doi.org/10.1080/15456870.2020.1684289

Stob, P. (2015). Talk like TED: The 9 public-speaking secrets of the world's top minds. *Quarterly Journal of Speech, 101*(1), 311–314.

Stokoe, E. (2009). "I've got a girlfriend": Police officers doing 'self-disclosure' in their interrogations of suspects. *Narrative Inquiry, 19*(1), 154–182.

Storm, B. C., Stone, S. M., & Benjamin, A. S. (2016). Using the internet to access information inflates future use of the internet to access other information. *Memory, 25*(6), 717–723. https://doi.org/10.1080/09658211.2016.1210171

Strauss, V. (2014, July 26). Students catch school superintendent plagiarizing part of commencement speech. *The Washington Post*. https://www.washingtonpost.com/news/answer-sheet/wp/2014/07/26/students-catch-school-superintendent-plagiarizing-part-of-commencement-speech/

Stroud, S. R., & Pye, D. (2013). Kant on unsocial sociability and the ethics of social blogging. In M. Drumwright (Ed.), *Ethical Issues in communication professions: New agendas in communication* (pp. 41–65). Routledge.

Suciu, P. (2020, April 8). During COVID-19 pandemic it isn't just fake news but seriously bad misinformation that is spreading on social media. *Forbes*. https://forbes.com/sites/petersuciu/2020/04/08/during-covid-19-pandemic-it-isnt-just-fake-news-but-seriously-bad-misinformation-that-is-spreading-on-social-media/#5d6111077e55

Sunstein, C. R., & Hastie, R. (2015). *Wiser: Getting beyond groupthink to make groups smarter.* Harvard Business Review Press.

Suter, E. A. (2016). Introduction: Critical approaches to family communication research: Representation, critique, and praxis. *Journal of Family Communication, 16*, 1–8.

Syal, S., & Anderson, A. K. (2013). It takes two to talk: A second-person neuroscience approach to language learning. *Behavioral and Brain Sciences, 36*(4), 439–440.

Syed, N. (2018, November 14). *Nabiha Syed: 2018 Salant lecture on freedom of the press* [Video]. Harvard Kennedy School's Shorenstein Center on Media, Politics, and Public Policy. https://shorensteincenter.org/nabiha-syed-2018-salant-lecture-freedom-press

Syrett, M., & Lammiman, J. (2002). *Creativity.* Capstone.

Szanto, T. (2017). Collaborative irrationality, akrasia, and groupthink: Social disruptions of emotion regulation. *Frontiers in Psychology, 7,* https://doi.org/10.3389/fpsyg.2016.02002

Tagliamonte, S. A., & Denis, D. (2008). Linguistic ruin? LOL! Instant messaging and teen language. *American Speech, 83*(1), 3–34

Tandoc, E., Lim, Z., & Ling, R. (2018). Defining "fake news": A typology of scholarly definitions. *Digital Journalism, 6*(2), 137–153.

Tanewong, S. (2019). Metacognitive pedagogical sequence for less-proficient Thai EFL listeners: A comparative investigation. *RELC Journal, 50*(1), 86–103.

Tang, K. N. (2019). *Leadership and change management.* Springer.

Tannen, D. (1981). Indirectness in discourse: Ethnicity as conversational style. *Discourse Processes, 4*(3), 221–238.

Tannen, D. (1990). *You just don't understand.* Ballantine Books.

Tannen, D. (1996). *Gender and discourse.* Oxford University Press.

Tapia, A., & Polonskaia, A. (2020). *The 5 disciplines of inclusive leaders.* Berrett-Koehler.

Tapia, J. E. (2001). Wilbur Samuel Howell: The trilogy of rhetoric, logic, and science. In J. Kuypers & A. King, *Twentieth- century roots of rhetorical studies* (pp. 103–122). Praeger.

Taraban, O., Heide, F., Woollacott, M., & Chan, D. (2016). The effects of a mindful listening task on mind-wandering. *Mindfulness, 8*(2), 433–443.

Tausczik, Y., & Huang, X. (2019). The impact of group size on the discovery of hidden profiles in online discussion groups. *ACM Transactions on Social Computing, 2*(3), 1–25.

Taylor, R. (2013). *Creativity at work.* Kogan Page.

TED Conferences. (n.d.). *History of TED.* https://ted.com/about/our-organization/history-of-ted

Teng, C., & Luo, Y. (2015). Effects of perceived social loafing, social interdependence, and group affective tone on students' group learning performance. *The Asia-Pacific Education Researcher, 24*(1), 259–269.

Teven, J. J., Richmond, V. P., McCroskey, J. C., & McCroskey, L. L. (2010). Updating relationships between communication traits and communication competence. *Communication Research Reports, 27*(3), 263–270.

The Washington Post. (2013, October 9). *Letter from Washington Redskins owner Dan Snyder to fans.* http://www.washingtonpost.com/local/letter-from-washington-redskins-owner-dan-snyder-to-fans/2013/10/09/e7670ba0-30fe-11e3-8627-c5d7de0a046b_story.html

Thomas, K. W. (1976). Conflict and conflict management. In M. Dunnette (Ed.), *Handbook of industrial and organizational psychology* (pp. 889–935). Rand-McNally.

Thomas, K. W., & Kilmann, R. H. (1974). The Thomas-Kilmann conflict mode instrument. CPP.

Thompson, J. (2009). Building collective communication competence in interdisciplinary research teams. *Journal of Applied Communication Research, 37*(3), 278–297.

Thompson, K., Leintz, P., Nevers, B., & Witkowski, S. (2004). The integrative listening model: An approach to teaching and learning listening. *The Journal of General Education, 53*(3/4), 225–246.

Thomsen, I. (2020, July 16). *The research is clear: White people are not more likely than black people to be killed by police.* News@Northeastern. https://news.northeastern.edu/2020/07/16/the-research-is-clear- white-people-are-not-more-likely-than-black-people-to-be-killed-by-police/

Thonney, T., & Montgomery, J. C. (2019). Defining critical thinking across disciplines: An analysis of community college faculty perspectives. *College Teaching, 67*(3), 169–176.

Thulin, E., & Vilhelmson, B. (2017). Does online co-presence increase spatial flexibility? On social media and young people's migration considerations. *Cybergeo.* https://doi.org/10.4000/cybergeo.28811

Tian, Q. (2013). Social anxiety, motivation, self-disclosure, and computer-mediated friendship: A path analysis of the social interaction in the blogosphere. *Communication Research, 40*(2), 237–260.

Tindale, C. W. (2011). Character and knowledge: Learning from the speech of experts. *Argumentation, 25*(3), 341–353.

Ting-Toomey, S. (1983). An analysis of verbal communication in high and low marital adjustment groups. *Human Communication Research, 9,* 306–319.

Tjosvold, D., Law, K., & Sun, H. F. (2003). Collectivistic and individualistic values: Their effects on group dynamics and productivity in China. *Group Decision and Negotiation, 12,* 243–263.

Todorov, A., Baron, S. G., & Oosterhof, N. N. (2008). Evaluating face trustworthiness: A model-based approach. *Social Cognitive and Affective Neuroscience, 3*(2), 119–127.

Toensing, G. C. (2014, April 11). *Navajo Nation officially joins fight against racist Redskins mascot.* Indian Country Today Media Network. http://indiancountrytodaymedianetwork.com/2014/04/11/navajo-nation-officially-joins-fight-against-redskins-mascot-154423

Torres, R., Gerhart, N., & Negahban, A. (2018). Epistemology in the era of fake news: An exploration of information verification behaviors among social networking site users. *ACM SIGMIS Database: The DATABASE for Advances in Information Systems, 49*(3), 78–97.

Tottie, G., & Hoffmann, S. (2006). Question tags in British and American English. *Journal of English Linguistics, 34*(4), 283–311.

Trager, G. L. (1958). Paralanguage: A first approximation. *Studies in Linguistics, 13,* 1–12.

Trager, G. L. (1960). Taos III: Paralanguage. *Anthropological Linguistics, 2,* 24–30.

Trager, G. L. (1961). The typology of paralanguage. *Anthropological Linguistics, 3*(1), 17–21.

Treasure, J. (2011). *5 ways to listen better* [Video]. TED Conferences. https://ted.com/talks/julian_treasure_5_ways_to_listen_better

Treasure, J. (2013, June 10). *How to speak so that people want to listen* [Video]. TED Conferences. https://ted.com/talks/julian_treasure_how_to_speak_so_that_people_want_to_listen

Treen, E., Atanasova, C., Pitt, L., Johnson, M., & Kietzmann, J. (2016). Evidence from a large sample on the effects of group size and decision-making time on performance in a marketing simulation game. *Journal of Marketing Education, 38*(2), 130–137.

Tsfati, Y., Boomgaarden, H. G., Strömbäck, J., Vliegenthart, R., Damstra, A., & Lindgren, E. (2020). Causes and consequences of mainstream media dissemination of fake news: Literature review and synthesis. *Annals of the International Communication Association, 44*(2), 157–173.

Tsikerdekis, M. (2013). The effects of perceived anonymity and anonymity states on conformity and groupthink in online communities: A Wikipedia study. *Journal of the American Society For Information Science & Technology, 64*(5), 1001–1015.

Turkle, S. (1996). Parallel loves: Working on identity in virtual space. In D. Grodin & T. R. Lindlof (Eds.), *Constructing the self in a mediated world* (pp. 156–175). SAGE.

Tuuli, M., & Rowlinson, S. (2009). Empowerment in project teams: A multilevel examination of the job performance implications. *Construction Management & Economics, 27*(5), 473–498.

Twiselton, K., Stanton, S. C. E., Gillanders, D., & Bottomley, E. (2020). Exploring the links between psychological flexibility, individual well-being, and relationship quality. *Personal Relationships, 27,* 880–906.

Umberson, D., & Montez, J. K. (2010). Social relationships and health: A flashpoint for health policy. *Journal of Health and Social Behavior, 51*(Suppl.), S54–S66.

Umberson, D., Williams, K., Powers, D. A., Liu, H., & Needham, B. (2006). You make me sick: Marital quality and health over the life course. *Journal of Health and Social Behavior, 47*(1), 1–16.

Uncapher, M., & Wagner, A. (2018). Minds and brains of media multitaskers: Current findings and future directions. *Proceedings of the National Academy of Sciences of the United States of America, 115*(40), 9889–9896.

United States Access Board. (1995, December 7). *Principles of universal design.* https://access-board.gov/guidelines-and-standards/communications-and-it/26-255-guidelines/825-principles-of-universal-design

University of Bristol. (2019, June 3). Feathers came first, then birds. *ScienceDaily.* https://sciencedaily.com/releases/2019/06/190603124542.htm

Vaidyanathan, R., Aggarwal, P., & Kozłowski, W. (2013). Interdependent self-construal in collectivist cultures: Effects on compliance in a cause-related marketing context. *Journal of Marketing Communications, 19*(1), 44–57.

Valine, Y. A. (2018). Why cultures fail: The power and risk of groupthink. *Journal of Risk Management in Financial Institutions, 11*(4), 301–307.

Van Damme, M., Anseel, F., Duyck, W., & Rietzschel, E. F. (2019). Strategies to improve selection of creative ideas: An experimental test of epistemic and social motivation in groups. *Creativity and Innovation Management, 28*(1), 61–71.

Vanden Abeele, M. M. P., & Postma-Nilsenova, M. (2018). More than just gaze: An experimental vignette study examining how phone-gazing and newspaper-gazing and phubbing-while-speaking and phubbing-while-listening compare in their effect on affiliation. *Communication Research Reports, 35*(4), 303–313.

Vandergrift, L. (2005). Relationships among motivation orientations, metacognitive awareness and proficiency in L2 listening. *Applied Linguistics, 26*(1), 70–89.

Vangelisti, A. L. (1993). Communication in the family: The influence of time, relational prototypes, and irrationality. *Communication Monographs, 60,* 42–54.

Vangelisti, A. L. (Ed.). (2013). *The Routledge handbook of family communication* (2nd ed.). Routledge.

Van Heekeren, M. (2020). The curative effect of social media on fake news: A historical re-evaluation. *Journalism Studies, 21,* 306–318.

VanLear, C. A., Koerner, A., & Allen, D. M. (2006). Relationship typologies. In A. L. Vangelisti & D. Perlman (Eds.), *The Cambridge handbook of personal relationships.* Cambridge University Press.

van Vilin, R. D., Jr. (2005). *Exploring the syntax-semantics interface.* Cambridge University Press.

Van Zant, A. B., & Berger, J. (2020). How the voice persuades. *Journal of Personality and Social Psychology, 118*(4), 661–682.

van Zijl, A. L., Vermeeren, B., Koster, F., & Steijn, B. (2021). Interprofessional teamwork in primary care: The effect of functional heterogeneity on performance and the role of leadership. *Journal of Interprofessional Care, 35*(1), 10–20, https://doi.org/10.1080/13561820.2020.1715357

Varnhagen, C. K., McFall, G. P., Pugh, N., Routledge, L., Sumida-MacDonald, H., & Kwong, T. E. (2010). Lol: New language and spelling in instant messaging. *Reading and Writing, 23*(6), 719–733.

Varpio, L., & Teunissen, P. (2021). Leadership in interprofessional healthcare teams: Empowering knotworking with followership. *Medical Teacher, 43*(1), 32–37. https://doi.org/10.1080/0142159X.2020.1791318

Vaughn, L. (2018). *The power of critical thinking: Effective reasoning about ordinary and extraordinary claims* (6th ed.). Oxford University Press.

Vaughn, M., & Parry, P. (2013). The statement of purpose speech: Helping students navigate the "sophomore slump". *Communication Teacher, 27*(4), 207–211.

Vecina, M., Chacón, F., Marzana, D., & Marta, E. (2013). Volunteer engagement and organizational commitment in nonprofit organizations: What makes volunteers remain within organizations and feel happy? *Journal of Community Psychology, 41*(3), 291–302.

Verouden, N., & Van Der Sanden, M. (2016). Is silence golden? Silence in interdisciplinary collaboration between scientists. *Journal of Science Communication, 15*(5), 1–4.

Vickery, A. (2018). "Listening enables me to connect with others": Exploring college students' (mediated) listening metaphors. *International Journal of Listening, 32*(2), 69–84.

Vickery, A. J., Keaton, S. A., & Bodie, G. D. (2015). Intrapersonal communication and listening goals: An examination of attributes and functions of imagined interactions and active-empathic listening behaviors. *Southern Communication Journal, 80*(1), 20–38.

Vigen, T. (2015). *Spurious correlations.* Hachette Books.

Vijayavalsalan, B. (2016). Mind mapping as a strategy for enhancing essay writing skills. *The New Educational Review, 45*(3), 137–150.

Vincenty, S. (2020, October 13). What Is cultural appropriation—And how can you avoid it? What to keep in mind when picking that Halloween costume. *O: The Oprah Magazine.* https://www.oprahmag.com/life/a34058415/what-is-cultural-appropriation/

Vingerhoets, A., & Scheirs, J. (2000). Gender differences in crying: Empirical findings and possible explanations. In A. H. Fisher (Ed.), *Gender and emotion: Social psychological perspectives* (pp. 143–165). Cambridge University Press.

Violanti, M. T., & Ray, C. A. (2021). *Follow, communicate, lead: Creating competent connections.* Cognella.

Vogels, E. A. (2020, September 20). *59% of U.S. parents with lower incomes say their child may face digital obstacles in schoolwork.* Pew Research Center. https://pewresearch.org/fact-tank/2020/09/10/59-of-u-s-parents-with-lower- incomes-say-their-child-may-face-digital-obstacles-in-schoolwork

Vogels, E. A., Perrin, A., Rainie, L., & Anderson, M. (2020, April 30). *53% of Americans say the internet has been essential during the COVID-19 outbreak.* Pew Research Center. https://www.pewresearch.org/internet/2020/04/30/53-of-americans-say-the-internet-has-been-essential-during-the-covid-19-outbreak/

Waddell, B. D., Roberto, M. A., & Yoon, S. (2013). Uncovering hidden profiles: Advocacy in team decision making. *Management Decision, 51*(2), 321–340.

Wagner, M., & Boczkowski, P. (2019). The reception of fake news: The interpretations and practices that shape the consumption of perceived misinformation. *Digital Journalism, 7*(7), 870–885.

Waisanen, D. (2015). Comedian-in-chief: Presidential jokes as enthymematic crisis rhetoric. *Presidential Studies Quarterly, 45*(2), 335–360.

Waldinger, R. J., & Schulz, M. S. (2010). What's love got to do with it? Social functioning, perceived health, and daily happiness in married octogenarians. *Psychology and Aging, 25,* 422–431.

Waldinger, R. J., & Schulz, M. S. (2016). The long reach of nurturing family environments: Links with midlife emotion-regulatory styles and late-life security in intimate relationships. *Psychological Science, 27,* 1443–1450.

Waldron, T. (2020, December 16). The Cleveland Indians will change their name. The Braves, Chiefs and Blackhawks are next. *Huffington Post.* https://www.huffpost.com/entry/cleveland-indians-name-change-braves-chiefs-blackhawks_n_5fd7ce92c5b663c37598ec2e

Walker, K. L., Hart, J. L., Magnuson, J., Hendricks, L., Khariwal, P., Friley, L. B., Jones, R., Koetter, S. B., Long, K., Nelson, G., & Mudd, M. (2011, March 23–27). An exploratory study investigating the verbal and nonverbal communication of later-life men [Paper presentation]. Annual Meeting of the Southern States Communication Association, Little Rock, AR, United States.

Walther, J., Liang, Y., DeAndrea, D., Tong, S., Carr, C., Spottswood, E., & Amichai-Hamburger, Y. (2011). The effect of feedback on identity shift in computer-mediated communication. *Media Psychology, 14*(1), 1–26.

Walther, J. B. (1996). Computer-mediated communication: Impersonal, interpersonal, and hyperpersonal interaction. *Communication Research, 23*(1), 3–44.

Walther, J. B. (2007). Selective self-presentation in computer-mediated communication: Hyperpersonal dimensions of technology, language, and cognition. *Computers in Human Behavior, 23*(5), 2538–2557.

Walther, J. B., & Whitty, M. T. (2021). Language, psychology, and new new media: The hyperpersonal model of mediated communication at twenty-five years. *Journal of Language and Social Psychology, 40*(1), 120–135. https://doi.org/10.1177/0261927X20967703

Waltman, M. S. (2003). Stratagems and heuristics in the recruitment of children into communities of hate: The fabric of our future nightmares. *Southern Communication Journal, 69*, 22–36.

Waltman, M. S., & Haas, J. W. (2007). Advertising hate on the internet. In D. W. Schumann & E. Thorson (Eds.), *Internet advertising: Theory and research* (2nd ed; pp. 395–424). Erlbaum.

Waltman, M. S., & Haas, J. W. (2010). *The communication of hate.* Peter Lang.

Wang, N., Roaché, D. J., & Pusateri, K. B. (2019). Associations between parents' and young adults' face-to-face and technologically mediated communication competence: The role of family communication patterns. *Communication Research, 46*, 1171–1196.

Wang, P., Rode, J. C., Shi, K., Luo, Z., & Chen, W. (2013). A workgroup climate perspective on the relationships among transformational leadership, workgroup diversity, and employee creativity. *Group & Organization Management, 38*(3), 334–360.

Wang, S., Moon, S., Kwon, K., Evans, C. A., & Stefanone, M. A. (2010). Face off: Implications of visual cues on initiating friendship on Facebook. *Computers in Human Behavior, 26*(2), 226–234.

Wang, Z., & Tchernev, J. M. (2012). The 'myth' of media multitasking: Reciprocal dynamics of media multitasking, personal needs, and gratifications. *Journal of Communication, 62*(3), 493–513.

Washington Military Department. (2018). *Map your neighborhood.* https://mil.wa.gov/map-your-neighborhood

Waters, R. D., & Lo, K. D. (2012). Exploring the impact of culture in the social media sphere: A content analysis of nonprofit organizations' use of Facebook. *Journal of Intercultural Communication Research, 41*(3), 297–319.

Watzlawick, P., Beavin, J. H., & Jackson, D. D. (1967). *Pragmatics of human communication: A study of interactional patterns, pathologies, and paradoxes.* Norton.

Waxer, P. H. (1977). Nonverbal cues for anxiety: An examination of emotional leakage. *Journal of Abnormal Psychology, 86*(3), 306–314.

Wazir, A. K., & Khan, H. G. A. (2018). Impact of participative leadership and employee voice through conscientiousness on organizational effectiveness: 3-time lagged study on banking sector of Pakistan. *Journal of Internet Banking and Commerce, 23*(3), 1–23.

Weaver, M. S., Hinds, P., Kellas, J. K., & Hecht, M. L. (2021). Identifying as a good parent: Considering the communication theory of identity for parents of children receiving palliative care. *Journal of Palliative Medicine, 24*(2), 305–309. https://doi.org/10.1089/jpm.2020.0131

Weber, P., & Wirth, W. (2014). When and how narratives persuade: The role of suspension of disbelief in didactic versus hedonic processing of a candidate film. *Journal of Communication, 64*(1), 125–144.

Weger, H., Bell, G., & Robinson, M. C. (2014). The relative effectiveness of active listening in initial interactions. *International Journal of Listening, 28*(1), 13–31.

Weir, K. (2018, March). Life-saving relationships. *Monitor on Psychology, 49*(3). http://www.apa.org/monitor/2018/03/life-saving-relationships

Weisel, J. J., & King, P. E. (2007). Involvement in a conversation and attributions concerning excessive self-disclosure. *Southern Communication Journal, 72*(4), 345–354.

Welch, S., & Mickelson, W. (2020). Listening across the life span: A listening environment comparison. *International Journal of Listening, 34*(2), 97–109.

Weller, A. (2020). Exploring practitioners' meaning of "ethics," "compliance," and "corporate social responsibility" practices: A communities of practice perspective. *Business & Society, 59*(3), 518–544.

Weng, Q., & Carlsson, F. (2015). Cooperation in teams: The role of identity, punishment, and endowment distribution. *Journal of Public Economics, 126*(C), 25–38.

Wessel-Tolvig, B., & Paggio, P. (2016). Revisiting the thinking-for-speaking hypothesis: Speech and gesture representation of motion in Danish and Italian. *Journal of Pragmatics, 99*, 39–61.

West, C., & Zimmerman, D. H. (1987). Doing gender. *Gender and Society, 1*, 125–151.

West, R., & Turner, J. H. (2020). *Interpersonal communication.* SAGE.

White paper: The plagiarism spectrum. (2012). Turnitin. https://www.turnitin.com/static/plagiarism-spectrum/

Whitman, M. V., & Mandeville, A. (2019). Blurring the lines: Exploring the work spouse phenomenon. *Journal of Management Inquiry, 30(3), 285–299.* https://doi.org/10.1177/1056492619882095

Whitmer, G. (2020, April 20). Michigan Governor Gretchen Whitmer press conference transcript April 20. *Rev.* https://rev.com/blog/transcripts/michigan-governor-gretchen-whitmer-press-conference-transcript-april-20

Widiyanti, E., Karsidi, R., Wijaya, M., & Utari, P. (2020). Identity gaps and negotiations among layers of young farmers: Case study in Indonesia. *Open Agriculture, 5*(1), 361–374. https://doi.org/10.1515/opag-2020-0041

Wiemann, J., & Knapp, M. (2006). Turn-taking in conversations. *Journal of Communication, 25*(2), 75–92.

Wilderom, C., Hur, Y., Wiersma, U., Van Den Berg, P., & Lee, J. (2015). From manager's emotional intelligence to objective store performance: Through store cohesiveness and sales-directed employee behavior. *Journal of Organizational Behavior, 36*(6), 825–844.

Wilkinson, F. C. (2015). Emotional intelligence in library response assistance teams: Competencies emerged? *College and Research Libraries, 76*(2), 188–204.

Williams, K. D., & Zadro, L. (2001). Ostracism: On being ignored, excluded, and rejected. In M. R. Leary (Ed.), *Interpersonal rejection* (pp. 21–54). Oxford University Press.

Williams, M. (2013, May 11). Tyson to grads: Future of exploration in your hands. *Rice University News & Media*. https://news.rice.edu/2013/05/11/tyson-to-grads-future-of-exploration-in-your-hands/

Williams, M. A. (2003). Arguing with style: How persuasion and the enthymeme work together in *On Invention, Book 3. Southern Communication Journal, 68*(2), 136–151.

Wilson, M. (2003). Discovery listening—improving perceptual processes. *ELT Journal, 57,* 335–343.

Winocur, R. (2009). Digital convergence as the symbolic medium of new practices and meanings in young people's lives. *Popular Communication, 7*(3), 179–187.

Winter, D. A., & Reed, N. (2020). Unprecedented times for many but not for all: Personal construct perspectives on the COVID-19 Pandemic. *Journal of Constructivist Psychology, 34*(3), 254–263. https://doi.org/10.1080/10720537.2020.1791291

Wiradhany, W., van Vugt, M. K., & Nieuwenstein, M. R. (2020). Media multitasking, mind-wandering, and distractibility: A large-scale study. *Attention, Perception & Psychophysics, 82*(3), 1112–1124. https://doi.org/10.3758/s13414-019-01842-0

Wise, A. F., Marbouti, F., Hsiao, Y., & Hausknecht, S. (2012). A survey of factors contributing to learners' 'listening' behaviors in asynchronous online discussions. *Journal of Educational Computing Research, 47*(4), 461–480.

Wise, A. F., Speer, J., Marbouti, F., & Hsiao, Y. (2013). Broadening the notion of participation in online discussions: Examining patterns in learners' online listening behaviors. *Instructional Science, 41*(2), 323–343.

Wiseman, J. P. (1986). Friendship: Bonds and binds in a voluntary relationship. *Journal of Social and Personal Relationships, 3,* 191–211.

Witt, P. L., Brown, K. C., Roberts, J. B., Weisel, J., Sawyer, C. R., & Behnke, R. R. (2006). Somatic anxiety patterns before, during, and after giving a public speech. *Southern Communication Journal, 71*(1), 87–100.

Woerkom, M., & Sanders, K. (2010). The romance of learning from disagreement: The effect of cohesiveness and disagreement on knowledge sharing behavior and individual performance within teams. *Journal of Business & Psychology, 25*(1), 139–149.

Wolfe, E. (2019, May). The history of video conferencing from 1870 to today. *Lifesize.* https://lifesize.com/en/video-conferencing-blog/history-of-video-conferencing

Wolvin, A. D., & Coakley, C. G. (2000). Listening education in the 21st century. *International Journal of Listening, 14,* 143–152.

Wood, J. (2002). A critical response to John Gray's Mars and Venus portrayals of men and women. *Southern Communication Journal, 2,* 201–210.

WSJ Staff. (2019, August 20). China resists charge by Twitter, Facebook of disinformation effort: Social media sites blocked hundreds of accounts they said were Beijing-linked attempts to undermine pro-democracy protestors in Hong Kong. *Wall Street Journal.* https://www.wsj.com/articles/china-resists-charge-by-twitter-facebook-of-disinformation-effort-11566339132

Xiangyu, Y., Huanhuan, L., Shan, J., Fei, P., & Zhongxin, L. (2014). Group laziness: The effect of social loafing on group performance. *Social Behavior & Personality: An International Journal, 42*(3), 465–471.

Xu, J. (2017). The mediating effect of listening metacognitive awareness between test-taking motivation and listening test score: An expectancy-value theory approach. *Frontiers in Psychology, 8,* 2201–2201. https://doi.org/10.3389/fpsyg.2017.02201

Yang, C., & Lee, Y. (2020). Interactants and activities on Facebook, Instagram, and Twitter: Associations between social media use and social adjustment to college. *Applied Developmental Science, 24*(1), 62–78. https://doi.org/10.1080/10888691.2018.1440233

Yang, Y., Lee, P. K. C, & Cheng, T. C. E. (2017). Leveraging selected operational improvement practices to achieve both efficiency and creativity: A multi-level study in frontline service operations. *International Journal of Production Economics, 191,* 298–310.

Yau, J. C., & Reich, S. M. (2019). "It's just a lot of work": Adolescents' self-presentation norms and practices on Facebook and Instagram. *Journal of Research on Adolescence, 29*(1), 196–209. https://doi.org/10.1111/jora.12376

Yeung, J. (2020, August 12). *New dinosaur species related to Tyrannosaurus rex discovered by scientists in England.* CNN. https://cnn.com/2020/08/12/uk/new-dinosaur-species-southampton-intl-hnk-sc-li-scn

Yong, K., Sauer, S. J., & Mannix, E. A. (2014). Conflict and creativity in interdisciplinary teams. *Small Group Research, 45*(3), 266–289.

You, J. W. (2020). Investigating the effects of achievement goals on team creativity and team achievement in learning communities at a South Korean university. *Higher Education.* https://sjsu-primo.hosted.exlibrisgroup.com/permalink/f/egdih2/TN_cdi_crossref_primary_10_1007_s10734_020_00545_y

Young, A., Oram, R., & Napier, J. (2019). Hearing people perceiving deaf people through sign language interpreters at work: On the loss of self through interpreted communication. *Journal of Applied Communication Research, 47*(1), 90–110.

Young, J. O. (2010). Cultural appropriation and the arts. John Wiley.

Yousafzai, M. (2014, December 10). *Malala Yousafzai—Nobel lecture.* https://nobelprize.org/nobel_prizes/peace/laureates/2014/yousafzai-lecture_en.html

Yue, C., Fong, P., & Li, T. (2019). Meeting the challenge of workplace change: Team cooperation outperforms team competition. *Social Behavior and Personality, 47*(7), 1–15.

Yule, G. (1996). *Pragmatics.* Oxford University Press.

Yung-Shui, W., & Tung-Chun, H. (2009). The relationship of transformational leadership with group cohesiveness and emotional intelligence. *Social Behavior & Personality: An International Journal, 37*(3), 379–392.

Zargaran, A., Ash, J., Kerry, G., Rasasingam, D., Gokani, S., Mittal, A., & Zargaran, D. (2018). Ethics of smartphone usage for medical image sharing. *Indian Journal of Surgery, 80*(3), 300–301.

Zentz, L. (2021). I am here and I matter. *Narrative Inquiry*. Advance online publication. https://doi.org/10.1075/ni.20117.zen

Zhang, G., Li, H., & Yan, S.. (2020, August 2). The vital few: Exploring the role of expertise in the process of team creativity. *The Journal of Creative Behavior, 55*(2), 464–475. https://doi.org/10.1002/jocb.466

Zhang, H., Dong, Y., Chiclana, F., & Yu, Shui. (2019). Consensus efficiency in group decision making: A comprehensive comparative study and its optimal design. *European Journal of Operational Research, 275*(2), 580–598.

Zhang, H., Li, F. & Reynolds, K. J. (2020). Creativity at work: Exploring role identity, organizational climate and creative team mindset. *Current Psychology.* https://doi-org.libaccess.sjlibrary.org/10.1007/s12144-020-00908-9

Zhang, L., Jiang, H., & Jin, T. (2020). Leader-member exchange and organisational citizenship behaviour: The mediating and moderating effects of role ambiguity. *Journal of Psychology in Africa, 30*(1), 17–22.

Zhang, N. N. (2009). *Coordination in syntax.* Cambridge University Press.

Zhang, R., & Fu, J. S. (2020). Privacy management and self-disclosure on social network sites: The moderating effects of stress and gender. *Journal of Computer-Mediated Communication, 25*(3), 236–251. https://doi.org/10.1093/jcmc/zmaa004

Zhu, M., Singh, S., & Wang, H. (2019). Perceptions of social loafing during the process of group development. *International Journal of Organization Theory & Behavior, 22*(4), 350–368.

Zhu, Y., & Stephens, K. (2019). Online support group participation and social support: Incorporating identification and interpersonal bonds. *Small Group Research, 50*(5), 593–622.

Zijl, A. L., Vermeeren, B., Koster, F., & Steijn, B. (2020). Interprofessional teamwork in primary care: The effect of functional heterogeneity on performance and the role of leadership. *Journal of Interprofessional Care, 35*(1), 1–11.

Zimmermann, K. A., & Emspak, J. (2017, June 27). *Internet history timeline: ARPANET to the World Wide Web.* Live Science. https://livescience.com/20727-internet-history.html

Zinko, R., Stolk, P., Furner, Z., & Almond, B. (2019). A picture is worth a thousand words: How images influence information quality and information load in online reviews. *Electronic Markets, 29*(7), 1–15.

Zinshteyn, M. (2016, April 25). The growing college-degree wealth gap. *The Atlantic.* https://www.theatlantic.com/education/archive/2016/04/the-growing-wealth-gap-in-who-earns-college-degrees/479688/

Zohoori, A. (2013). A cross-cultural comparison of the HURIER Listening Profile among Iranian and U.S. students. *International Journal of Listening, 27*(1), 50–60.

INDEX

CPSIA information can be obtained
at www.ICGtesting.com
Printed in the USA
LVHW021916230623
750640LV00006B/11